EARTH SCIENCE

EARTH SCIENCE

FIFTH EDITION

Edward J. Tarbuck
Frederick K. Lutgens

ILLINOIS CENTRAL COLLEGE

MERRILL PUBLISHING COMPANY
A BELL & HOWELL INFORMATION COMPANY
COLUMBUS   TORONTO   LONDON   MELBOURNE

Published by Merrill Publishing Company
A Bell & Howell Information Company
Columbus, Ohio 43216

This book was set in Usherwood

Administrative Editor: David Gordon
Production Coordinator: Rex Davidson
Art Coordinator: Lorraine Woost
Cover Designer: Cathy Watterson
Text Designer: Cynthia Brunk
Page layout: Ellen Robison

Illustrations by Dennis Tasa, Tasa
Graphic Arts, Inc.

Library of Congress Catalog Card Number: 87–61300
International Standard Book Number: 0–675–20748–7
Printed in the United States of America
   3 4 5 6 7 8 9—92 91 90 89 88

PHOTO CREDITS

Cover Photo and Title Page: Oak Creek and Cathedral
Rock. Red Rock, Arizona. (Photo by Larry Ulrich)

Part One: Shoshone Point, Grand Canyon. (Photo by
Stephen Trimble)

Part Two: Surf from lighthouse. Point Reyes,
California. (Photo by Stephen Trimble)

Part Three: Lightning over the Saguaro National
Monument, Arizona. (Photo by Ralph H. Wetmore, II)

Part Four: The Pleiades star cluster in Taurus. (Hale
Observatories photo. Copyright by the California
Institute of Technology and Carnegie Institution of
Washington)

PREFACE

In recent years, there has been an increased awareness and concern for our physical environment. Probes of our solar system, exploration of the bottom of the sea, and the development of a revolutionary theory in geology have called for new explanations of the earth's long and complex history. These events and many more have helped to include the earth sciences among the most dynamic and exciting disciplines. To help the beginning student share in this excitement and gain a basic understanding of the physical environment, we have attempted to write a text that is highly usable—a tool for learning the basic concepts of earth science. To achieve this goal, the Fifth Edition of *Earth Science,* like its four predecessors, is written concisely, contains useful learning aids, and is well illustrated. Each chapter begins with a brief overview, and concludes with review questions and a list of key terms. The key terms are in boldface type in the body of the text for easy reference. Furthermore, careful attention has been given to the line drawings and photographs, which constitute an important part of the text by serving to clarify concepts and make them less abstract.

Earth Science is a broad and nonquantitative survey at the introductory level of topics in geology, oceanography, meteorology, and astronomy. The Fifth Edition of *Earth Science* has been thoroughly revised and updated. Extensive rewriting has made many discussions more timely and more readable. The Fifth Edition contains substantial new material including a new section on the nature of scientific inquiry in the Introduction. Additional treatment of environmental topics includes expanded discussions of geothermal energy, earthquake destruction and prediction, and volcanic hazards. A major new section on shoreline erosion problems has been added. The treatment of plate tectonics has been brought up to date, and there has been a substantial revision of the discussion on mountain building. Chapter 9, Geologic Time and Earth History, has been strengthened with expanded sections on radiometric dating, relative dating, and unconformities. An entirely new section dealing with life of the geologic past has been added.

Among the changes to the unit on the atmosphere is new material on tornadoes as well as an updating of discussions on human impact on global climate and intentional weather modification. The astronomy unit includes rewritten discussions on optical telescopes and H-R diagrams in addition to an updated treatment of lunar history and new data from *Voyager 2* on Uranus and its satellite.

In previous editions of *Earth Science* special attention was given to the quality of photographs and artwork. This emphasis has been maintained in the Fifth Edition, which includes more than 60 new color photographs. These images were selected not only to instruct students, but also to heighten their interest. Furthermore, because we feel a carefully planned and executed art program is critical in a text such as ours, the already excellent art program has been strengthened. There are 115 new or substantially redrawn pieces of line art.

Whereas student use of the text is a primary concern, the book's adaptability to the needs and desires of the instructor is equally important. Realizing the broad diversity of earth science courses in both content and approach, we have continued to use a relatively nonintegrated format to allow maximum flexibility for the instructor. Each of the four major units stands alone; hence, they can be taught in any order. A unit can be omitted entirely without appreciable loss of continuity, and portions of some chapters may be interchanged or excluded at the instructor's discretion. For example, some instructors may wish to include the section on the sun, which appears as a portion of Chapter 18, with the discussion of stars in Chapter 20.

The authors wish to thank the many individuals, institutions, and state and federal agencies that provided information, photographs, and illustrations for use in this text. Special gratitude goes to those colleagues who were kind enough to review all or part of the manuscript. Over the years their critical comments have greatly improved this text. Our thanks go to Diane Weber who typed much of the manuscript, Dennis Tasa for his imaginative production of the line art, and to Rex Davidson, our production editor, who along with the many other fine people at Merrill Publishing skillfully transformed our manuscript into a finished product. Finally, a special debt of gratitude goes to our wives, Karen and Nancy, for their patience, encouragement, and help.

EJT
FKL

CONTENTS

PART TWO
THE OCEANS

INTRODUCTION

A view of the earth from space affords us a unique perspective of our planet (Figure I.1). At first, it may strike us that the earth is a somewhat fragile-appearing sphere surrounded by the blackness of space. In fact, it is just a speck of matter in an infinite universe. As we look more closely, it becomes apparent that the earth is much more than just rock and soil (Figure I.2). Indeed, the most conspicuous features are not the continents but the swirling

FIGURE I.1
View of the earth that greeted *Apollo 8* astronauts as their spacecraft came from behind the moon. (Courtesy of NASA)

FIGURE I.2
View of the earth from *Apollo 17.* (Courtesy of NASA)

clouds suspended above the surface and the vast global ocean. From such a vantage point, we can appreciate why the earth is traditionally divided into three major parts: the solid lithosphere, the liquid hydrosphere, and the gaseous atmosphere. However, the earth is not dominated by rock, water, or air alone. Rather, it is characterized by continuous interaction as air comes into contact with rock, rock with water, and water with air.

THE EARTH'S STRUCTURE

The solid earth, or **lithosphere,** may be divided into three principal units: the dense **core;** the less dense **mantle;** and the **crust,** which is the light and very thin outer skin of the earth (Figure I.3).*

---

*In recent years, since the development of the theory of plate tectonics (Chapter 6), the term *lithosphere* is also used more specifically to denote the rigid outer layer of the earth, which includes the crust and upper mantle.

The crust is not a layer of uniform thickness; rather, it is characterized by many irregularities. It is thinnest beneath the oceans and thickest where continents exist. Although the crust may seem insignificant when compared with the other units of the solid earth, it was created by the same general processes that were responsible for the earth's present structure. Thus, the crust is important in understanding the history and nature of our planet.

The **hydrosphere** is the water portion of our planet. This dynamic mass of liquid is continuously on the move, from the oceans to the air, to the land, and back again. The global ocean is obviously the most prominent feature of the hydrosphere, blanketing 71 percent of the earth's surface and accounting for about 97 percent of the earth's water. However, the hydrosphere also includes the fresh water found in streams, lakes, and glaciers, as well as that found in the ground. Although these latter sources constitute just a tiny

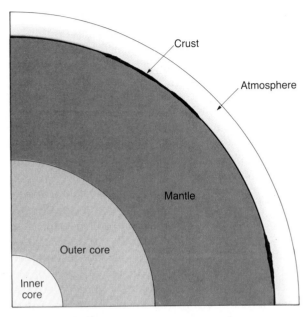

FIGURE I.3
Cross-sectional view of the earth's layered structure.
The inner core, outer core, and mantle are drawn to
scale, but the crust's thickness is exaggerated by
about three times.

fraction of the total, they are much more important than their meager percentage indicates, for
they are responsible for sculpturing and creating
many of our planet's varied landforms.

The earth is surrounded by a life-giving gaseous envelope called the **atmosphere.** This blanket
of air, hundreds of kilometers thick, is an integral
part of the planet. It not only provides the air that
we breathe but also acts to protect us from the
sun's intense heat and dangerous radiation. The
energy exchanges that continually occur between
the atmosphere and the earth's surface and between the atmosphere and space produce the effects we call weather.

THE EARTH SCIENCES

Earth science is a name for all the sciences that
collectively seek to understand the earth and its
neighbors in space. *Geology,* which literally means
"the study of the earth," examines the origin and
development of the solid earth, as well as its composition and the processes that operate beneath
and upon its surface. In many instances, it requires significant consideration of the hydrosphere and atmosphere. *Meteorology* and *climatology* are concerned with the activities and structure
of the atmosphere, and *oceanography* deals with
the dynamics of the oceans. At the boundary between the ocean and the atmosphere, oceanography and meteorology merge, and in a similar
manner, oceanography merges with geology at the
ocean floor.

The study of the earth is not confined to investigations of the lithosphere, hydrosphere, and atmosphere and to their many interactions and interrelationships. The earth sciences also attempt
to relate our planet to the larger universe. Because
the earth is related to all of the other objects in
space, the science of *astronomy* is very useful in
probing the origins of our own environment. Since
we are so closely acquainted with the planet on
which we live, it is easy to forget that the earth is
just a tiny object in a vast universe. Indeed, the
earth is subject to the same physical laws that govern the great many other objects that populate the
infinite expanses of space. Thus, to understand
the theories of the earth's origin, it is necessary to
learn something about the other members of our
solar system. Moreover, it is necessary to view the
solar system as a part of the great assemblage of
stars that comprise our galaxy, which, in turn, is
but one of many galaxies.

THE NATURE OF SCIENTIFIC INQUIRY

As members of a modern society, we are constantly reminded of the significant benefits derived from scientific investigations. What exactly is
the nature of this inquiry?

All science is based on the assumption that the
natural world behaves in a consistent and predictable manner. This implies that the physical laws
which govern the smallest atomic particles also
operate in the largest, most distant galaxies. Evidence for the existence of these underlying patterns can be found in the physical world as well as
the biological world. For example, the same biochemical processes and the same genetic codes

that are found in bacterial cells are also found in human cells. The overall goal of science is to discover the underlying patterns in the natural world and then to use this knowledge to make predictions about what should or should not be expected to happen given certain facts or circumstances.

The development of new scientific knowledge involves some basic, logical processes that are universally accepted. To determine what is occurring in the natural world, scientists collect scientific *facts* through observation and measurement (Figure I.4). These data are essential to science and serve as the springboard for the development of scientific theories and laws.

Once a set of scientific facts (or principles) that describe a natural phenomenon are gathered, in-

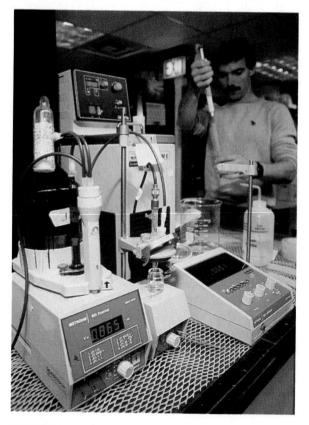

FIGURE I.4
Scientist analyzing sea-floor samples collected by the drilling ship *JOIDES Resolution*. (Courtesy of Ocean Drilling Program)

vestigators try to explain how or why things happen in the manner observed. They can do this by constructing a tentative (or untested) explanation, which we call a scientific *hypothesis*. Often several hypotheses are advanced to explain the same factual evidence. For example, there are currently five major hypotheses that have been proposed to explain the origin of the earth's moon. Until recently, the most widely held hypothesis argued that the moon and the earth formed simultaneously from the same cloud of nebular dust and gases. A newer hypothesis suggests that a Mars-sized body impacted the earth, thereby ejecting a huge quantity of material into earth orbit that eventually accumulated into the moon. Both hypotheses will be submitted to rigorous testing that will undoubtedly result in the modification or rejection of one, or perhaps both, of these proposals. The history of science is littered with discarded hypotheses. One of the best known is the idea that the earth was at the center of the universe, a proposal that was supported by the apparent daily motion of the sun, moon, and stars around the earth.

When a hypothesis has survived extensive scrutiny and when competing hypotheses have been eliminated, a hypothesis may be elevated to the status of a scientific *theory*. A scientific theory is a well-tested and widely accepted view that scientists agree best explains certain observable facts. However, scientific theories, like scientific hypotheses, are accepted only provisionally. It is always possible that a theory which has withstood previous testing may eventually be disproven. As theories survive more testing, they are regarded with higher levels of confidence. Theories that have withstood extensive testing, as for example, the theory of plate tectonics or the theory of evolution, are held with a very high degree of confidence.

Some concepts in science are formulated into scientific laws. A scientific *law* is a generalization about the behavior of nature from which there has been no known deviation after numerous observations or experiments. Scientific laws are generally narrower in scope than theories and can usually be expressed mathematically. Examples include Newton's laws of motion and the laws of thermodynamics. Scientific laws describe what happens

in nature, but they do not explain how or why things happen this way. Also, unlike hypotheses, which are inventions of the mind to explain scientific facts, scientific laws are based on observations and measurements and thus are rarely overthrown or seriously modified. Nevertheless, even scientific laws are not necessarily "perpetual truths." As the mathematician Jacob Bronowski so ably stated, "Science is a great many things, but in the end they all return to this: Science is the acceptance of what works and the rejection of what does not."

The processes just described, in which scientists gather facts through observations and formulate scientific hypotheses, theories, and laws, is called the *scientific method*. Contrary to popular belief, the scientific method is not a standard recipe that scientists apply in a routine manner to unravel the secrets of our natural world. Neither is this process a haphazard one. Some scientific knowledge is gained through the following steps: (1) the collection of scientific facts (or laws) through observation and measurement (Figure I.5); (2) the development of a working hypothesis to explain these facts; (3) construction of experiments to validate the hypothesis; and (4) the acceptance, modification, or rejection of the hypothesis based on extensive testing. Other scientific discoveries represent purely theoretical ideas, which stood up to extensive examination. Still other scientific advancements have been made when a totally unexpected happening occurred during an experiment. These so-called serendipitous discoveries are more than pure luck; for as Louis Pasteur said, "In the field of observation, chance favors only the prepared mind." Since scientific knowledge is acquired through several avenues, it might be best to describe the nature of scientific inquiry as the *methods* of science, rather than *the* scientific method.

PART ONE

THE SOLID EARTH

1

MINERALS AND ROCKS

The study of geology appropriately begins with the study of rocks and minerals, because a knowledge of the materials that make up the earth is basic to our understanding of much of the rest of geology. In the pages that follow we shall examine atoms and elements, the building blocks of minerals, and then investigate the three major groups of rocks. Our inquiry shall attempt to answer some important questions: How do rocks form? How are minerals and rocks identified? What characteristics differentiate one group of rocks from another? What can the study of rocks and minerals tell us about the nature and history of our planet?

Rock outcrop near Glen Canyon, Arizona. (Photo by Stephen Trimble)

MINERALS

Although the outer layer of the earth, called the crust, is very thin, it is of supreme importance to us. We depend on it for fossil fuels and for such diverse minerals as talc for baby powder, salt to flavor food, and gold for world trade. In fact, on occasion, the availability or absence of certain minerals and rocks has altered history.

In addition to the economic uses of rocks and minerals, all of the processes studied by geologists are in some way dependent upon the properties of these basic earth materials. Events such as volcanic eruptions, mountain building, weathering and erosion, and even earthquakes involve rocks and minerals. Consequently, a basic knowledge of earth materials is essential to the understanding of all geologic phenomena.

The term *mineral* has many meanings. To some, minerals are considered to be nutrients or dietary supplements. Others consider minerals to be rare ores or precious gems. These common perceptions are inaccurate. Minerals are actually very common substances. Sand and soil are but two common examples of substances that consist largely of minerals. In fact, a **mineral** is any naturally occurring inorganic solid that is characterized by a definite chemical composition and by a specific and regular arrangement of atoms.

By contrast, a **rock** can be defined simply as an aggregate of one or more minerals. Here the term *aggregate* implies that the minerals are found together as a *mixture* in which the properties of the individual minerals are retained. Although most rocks are composed of more than one mineral, certain minerals are commonly found by themselves in large quantities. In these instances they are considered to be both a mineral and a rock. A common example is the mineral calcite, which frequently is the dominant constituent in large rock units, where it is given the name *limestone*.

STRUCTURE OF MINERALS

Minerals, like all matter, are made of **elements.** At present, over 100 elements are known, a dozen and a half of which have been produced only in the laboratory. Some minerals such as gold and sulfur are made entirely of one element, but most

are a combination of two or more elements joined to form a chemically stable **compound.** In order to better understand how elements combine to form compounds, we must first consider the **atom,** the smallest part of matter that still retains the characteristics of an element, because it is this extremely small particle that does the combining.

Individual atoms are far too small to be observed directly; therefore, our concept of atomic structure has come from experimental evidence and mathematical models. A simplified model of the structure of an atom is shown in Figure 1.1. Each atom has a central region, called the **nucleus,** which contains very dense positively charged **protons** and equally dense neutral particles called **neutrons.** Orbiting the nucleus are negatively charged particles known as **electrons.** Unlike the orderly orbiting of the planets around the sun, electrons move so rapidly that their positions cannot be pinpointed. Hence, a more realistic picture of the positions of electrons can be obtained by envisioning a cloud of electrons surrounding the nucleus.

The number of protons in the nucleus determines the **atomic number** and name of the element. For example, all atoms with six protons are carbon atoms, all those with eight protons are oxy-

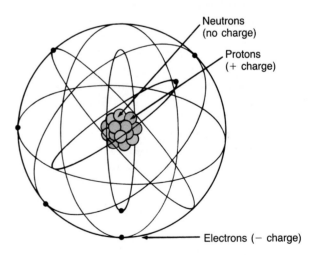

FIGURE 1.1
Simplified model of an atom. An atom consists of a central nucleus composed of protons and neutrons that is encircled by electrons.

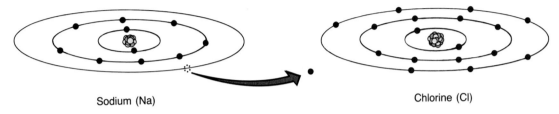

Sodium (Na)   Chlorine (Cl)

FIGURE 1.2
Chemical bonding of sodium and chlorine to produce sodium chloride. Through
the transfer of one electron from sodium to chlorine, sodium becomes a positive
ion and chlorine a negative ion.

gen atoms, and so forth. Since all atoms have the same number of electrons as protons, the atomic number also equals the number of electrons surrounding the nucleus. Moreover, since neutrons have no charge, the positive charge of the protons is exactly balanced by the negative charge of the electrons. Consequently, every atom is an electrically neutral particle. Elements can be considered to be a large collection of electrically neutral atoms, all having the same atomic number.

Chemical bonds develop when atoms of two or more elements join to form a compound. When the atoms separate, the bonds are broken and the compound is destroyed. Through experimentation it has been learned that the forces bonding the atoms are electrical in nature. Further, it is known that chemical bonding results in a change in the electronic structures of the bonded atoms.

When an atom combines chemically, it either gains, loses, or shares electrons with another atom. An atom that gains electrons becomes negatively charged because it has more electrons than protons, whereas atoms that lose electrons become positively charged. Atoms that have an electrical charge because of a gain or loss of electrons are called **ions.** Simply stated, oppositely charged ions attract one another and produce a neutral chemical compound.

An example of chemical bonding involving sodium (Na) and chlorine (Cl) to produce sodium chloride (NaCl, common table salt) is shown in Figure 1.2. When the sodium atom loses one electron it becomes a positive ion, and when the chlorine atom gains one electron it becomes a negative ion. These opposite charges act as a bond, holding

the atoms together. Here it is interesting to note that chlorine is a green, poisonous gas and sodium is a silvery metal, which, if held in your hand, would give a severe chemical burn. Together, however, these atoms produce the compound sodium chloride, which looks nothing like either and is a requirement for human life. This example also illustrates a difference between a rock and a mineral. The mineral sodium chloride (called halite) is a chemical compound and has properties unique to it and very different from the elements which make it up. Rocks, on the other hand, are *mixtures* of minerals, with each mineral retaining its own characteristics.

Although electrons play an active role in chemical reactions, they do not contribute significantly to the weight (mass) of an atom. It follows then that the weight of an atom is centered in the nucleus. The **mass number** of an atom is obtained by totaling the number of neutrons and protons in the nucleus. It is not uncommon, however, for atoms of the same element to have varying numbers of neutrons, and therefore, to have different mass numbers. Such atoms are called **isotopes** of that element. For example, carbon has two well-known isotopes, one having a mass number of 12 (carbon-12), the other a mass number of 14 (carbon-14). Recall that all atoms of the same element must have the same number of protons (atomic number) and that carbon always has six protons. Hence, carbon-12 must have six neutrons to give it a mass number of 12, whereas carbon-14 must have eight neutrons to give it a mass number of 14. The term commonly used to express the average of the atomic masses of isotopes for a given element is **atomic weight.** The atomic weight of carbon is much closer to 12 than 14, because carbon-12 is the more common isotope. Note that in a chemical sense all isotopes of the same element are nearly identical. To distinguish among them would be like trying to differentiate individual members from a group of similar objects, all having the same shape, size, and color, with only some being slightly heavier.

Although the vast majority of atoms are stable, many elements do have isotopes that are unstable. Unstable isotopes such as carbon-14 go through a process of natural disintegration called **radioactivity,** which occurs when the forces that bind the nucleus are not strong enough. The rate at which the unstable nuclei break apart (decay) is measurable and makes such elements useful

"clocks" in dating the events of earth history. A discussion of radioactivity and its application in dating events of the geologic past can be found in Chapter 9.

PHYSICAL PROPERTIES OF MINERALS

Minerals are naturally occurring solids formed by inorganic processes. Each mineral has an orderly arrangement of atoms (crystalline structure) and a definite chemical composition which give it a unique set of physical properties. Since the internal structure and chemical composition of a mineral are difficult to determine without the aid of sophisticated tests and apparatus, the more easily recognized physical properties are used in identi-

A.

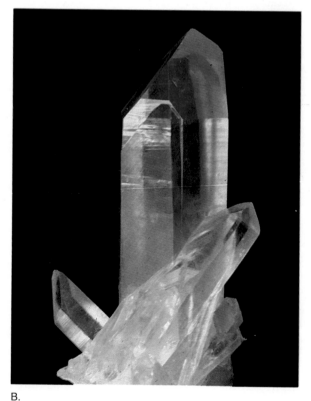

B.

FIGURE 1.3
Crystal form is the external expression of a mineral's orderly internal structure.
A. Pyrite crystals. (Courtesy of JLM Visuals). **B.** Quartz crystals. (Photo by E. J. Tarbuck)

fication. A discussion of some diagnostic physical properties follows.

Crystal Form   Most people think of a crystal as a rare commodity, when in fact most inorganic solid objects are composed of crystals. The reason for this misconception is that most crystals do not exhibit their crystal form. The **crystal form** is the external expression of a mineral that reflects the orderly internal arrangement of atoms. Figure 1.3A illustrates the characteristic form of the iron-bearing mineral pyrite. Whenever a mineral forms without space restrictions, individual crystals with well-formed crystal faces will develop. Some crystals, such as those of the mineral quartz, have a very distinctive crystal form that can be helpful in identification (Figure 1.3B). However, most of the time crystal growth is interrupted because of competition for space, resulting in an intergrown mass of crystals, none of which exhibits its crystal form.

Color   Although color is the most obvious feature of a mineral, it is possibly the least reliable. Slight impurities in the common mineral quartz, for example, give it a variety of colors, including pink, purple (amethyst), white, and even black. For other minerals, such as sulfur, color is an excellent diagnostic property.

Streak   Streak is the color of a mineral in its powdered form and is obtained by rubbing a mineral across a plate of unglazed porcelain. Whereas the color of a mineral often varies from sample to sample, the streak usually does not and is therefore the more reliable property.

Luster   Luster is the appearance or quality of light reflected from the surface of a mineral. Minerals that have the appearance of metals, regardless of color, are said to have a metallic luster. Minerals with a nonmetallic luster are described by various adjectives, including vitreous (glassy), pearly, silky, resinous, and earthy (dull).

Hardness   One of the most useful diagnostic properties of a mineral is **hardness,** the resistance of a mineral to abrasion or scratching. This is a relative property that is determined by rubbing a mineral of unknown hardness against one of known hardness, or vice versa. A numerical value

can be obtained by using **Mohs scale** of hardness, which consists of ten minerals arranged in order from 1 (softest) to 10 (hardest) as follows:

Hardness	Mineral
1	Talc
2	Gypsum
3	Calcite
4	Fluorite
5	Apatite
6	Orthoclase
7	Quartz
8	Topaz
9	Corundum
10	Diamond

Any mineral of unknown hardness can be compared to these or to other objects of known hardness. For example, a fingernail has a hardness of 2.5, a copper penny 3, and a piece of glass 5.5. The mineral gypsum, which has a hardness of 2, can be easily scratched with your fingernail. On the other hand, the mineral calcite, which has a hardness of 3, will scratch your fingernail but will not scratch glass. Quartz, the hardest of the common minerals, will scratch a glass plate with ease.

Cleavage   Cleavage is the tendency of a mineral to break along planes of weak bonding. Minerals that possess cleavage are identified by the smooth, flat surfaces produced when the mineral is broken. The simplest type of cleavage is exhibited by the micas (Figure 1.4).

FIGURE 1.4
Sheet-type cleavage common to the micas. (Courtesy of Ward's Natural Science Establishment, Inc., Rochester, N.Y.)

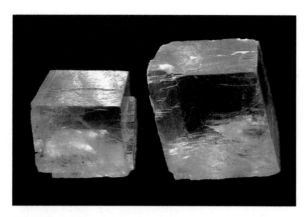

FIGURE 1.5
Smooth surfaces produced when a mineral with cleavage is broken. These samples exhibit three planes of cleavage (six sides.) The mineral on the left (halite) has cleavage planes which meet at 90-degree angles, whereas the mineral on the right (calcite) has cleavage planes which meet at 75-degree angles. (Photo by E. J. Tarbuck)

Because the micas have excellent cleavage in one direction, they break to form thin, flat sheets. Some minerals have several cleavage planes which produce smooth surfaces when broken, while others exhibit poor cleavage, and still others have no cleavage at all. When minerals break evenly in more than one direction, cleavage is described by the number of planes exhibited and the angles at which they meet (Figure 1.5).

Cleavage should not be confused with crystal form. When a mineral exhibits cleavage, it will break into pieces that have the same configuration as the original sample. By contrast, the quartz crystals shown in Figure 1.3 do not have cleavage, and if broken, would shatter into shapes that do not resemble each other or the original crystals.

Fracture Minerals that do not exhibit cleavage are said to **fracture** when broken. Those that break into smooth curved surfaces resembling broken glass have a conchoidal fracture. Others break into splinters or fibers, but most fracture irregularly.

Specific Gravity **Specific gravity** is a number which represents the ratio of the weight of a min-

eral to the weight of an equal volume of water. For example, if a mineral weighs three times as much as an equal volume of water, its specific gravity is 3. With a little practice, you can estimate the specific gravity of minerals by hefting them in your hand. The average specific gravity for minerals is around 2.7, but some metallic minerals have a specific gravity two or three times greater.

THE SILICATES

Over two thousand minerals are presently known and new ones are still being discovered. Fortunately for those who study minerals, no more than two dozen are abundant. Collectively, these few make up most of the rocks of the earth's crust and as such are classified as the *rock-forming minerals*. It is also interesting to note that only eight elements compose the bulk of these minerals and represent over 98 percent (by weight) of the continental crust (Table 1.1). The two most abundant elements are silicon and oxygen, which combine to form the framework of the most common mineral group, the **silicates.** Every silicate mineral contains oxygen and silicon, and except for quartz, one or more additional elements are needed to acquire electrical neutrality.

TABLE 1.1
Relative abundance of the most common elements in the earth's crust.

Element	Approximate Percentage by Weight
Oxygen (O)	46.6
Silicon (Si)	27.7
Aluminum (Al)	8.1
Iron (Fe)	5.0
Calcium (Ca)	3.6
Sodium (Na)	2.8
Potassium (K)	2.6
Magnesium (Mg)	2.1
All others	1.5
Total	100.0

SOURCE: Data from Brian Mason.

All silicate minerals have the same basic building block, the **silicon-oxygen tetrahedron.** This structure is composed of four oxygen atoms with a smaller silicon atom positioned in the space between them (Figure 1.6). In some minerals the tetrahedra are joined into chains, sheets, or three-dimensional networks by sharing oxygen atoms (Figure 1.7). These larger silicate structures are then connected to one another by other elements. The primary elements that join silicate structures are iron (Fe), magnesium (Mg), potassium (K), sodium (Na), and calcium (Ca). Because iron and magnesium are nearly the same size, they can readily substitute for each other without changing the structure of the mineral. This also holds true for calcium and sodium, which can occupy the same site in a crystal structure. Aluminum (Al) is unique as a substitute for silicon in the tetrahedron.

The main groups of silicate minerals and common examples of each are given in Figure 1.8. The feldspars are by far the most abundant group, comprising over 50 percent of the crust. Quartz, the second most common mineral in the continental crust, is the only one made completely of silicon and oxygen. Notice in Figure 1.8 that each

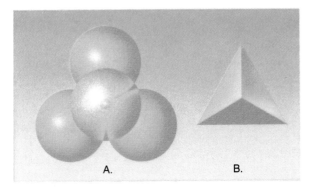

FIGURE 1.6
Top view of the silicon-oxygen tetrahedron. **A.** The four large spheres represent oxygen atoms, and the one dark sphere represents a silicon atom. **B.** Diagrammatic representation of the tetrahedron using four points to represent the positions of the oxygen atoms.

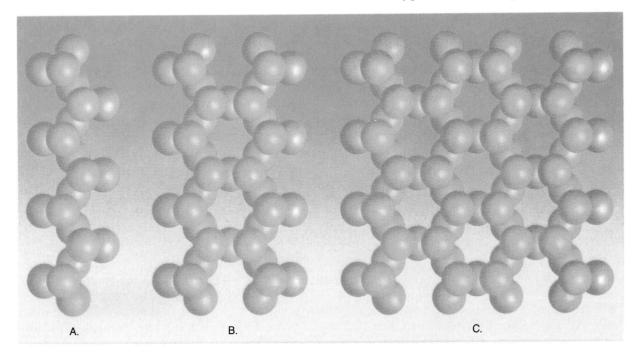

FIGURE 1.7
Three types of silicate structures. **A.** Single chains. **B.** Double chains. **C.** Sheet structures.

Mineral		Idealized Formula	Cleavage	Silicate Structure	
Olivine		$(Mg,Fe)_2SiO_4$	None	Single tetrahedron	
Pyroxene group		$(Mg,Fe)SiO_3$	Two planes at right angles	Chains 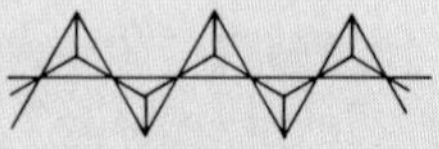	
Amphibole group		$(Ca_2Mg_5)Si_8O_{22}(OH)_2$	Two planes at 60° and 120°	Double chains 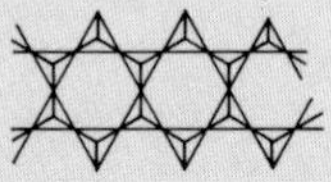	
Micas	Muscovite	$KAl_3Si_3O_{10}(OH)_2$	One plane	Sheets 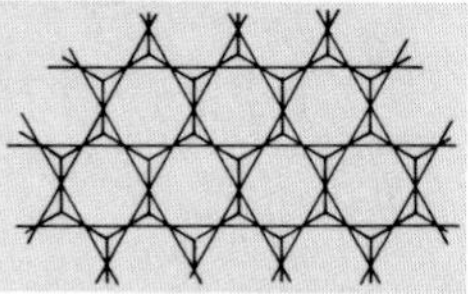	
	Biotite	$K(Mg,Fe)_3Si_3O_{10}(OH)_2$			
Feld-spars	Orthoclase	$KAlSi_3O_8$	Two planes at 90°	Three-dimensional networks 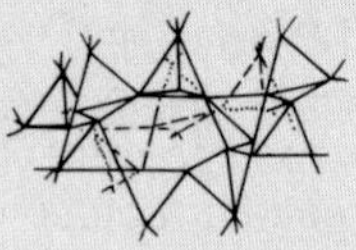	
	Plagioclase	$(Ca,Na)AlSi_3O_8$			
Quartz		SiO_2	None		

FIGURE 1.8

Common silicate minerals. Note that the complexity of the silicate structure increases down the chart.

mineral group has a particular silicate structure. A relationship exists between the internal structure of a mineral and the cleavage it exhibits. Because the silicon-oxygen bonds are strong, silicate minerals tend to cleave between the silicon-oxygen structures rather than across them. For example, the micas have a sheet structure and tend to cleave into flat plates (see Figure 1.4). Quartz, which has equally strong silicon-oxygen bonds in all directions, has no cleavage.

Most silicate minerals form when molten rock cools and crystallizes. This process can occur at or near the earth's surface, or at great depths where temperatures and pressures are very high. The environment during crystallization and the chemical composition of the molten rock to a large degree determine the minerals that are produced. In addition, some silicate minerals are stable at the earth's surface and represent the weathered products of pre-existing silicate minerals. Still other silicate minerals are formed under the extreme pressures associated with metamorphism.

The various silicate minerals can be divided into two groups based on their chemical makeup. The so-called dark silicates contain ions of iron and/or magnesium. Examples include hornblende and biotite, which, like other dark silicate minerals, are dark in color and have a specific gravity greater than 3. By contrast, quartz and feldspar are light colored and have an average specific gravity of 2.7. These observed differences are mainly attributable to the presence or absence of iron.

NONSILICATE MINERALS

Although many mineral groups are important economically, they are considered scarce compared to the silicates. Table 1.2 lists examples of oxides, sulfides, sulfates, halides, carbonates, and native elements of economic value. This group includes ores of metals such as hematite (iron), sphalerite (zinc), and galena (lead); the native elements gold, silver, and carbon (diamonds); and a host of others such as fluorite, corundum, and malachite.

TABLE 1.2
Common nonsilicate mineral groups.

Group	Member	Formula	Economic Use
Oxides	Hematite	Fe_2O_3	Ore of iron
	Magnetite	Fe_3O_4	Ore of iron
	Corundum	Al_2O_3	Abrasive, gemstone
	Ice	H_2O	
Sulfides	Galena	PbS	Ore of lead
	Sphalerite	ZnS	Ore of zinc
	Pyrite	FeS_2	Sulfur
	Chalcopyrite	$CuFeS_2$	Ore of copper
Sulfates	Gypsum	$CaSO_4 \cdot 2H_2O$	Used for plaster
	Anhydrite	$CaSO_4$	Used for plaster
Native elements	Gold	Au	Trade, electronics, medical
	Copper	Cu	Used as an electrical conductor
	Diamond	C	Gemstone, abrasive
	Sulfur	S	Used in numerous chemicals
	Graphite	C	Pencil lead and dry lubricant
Halides	Halite	NaCl	Common salt
	Fluorite	CaF_2	Used in steel making, chemicals, ceramics
Carbonates	Calcite	$CaCO_3$	Portland cement, agricultural lime
	Dolomite	$CaMg(CO_3)_2$	Portland cement, agricultural lime
	Malachite	$Cu_2(OH)_2CO_3$	Ore of copper

An important rock-forming group is the *carbonates,* which includes the mineral calcite ($CaCO_3$). This mineral is the major constituent of limestone and marble. Limestone is used commercially for road aggregate, building stone, and as the main ingredient in portland cement.

Two other nonsilicate minerals frequently found in rocks are *halite* and *gypsum*. Both minerals are commonly found in thick layers, where they accumulated when the waters of a saline lake or an ancient sea evaporated. Halite is the mineral name for common table salt. Gypsum is the mineral raw material from which plaster and other similar building materials are made.

THE ROCK CYCLE

The **rock cycle** (Figure 1.9) is one means of viewing many of the interrelationships of geology. By studying the rock cycle we may ascertain the origin of the three basic rock types and gain some insight into the role of various geologic processes in transforming one rock type into another. The concept of the rock cycle, which may be considered as a basic outline of physical geology, was initially proposed in the late eighteenth century by James Hutton, a founding father of modern geology. This rock cycle, shown in Figure 1.9, uses arrows to indicate chemical and physical processes, and boxes to represent earth materials.

The first rock type, **igneous rock,** originates when molten material called **magma** cools and solidifies. This process, called **crystallization,** may occur either beneath the earth's surface or, following a volcanic eruption, at the surface. Initially, or shortly after forming, the earth's outer shell is believed to have been molten. As this molten material gradually cooled and crystallized, it

FIGURE 1.9
The rock cycle. Originally proposed by James Hutton, the rock cycle illustrates the role of the various geologic processes which act to transform one rock type into another.

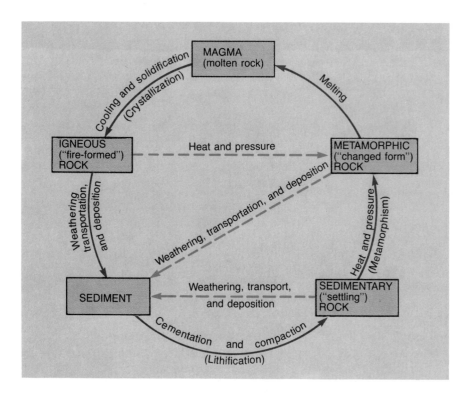

generated a primitive crust that consisted entirely of igneous rocks.

If igneous rocks are exposed at the earth's surface, they will undergo **weathering,** in which the day-in and day-out influences of the atmosphere slowly disintegrate and decompose the rock. The resulting material will be picked up, transported, and deposited by any of a number of erosional agents—gravity, running water, glaciers, wind, or waves. Once this material, called **sediment,** is deposited, usually as horizontal beds in the ocean, it will undergo **lithification,** a term meaning "conversion into rock." Sediment is lithified when compacted by the weight of overlying layers or when cemented as percolating groundwater fills the pores with mineral matter. If the resulting **sedimentary rock** is buried deep within the earth or involved in the dynamics of mountain building, it will be subjected to great pressures and heat. The sedimentary rock will react to the changing environment and turn into the third rock type, **meta-morphic rock.** When metamorphic rock is subjected to still greater heat and pressure, it will melt to create magma, which will eventually solidify as igneous rock.

The full cycle just described does not always take place. "Shortcuts" in the cycle are indicated by dashed lines in Figure 1.9. Igneous rock, for example, rather than being exposed to weathering and erosion at the earth's surface, may be subjected to the heat and pressure found far below and change to metamorphic rock. On the other hand, metamorphic and sedimentary rocks, as well as sediment, may be exposed at the surface and turned into new raw materials for sedimentary rock.

As you study the remaining pages of this chapter, which discuss details of each of the three basic rock types, remember the rock cycle. Although rocks may seem to be unchanging masses, the rock cycle shows that they are not. The changes, however, take time—great amounts of time.

IGNEOUS ROCKS

In our discussion of the rock cycle, it was pointed out that igneous rocks form when magma cools and crystallizes. This molten rock, which originates at depths as great as 200 kilometers within the earth, consists primarily of the elements found in silicate minerals, along with some gases, particularly water vapor, which are confined within the magma by the pressure of the surrounding rocks. Because the magma body is less dense than the surrounding rocks, it works its way toward the surface, and on occasion breaks through, producing a volcanic eruption (Figure 1.10).

The spectacular explosions that sometimes accompany an eruption are produced by the gases (volatiles) escaping as the confining pressure lessens near the surface. Sometimes blockage of the vent coupled with surface water seepage into the magma chamber can produce catastrophic explosions. Along with ejected rock fragments, a volcanic eruption often generates extensive lava flows. **Lava** is similar to magma, except that most of the gaseous component has escaped. The rocks which result when lava solidifies are classified as **extrusive,** or **volcanic.** The magma not able to reach the surface eventually crystallizes at depth. Igneous rocks produced in this manner are termed **intrusive,** or **plutonic,** and would never be observed if not for the processes of erosion stripping away the overlying rocks.

FIGURE 1.10
Lava fountain along the east rift zone of Hawaii's Kilauea Volcano, 1983. (Photo by J. D. Griggs, U.S. Geological Survey)

CRYSTALLIZATION OF MAGMA

Magma is a hot liquid; therefore, the ions that compose it move about freely and are said to be unordered. However, as magma cools, the random movements of the ions slow and the ions begin to arrange themselves into orderly patterns. This process is called *crystallization.* Usually all of the molten material does not solidify at the same time. Rather, as it cools, numerous small crystals develop. In a systematic fashion, ions are added to these centers of crystal growth. When the crystals grow large enough for their edges to meet, their growth ceases and crystallization continues elsewhere. Eventually, all of the liquid is transformed into a solid mass of interlocking crystals.

The rate of cooling strongly influences the crystallization process, in particular the size of the crystals. When a magma cools very slowly, relatively few centers of crystal growth develop. Slow cooling also allows ions to migrate over relatively great distances. Consequently, slow cooling results in the formation of rather large crystals. On the other hand, when cooling occurs quite rapidly, the ions quickly lose their motion and readily combine. This results in the development of large numbers of tiny crystals that all compete for the available ions. Therefore, the outcome of rapid cooling is the formation of a solid mass composed of very small intergrown crystals.

When the molten material is quenched instantly, there is not sufficient time for the ions to arrange themselves into a crystalline network. Therefore, the solids produced in this manner consist of randomly distributed ions. Rocks that consist of unordered atoms are referred to as *glass* and are quite similar to ordinary manufactured glass.

TEXTURE AND MINERAL COMPOSITION

A large variety of igneous rocks exist and are differentiated on the basis of their texture and mineral content. The term **texture,** when applied to an igneous rock, is used to describe the overall appearance of the rock based on the size and arrangement of its interlocking crystals. Texture is a very important characteristic since it reveals a great deal about the environment in which the rock formed. From our discussion of crystallization, we learned that rapid cooling produces small crystals, whereas very slow cooling results in the formation of much larger crystals. As we might

expect, the rate of cooling is quite slow in magma chambers lying deep within the crust, whereas a thin layer of lava extruded upon the earth's surface may chill in a matter of hours, and small molten blobs ejected into the air during a violent eruption can solidify almost instantly.

Igneous rocks that form at the earth's surface or as small masses within the upper crust have a very **fine grained texture,** with the individual crystals being too small to be seen with the unaided eye (Figure 1.11A). A common feature in many fine-grained igneous rocks are the voids left by escaping gases (Figure 1.12). These spherical or elongated openings are called *vesicles* and are limited to the outer portion of lava flows. Only in the outer zone of a lava flow can cooling occur rapidly enough to "freeze" the lava and preserve the openings produced by the escaping gas.

When large masses of magma solidify far below the surface, they form igneous rocks that exhibit a **coarse-grained texture.** These coarse-grained rocks have the appearance of a mass of intergrown crystals, which are roughly equal in size and large enough that the individual minerals can be identified with the unaided eye (Figure 1.11B).

A large mass of magma located at depth may require tens of thousands, even millions, of years

A.

B.

FIGURE 1.11
A. Rhyolite exhibits a fine-grained texture. **B.** Granite is a common igneous rock that has a coarse-grained texture. (Photos by E. J. Tarbuck)

FIGURE 1.12
Vesicular texture. Vesicles form as gas bubbles escape near the top of a lava flow. (Photo by E. J. Tarbuck)

to solidify. Since all minerals within a magma do not crystallize at the same rate or at the same time during cooling, it is possible for some to become quite large before others even start to form. If magma containing some large crystals were to change environments by erupting at the surface, for example, the molten portion of the lava would cool quickly. The resulting rock, which has large crystals embedded in a matrix of smaller crystals, is said to have a **porphyritic texture** (Figure 1.13).

During some volcanic eruptions, molten rock is ejected into the atmosphere, where it is quenched very quickly. Rapid cooling of this type may generate rock with a **glassy texture.** As was indicated, glass results when the ions do not have sufficient time to unite into an orderly crystalline structure. *Obsidian,* a common type of natural glass, is similar in appearance to a dark chunk of manufactured glass (Figure 1.14).

The mineral makeup of an igneous rock is ultimately determined by the chemical composition of the magma from which it crystallized. Such a large variety of igneous rocks exists that it is logical to assume an equally large variety of magmas must also exist. However, geologists have found that various eruptive stages of the same volcano often extrude lavas exhibiting somewhat different mineral compositions, particularly if an extensive

time period separated the eruptions. Evidence of this type led geologists to examine the possibility that a single magma might produce rocks of varying mineral content.

A pioneering investigation into the crystallization of magma was carried out by N. L. Bowen in the first quarter of this century. Bowen discovered that as magma cools in the laboratory, certain

FIGURE 1.13
Andesite porphyry. Notice the two distinctively different sizes of crystals. (Photo by E. J. Tarbuck)

FIGURE 1.14
Obsidian, a glassy volcanic rock. (Courtesy of Ward's Natural Science Establishment, Inc., Rochester, N.Y.)

minerals crystallize first. At successively lower temperatures, other minerals begin to crystallize as shown in Figure 1.15.

Bowen also demonstrated that if a mineral remains in the melt after crystallization, it will react with the remaining melt to produce the next mineral in the sequence shown in Figure 1.15. For this reason, this arrangement of minerals became known as *Bowen's reaction series*. On the upper left branch of this reaction series, olivine, the first mineral to form, will react with the remaining melt to become pyroxene. This reaction will continue until the last mineral in the series, biotite, is formed. The right branch of the reaction series is a continuum in which the earliest-formed calcium-rich feldspar crystals react with the sodium ions contained in the melt to become progressively more sodium rich. Ordinarily, these reactions are not complete so that various amounts of several of these minerals may exist at any given time.

During the last stage of crystallization, after most of the magma has solidified, the remaining melt will form the minerals quartz, muscovite, and potassium feldspar. Although these minerals crystallize in the order shown, this sequence is not a true reaction series.

Bowen demonstrated that minerals crystallize from magma in a systematic fashion. But how does Bowen's reaction series account for the great diversity of igneous rocks? Apparently, at one or more stages in the crystallization process, a separation of the solid and liquid components of a magma frequently occurs. This can happen, for example, if the earlier-formed minerals are heavier than the liquid portion and settle to the bottom of the magma chamber as shown in Figure 1.16A. This settling is thought to occur frequently with the dark silicates, such as olivine. When the remaining melt crystallizes, either in place or in a new location if it migrates out of the chamber, it

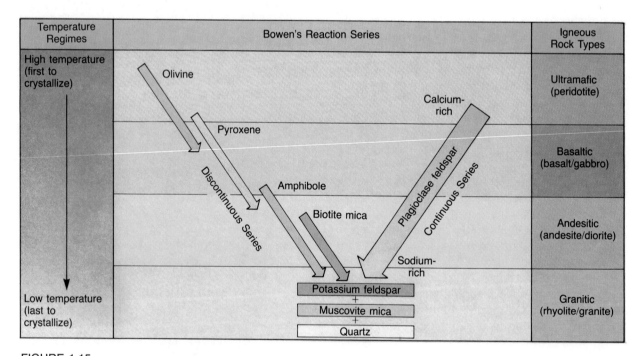

FIGURE 1.15

Bowen's reaction series shows the sequence in which minerals crystallize from a magma. Compare this figure to the mineral composition of the rock groups in Table 1.3. Note that each rock group consists of minerals that crystallize at the same time.

FIGURE 1.16

FIGURE 1.16
Separation of minerals by fractional crystallization.
A. Illustration of how the earliest-formed minerals can be separated from a magma by settling. **B.** The remaining melt could migrate to a number of locations and, upon further crystallization, generate rocks having a composition much different from the parent magma.

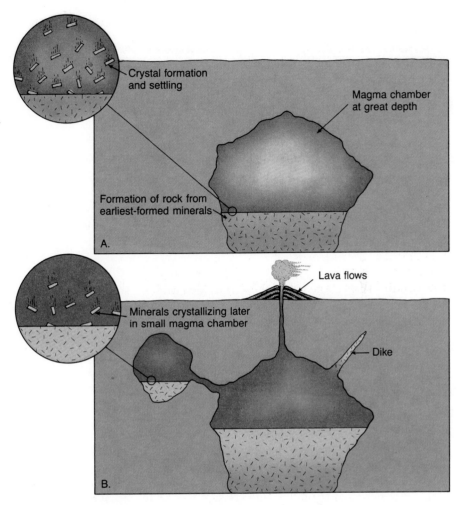

will form a rock with a chemical composition much different from the original magma (Figure 1.16B).

The process involving the segregation of minerals by differential crystallization and separation is called **fractional crystallization.** At any stage in the crystallization process the melt might be separated from the solid portion of the magma. Consequently, fractional crystallization can produce igneous rocks having a wide range of compositions.

NAMING IGNEOUS ROCKS

As was stated previously, igneous rocks are most often classified, or grouped, on the basis of their texture and mineral composition. The various ig-

neous textures result from different cooling histories, while the mineral composition of an igneous rock is the consequence of the chemical makeup of the parent magma and the environment of crystallization. As we might expect from the results of Bowen's work, minerals that crystallize under similar conditions are most often found together composing the same igneous rock. A general classification scheme based on texture and mineral composition is provided in Table 1.3.

The rocks on the right side of Table 1.3 consist of the first minerals to crystallize and are higher in iron (Fe) and magnesium (Mg) and lower in silica (SiO_2) than later ones. Their iron content makes them darker in color and slightly more dense than the other rocks. The term **mafic,** derived from

TABLE 1.3
Common igneous rocks.

	Felsic (granitic)	Intermediate (andesitic)	Mafic (basaltic)	Ultramafic
Intrusive (coarse-grained)	Granite	Diorite	Gabbro	Peridotite
Extrusive (fine-grained)	Rhyolite	Andesite	Basalt	None
Mineral Composition	Quartz Potassium feldspar Sodium feldspar	Hornblende Sodium feldspar Calcium feldspar	Calcium feldspar Pyroxene	Olivine Pyroxene
Minor Mineral Constituents	Muscovite Biotite Hornblende	Biotite Pyroxene	Olivine Hornblende	Calcium feldspar

magnesium and Fe, the chemical symbol for iron, is used for rocks of this composition. Geologists also employ the term *basaltic* (after basalt, a common rock of this composition) to indicate any igneous rock having this composition. *Basalt* is a dark green to black, fine-grained rock composed primarily of pyroxene and calcium-rich feldspar. Basalt is the most common extrusive igneous rock. Many volcanic islands, such as the Hawaiian Islands and Iceland, are composed mainly of basalt. Further, the upper layers of the oceanic crust consist primarily of basalt.

The rocks on the left side of Table 1.3 contain the last minerals to crystallize and consist mainly of **fel**dspar and **si**lica (quartz); consequently the term **felsic** is applied to them. The felsic rocks are also commonly referred to as having a *granitic* composition. *Granite* is perhaps the best-known igneous rock (see Figure 1.11B). This is partly because of its natural beauty, which is enhanced when it is polished, and partly because of its abundance. Slabs of polished granite are commonly used for tombstones, monuments, and as building stones. Granite is often produced by the processes which generate mountains. Because granite is a by-product of mountain building and is very resistant to weathering and erosion, it frequently forms the core of eroded mountains. For example, Pikes Peak in the Rockies, Mount Rushmore in the Black Hills, the White Mountains of New Hampshire, Stone Mountain, Georgia, and Yo-

semite National Park in the Sierra Nevada are all areas where large quantities of granite are exposed at the surface.

This discussion has dealt with only two mineral compositions, yet it is important to note that gradations between these types also exist (Table 1.3). For example, an abundant extrusive igneous rock called *andesite* has a mineral composition between that of rocks with a granitic composition and those with a basaltic composition. Andesite is a fine-grained rock of volcanic origin, whose name derives from the Andes Mountains, where numerous volcanoes are composed of this rock type. In addition to the Andes volcanoes, many volcanic structures encircling the Pacific Ocean exhibit this andesitic composition. Another important igneous rock called *peridotite* contains mostly olivine and thus falls near the very beginning of Bowen's reaction series. Peridotite is believed to be a major constituent of the upper mantle.

SEDIMENTARY ROCKS

The products of weathering constitute the raw materials for sedimentary rocks. The word *sedimentary* indicates the nature of these rocks, for it is derived from the Latin *sedimentum,* which means "settling," a reference to solid material settling out of a fluid. Most but not all sediment is deposited in this fashion. Weathered debris is constantly being swept from bedrock and carried away by water and ice. Eventually the material is deposited in lakes, river valleys, seas, and countless other places. The particles in a desert sand dune, the mud on the floor of a swamp, the gravels in a stream bed, and even household dust are examples of sediment produced by this never-ending process. Since the weathering of bedrock and the transport and deposition of the weathering products are continuous, sediment is found almost everywhere. As piles of sediment accumulate, the materials near the bottom are compacted by the weight of the overlying layers. Over long periods, these sediments are cemented together by mineral matter deposited in the spaces between particles to form solid rock.

Geologists estimate that sedimentary rocks account for only about 5 percent (by volume) of the earth's outer 16 kilometers (10 miles). However, the importance of this group of rocks is far greater than this percentage would imply. If we were to sample the rocks exposed at the earth's surface, we would find that the great majority are sedimentary (Figure 1.17). Indeed, about 75 percent of all rock outcrops on the continents are sedimentary. Therefore, we can think of sedimentary rocks as comprising a relatively thin and somewhat discontinuous layer in the uppermost portion of the crust. This fact is readily understood when we consider that sediment accumulates at the surface of the earth.

FIGURE 1.17
Sedimentary rocks exposed near Canyonlands National Park, Utah. About 75 percent of all rock outcrops on the continents are sedimentary. (Photo by Stephen Trimble)

Since sediments accumulate at the earth's surface, the rock layers that they eventually form contain evidence of past events at the surface. By their very nature, sedimentary rocks contain indications of past environments in which their particles were deposited and in some cases, clues to the mechanisms involved in their transport. Furthermore, it is sedimentary rocks that contain fossils, which are vital tools in the study of the geologic past. Thus, it is largely from this group of rocks that geologists must reconstruct the details of earth history.

Finally, it should be mentioned that many sedimentary rocks are very important economically. Coal, for example, is classified as a sedimentary rock, whereas our other major energy resources, petroleum and natural gas, are found in association with sedimentary rocks. Still others represent major sources of iron, aluminum, manganese, and fertilizer as well as numerous materials essential to the construction industry.

CLASSIFICATION

Materials accumulating as sediment have two principal sources. First, sediments may be accumulations of materials that originate and are transported as solid particles derived from weathering. Deposits of this type are termed *detrital* and the sedimentary rocks that they form are called **detrital sedimentary rocks.** The second major source of sediment is soluble material produced largely by chemical weathering. When these dissolved substances are precipitated by either inorganic or organic processes, the material is known as chemical sediment and the rocks formed from it are called **chemical sedimentary rocks.**

Detrital Sedimentary Rocks Though a wide variety of minerals and rock fragments may be found in detrital rocks, clay minerals and quartz are the chief constituents of most sedimentary rocks in this category. As we shall see in Chapter 2, clay minerals are the most abundant product of the chemical weathering of silicate minerals, especially the feldspars. Quartz, on the other hand, is plentiful due to its extreme resistance to chemical weathering. Thus, when igneous rocks such as

TABLE 1.4
Particle size classification for detrital rocks.

Sediment Name	Size Range (mm)	Detrital Rock Name
Gravel	>2	Conglomerate or breccia
Sand	1/16–2	Sandstone
Silt	1/256–1/16	Siltstone
Clay	< 1/256	Shale

granite are attacked by weathering processes, individual quartz grains are set free.

Particle size is the primary basis for distinguishing among various detrital sedimentary rocks. Table 1.4 presents the size categories for particles making up detrital rocks. When gravel-sized particles predominate, the rock is called *conglomerate* if the sediment is rounded (Figure 1.18A) and *breccia* if the pieces are angular (Figure 1.18B). Angular fragments indicate that the particles were not transported very far from their source prior to deposition. *Sandstone* is the name given rocks when sand-sized grains prevail (Figure 1.18C), whereas *shale,* the most common sedimentary rock, is made of very fine grained sediment (Figure 1.18D). *Siltstone,* another rather fine grained rock, is sometimes difficult to differentiate from rocks such as shale that are composed of even smaller clay-sized sediment.

The size of the particles in a detrital rock can often be related to the energy of the transporting medium. Currents of water or air sort the particles by size; the stronger the current, the larger the particle size carried. Gravels, for example, are moved by swiftly flowing rivers as well as by rockslides and glaciers. Less energy is required to transport sand, thus it is common to such features as windblown dunes, as well as some river deposits and beaches. Since silts and clays settle very slowly, accumulations of these materials are generally associated with the quiet waters of a lake, lagoon, swamp, or marine environment.

Although detrital sedimentary rocks are classified primarily by the size of their particles there are exceptions. In certain cases the mineral composition is important to its classification. For ex-

A.

B.

C.

D.

FIGURE 1.18
Common detrital sedimentary rocks. **A.** Conglomerate. **B.** Breccia. **C.** Sandstone.
D. Shale. (Photos by E. J. Tarbuck)

ample, most sandstones are predominantly quartz, but when appreciable quantities of feldspar are present, the rock is called *arkose*. In addition, rocks consisting of detrital sediments are rarely composed of grains of just one size. Consequently, a rock containing quantities of both sand and silt can be correctly classified as sandy siltstone or silty sandstone, depending upon which particle size dominates.

Chemical Sedimentary Rocks In contrast to detrital rocks, which form from the solid products of

weathering, chemical sediments derive from material that is carried in solution to lakes and seas. This material does not remain dissolved in the water indefinitely. Rather, some of it precipitates to form chemical sediments. This precipitation may occur directly as the result of inorganic processes or indirectly as the result of the life processes of water-dwelling organisms. Sediment formed in this second way is said to have a *biochemical* origin.

An example of a deposit resulting from inorganic chemical processes is the salt left behind as a body of salt water evaporates. In contrast, many water-dwelling animals and plants extract dissolved mineral matter to form shells and other hard parts. After the organisms die, their skeletons may accumulate on the floor of a lake or ocean.

Limestone is the most abundant chemical sedimentary rock. It is composed chiefly of the mineral calcite ($CaCO_3$) and forms by either inorganic means or as the result of biochemical processes. Limestones having a biochemical origin are by far the most common. As much as 90 percent of the world's limestone may have originated as accumulations of biochemical sediment.

Although most limestone is the product of biological processes, this origin is not always evident because shells and skeletons may undergo considerable change before being converted to rock. However, one easily identified biochemical limestone is *coquina,* a coarse rock composed of poorly cemented shells and shell fragments. Another less obvious, but nevertheless familiar example is *chalk,* a rock made up almost entirely of the hard parts of microscopic organisms that are no larger than the head of a pin (Figure 1.19).

Limestones having an inorganic origin form when evaporation and/or high water temperatures increase the concentration of dissolved calcium carbonate to the point that it precipitates. *Travertine,* the type of limestone commonly seen decorating caverns is one example and is deposited when groundwater containing calcium carbonate evaporates.

Dissolved silica (SiO_2) precipitates to form varieties of microcrystalline quartz. These include chert (light color), flint (dark), jasper (red), and agate (banded). These chemical sedimentary rocks may have either an inorganic or biochemical origin, but the mode of origin is usually difficult to determine.

Evaporation very often triggers deposition of chemical precipitates. Minerals commonly precipitated in this fashion include halite, the chief component of *rock salt,* and gypsum, the main ingredient of *rock gypsum.*

In the geologic past, many areas that are now dry land were covered by shallow arms of the sea that had only narrow connections to the open ocean. Under these conditions, water continually moved into the bay to replace water lost by evaporation. Eventually the waters of the bay became saturated and salt deposition began. Such deposits are called **evaporites.** Today such deposits serve as an important source of many chemicals. Similar deposits may be seen in such places as Death Valley, California. Here, following rains or periods of snowmelt in the mountains, streams flow from surrounding mountains into an enclosed basin. As the water evaporates, *salt flats* form from dissolved materials left behind as a white crust on the ground (Figure 1.20).

Coal is difficult to classify because it is quite different from other sedimentary rocks. Nevertheless, it is often grouped with biochemical sedi-

FIGURE 1.19
The White Chalk Cliffs of Dover. (Photo by Jerome Wyckoff)

FIGURE 1.20
These salt flats (composed mainly of gypsum and rock salt) are examples of evaporite deposits and are common in basins located in the arid Southwest. (Photo by John S. Shelton)

mentary rocks. However, unlike other rocks in this category, which are calcite rich or silica rich, coal is made of organic matter. Close examination of a piece of coal under a microscope or magnifying glass often reveals the presence of various plant structures such as leaves, bark, and wood that have been chemically altered but are nevertheless still identifiable. This supports the conclusion that coal is the end product of the burial of large amounts of plant material over extended periods.

Under certain circumstances a great volume of plants may accumulate rather than totally decompose, which is normally the case. One such environment is an oxygen-poor swamp. At various times during earth history such environments have been relatively common. With each successive stage in coal formation, higher temperatures and pressures drive off impurities and volatiles as follows:

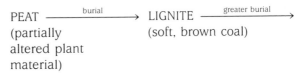

PEAT ——————burial——————→ LIGNITE ——————greater burial——————→
(partially   (soft, brown coal)
altered plant
material)

BITUMINOUS ——————metamorphism——————→ ANTHRACITE
(soft, black coal)   (hard, black coal)

Bituminous coal is by far the most important coal resource. *Anthracite* is formed when bituminous coal undergoes metamorphism, and although it burns more cleanly, it is not as widespread and is more expensive to mine.

Organic matter is present in small quantities in many other sedimentary rocks, both detrital and chemical. When present, it usually colors the rock black. In such cases the term *carbonaceous* is often applied. Thus, a black, organic-rich shale would be called carbonaceous shale.

In summary, the classification of sedimentary rocks is a difficult matter. As was pointed out, many detrital sedimentary rocks are a mixture of more than one particle size. Furthermore, many of the sedimentary rocks classified as chemical also contain at least small quantities of detrital sediment. Many limestones, for example, contain varying amounts of mud or sand, giving them a "sandy" or "shaly" quality. On the other hand, since practically all detrital rocks are cemented with material that was originally dissolved in water, they too are far from being "pure."

LITHIFICATION

Lithification refers to the processes by which unconsolidated sediments are transformed into solid sedimentary rocks. One of the most common processes affecting sediments is *compaction.* As sediments accumulate through time, the weight of overlying material compresses the deeper sediments. As the grains are pressed closer and closer, there is a considerable reduction in pore space. For example, when clays are buried beneath several thousand meters of material, the volume of the clay may be reduced by as much as 40 percent. Since sands and other coarse sediments are only slightly compressible, compaction is most significant as a lithification process in fine-grained sedimentary rocks such as shale.

Cementation is another important means by which sediments are converted to sedimentary rocks. The cementing materials are carried in solution by water percolating through the open spaces between particles. Through time, the cement precipitates onto the sediment grains, fills the open spaces, and joins the particles. Calcite, silica, and iron oxide are the most common cements. The identification of the cementing material is a relatively simple matter. Calcite cement will effervesce with dilute hydrochloric acid. Silica is the hardest cement and thus produces the hardest sedimentary rocks. When a sedimentary rock has an orange or red color, this usually means iron oxide is present.

Although most sedimentary rocks are lithified by compaction, cementation, or a combination of both, some are made of interlocking crystals. This type of lithification is confined largely to certain chemical sedimentary rocks.

FEATURES OF SEDIMENTARY ROCKS

Sedimentary rocks are particularly important in the interpretation of earth history. These rocks form at the earth's surface, and as layer upon layer of sediment accumulates, each records the nature of the environment at the time the sediment was deposited. These layers, called **strata,** or **beds,** are the single most characteristic feature of sedimentary rocks (see Figure 1.17).

As geologists examine sedimentary rocks, much can be deduced. A conglomerate, for example, may indicate a high-energy environment, such as a rushing stream, where only the coarse materials can settle out. If the rock is arkose, it may signify a dry climate, where little chemical alteration of feldspar is possible. Carbonaceous shale is a sign of a low-energy, organic-rich environment such as a swamp or lagoon. Other features found in some sedimentary rocks also give clues to past environments (Figure 1.21).

Fossils, the evidence or remains of prehistoric life, are perhaps the most important inclusions found in some sedimentary rock. Knowing the nature of the life forms that existed at a particular time may help to answer many questions about the environment. Was it land or ocean? A lake or swamp? Was the climate hot or cold, rainy or dry? Was the ocean water shallow or deep, turbid or clear? Fossils are important tools used in interpreting the geologic past (more on this in Chapter 9).

METAMORPHIC ROCKS

Metamorphism involves the transformation of pre-existing rock. Metamorphic rocks can form from igneous, sedimentary, or even from other metamorphic rocks. The term for this process is very appropriate because it literally means to "change form." The agents of change include heat, pressure, and chemically active fluids, while the changes that occur are textural as well as mineralogical.

A.

B.

FIGURE 1.21

A. Ripple marks may indicate a beach or stream channel environment. (Photo by Stephen Trimble) B. Mud cracks form when wet mud or clay dries and shrinks, perhaps signifying a tidal flat or desert basin. (Photo by Garrett Deckert) C. The cross-bedding in this sandstone indicates that it was once a sand dune. (Photograph used by permission of Dennis Tasa)

C.

FIGURE 1.22
Ancient deformed metamorphic rocks exposed near the Colorado National Monument. (Photo by Stephen Trimble)

In some instances rocks are only slightly changed, becoming more compact. In other cases the transformation is so complete that the identity of the original rock cannot be determined. In high-grade metamorphism, such features as bedding planes, fossils, and vesicles that may have existed in the parent rock are completely destroyed. Further, when subjected to intense heat and directional pressure, these rocks behave plastically and may bend into intricate folds (Figure 1.22). In the most extreme metamorphic environments, the temperatures approach those at which rocks melt. However, during metamorphism the deformed material must remain solid, for once melting occurs, we have entered the realm of igneous activity.

The process of metamorphism takes place when rock is subjected to conditions unlike those in which it originally formed; the rock becomes unstable and gradually changes until a state of equilibrium with the new environment is reached. The changes occur at the temperatures and pressures in the region extending from a few kilometers below the earth's surface to the crust-mantle boundary. Since the formation of metamorphic rocks is completely hidden from view (which is not the case for many sedimentary and some igneous rocks), metamorphism is undoubtedly one of the most difficult processes for geologists to study.

Metamorphism most often occurs in one of two settings. First, during mountain building great quantities of rock are subjected to the intense stresses and high temperatures associated with large-scale deformation. The end result may be extensive areas of metamorphic rocks that are said to have undergone **regional metamorphism.** The greatest volume of metamorphic rock is produced in this fashion. Second, when rock is in contact or close proximity to a mass of magma, **contact metamorphism** takes place. In this circumstance the changes are caused primarily by the high temperatures of the molten material, which in effect "bake" the surrounding rock.

AGENTS OF METAMORPHISM

As stated earlier, the agents of metamorphism include heat, pressure, and chemically active fluids.

Perhaps the most important metamorphic agent is heat. Rocks formed near the earth's surface may be subjected to intense heat when they are intruded by molten material rising from below. Since temperature increases with depth, rocks that originate in a surface environment may also be subjected to extreme temperatures if they are subsequently buried deep within the earth. When buried to a depth of only a few kilometers, certain minerals such as clay become unstable and begin to recrystallize into different minerals that are stable in this environment. Other minerals, particularly those in crystalline igneous rocks, are stable at relatively high temperatures and pressures and therefore require burial to 20 kilometers or more before metamorphism will occur.

Pressure, like temperature, also increases with depth. Buried rocks are subjected to the force exerted by the load above. This confining pressure is analogous to air pressure where the force is applied equally in all directions. In addition to the pressure exerted by the load of material above, rocks are also subjected to *stress* during the process of mountain building. In this situation the applied force is directional, and the material is squeezed as if it had been placed in a vise. Rock located at great depth is quite warm and behaves plastically during deformation. This accounts for its ability to flow and bend into intricate folds.

Chemically active fluids, most commonly water containing ions in solution, also influence the metamorphic process. Some water is contained in the pore spaces of virtually every rock. Water that surrounds the crystals acts as a catalyst by aiding the migration of ions. In some instances the minerals recrystallize into more stable configurations. In other cases, ion exchange among minerals results in the formation of completely new minerals.

TEXTURAL AND MINERALOGICAL CHANGES

The degree of metamorphism is reflected in the rock's texture and mineralogy. When rocks are subjected to very low grade metamorphism, they become more compact and thus more dense. A common example is the metamorphic rock *slate,* formed from the further compaction of shale (Table 1.5). Under more extreme conditions, pressure causes certain minerals to recrystallize. As described earlier, water is believed to play a very important role in the recrystallization process by aiding the migration of ions. In general, recrystallization encourages the growth of larger crystals. Consequently, many metamorphic rocks consist of visible crystals, much like coarse-grained igneous rocks. The crystals of some minerals, such as micas, which have a sheet structure, and horn-

blende, which has an elongated structure, will recrystallize with a preferred orientation. The new orientation will be essentially perpendicular to the direction of the compressional force. The resulting mineral alignment usually gives the rock a layered or banded appearance termed **foliation** (Figure 1.23).

Not all metamorphic rocks have a foliated texture. Such rocks are said to exhibit a **nonfoliated** texture. Metamorphic rocks composed of only one mineral which forms equidimensional crystals are

TABLE 1.5
Description of common metamorphic rocks.

FOLIATED TEXTURE

Slate Very fine grained rock composed primarily of microscopic flakes of mica. Slate is produced by the low-grade metamorphism of shale (Figure 1.24A).

Schist Among the most abundant foliated metamorphic rocks, schists are composed in large part of visible flakes of platy minerals. As with slate, schist often forms from shale, but in the case of schist, the metamorphism was more intense. Often included are new minerals formed during metamorphism and unique to metamorphic rocks. Schists are named based upon their mineral composition (Figure 1.24B).

Gneiss (pronounced "nice") Most often has the composition of granite, although other compositions are possible. The outstanding feature of gneiss is a coarsely banded or streaked appearance, reflecting the characteristic alternating layers of light and dark minerals (Figure 1.23, right).

NONFOLIATED TEXTURE

Marble Results when limestone is metamorphosed. The large interlocking crystals of calcite so characteristic of marble form when the much smaller grains in limestone are recrystallized. When pure, marble is white, but it is often colored by impurities (Figure 1.24C).

Quartzite Common metamorphic rock formed from quartz sandstone. Quartzite may closely resemble marble in appearance but is much harder.

as a rule not visibly foliated. For example, when a fine-grained limestone is metamorphosed, the small calcite crystals combine to form relatively large interlocking crystals. The resulting rock has an appearance similar to a coarse-grained igneous rock. This metamorphic equivalent of limestone is called *marble* (Table 1.5).

In some environments, new materials are actually introduced during the metamorphic process. For example, host rock adjacent to a large magma

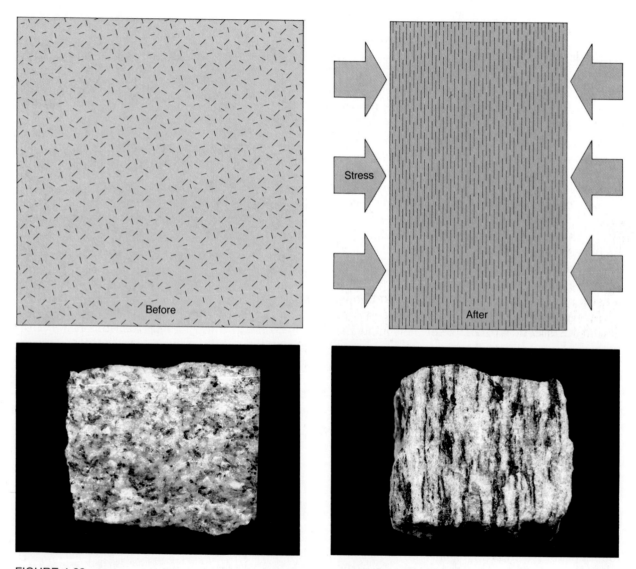

FIGURE 1.23
Under the pressures of metamorphism, some mineral grains become reoriented and aligned at right angles to the stress. The resulting linear orientation of mineral grains gives the rock a foliated texture. If the coarse-grained igneous rock (granite) on the left underwent intense metamorphism, it could end up closely resembling the metamorphic rock on the right (gneiss). (Photos by E. J. Tarbuck)

body would be altered by ion-rich *hydrothermal* (hot water) *solutions* released during the latter stages of crystallization. Many metallic ore deposits are formed by the precipitation of minerals from hydrothermal solutions.

In summary, the metamorphic process causes many changes in rocks, including increased density, growth of larger crystals, reorientation of the mineral grains into a layered or banded appearance known as foliation, and the transformation of low-temperature minerals into high-temperature minerals. Further, the introduction of ions generates new minerals, some of which are economically important.

FIGURE 1.24
A. Slate. B. Mica schist.
C. Pink marble. (Photos by
E. J. Tarbuck)

REVIEW QUESTIONS

1   List the three main particles of an atom and explain how they differ from one another.

2   If the number of electrons in an atom is 35 and the mass number is 80, calculate the following:
   (a) The number of protons.
   (b) The atomic number.
   (c) The number of neutrons.

3   What occurs in an atom to produce an ion?

4   What is an isotope?

5   Although all minerals have an orderly internal arrangement of atoms (crystalline structure), most mineral samples do not demonstrate their crystal form. Explain.

6   Why might it be difficult to identify a mineral by its color?

7   If you found a glassy-appearing mineral while rock hunting and had hopes that it was a diamond, what simple test might help you make a determination?

8   Explain the economic use of corundum as given in Table 1.2 in terms of Mohs hardness scale.

9   Gold has a specific gravity of about 20. If a 25-liter container of water weighs about 25 kilograms, how much would a 25-liter container of gold weigh?

10   What is the difference between silicon and silicate?

11   Explain the statement "One rock is the raw material for another" using the rock cycle.

12   If a lava flow at the earth's surface had a mafic composition, what rock type would the flow likely be (see Table 1.3)? What igneous rock would form from the same magma if it did not reach the surface but instead crystallized at great depth?

13   What does a porphyritic texture indicate about an igneous rock?

14   How are granite and rhyolite different? The same? (See Table 1.3.)

15   Relate the classification of igneous rocks to Bowen's reaction series.

16   What minerals are most common in detrital sedimentary rocks? Why are these minerals so abundant?

17   What is the primary basis for distinguishing among various detrital sedimentary rocks?

18   Distinguish between the two categories of chemical sedimentary rocks.

19   What are evaporite deposits? Name a rock that is an evaporite.

20   Compaction is an important lithification process with which sediment size?

21   What is probably the single most characteristic feature of sedimentary rocks?

22   What is metamorphism? What are the *agents of change*?

23   Distinguish between regional and contact metamorphism.

24   What feature would easily distinguish schist and gneiss from quartzite and marble?

25   How do metamorphic rocks differ from the igneous and sedimentary rocks from which they formed?

KEY TERMS

Review your understanding of important terms in this chapter by defining and explaining the importance of each of the following. Terms are listed in order of occurrence within the chapter.

mineral	Mohs scale	texture
rock	cleavage	fine-grained texture
element	fracture	coarse-grained texture
compound	specific gravity	porphyritic texture
atom	silicate	glassy texture
nucleus	silicon-oxygen tetrahedron	fractional crystallization
proton		
neutron	rock cycle	mafic
electron	igneous rock	felsic
atomic number	magma	detrital sedimentary rock
ion	crystallization	
mass number	weathering	chemical sedimentary rock
isotope	sediment	
atomic weight	lithification	evaporite deposit
radioactivity	sedimentary rock	strata (beds)
crystal form	metamorphic rock	regional metamorphism
streak	lava	
luster	extrusive (volcanic)	contact metamorphism
hardness	intrusive (plutonic)	
		foliated texture
		nonfoliated texture

2

WEATHERING, SOIL, AND MASS WASTING

The earth's surface is constantly changing. Rock is disintegrated and decomposed, moved to lower elevations by gravity, and carried away by water, wind, or ice. In this manner the earth's physical landscape is sculptured. This chapter focuses on the first two steps of this never-ending process—weathering and mass wasting—probing into how and why rock disintegrates and decomposes and what mechanisms act to move it downslope. Soil, an important product of the weathering process and a vital resource, is also examined.

Colorado National Monument. Weathering accentuates differences in rocks to produce some of our most spectacular scenery. (Photo by Stephen Trimble)

To the casual observer the face of the earth may appear to be without change, unaffected by time. For that matter, less than 200 years ago most people believed that mountains, lakes, and deserts were permanent features of an earth that was thought to be no more than a few thousand years old. Today, however, we know that mountains eventually succumb to weathering and erosion and are washed into the sea, lakes fill with sediment and vegetation or are drained by streams, and deserts come and go as relatively minor climatic changes occur.

The earth is indeed a dynamic body. Volcanic and tectonic activities* are elevating parts of the earth's surface, while opposing processes are continually removing materials from higher elevations and moving them to lower elevations. The latter processes include:

1 **Weathering**—disintegration and decomposition of rock at or near the surface of the earth.
2 **Erosion**—incorporation and transportation of material by a mobile agent, usually water, wind, or ice.
3 **Mass wasting**—transfer of rock material downslope under the influence of gravity.

We will first turn our attention to the process of weathering and the products generated by this activity. However, weathering cannot be easily separated from the other two processes because as weathering breaks rocks apart, it facilitates the movement of rock debris by erosion and mass wasting. On the other hand, the transport of material by erosion and mass wasting furthers the disintegration and decomposition of rock.

WEATHERING

All materials are susceptible to weathering. Consider, for example, the fabricated product concrete, which closely resembles the sedimentary rock called conglomerate. A newly-poured concrete sidewalk has a smooth, fresh look. However, not many years later the same sidewalk will ap-

pear chipped, cracked, and rough, with pebbles exposed at the surface. If a tree is nearby, its roots may heave and buckle the concrete as well. The same natural processes that eventually destroy a concrete sidewalk also act to disintegrate rock.

Why does rock weather? Simply, weathering is the response of earth materials to a changing environment. For instance, after millions of years of uplift and erosion, the rocks overlying a large intrusive igneous body may be removed, exposing it at the surface. This mass of crystalline rock, which formed in a high-temperature, high-pressure environment perhaps several kilometers below ground, is now subjected to a very different and comparatively hostile surface environment. In response, this rock mass will gradually change until it is once again in equilibrium, or balance, with its new environment. This transformation of rock is what we call weathering.

In the following sections we will discuss the various types of mechanical and chemical weathering. Although we will consider these two processes separately, keep in mind that they usually work simultaneously in nature.

MECHANICAL WEATHERING

When a rock undergoes **mechanical weathering** it is broken into smaller and smaller pieces, each retaining the characteristics of the original material. The end result is many small pieces from a single large one. Figure 2.1 shows that breaking a rock into smaller pieces increases the surface area available for chemical attack. An analogous situation occurs when sugar is added to a liquid. In this situation, a cube of sugar will dissolve much slower than an equal volume of granules because of the vast difference in surface area. Hence, by breaking rocks into smaller pieces, mechanical weathering increases the amount of surface area available for chemical weathering.

In nature four important physical processes lead to the fragmentation of rock: frost wedging, expansion resulting from unloading, thermal expansion, and organic activity.

Frost Wedging Alternate freezing and thawing of water is one of the most important processes of mechanical weathering. Water has the unique

---

*Tectonic activities** are activities that result in the deformation of the earth's crust.

property of expanding about 9 percent as it freezes. This increase in volume occurs because as ice forms, the water molecules arrange themselves into a very open crystalline structure. As a

result, when water freezes it expands and exerts a tremendous outward force. This can be verified by filling a container with water and freezing it. If sufficient volume does not exist within the container, the formation of ice will cause it to shatter.

In nature, water works its way into cracks or voids in rock and, upon freezing, expands and wedges the rock apart. This process is appropriately called **frost wedging** (Figure 2.2). Frost wedging is most pronounced in mountainous regions in the middle latitudes where a daily freeze-thaw cycle often exists. Here sections of rock are wedged loose and may tumble into large piles called **talus slopes** that often form at the base of steep rock outcrops (Figure 2.2).

Unloading   When large igneous bodies, particularly those composed of granite, are exposed by erosion, concentric slabs begin to break loose. The process generating these onionlike layers is called **sheeting** and is thought to occur, at least in part, because of the great reduction in pressure when

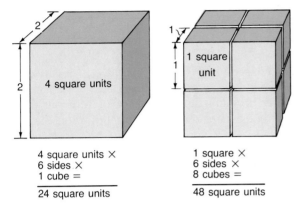

FIGURE 2.1
Mechanical weathering increases the surface area available for chemical attack.

FIGURE 2.2
Frost wedging. As water freezes it expands, exerting a force great enough to break rock. Here, as the rock fragments accumulate at the base of the cliff, a talus slope is formed.

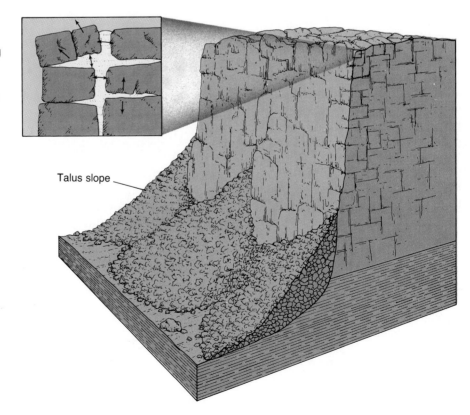

FIGURE 2.3
Summit of Half Dome, an exfoliation dome in Yosem-ite National Park. (Photo by Stephen Trimble)

the overlying rock is stripped away. Accompany-ing the unloading, the outer layers expand more than the rock below, and thus separate from the rock body. The fractures typically develop parallel to the surface topography and give the exhumed igneous body a domed shape. Continued weather-ing eventually causes the slabs produced by sheeting to separate and spall off these large structures known as **exfoliation domes.** Excellent examples of exfoliation domes include Stone Mountain, Georgia, and Half Dome and Liberty Cap in Yosemite National Park (Figure 2.3).

Mine shafts provide us with a view of how rocks behave once the confining pressure is removed. Large rock slabs have been known to explode off the walls of newly cut mine shafts because of the reduced pressure. Evidence of this type, plus the fact that fracturing occurs parallel to the floor of a rock quarry when large blocks are removed, strongly supports the process of unloading as the cause of sheeting.

Although many fractures are created by expan-sion, others are produced by contraction during the crystallization of magma and still others by tectonic forces during mountain building. Frac-tures produced by these activities generally form a definite pattern and are called *joints.* Joints are important rock structures which allow water to penetrate to depth and start the process of weath-ering long before the rock is exposed at the sur-face.

Thermal Expansion The daily cycle of tempera-ture change is thought to weaken rocks, particu-larly in hot, dry regions where daily variations often exceed 30°C (54°F). Heating a rock causes it to expand, and cooling causes it to contract. Re-peated swelling and shrinking of minerals with different expansion and contraction rates should logically exert some stress on the rock's outer shell. Although this process was once thought to be of major importance in the disintegration of rock, laboratory experiments have not substanti-ated this idea. In one test rocks were heated to temperatures much higher than those normally experienced on the earth's surface and then were cooled. This procedure was repeated many times to simulate hundreds of years of weathering, but the rocks showed little apparent change. Addi-tional work is needed before the impact of daily temperature variations on weathering can be de-termined.

Organic Activity Weathering is also accom-plished by the activities of organisms, including plants, burrowing animals, and humans. Plant roots in search of minerals and water grow into fractures, and as the roots grow they wedge the rock apart. Burrowing animals further break down rock by moving fresh material to the surface, where physical and chemical processes can more effectively attack it. Decaying organisms also pro-duce acids, which contribute to chemical weather-ing. Where rock has been blasted in search of min-erals or for road construction, the impact is quite noticeable, but on a worldwide scale human activ-ity probably ranks behind burrowing animals in earth-moving accomplishments.

CHEMICAL WEATHERING

Chemical weathering involves the complex proc-esses that alter the internal structures of minerals by removing and/or adding elements. Water is by far the most important agent of chemical weathering. Although pure water is nonreactive, a small amount of dissolved material is generally all that is needed to activate it. Oxygen dissolved in water will *oxidize* some materials. For example, when an iron nail is found in the soil it will have a coating of rust (iron oxide), and if the time of exposure has been long, the nail will be so weak that it can be broken as easily as a toothpick. When rocks containing iron-rich minerals oxidize, a yellow to reddish brown rust will appear on the surface.

Carbon dioxide (CO_2) dissolved in water (H_2O) forms carbonic acid (H_2CO_3), the same weak acid produced when soft drinks are carbonated. Rain dissolves some carbon dioxide as it falls through the atmosphere, and additional amounts released by decaying organic matter are acquired as the water percolates through the soil. Carbonic acid ionizes to form the very reactive hydrogen ion (H^+) and the bicarbonate ion (HCO_3^-).

To illustrate how rock chemically weathers when attacked by carbonic acid, we will consider the weathering of granite, the most abundant continental rock. Recall that granite consists mainly of quartz and potassium feldspar. The weathering of the potassium feldspar component of granite takes place as follows:

$$2KAlSi_3O_8 + 2(H^+ + HCO_3^-) + H_2O \longrightarrow$$
potassium   carbonic acid   water
feldspar

$$Al_2Si_2O_5(OH)_4 + 2KHCO_3 + 4SiO_2$$
clay mineral   potassium   silica
                  bicarbonate

In this reaction, the hydrogen ions (H^+) attack and replace potassium ions (K^+) in the feldspar structure, thereby disrupting the crystalline network. Once removed, the potassium is available as a nutrient for plants or becomes the soluble salt potassium bicarbonate ($KHCO_3$), which may be incorporated into other minerals or carried to the ocean in dissolved form by streams. The most abundant products of the chemical breakdown of feldspar are residual clay minerals. Because clay minerals are the end product of weathering, they are very stable under surface conditions. Consequently, clay minerals make up a high percentage of the inorganic material in soil. The most abundant sedimentary rock, shale, is also composed of

clay minerals. Some of the silica removed from the feldspar structure will go into solution with the groundwater (water beneath the earth's surface). This silica will eventually precipitate to produce nodules of chert or flint, fill in the pore spaces between sediment grains, or be carried to the ocean, where microscopic animals such as diatoms will remove it to build silica shells.

To summarize, the weathering of potassium feldspar generates a residual clay mineral, a soluble salt (potassium bicarbonate), and some silica which enters into solution.

Quartz, the other main component of granite, is very resistant to chemical weathering; hence, it remains substantially unaltered when attacked by weakly acidic solutions. As a result, when granite weathers, the feldspar crystals dull and slowly turn to clay, releasing the once-interlocked quartz grains, which still retain their fresh, glassy appearance. Although some quartz remains in the soil, much is transported to the sea, where it becomes the main constituent of sandy beaches and in time may be lithified to form the sedimentary rock sandstone.

Table 2.1 lists the weathered products of some of the most common silicate minerals. Remember that silicate minerals make up most of the earth's crust and that these minerals are essentially

TABLE 2.1
Products of weathering.

Mineral	Residual Products	Material in Solution
Quartz	Quartz grains	Silica
Feldspars	Clay minerals	Silica K^+, Na^+, Ca^{2+}
Hornblende	Clay minerals Limonite Hematite	Silica Ca^{2+}, Mg^{2+}
Olivine	Limonite Hematite	Silica Mg^{2+}

composed of only eight elements. When chemically weathered, the silicate minerals yield sodium, calcium, potassium, and magnesium ions that form soluble products which may be removed by groundwater. The element iron combines with oxygen, producing relatively insoluble iron oxides, most notably hematite and limonite, which give soil a reddish-brown or yellowish color. Under most conditions the three remaining elements, aluminum, silicon, and oxygen, join with water to produce residual clay minerals. However, even the highly insoluble clay minerals are very slowly removed by subsurface water.

As chemical weathering alters the internal structure of minerals, physical changes are also occurring. For example, when angular rock fragments produced by regular joint systems are attacked by chemical weathering, the fragments take on a spherical shape. Any process that tends to give the weathered rock a spherical shape is called **spheroidal weathering.** One type of spheroidal weathering resembles sheeting, except that much smaller rocks are involved. As the minerals in the rock weather to clay, they increase in size

FIGURE 2.4
Successive shells are loosened as the weathering process continues to penetrate ever deeper into the rock. (Photo by Kenneth Hasson)

because of the addition of water into their structure. This increase in bulk exerts an outward force that is thought to cause concentric layers of rock to break loose and fall off (Figure 2.4). Hence chemical weathering can produce forces great enough to cause mechanical weathering.

RATES OF WEATHERING

The rate at which rock weathers depends on many factors. We have already seen that particle size influences the rate of weathering. The mineral makeup of a rock is also a very important factor, which can be demonstrated by comparing headstones carved from different rock types. A headstone made of granite, a rock composed of silicate minerals, is relatively resistant to chemical weathering as we can see by examining the inscriptions on the headstone in Figure 2.5A. This is not true of the marble headstone, which shows signs of extensive chemical alteration over a relatively short period (Figure 2.5B). Recall that marble is composed of calcite, which readily dissolves even in a weakly acidic solution.

The silicates, the most abundant mineral group, weather in essentially the same order as their order of crystallization. The minerals that crystallized first formed at much higher temperatures than those that crystallized last. Consequently, these early-formed minerals are not as stable at the earth's surface, where the temperature and pressure are radically different from the environment where they formed. By examining Bowen's reaction series (see Figure 1.15), we can see that olivine crystallizes first and is therefore the least resistant to weathering, while quartz, which crystallizes last, is the most resistant.

Climatic factors, particularly temperature and moisture, are of primary significance to the rate of rock weathering. The optimum environment for chemical weathering is a combination of warm temperatures and abundant moisture. In polar regions and at high altitudes, chemical weathering is reduced because frigid temperatures keep the available moisture locked up as ice, whereas in arid regions there is insufficient moisture to foster much chemical weathering. A classic example of how climate affects the rate of weathering was

FIGURE 2.5
An examination of head-
stones reveals the rate of
chemical weathering on diverse rock types. The granite headstone (left) was
erected a few years before
the marble headstone (right),
whose inscription date of
1892 is nearly illegible.
(Photos by E. J. Tarbuck)

A.

B.

provided when Cleopatra's Needle, a granite obelisk, was moved from Egypt to New York City. After
withstanding approximately 3500 years of exposure in the dry climate of Egypt, the hieroglyphics
were almost completely removed from the wind-

ward side in less than 75 years in the wet and
chemical-laden air of New York City (Figure 2.6).

The sum of these factors determines the type
and rate of rock weathering for a given region.
However, there is generally enough variation, even

FIGURE 2.6
Chemical weathering of
Cleopatra's Needle, a granite obelisk. A. Before it was
removed from Egypt. (Courtesy of The Metropolitan
Museum of Art). B. After a
span of 75 years in New
York City's Central Park.
After having survived intact
for about 35 centuries in
Egypt, the windward side
has been almost completely
defaced in less than a century. (Courtesy of New York
City Parks)

A.

B.

within a relatively small area, for the rocks to ex-hibit some differential weathering. Differential weathering and subsequent erosion are responsible for many unusual and often spectacular landforms (see chapter-opening photo). Included are features such as natural bridges like those found in Arches National Park and sculptured rock pinnacles such as those found in Bryce Canyon National Park.

SOIL

Soil has accurately been called "the bridge between life and the inanimate world." All life owes its existence to a dozen or so elements that must ultimately come from the earth's crust. Once weathering and other processes create soil, plants carry out the intermediary role of assimilating the necessary elements and making them available to animals and people.

With few exceptions, the earth's land surface is covered by **regolith,** the layer of rock and mineral fragments produced by weathering. Some would call this material soil, but soil is more than an accumulation of weathered debris. **Soil** is a combi-

nation of mineral and organic matter, water, and air—that portion of the regolith that supports the growth of plants. Although the proportions vary, the major components do not (Figure 2.7). About one-half of the total volume of a good quality surface soil is a mixture of disintegrated and decomposed rock (mineral matter) and **humus,** the decayed remains of animal and plant life (organic matter). The remaining half consists of pore spaces, where air and water circulate.

Although the mineral portion of the soil is usually much greater than the organic portion, humus is a very significant component. In addition to being an important source of plant nutrients, humus enhances the soil's ability to retain water. Since plants require air and water to live and grow, the portion of the soil consisting of pore spaces that allow for the circulation of these fluids is as vital as the solid soil constituents. Soil water is not "pure" water; instead, it is a complex solution containing many soluble nutrients. Soil water not only provides the necessary moisture for the chemical reactions that sustain life, it also supplies plants with nutrients in a form they can use. The pore spaces not filled with water contain air. This air is the source of necessary oxygen and carbon dioxide for most microorganisms and plants that live in the soil.

SOIL TEXTURE AND STRUCTURE

Soil texture refers to the relative portions of different particle sizes in a soil. Texture is a very basic soil property because it strongly influences the soil's ability to retain and transmit water and air, both of which are essential to plant growth. Sandy soils may drain too rapidly, whereas the pore spaces of clay-rich soils may be too small for adequate drainage. Moreover, when the clay and silt content is very high, plant roots may have difficulty penetrating the soil.

Soils rarely consist of particles of only one size; therefore, textural categories have been established based upon the varying proportions of clay, silt, and sand. The standard system of classes used by the U.S. Department of Agriculture is shown in Figure 2.8. For example, point A on this

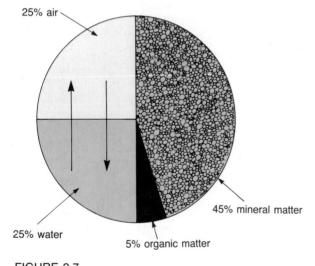

25% air

25% water

5% organic matter

45% mineral matter

FIGURE 2.7
Composition (by volume) of a soil in good condition for plant growth. Although the percentages vary, each soil is composed of mineral and organic matter, water, and air.

FIGURE 2.8

The texture of any soil can be represented by a point on this soil texture diagram. Soil texture is one of the most significant factors used to estimate agricultural potential and engineering characteristics. (After U.S. Department of Agriculture)

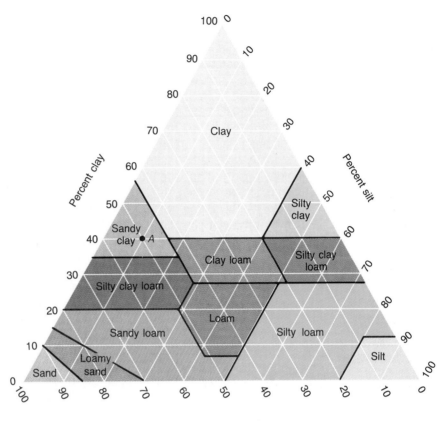

Percent sand

triangular diagram represents a soil composed of 10 percent silt, 40 percent clay, and 50 percent sand. Such a soil is called a *sandy clay*. The soils called *loam,* which occupy the central portion of the diagram, are those in which no single particle size predominates over the other two. Loam soils are best suited to support plant life because they generally have better moisture characteristics and nutrient storage ability than soils composed predominantly of clay or coarse sand.

Soil particles are seldom completely independent of one another. Rather, they usually form clumps called *peds* that give soils a particular structure. Four basic soil structures are recognized: platy, prismatic, blocky, and spheroidal. Soil structure is important because it influences the ease of a soil's cultivation as well as the susceptibility of a soil to erosion. In addition, soil structure affects the porosity and permeability of

soil (that is, the ease with which water can penetrate). This, in turn, influences the movement of nutrients to plant roots. Prismatic and blocky peds usually allow for moderate water infiltration, whereas platy and spheroidal structures are characterized by slower infiltration rates.

CONTROLS OF SOIL FORMATION

Soil is the product of the complex interplay of several factors, including parent material, time, climate, plants and animals, and slope. Although all of these factors are interdependent, it will be helpful to examine their roles separately.

Parent Material   The **parent material** from which a soil has evolved may either be underlying bedrock or a layer of unconsolidated deposits. Soils formed on bedrock are termed **residual soils,**

whereas those developed on unconsolidated sediment are called **transported soils** (Figure 2.9).

The nature of the parent material influences soils in two ways. First, the type of parent material to some degree will affect the rate of weathering and thus the rate of soil formation. For example, the mineral composition of the parent material will influence the rate of chemical weathering. Also, since unconsolidated deposits are already partly weathered, soil development on such material will likely progress more rapidly than when bedrock is the parent material. Second, the chemical makeup of the parent material will affect the fertility of the soil. If it lacks the necessary elements for plant growth, its usefulness is obviously diminished.

At one time it was thought that the parent material was the primary factor causing differences among soils. Today soil scientists realize that other factors, especially climate, are more important. In fact, it has been found that similar soils are often produced from different parent materials and that dissimilar soils have developed from the same parent material. Such discoveries reinforce the importance of the other soil-forming factors.

Time If weathering has been going on for a comparatively short time, the character of the parent material determines to a large extent the characteristics of the soil. As the weathering process continues, the influence of parent material on soil is overshadowed by the other soil-forming factors. Therefore time is an important control of soil formation. It is not possible to list the length of time required for various soils to evolve, because the soil-forming processes act at varying rates under different circumstances. However, it is safe to say that the longer a soil has been forming, the thicker it becomes and the less it resembles the parent material from which it formed.

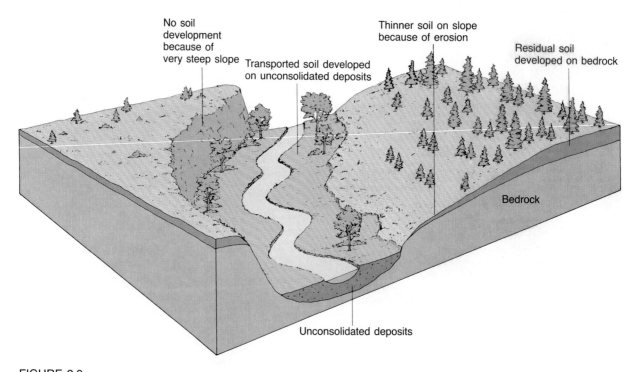

FIGURE 2.9
Parent material. The parent material for residual soils is the underlying bedrock, whereas transported soils form on unconsolidated deposits. Also note that soils are thinner, or nonexistent, on the slopes.

Climate Climate is considered to be the most important control of soil formation. It determines whether chemical or mechanical weathering will predominate and also greatly influences the rate and depth of weathering. For instance, a hot and wet climate may produce a thick layer of chemically weathered soil in the same amount of time that a cold and less humid climate produces a thin mantle of mechanically weathered debris. Furthermore, the amount of precipitation influences the degree to which various materials are removed from the soil, thereby affecting soil fertility. Finally, climatic conditions are an important control on the type of plant and animal life present.

Plants and Animals The chief function of plants and animals is to supply organic matter to the soil. Certain bog soils are composed almost entirely of organic matter, whereas desert soils may contain only a very small percentage. Although the quantity of organic matter present varies substantially among soils, no soil completely lacks it.

The primary source of organic matter is plants, although animals and the uncountable numbers of microorganisms also contribute. When organic matter is decomposed, it supplies important nutrients for plants, as well as food for the animals and microorganisms living in the soil. Consequently, soil fertility is in part related to the amount of organic matter present. Furthermore, the decay of plant and animal remains causes various organic acids to form. These complex acids hasten the weathering process. Organic matter also has a high water-holding ability and thus aids water retention in a soil.

Microorganisms, including fungi, bacteria, and the single-celled protozoa, play the active role in the decay of plant and animal remains. The end product is humus, a material that no longer resembles the plants and animals from which it formed. In addition, certain microorganisms aid soil fertility because they have the ability to *fix* (change) atmospheric nitrogen into soil nitrogen.

Earthworms and other burrowing animals act to mix the mineral and organic portions of a soil. Earthworms, for example, feed on the organic matter in the soil and thoroughly mix soils in which they live, often moving and enriching many tons per acre each year. Burrows and holes also aid the passage of water and air through the soil.

Slope Slope significantly affects the amount of erosion and the water content of soil. On steep slopes soils are often poorly developed. In such situations the quantity of water soaking in is slight, and as a result, the moisture content of the soil may not be sufficient for vigorous plant growth. Further, because of accelerated erosion on steep slopes, the soils there are thin, or in some cases, nonexistent (Figure 2.9). On the other hand, poorly drained and waterlogged soils found in bottomlands have a much different character. Such soils are usually very thick and very dark, the dark color resulting from the large quantity of organic matter that accumulates because saturated conditions retard the decay of vegetation. The optimum slope for soil development is a flat-to-undulating, upland surface. Here we find good drainage, minimum erosion, and sufficient infiltration of water into the soil.

Slope orientation, the direction the slope faces, is another aspect worthy of mention. In the midlatitudes a south-facing slope will receive a great deal more sunlight than a north-facing slope. In fact, a steep north-facing slope may receive no direct sunlight at all. The difference in the amount of solar radiation received will cause differences in soil temperature and moisture, which in turn may influence the nature of the vegetation and the character of the soil.

Although this section has dealt separately with each of the soil-forming factors, remember that all of them work together to form soil. No single factor is responsible for a soil being as it is, but rather the combined influence of parent material, time, climate, plants and animals, and slope.

THE SOIL PROFILE

Since soil-forming processes operate from the surface downward, variations in composition, texture, structure, and color gradually evolve at varying depths. These vertical differences, which usually become more pronounced as time passes, divide the soil into zones or layers known as **horizons.** If you were to dig a trench in soil, you would see that its walls are layered. Such a vertical section

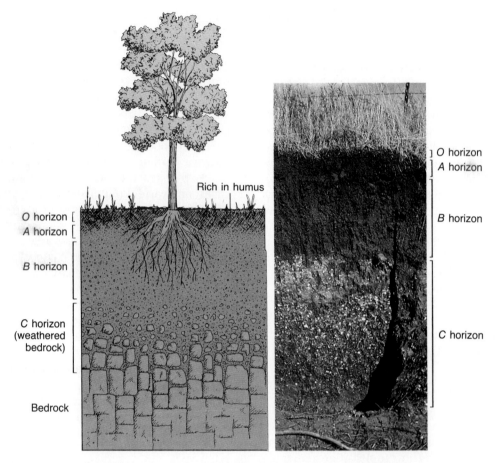

Rich in humus

O horizon [
A horizon [

B horizon

C horizon
(weathered
bedrock)

Bedrock

] O horizon
] A horizon

B horizon

C horizon

FIGURE 2.10
Soil profile. Mature soils are characterized by a series of horizontal layers called
horizons, which comprise the soil profile.

through all of the soil horizons constitutes the **soil profile** (Figure 2.10).

Four basic horizons are identified and from top to bottom are designated as *O, A, B,* and *C,* respectively. The three upper layers may be further divided.

Unlike the layers beneath it which consist mainly of mineral matter, the *O* horizon consists largely of organic material. The upper portion of this horizon is primarily plant litter such as loose leaves and other organic debris that are still recognizable. By contrast, the lower portion of the *O* horizon is made up of partly decomposed organic matter (humus) in which plant structures can no longer be identified.

Underlying the organic-rich *O* horizon is the *A* horizon. This zone is largely mineral matter, yet biological activity is high and humus is generally present—up to 30 percent in some instances. As water percolates downward from the surface through the *A* horizon, finer particles are carried away with it. This washing out of fine soil components is termed **eluviation.** As a consequence of eluviation, the texture of the *A* horizon gradually becomes coarser, because a portion of the fine particles is removed. Water percolating downward also dissolves soluble inorganic soil components and carries them to deeper zones. This depletion of soluble materials from the upper soil is termed **leaching.**

Immediately below the *A* horizon is the *B* horizon, or *subsoil*. Much of the material removed from the *A* horizon by eluviation is deposited in the *B* horizon, which is often referred to as the *zone of accumulation*. The accumulation of fine clay particles derived from the *A* horizon enhances water retention in the subsoil. However, in extreme cases, clay accumulation can form an extremely dense and impermeable layer called *hardpan*. Since the *B* horizon has an intermediate position in the soil profile, it may be considered, at least in part, a transitional zone. For example, living organisms and organic matter are more abundant in the *B* than in the *C* horizon, but considerably less so than in the *A* horizon. The *O, A,* and *B* horizons together constitute the **solum,** or "true soil." It is in the solum that the soil-forming processes are active and that living roots and other plant and animal life are largely confined.

Below the solum is the *C* horizon, a layer characterized by partially altered parent material and little if any organic matter. While the parent material may be so dramatically altered in the solum that its original character is not recognizable, it is easily identifiable in the *C* horizon.

The boundaries between soil horizons may be very sharp, or the horizons may blend gradually from one to another. Furthermore, some soils lack horizons altogether. Such soils are called *immature* because soil building has been going on for only a short time. Immature soils are also characteristic of steep slopes where erosion continually strips away the soil, preventing full development.

SOIL TYPES

In the discussion which follows we shall briefly examine some common soil types. As you read, notice that the characteristics of each of the soil types are primarily manifestations of the prevailing climatic conditions. A summary of the characteristics of the soils discussed in this section is provided in Table 2.2.*

The term **pedalfer** gives a clue to the basic characteristic of this soil type. The word is derived

from the Greek **ped**on, meaning "soil," and the chemical symbols **Al** (aluminum) and **Fe** (iron). Pedalfers are characterized by an accumulation of iron oxides and aluminum-rich clays in the *B* horizon. In mid-latitude areas where the annual rainfall exceeds 63 centimeters (25 inches) most of the soluble materials, such as calcium carbonate, are leached from the soil and carried away by underground water. The less soluble iron oxides and clays are carried from the *A* horizon and deposited in the *B* horizon, giving it a brown to red-brown color. These soils are best developed under forest vegetation where large quantities of decomposing organic matter provide the acid conditions necessary for leaching. In the United States pedalfers are found east of a line extending from northwestern Minnesota to south-central Texas.

Pedocal is derived from the Greek **ped**on, meaning "soil," and the first three letters of **cal**cite (calcium carbonate). As the name implies, pedocals are characterized by an accumulation of calcium carbonate. This soil type is found in the drier western United States in association with grassland and brush vegetation. Here rainwater percolating through the soil often evaporates before it can remove soluble materials, chiefly calcium carbonate. The result is a whitish accumulation called *caliche*. In addition, since chemical weathering is less intense in drier areas, pedocals generally contain a smaller percentage of clay minerals than pedalfers.

In the hot and wet climates of the tropics, soils called **laterites** develop. Since chemical weathering is intense under such climatic conditions, these soils are usually deeper than soils developing over a similar period of time in the mid-latitudes. Not only does leaching remove the soluble materials such as calcite, but the great quantities of percolating water also remove much of the silica, with the result that oxides of iron and aluminum become concentrated in the soil. The iron gives the soil a distinctive red color. When dried, laterites are very hard. In fact, some people use this soil for making bricks. If the parent rock contained appreciable amounts of aluminum, the product of weathering is an aluminum-rich accumulation called *bauxite*. Bauxite is the primary ore of aluminum.

---

*For more information on different soil groups and their distribution, see Appendix G.

TABLE 2.2
Summary of soil types.

	Temperate humid (>63 cm rainfall)	Temperate dry (<63 cm rainfall)	Tropical (heavy rainfall)	Extreme arctic or desert
Climate	Temperate humid (>63 cm rainfall)	Temperate dry (<63 cm rainfall)	Tropical (heavy rainfall)	Extreme arctic or desert
Vegetation	Forest	Grass and brush	Grass and trees	Almost none, so no humus develops
Typical Area	Eastern U.S.	Western U.S.		
Soil Type	Pedalfer	Pedocal	Laterite	
Topsoil	Sandy; light colored; acid	Commonly enriched in calcite, whitish color	Enriched in iron (and aluminum); brick red color	No real soil forms because there is no organic material. Chemical weathering is very slow
Subsoil	Enriched in aluminum, iron, and clay; brown color	Enriched in calcite; whitish color	All other elements removed by leaching	
Remarks	Extreme development in conifer forests, because abundant humus makes groundwater very acid. Produces light gray soil because of removal of iron	Caliche is name applied to the accumulation of calcite	Apparently bacteria destroy humus, so no acid is available to remove iron	

(Zones not developed — Topsoil and Subsoil, Tropical column)

Since bacterial activity is very high in the tropics, laterites contain practically no humus. This fact, coupled with the highly leached and bricklike nature of these soils, make laterites poor for growing crops. The infertility of these soils has been borne out repeatedly in tropical countries where cultivation has been expanded into such areas.

In cold or dry climates soils are generally very thin and poorly developed. The reasons for this are fairly obvious. Chemical weathering progresses very slowly in such climates, and the scanty plant life yields very little organic matter.

MASS WASTING

Landslides are spectacular examples of a common geologic process called mass wasting. Mass wasting refers to the downslope movement of rock, regolith, and soil under the direct influence of gravity. Once weathering weakens and breaks rock apart, mass wasting transfers the debris downslope, where a stream, acting as a conveyor belt, carries it away. Although there may be many intermediate stops along the way, the sediment is eventually transported to its ultimate destination, the sea. It is the combined effects of mass wasting and running water that produce stream valleys, which are the most common and conspicuous landforms at the earth's surface. If streams alone were responsible for creating the valleys in which they flow, valleys would be very narrow features. However, the fact that most river valleys are much wider than they are deep is a strong indication of the significance of mass wasting processes in supplying material to streams.

Although gravity is the controlling force of mass wasting, other factors play an important part in bringing about the downslope movement of material. Water is one of these factors. When the pores in sediment become filled with water, the cohesion between the particles is destroyed, allowing them to slide past one another with relative ease. For example, when sand is slightly moist, it may stick together quite well. However, if more water is added, filling the openings between the grains, the sand will ooze out in all directions. Thus, saturation reduces the internal resistance of materials, which are then easily set in motion by the force of gravity. When clay is wetted, it becomes very slick—another example of the "lubricating" effect of water. Water also adds considerable weight to a mass of material. The added weight in itself may be enough to cause the material to slide or flow downslope.

Oversteepening of slopes is another cause for many mass movements. Loose, undisturbed particles assume a stable slope called the **angle of repose,** the steepest angle at which material remains stable. Depending upon the size and shape of the particles, the angle varies from 25 to 40 degrees. The larger, more angular particles maintain the steepest slopes. If the angle is increased, the rock debris will adjust by moving downslope. There are many situations in nature where this takes place. A stream undercutting a valley wall and waves pounding against the base of a cliff are but two familiar examples. Furthermore, through their activities, people often create oversteepened and unstable slopes that become prime sites for mass wasting.

CLASSIFICATION OF MASS WASTING PROCESSES

There is a broad complex of activities that geologists call mass wasting. Generally the different types are divided and described on the basis of the type of material involved, the kind of motion that is displayed, and by the velocity of the movement.

The classification of mass wasting processes based on the material involved in the movement depends upon whether the descending mass began as unconsolidated material or as bedrock. If soil and regolith dominate, terms such as "debris," "mud," or "earth" are used in the description. On the other hand, when a mass of bedrock breaks loose and moves downslope, the term "rock" may be part of the description.

In addition to the type of material involved in a mass wasting event, the way in which material moves may also be important. Generally the kind of motion is described as either a fall, a slide, or a flow.

When the movement involves the free-fall of detached individual pieces of any size, it is termed a *fall*. Fall is a common form of movement on slopes that are so steep that loose material cannot remain on the surface. The rock may fall directly to the base of the slope or move in a series of leaps and bounds over other rocks along the way. Many falls result when freeze and thaw cycles or the action of plant roots loosen rock to the point that gravity takes over.

Many mass wasting processes are described as *slides*. Slides occur whenever material remains fairly coherent and moves along a well-defined surface. Sometimes the surface is a joint, a fault, or a bedding plane that is approximately parallel to the slope. However, in the case of the movement called slump, the descending mass moves along a curved surface of rupture. A note of clarification is appropriate at this point. Sometimes the word "slide" is used as a synonym for the word "landslide." It should be pointed out that although many people, including geologists, use the term, the word "landslide" has no specific definition in geology. Rather it should be considered as a popular nontechnical term to describe all perceptible forms of mass wasting, including those in which sliding does not occur.

The third type of movement common to mass wasting is termed *flow*. Flow occurs when material moves downslope as a viscous fluid. Most flows are saturated with water and typically move as lobes or tongues.

When mass wasting events make the news, a large quantity of material has in all likelihood moved rapidly downslope and had a disastrous

FIGURE 2.11
Debris deposited atop Sherman Glacier in Alaska by a rock avalanche. The event was triggered by a tremendous earthquake in March, 1964. (Photo by Austin Post, U.S. Geological Survey)

effect upon people and property. Indeed, during events called *rock avalanches,* rock and debris can move downslope at speeds well in excess of 200 kilometers (125 miles) per hour. Many researchers believe that rock avalanches, such as the one that produced the scene in Figure 2.11, must literally "float on air" as they move downslope. That is, high velocities result when air becomes trapped and compressed beneath the falling mass of debris, allowing it to move as a buoyant, flexible sheet across the surface.

Most mass movements, however, do not move with the speed of a rock avalanche. In fact, a great deal of mass wasting is imperceptibly slow. One process that we will examine later, termed creep, results in particle movements that are usually measured in millimeters or centimeters per year. Thus, as you can see, rates of movement can be spectacularly sudden or exceptionally gradual. Although various types of mass wasting are often classified as either rapid or slow, such a distinction is highly subjective because there is a wide

FIGURE 2.12
Slump occurs when material
slips downslope en masse
along a curved surface of
rupture.

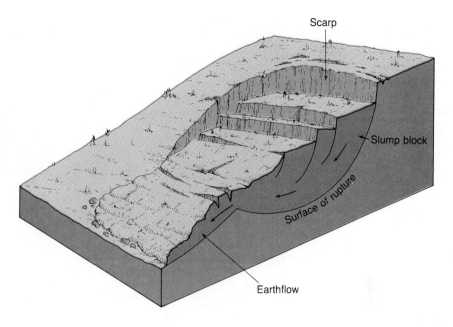

range of rates between the two extremes. Even the velocity of a single process at a particular site can vary considerably from one time to another.

SLUMP

Slump refers to the downward slipping of a mass of rock or unconsolidated material moving as a unit along a curved surface (Figure 2.12). Usually the slumped material does not travel spectacularly fast nor very far. This is a common form of mass wasting, especially in thick accumulations of cohesive materials such as clay. The surface of rupture beneath the slump block is characteristically spoon-shaped and concave upward or outward. As the movement occurs, a crescent-shaped scarp (cliff) is created at the head and the block's upper surface is tilted backwards.

Slump commonly occurs because a slope has been oversteepened. The material on the upper portion of a slope is held in place by the material at the bottom of the slope. As this anchoring material at the base is removed, the material above is made unstable and reacts to the pull of gravity. Slumping may also occur when a slope is over-

loaded, causing internal stress on the material below. This type of slump often occurs where weak, clay-rich material underlies layers of stronger, more resistant rock such as sandstone. The seepage of water through the upper layers reduces the strength of the clay below and slope failure results.

ROCKSLIDE

Rockslides occur when blocks of bedrock break loose and slide downslope. Such events are among the fastest and most destructive mass movements. Usually rockslides take place in a geologic setting where the rock strata are inclined, or joints and fractures exist, parallel to the slope. If the rock is undercut at the base of the slope, it loses support and the rock eventually gives way. Sometimes the rockslide is triggered when rain or melting snow lubricates the underlying surface to the point where friction is no longer sufficient to hold the rock unit in place. As a result, rockslides tend to be most prevalent during the spring, when heavy rains and melting snow are greatest. Earthquakes are another mechanism that triggers

FIGURE 2.13
The scar, 2.4 kilometers (1.5 miles) long and 0.8 kilometer (0.5 mile) wide, left on the side of Sheep Mountain by the Gros Ventre rockslide. (Photo by Stephen Trimble)

rockslides and other mass movements. The 1811 earthquake at New Madrid, Missouri, for example, caused slides in an area of more than 13,000 square kilometers (5000 square miles) along the Mississippi River valley. In terms of property damage, the most devastating effects of the famous 1964 Alaska earthquake were the slides in Anchorage, some 130 kilometers (80 miles) from the center of the quake.

The Gros Ventre River flows west from the Wind River Range in northwestern Wyoming, through Grand Teton National Park, eventually emptying into the Snake River. On June 23, 1925, a classic rockslide took place in its valley, just east of the small town of Kelly. In the span of only a few min-

utes a great mass of sandstone, shale, and soil crashed down the south side of the valley, carrying with it a dense pine forest. The volume of debris, estimated at 38 million cubic meters (50 million cubic yards) created a 67–75-meter-(220–250-foot-) high dam on the Gros Ventre River (Figure 2.13). The river was completely blocked, creating a lake, which filled so quickly that a house that had been 18 meters (60 feet) above the river was floated off its foundation 18 hours after the slide. In 1927 the lake overflowed the dam, partially draining the lake and resulting in a devastating flood downstream.

Why did the Gros Ventre rockslide take place? Figure 2.14 is a diagrammatic cross-sectional view

FIGURE 2.14
Cross-sectional view of the Gros Ventre rockslide. The slide occurred when the tilted and undercut sandstone bed could no longer maintain its position atop the saturated bed of clay. (After W. C. Alden, "Landslide and Flood at Gros Ventre, Wyoming." *Transactions* (AIME) 76 (1928):348)

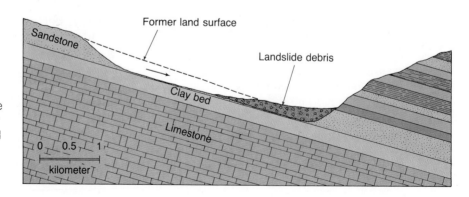

of the geology of the valley. Notice the following points: (1) The sedimentary strata in this area dip (tilt) 15–21 degrees; (2) Underlying the bed of sandstone is a relatively thin layer of clay; and (3) At the bottom of the valley the river had cut through much of the sandstone layer. During the spring of 1925 water from the heavy rains and melting snow seeped through the sandstone, saturating the clay below. Since much of the sandstone layer had been cut through by the Gros Ventre River, the layer had virtually no support at the bottom of the slope. Eventually the sandstone could no longer hold its position on the wetted clay, and gravity pulled the mass down the side of the valley. The circumstances at this location were such that a rockslide was inevitable.

MUDFLOW

Mudflow is a relatively rapid type of mass wasting that involves a flowage of debris containing a large amount of water. Mudflows are most characteristic of semiarid mountainous regions and, because of their high water content and the predominance of fine particles, they tend to follow canyons and gullies. Although rains in semiarid regions are in-frequent, they are typically heavy. When a cloudburst or rapidly melting mountain snow cause a sudden flood, large quantities of soil and regolith are washed into nearby stream channels because there is usually little or no vegetation to anchor the surface material. This creates a flowing tongue of well-mixed mud, soil, rock, and water. Its consistency may range from that of wet concrete to a soupy mixture not much thicker than muddy water. The rate of flow therefore depends not only on the slope but on the water content as well. When dense, mudflows are capable of carrying or pushing large boulders, trees, and even houses with relative ease. Mudflows periodically are a serious hazard in dry mountainous areas such as portions of southern California. Here construction of homes on canyon hillsides and the removal of native vegetation by brush fires and other means have increased the frequency of these destructive events (Figure 2.15).

Mudflows are also common on the slopes of some volcanoes, in which case they are termed **lahars.** The word originated in Indonesia, a volcanic region that has experienced many of these often destructive events. Lahars result when highly unstable layers of ash and debris become

FIGURE 2.15
Severe damage resulted when a mudflow buried the lower portion of this house located near the mouth of a canyon in southern California. (Photo by James E. Patterson)

saturated with water and flow down steep volcanic slopes, generally following existing stream channels. Some are initiated when heavy rainfalls erode volcanic deposits. Others are triggered when large volumes of ice and snow are suddenly melted by heat flowing to the surface from within the volcano or by the near-molten debris emitted during a violent eruption.

In November, 1985, Nevado del Ruiz, a large volcano in the Andes Mountains of Colombia, erupted. The eruption melted much of the snow and ice that capped the peak, producing a torrent of hot viscous mud, ash, and debris. The lahar moved outward from the volcano following the valleys and three rain-swollen rivers that radiate from the peak. The flow that moved down the valley of the Lagunilla River was the most destructive, devastating the town of Armero, 48 kilometers (30 miles) from the mountain. Most of the more than 20,000 deaths caused by the event were in this once-thriving agricultural community. Death and property damage also occurred in 13 other villages within the 180-square-kilometer disaster area. Although a great deal of material was explosively ejected from Nevado del Ruiz, it was the lahars triggered by this eruption that made this such a devastating natural disaster.

EARTHFLOW

Unlike mudflows, which are usually confined to channels in semiarid regions, **earthflows** most often form on hillsides in humid areas as the result of excessive rainfall. When water saturates clay-rich regolith on a hillslope, the material may break away and flow a short distance downslope, leaving a scar on the hillside (Figure 2.16). Depending upon the steepness of the slope and the material's consistency, the speed of an earthflow may vary from a few meters per hour to several meters per minute. However, since earthflows are quite viscous, they generally move more slowly than the more fluid mudflows described in the preceding section. In addition to occurring as isolated hillside phenomena, earthflows commonly take place in association with large slumps. In this situation, they may be seen as tonguelike flows at the base of the slump block (see Figure 2.12).

CREEP

Movements such as rockslides and rock avalanches are certainly the most spectacular and catastrophic forms of mass wasting. Since these events have been known to kill thousands, they

FIGURE 2.16
This earthflow occurred in clay-rich material on a newly formed slope along a recently constructed highway. Notice the small slump at the head of the earthflow. (Photo by E. J. Tarbuck)

deserve intensive study so that, through more effective prediction, timely warnings and better controls can help save lives. However, because of
their large size and spectacular nature they give
us a false impression of their importance as a
mass wasting process. Indeed, sudden movements are responsible for moving less material
than the slow and far more subtle action of creep.
Whereas rapid types of mass wasting are characteristic of mountains and steep hillsides, creep can
take place on gentle slopes and is thus much more
widespread.

Creep is a type of mass wasting that involves
the gradual downhill movement of soil and regolith. One of the primary causes of creep is the alternate expansion and contraction of surface material caused by freezing and thawing or wetting
and drying. As shown in Figure 2.17, freezing or
wetting lifts the soil at right angles to the slope,
and thawing or drying allows the particles to fall
back to a slightly lower level. Each cycle therefore
moves the material a short distance downhill.
Creep may also be initiated if the ground becomes
saturated with water. Following a heavy rain or
snowmelt, a waterlogged soil may lose its internal
cohesion, allowing gravity to pull the material
downslope. Although the movement is imperceptibly slow, its effects are recognizable. Creep
causes fences and telephone poles to tilt, and tree
trunks will often be bent as a consequence of this
movement.

SOLIFLUCTION

In the frigid zones found at high latitudes **solifluction** is of major importance. In such regions only
the ice in the upper few meters of regolith melts
during the spring and summer, while the ground
below remains permanently frozen. Since the
water in the upper zone has nowhere to go, this

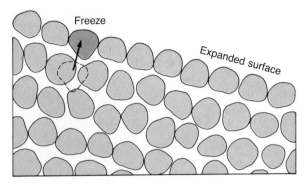

A.

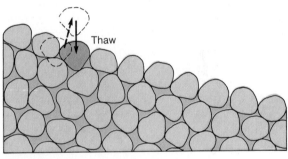

B.

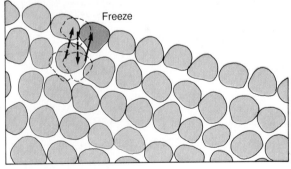

C.

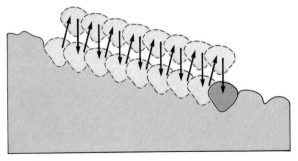

D.

FIGURE 2.17
The repeated expansion and contraction of the surface material causes a net downslope migration of
rock particles—a process called creep.

layer remains saturated and slowly flows down even the gentlest slopes (Figure 2.18). In this manner the overburden is removed and the unweath-ered rock below is exposed. When the newly exposed rock is eventually weathered, it too will be removed by solifluction.

FIGURE 2.18
Solifluction lobes northeast of Fairbanks, Alaska. (Photo by James E. Patterson)

REVIEW
QUESTIONS

1   If two identical rocks were weathered, one mechanically and the other chemically, how would the products of weathering for the two rocks differ?

2   How does mechanical weathering add to the effectiveness of chemical weathering?

3   Describe the formation of an exfoliation dome. Give an example of such a feature.

4   Granite and basalt are exposed at the surface in a hot, wet region.
    (a)   Which type of weathering will predominate?
    (b)   Which of the rocks will weather most rapidly? Why?

5   Heat speeds up a chemical reaction. Why then does chemical weathering proceed slowly in a hot desert?

6   How is carbonic acid (H_2CO_3) formed in nature? What results when this acid reacts with potassium feldspar?

7   What is the difference between soil and regolith?

8   Using the soil texture diagram (Figure 2.8), name the soil that consists of 60 percent sand, 30 percent silt, and 10 percent clay.

9 What factors might cause different soils to develop from the same parent material, or similar soils to form from different parent materials?

10 Which of the controls of soil formation is most important? Explain.

11 How can slope affect the development of soil? What is meant by the term *slope orientation*?

12 List the characteristics associated with each of the horizons in a well-developed soil profile. Which of the horizons constitute the solum? Under what circumstances do soils lack horizons?

13 Distinguish between pedalfers and pedocals.

14 Soils formed in the humid tropics and the Arctic both contain little organic matter. Do both lack humus for the same reasons?

15 What role does mass wasting play in sculpturing the earth's landscape?

16 What is the controlling force of mass wasting? What other factors are important?

17 Distinguish among fall, slide, and flow.

18 Why can rock avalanches move at such great speeds?

19 What factors led to the massive rockslide at Gros Ventre, Wyoming (Figure 2.14)?

20 What type of mass wasting event killed thousands of people in the vicinity of the Colombian volcano Nevado del Ruiz in 1985?

21 Compare and contrast mudflow and earthflow.

22 Since creep is an imperceptibly slow process, what evidence might indicate to you that this phenomenon is affecting a slope? Describe the mechanism that creates this slow movement (Figure 2.17).

23 Why is solifluction only a summertime phenomenon?

KEY TERMS

weathering	soil	pedalfer
erosion	humus	pedocal
mass wasting	soil texture	laterite
mechanical weathering	parent material	angle of repose
frost wedging	residual soil	slump
talus slope	transported soil	rockslide
sheeting	horizon	mudflow
exfoliation dome	soil profile	lahar
chemical weathering	eluviation	earthflow
spheroidal weathering	leaching	creep
regolith	solum	solifluction

3

RUNNING WATER
AND GROUNDWATER

The Lower Falls of the Grand Canyon of the Yellowstone. An excellent example of a valley in the youthful stage of development. (Photograph used by permission of Dennis Tasa)

All the rivers run into the sea; yet the sea is not full; unto the place from whence the rivers come, thither they return again. (Ecclesiastes 1:7)

As the perceptive writer of Ecclesiastes indicated, water is continually on the move, from the ocean to the land and back again. This chapter deals with that part of the water cycle which involves the return of water to the sea, with some water traveling quickly via a rushing stream, and some moving more slowly below the surface. We shall examine the factors that control the movement of water, as well as look at how water sculptures the landscape and at the features that result. To a great extent, the Grand Canyon, Niagara Falls, Old Faithful, and Mammoth Cave all owe their existence to the activities of water on its way to the sea.

THE HYDROLOGIC CYCLE

The amount of water on earth is immense, an estimated 1.36 billion cubic kilometers (326 million cubic miles). Of this total, the vast bulk—97.2 percent—is part of the world ocean. Icecaps and glaciers account for another 2.15 percent, leaving only 0.65 percent to be divided among lakes, streams, subsurface water, and the atmosphere. Although the percentage of the earth's total water found in each of the latter sources is but a small fraction of the total inventory, the absolute quantities are great.

An adequate supply of water is vital to life on earth. With increasing demands on this finite resource, scientists have given a great deal of attention to the exchanges of water among the oceans, the atmosphere, and the continents. This unending circulation of the earth's water supply has come to be called the **hydrologic cycle.** It is a gigantic system powered by energy from the sun in which the atmosphere provides the vital link between the oceans and continents. Water from the oceans, and to a much lesser extent from the continents, is constantly evaporating into the atmosphere. Winds transport the moisture-laden air, often great distances, until the complex processes of cloud formation are set in motion that eventually result in precipitation. The precipitation that falls into the ocean has ended its cycle and is ready to begin another. The water that falls on the continents, however, must still make its way back to the ocean.

What happens to precipitation once it has fallen on land? A portion of the water soaks into the ground, some of it moving downward, then laterally, finally seeping into lakes, streams, or directly into the ocean. When the rate of rainfall is greater than the earth's ability to absorb it, the additional water flows over the surface into lakes and streams. Much of the water which soaks in **(infiltration)** or runs off **(runoff)** eventually finds its way back to the atmosphere because of evaporation from the soil, lakes, and streams. Also, some of the water that infiltrates the ground surface is absorbed by plants, which then release it into the atmosphere. This process is called **transpiration.** Each year a field of crops may transpire an amount of water equivalent to a layer 60 centimeters (2 feet) deep over the entire field, whereas a forest may pump twice this amount into the atmosphere.

When precipitation falls at high elevations or high latitudes, the water may not immediately soak in, run off, or evaporate. Instead it may become part of a snowfield or a glacier. Glaciers store large quantities of water on land. If present-day glaciers were to melt and release their storage of water, sea level would rise by several tens of meters and submerge many heavily populated coastal areas. As we shall see in Chapter 4, over the past two million years, huge continental glaciers have formed and melted on several occasions.

A diagram of the earth's water balance, a quantitative view of the hydrologic cycle, is shown in Figure 3.1. Although the amount of water vapor in the air at any one time is but a minute fraction of the earth's total water supply, the absolute quantities that are cycled through the atmosphere over a one-year period are immense—some 380,000 cubic kilometers—enough to cover the earth's surface to a depth of about one meter (39 inches). Since the total amount of water vapor in the atmosphere remains about the same, average annual precipitation over the earth must be equal to the quantity of water evaporated. However, for all of the continents taken together, precipitation exceeds evaporation. Conversely, over the oceans, evaporation exceeds precipitation. Since the level of the world ocean is not dropping, runoff from land areas must balance the deficit of precipitation over the oceans.

The hydrologic cycle represents the continuous movement of water from the oceans to the atmosphere, from the atmosphere to the land, and from the land back to the sea. The wearing down of the earth's land surface is primarily attributable to the last of the steps in this cycle.

RUNNING WATER

Running water is of great importance to people. We depend upon rivers for energy, travel, and irrigation; their fertile floodplains have fostered human progress since the dawn of civilization. Moreover, as the dominant agent of landscape al-

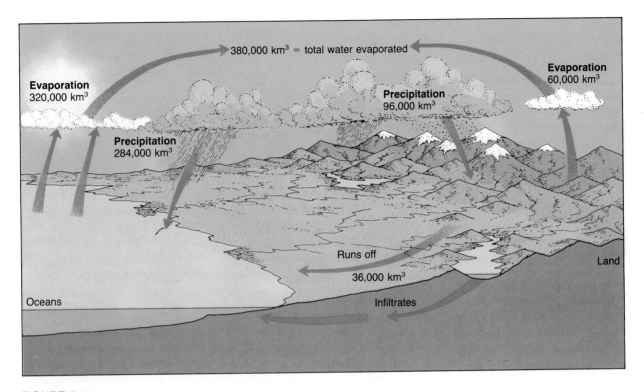

FIGURE 3.1
The earth's water balance. About 320,000 cubic kilometers of water are evaporated each year from the oceans, while evaporation from the land (including lakes and streams) contributes 60,000 cubic kilometers of water. Of this total of 380,000 cubic kilometers of water, about 284,000 cubic kilometers fall back to the ocean, and the remaining 96,000 cubic kilometers fall onto the earth's land surface. Since 60,000 cubic kilometers of water evaporate from the land, 36,000 cubic kilometers of water remain to erode the land during the journey back to the oceans.

teration, streams have sculptured much of our physical environment.

Although people have always depended to a great extent on running water, its source eluded them for centuries. It was not until the sixteenth century that they first realized streams were supplied by runoff and underground water, which ultimately had their sources as rain and snow. Runoff initially flows in broad sheets that enter small rills, which carry it to a stream. The word *stream* is used to denote channelized flow of any size, from the smallest brook to the mighty Amazon. Although the terms *river* and *stream* are used synonymously, the term *river* is often preferred when

describing a main stream into which several tributaries flow.

STREAMFLOW

Flowing water makes its way to the sea under the influence of gravity. The time required for the journey depends upon the velocity of the stream, which is measured in terms of the distance the water travels in a given unit of time. Some sluggish streams travel at less than 0.8 kilometer (0.5 mile) per hour, while a few rapid ones reach speeds as high as 32 kilometers (20 miles) per hour. Velocities are determined at gauging

FIGURE 3.2
Influence of channel shape on velocity. Although the cross-sectional area of all three channels is the same, the semicircular channel has less water in contact with the channel, and hence less frictional drag. As a result, the water will flow more rapidly in this channel, all other factors being equal.

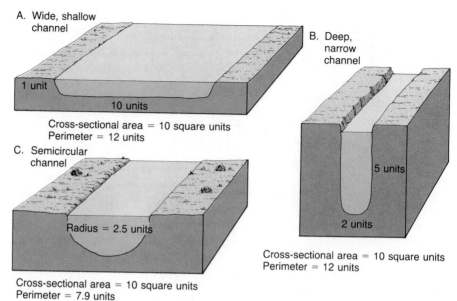

A. Wide, shallow channel

1 unit

10 units

Cross-sectional area = 10 square units
Perimeter = 12 units

C. Semicircular channel

Radius = 2.5 units

Cross-sectional area = 10 square units
Perimeter = 7.9 units

B. Deep, narrow channel

5 units

2 units

Cross-sectional area = 10 square units
Perimeter = 12 units

stations where measurements are taken at several locations across the channel and then averaged. Along straight stretches the highest velocities are near the center of the channel just below the surface, where friction is lowest. But when a stream curves, its zone of maximum speed shifts toward its outer bank.

The ability of a stream to erode and transport materials is directly related to its velocity; thus, it is a very important characteristic. Even slight variations in velocity can lead to significant changes in the load of sediment transported by the water. Several factors determine the velocity of a stream and therefore control the amount of erosional work a stream may accomplish. These factors include the following: (1) Gradient, (2) Shape, size, and roughness of the channel, and (3) Discharge.

Certainly one of the most obvious factors controlling stream velocity is the **gradient,** or slope, of a stream channel. Gradient is typically expressed as the vertical drop of a stream over a fixed distance. Gradients may vary considerably from one stream to another as well as along the course of a given stream. Portions of the lower Mississippi River, for example, have gradients of 10 centimeters per kilometer and less. By way of contrast, some mountain stream channels decrease in elevation at a rate of more than 40 meters per kilome-

ter, 400 times more abruptly than the lower Mississippi. The higher the gradient, the more energy available for streamflow. If two streams were identical in every respect except gradient, the stream with the higher gradient would obviously have the greater velocity.

The cross-sectional shape of a channel determines the amount of water in contact with the channel and hence affects the frictional drag. The most efficient channel is one with the least perimeter for its cross-sectional area. Figure 3.2 compares three types of channels. Although the cross-sectional area of all three is identical, the semicircular shape has less water in contact with the channel than the others and therefore less frictional drag. As a result, if all other factors are equal, the water will flow more rapidly in the semicircular channel.

The size and roughness of the channel also affect the amount of friction. An increase in the size of a channel reduces the ratio of perimeter to cross-sectional area and therefore increases the efficiency of flow. The effect of roughness is obvious. A smooth channel promotes a more uniform flow, whereas an irregular channel filled with boulders creates enough turbulence to significantly retard the stream's forward motion.

The **discharge** of a stream is the amount of

water flowing past a certain point in a given unit of time. This is usually measured in cubic meters per second or cubic feet per second. Discharge is determined by multiplying a stream's cross-sectional area by its velocity.

The largest river in North America, the Mississippi, discharges an average of 17,715 cubic meters (625,340 cubic feet) per second. Although this is a huge quantity of water, it is nevertheless dwarfed by the mighty Amazon, the world's largest river. Draining a rainy region that is nearly three-quarters the size of the conterminous United States, the Amazon discharges 10 times more water than the Mississippi. In fact, the flow of the Amazon accounts for about 15 percent of all the fresh water discharged into the ocean by all of the world's rivers. Just one day's discharge would supply the water needs of New York City for 9 years!

The discharges of rivers are far from constant. This is true because of such variables as rainfall and snowmelt. If discharge changes, the factors noted earlier must also change. When discharge increases, the width or depth of the channel must increase or the water must flow faster, or some combination of these factors must change. Indeed, measurements show that when the amount of water in a stream increases, the width, depth, and velocity all increase. In order to handle the additional water, the stream will increase the size of its channel by widening and deepening it. As we saw earlier, when the size of the channel increases, proportionally less of the water is in contact with the bed and banks of the channel. This means that friction, which acts to retard the flow,

is relatively decreased. The less friction, the more swiftly the water will flow.

CHANGES DOWNSTREAM

One useful way of studying a stream is to examine its *longitudinal profile*. Such a profile is simply a cross-sectional view of a stream from its source area (called the *head* or *headwaters*) to its *mouth,* the point downstream where the river empties into another water body. By examining Figure 3.3, you can see that the most obvious feature of a typical longitudinal profile is a constantly decreasing gradient from the head to the mouth. Although many local irregularities may exist, the overall profile is a smooth concave-upward curve.

The longitudinal profile shows that the gradient decreases downstream. To see how other factors change in a downstream direction, observations and measurements must be made. When data are collected from successive gauging stations along a river, they show that discharge increases toward the mouth. This should come as no surprise since, as we move downstream, more and more tributaries contribute water to the main channel. Furthermore, in most humid regions, additional water is continually being added from the groundwater supply. Since this is the case, the width, depth, and velocity must change in response to the increased volume of water carried by the stream. Indeed, the downstream changes in these variables have been shown to vary in a manner similar to what occurs when discharge increases at one

FIGURE 3.3
A longitudinal profile is a cross section along the length of a stream. Note the concave-upward curve of the profile, with a steeper gradient upstream and a gentler gradient downstream.

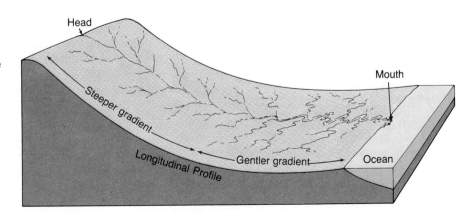

place; that is, width, depth, and velocity all increase systematically.

The observed increase in velocity that occurs downstream contradicts our impressions about wild, rushing mountain streams and wide, placid rivers. The mental picture that we may have of "old man river just rollin' along" is just not so. Although a mountain stream may have the appearance of a raging torrent, its average velocity is often less than for the river near its mouth. The difference is primarily attributable to the greater efficiency of the larger channel in a downstream direction.

In the headwaters region where the gradient is steep, the water must flow in a relatively small and often boulder-strewn channel. The small channel and rough bed create great drag and inhibit movement by sending water in all directions with almost as much backward motion as forward motion. However, downstream, the material on the bed of the stream becomes much smaller, offering less resistance to flow, and the width and depth of the channel increase to accommodate the greater discharge. These factors, especially the wider and deeper channel, permit the water to flow more freely and hence more rapidly.

In summary, we have seen that there is an inverse relationship between gradient and discharge. Where gradient is high, discharge is small, and where discharge is great, gradient is small. Stated another way, a stream can maintain a higher velocity near its mouth even though it has a lower gradient than upstream because of the greater discharge, larger channel, and smoother bed.

BASE LEVEL AND GRADED STREAMS

An important control over streamflow is base level. **Base level** is the lowest point to which a stream can erode its channel. Two general types of base level exist. Sea level is considered the **ultimate base level,** since it represents the lowest level to which stream erosion could lower the land. **Temporary,** or **local, base levels** include lakes, resistant rock, and main streams which act as base lev-

els for their tributaries. All have the capacity to limit a stream at a certain level. For example, when a stream enters a lake, its velocity quickly approaches zero and its ability to erode decreases. Thus the lake prevents the stream from eroding below its level at any point upstream from the lake (Figure 3.4). However, since the outlet of the lake can cut downward and drain the lake, the lake is only a temporary hindrance to the stream's ability to downcut its channel.

Any change in base level will cause a corresponding readjustment of stream activities. When a dam is built along a stream course, the reservoir which forms behind it raises the base level of the stream (Figure 3.5). Upstream from the dam the stream gradient is reduced, lowering its velocity and, hence, its sediment-transporting ability. The stream, now unable to transport all of its load, will deposit material, thereby building up its channel. This process continues until the stream again has a gradient sufficient to carry its load. The profile of the new channel would be similar to the old, except that it would be somewhat higher.

If, on the other hand, the base level were lowered, either by an uplift of the land or by a drop in base level, the stream would readjust. The stream, now above base level, would have excess energy and would downcut its channel to establish a balance with its new base level. Erosion would first progress near the mouth, then work upstream until the stream profile was adjusted along its full length.

The observation that streams adjust their profile for changes in base level led to the concept of a graded stream. A **graded stream** has the correct slope and other channel characteristics necessary to maintain just the velocity required to transport the material supplied to it. On the average, a graded system is not eroding or depositing material but is simply transporting it. Once a stream has reached this state of equilibrium, it becomes a self-regulating system in which a change in one characteristic causes an adjustment in the others to counteract the effect. Referring again to our example of a stream adjusting to a lowering of its base level, the stream would not be graded while cutting its new channel but would achieve this state after downcutting had ceased.

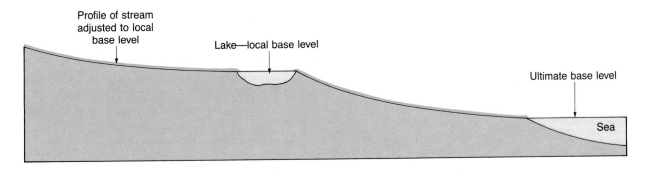

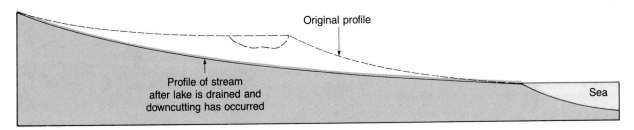

FIGURE 3.4
The lake acts as local base level. As long as the lake is present, the stream cannot erode below the level of the lake. If the lake is drained, the stream will
downcut toward sea level (ultimate base level).

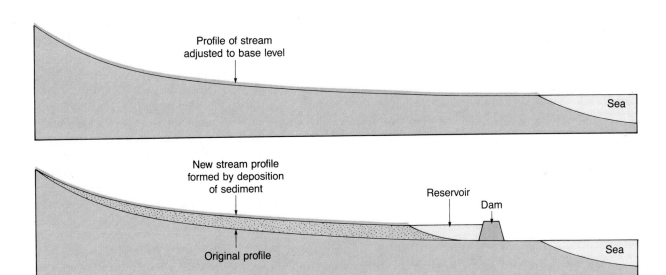

FIGURE 3.5
When a dam is built and a reservoir forms, the stream's base level is raised. This
reduces the stream's velocity and leads to deposition and a reduction of the gradient upstream from the reservoir.

WORK OF STREAMS

The work of streams includes erosion, transportation, and deposition. These activities go on simultaneously in all stream channels, even though they are presented individually here.

EROSION

Although much of the material carried by streams has been brought in by underground water, overland flow, and mass wasting, streams do contribute to their load by eroding their own channels. If a channel is composed of bedrock, most of the erosion is accomplished by the abrasive action of water armed with sediment, a process analogous to sandblasting. Pebbles caught in eddies serve as cutting tools and bore circular holes called **potholes** into the channel floor. In channels consisting of unconsolidated material, considerable lifting can be accomplished by the impact of water alone.

TRANSPORTATION

Once streams acquire their load of sediment, they transport it in three ways: (1) In solution (**dissolved load**); (2) In suspension (**suspended load**); and (3) Along the bottom (**bed load**).

The dissolved load is brought to the stream by groundwater and to a lesser degree is acquired directly from soluble rock along the stream's course. The quantity of material carried in solution is highly variable and depends upon such factors as climate and the geologic setting. Usually the dissolved load is expressed as parts of dissolved material per million parts of water (parts per million, or ppm). Although some rivers may have a dissolved load of 1000 ppm or more, the average figure for the world's rivers is estimated to be between 115 and 120 ppm. Almost 4 billion metric tons of dissolved mineral matter are supplied to the oceans each year by streams.

Most, but not all, streams carry the bulk of their load as suspended load. Usually only fine sand-, silt-, and clay-sized particles can be carried this way, but during floodstage much larger particles are carried as well. Also during floodstage, the total quantity of material carried in suspension increases dramatically, as can be verified by persons whose homes have been sites for the deposition of this material.

A portion of a stream's load of solid material consists of sediment that is too large to be carried in suspension. These coarser particles move along the bottom of the stream and constitute the bed load. In terms of the erosional work accomplished by a downcutting stream, the grinding action of the bed load is of great importance.

The particles composing the bed load move along the bottom by rolling, sliding, and saltation. Sediment moving by **saltation** appears to jump or skip along the stream bed. This occurs as particles are propelled upward by collisions or sucked upward by the current and then carried downstream a short distance until gravity pulls them back to the bed of the stream. Particles that are too large or heavy to move by saltation either roll or slide along the bottom, depending upon their shape.

Unlike the suspended and dissolved loads, which are constantly in motion, the bed load is in motion only intermittently, when the force of the water is sufficient to move the larger particles. Although the bed load may constitute up to 50 percent of the total load of a few streams, it usually does not exceed 10 percent of a stream's total load. For example, consider the distribution of the 750 million tons of material carried to the Gulf of Mexico by the Mississippi River each year. Of this total, it is estimated that approximately 500 million tons are carried in suspension, 200 million tons in solution, and the remaining 50 million tons as bed load. Estimates of a stream's bed load, however, should be viewed cautiously because this fraction of the load is very difficult to measure accurately. Not only is the bed load more inaccessible than the suspended and dissolved loads, but it moves primarily during periods of flooding when the bottom of a stream channel is most difficult to study.

A stream's ability to carry its load is established using two criteria. First, the **competence** of a stream is a measure of the maximum size of particles it is capable of transporting. The stream's velocity determines its competence. If the velocity of a stream doubles, the impact force of water increases four times; if the velocity triples, the force increases nine times; and so forth. Hence, the huge boulders which are visible during the low-water stage and seem immovable can, in fact, be transported during floodstage because of the stream's increased velocity. Second, the maximum load a stream can carry is termed its **capacity.** The capacity of a stream is directly related to its discharge. The greater the amount of water flowing in the stream, the greater the stream's capacity for hauling sediment.

DEPOSITION

Whenever a stream's velocity is reduced, its competence is also lowered. Consequently, some of the suspended particles begin to settle out. Sediment deposited in this manner is termed **alluvium.** Although some material temporarily settles

in the channel, most eventually reaches the ocean. When a stream enters the relatively still waters of an ocean or lake, its forward motion is quickly lost, and the resulting deposits form a **delta.** The finer silts and clays will settle out some distance from the mouth into nearly horizontal layers called *bottomset beds* (Figure 3.6A). Prior to the accumulation of bottomset beds, *foreset beds* begin to form. These beds are composed of coarse sediment, which is dropped almost immediately upon entering a lake or ocean, forming sloping layers. The foreset beds are usually covered by thin, horizontal *topset beds* deposited during floodstage. As the delta grows outward, the gradient of the river is continually lowering, causing the stream to search for a shorter route to base level, a process illustrated in Figure 3.6B. This illustration

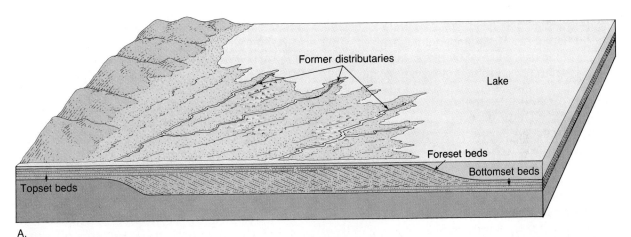

 A.

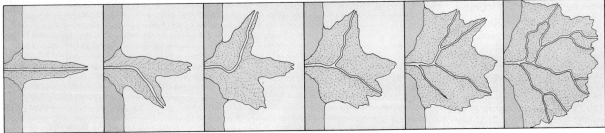

B.

FIGURE 3.6
A. Structure of a simple delta that forms in the relatively quiet waters of a lake.
B. Growth of a simple delta. Once a stream has extended its channel, the reduced gradient causes it to find a shorter distance to its base level. (After Ward's Natural Science Establishment, Inc., Rochester, N.Y.)

also shows the main channel dividing into several smaller ones called **distributaries.** Most deltas are characterized by these shifting channels that act in an opposite way to that of tributaries. Rather than carrying water into the main channel, distributaries carry water away from the main channel in varying paths to base level. After numerous shifts of the channel, the simple delta grows into the idealized triangular shape of the Greek letter delta (Δ), for which it was named. Note, however, that many deltas do not exhibit this idealized shape. Differences in the configurations of shorelines and variations in the nature and strength of wave activity result in many shapes.

Many large rivers have deltas extending over thousands of square kilometers. The Mississippi delta began forming millions of years ago near the present-day town of Cairo, Illinois, and has since advanced nearly 1600 kilometers (1000 miles) to the south. New Orleans rests where there was ocean less than 5000 years ago. Figure 3.7 shows that portion of the Mississippi delta that has been built over the past 5000–6000 years. As shown, the delta is actually a series of seven coalescing

subdeltas. Each formed when the river left its existing channel in favor of a shorter, more direct path to the Gulf of Mexico. The individual subdeltas interfinger and partially cover one another to produce a very complex structure. The present subdelta, called a *bird-foot* delta because of the configuration of its distributaries, has been built by the Mississippi in the last 500 years.

Alluvial fans are features similar to deltas which form on land. When mountain streams reach a plain, their gradient is abruptly reduced and they immediately dump much of their load. Usually the coarse material is dropped near the base of the slope, while finer material is carried farther out on the plain.

Rivers that occupy valleys with broad, flat valley floors on occasion build a landform called a **natural levee** that parallels its channel. Natural levees are built by successive floods over many years. When a stream overflows its banks, its velocity immediately diminishes, leaving coarse sediment deposited in strips bordering the channel (Figure 3.8). As the water spreads out over the valley, a lesser amount of fine sediment is deposited over

FIGURE 3.7
During the past 5000-6000 years, the Mississippi River has built a series of seven coalescing subdeltas. The numbers indicate the order in which the subdeltas were deposited. The present bird-foot delta (number 7) represents the activity of the past 500 years. Without ongoing human efforts, the present course will shift and follow the path of the Atchafalaya River (inset). (After C. R. Kolb and J. R. Van Lopik, *Depositional Environments of the Mississippi River Deltaic Plain*, p. 22. Copyright © 1966 by the Houston Geological Society)

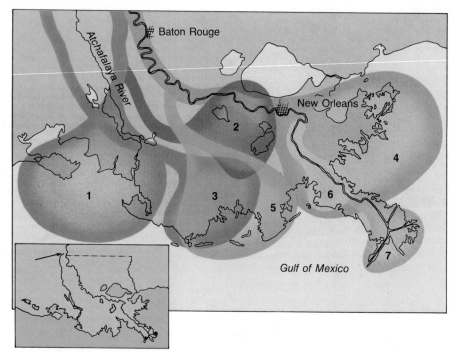

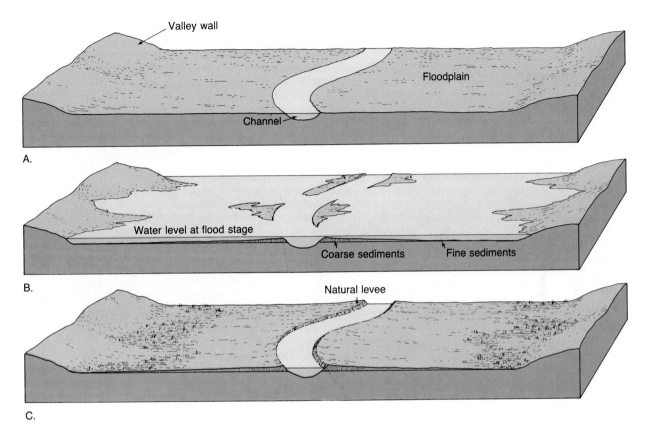

Valley wall

Floodplain

Channel

A.

Water level at flood stage

Coarse sediments   Fine sediments

B.

Natural levee

C.

FIGURE 3.8
Formation of natural levees. After repeated flooding, streams may build very
gently sloping levees.

the valley floor. This uneven distribution of material produces the very gentle slope of the natural levee. The natural levees of the lower Mississippi rise 6 meters (20 feet) above the valley floor. The area behind the levee is characteristically poorly drained for the obvious reason that water cannot flow up the levee and into the river. Marshes called **back swamps** result. A tributary stream that attempts to enter a river with natural levees often has to flow parallel to the main stream until it can breach the levee. Such streams are called **yazoo tributaries** after the Yazoo River, which parallels the Mississippi for over 300 kilometers.

Sometimes artificial levees are built along rivers as a means of flood control. These are usually easy to distinguish from natural levees because their slopes are much steeper. When a river is con-fined by levees during periods of high water, it deposits material in its channel as the discharge diminishes. This is sediment that otherwise would have been dropped on the floodplain. Thus, each time there is a high flow, deposits are left on the river bed and the bottom of the channel is built up. With the buildup of the bed, less water is required to overflow the original levee. As a result, the height of the levee must be raised to protect the floodplain. For this reason, many levees along the lower Mississippi River have had to be raised to cope with the increasing height of the water in the channel. As you can see, artificial levees are not a permanent solution to the problem of flooding. If protection is to be maintained, the structures must be heightened periodically, a process that cannot go on indefinitely.

FIGURE 3.9
The smaller American Falls at Niagara Falls. The river plunges over the falls and erodes the shale beneath the more resistant Lockport Dolomite. As a section of dolomite is undercut, it loses support and breaks off. (Photo by James E. Patterson)

STREAM VALLEYS

Stream valleys can be divided into two general types. Narrow V-shaped valleys and wide valleys with flat floors exist as the ideal forms, with many gradations between. In arid regions narrow valleys often have nearly vertical walls, while in humid regions the effect of mass wasting and slope erosion caused by heavy rainfall produce the typical V-shaped valley (see chapter-opening photo). The type of valley gives an indication of what the stream has been doing. A narrow V-shaped valley indicates that the primary work of the stream has been downcutting toward base level. On the other hand, streams with flat-floored valleys have been widened by lateral (side-to-side) erosion.

The most prominent features of a narrow valley are **rapids** and **waterfalls.** Both occur where the stream profile drops rapidly, a situation usually caused by variations in the erodibility of the bedrock into which the stream channel is cutting. Resistant beds create rapids by acting as a temporary

base level upstream while allowing downcutting to continue downstream. Once erosion has eliminated the resistant rock, the stream profile smooths out again. Waterfalls are places where the stream profile makes a vertical drop. One type of waterfall is exemplified by Niagara Falls (Figure 3.9). Here the falls are supported by a resistant bed of dolomite that is underlain by a less resistant shale. As the water plunges over the lip of the falls it erodes the less resistant shale, undermining a section of dolomite, which eventually breaks off. In this manner the waterfall retains its vertical cliff while slowly but continually retreating upstream. Since its formation Niagara Falls has retreated approximately 11 kilometers (7 miles) upstream.

Once a stream has cut its channel closer to base level it begins to reach a graded condition, and downward erosion becomes less dominant. At this point more of the stream's energy is directed from side to side. The reason for this change is not fully understood, but the reduced gradient probably is an important factor. Nevertheless it does occur, and the result is a widening of the valley as the river cuts away first at one bank and then at the other (Figure 3.10). In this manner the flat valley floor, or **floodplain,** is produced. It is appropriately named because the river is confined to its channel, except during floodstage, when it overflows its banks and inundates the floodplain.

When a river erodes laterally, creating a floodplain as just described, it is called an *erosional floodplain*. Floodplains can be depositional in nature as well. *Depositional floodplains* are produced by a major fluctuation in conditions, such as a

change in base level. The floodplain in Yosemite Valley is one such feature, and was produced when a glacier gouged the former stream valley about 300 meters (985 feet) deeper than it had been. After the glacial ice melted, the stream readjusted to its former base level by refilling the valley with alluvium.

Streams that flow upon floodplains, whether erosional or depositional, move in sweeping bends called **meanders.** Meanders continually change position by eroding sideways and slightly downstream. The sideways movement occurs because the maximum velocity of the stream shifts toward

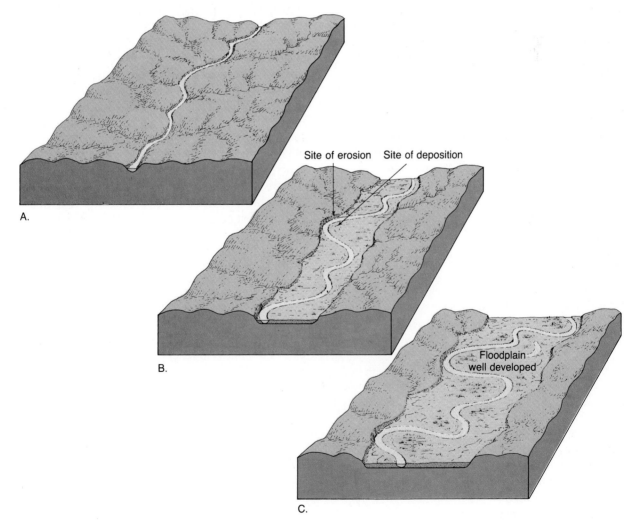

FIGURE 3.10
Stream eroding its floodplain.

FIGURE 3.11
Lateral movement of mean-
ders. By eroding its outer
bank and depositing material
on the inside of the bend, a
stream is able to shift its
channel.

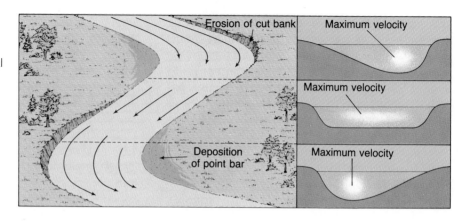

the outside of the bend, causing erosion of the
outer bank (Figure 3.11). At the same time the
reduced current at the inside of the meander results in the deposition of coarse sediment, especially sand. Thus by eroding its outer bank and
depositing material along its inner bank, a stream
moves sideways without changing its channel
size. Due to the slope of the channel, erosion is
more effective on the downstream side of a meander. Therefore, in addition to growing laterally, the
bends also gradually migrate down the valley.
Sometimes the downstream migration of a meander is slowed when it reaches a more resistant portion of the floodplain. This allows the next meander upstream to "catch up." Gradually the neck of
land between the meanders is narrowed. When
they get close enough, the river may erode
through the narrow neck of land to the next loop

(Figure 3.12). The new, shorter channel segment
is called a **cutoff** and, because of its shape, the
abandoned bend is called an **oxbow lake.**

One method of flood control is to straighten a
channel by creating artificial cutoffs. The idea is
that by shortening the stream, the gradient and
hence the velocity are increased. By increasing
velocity, the larger discharge associated with
flooding can be dispersed more rapidly. Since
1932 about 16 artificial cutoffs have been constructed on the lower Mississippi for the purpose
of increasing the channel efficiency and reducing
the threat of flooding. The program has been
somewhat successful in reducing the height of the
river in flood. However, since the river's tendency
to meander still exists, preventing the river from
returning to its previous condition has been difficult.

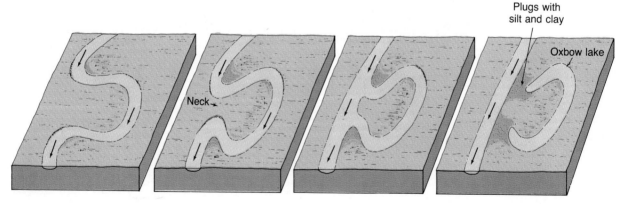

FIGURE 3.12
Formation of a cutoff and oxbow lake.

Artificial cutoffs increase a stream's velocity and may also accelerate erosion of the bed and banks of the channel. A case in point is the Blackwater River in Missouri, whose meandering course was shortened in 1910. Among the many effects of this project was a dramatic increase in the width of the channel caused by the increased velocity of the stream. One bridge collapsed because of bank erosion in 1930. Over the next 17 years the same bridge was replaced on three more occasions, each time with a wider span.

DRAINAGE SYSTEMS AND PATTERNS

A stream is just one small component in a much larger system. Each system consists of a **drainage basin,** the land area that contributes water to the stream. The drainage basin of one stream is separated from another by an imaginary line called a **divide.** Divides range from a ridge separating two small gullies to *continental divides,* which split continents into enormous drainage basins. For example, the continental divide that runs somewhat north-south through the Rocky Mountains separates the drainage which flows west to the Pacific Ocean from that which flows to the Gulf of Mexico. Although divides separate the drainage of two streams, if they are tributaries of the same river, they are both a part of that larger drainage system.

All drainage systems are made up of an interconnected network of streams which together form particular patterns. The nature of a drainage pattern can vary greatly from one type of terrain to another, primarily in response to the kinds of rock on which the streams developed or the structural pattern of faults and folds.

Certainly the most commonly encountered drainage pattern is the **dendritic** pattern (Figure 3.13A). This pattern is characterized by an

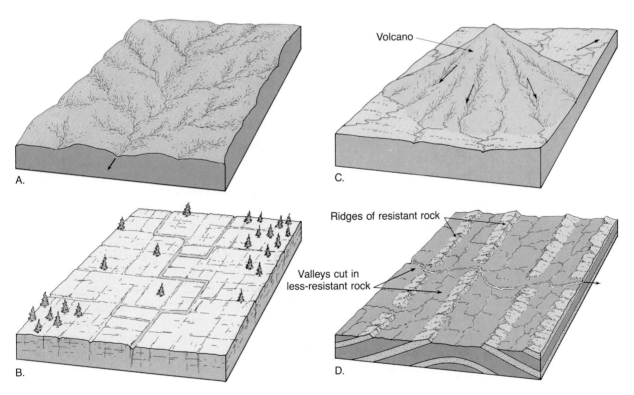

FIGURE 3.13
Drainage patterns **A.** Dendritic. **B.** Rectangular. **C.** Radial. **D.** Trellis.

irregular branching of tributary streams that resembles the branching pattern of a deciduous tree. In fact, the word *dendritic* means "treelike." The dendritic pattern forms where the underlying material is relatively uniform. Since the surface material is essentially uniform in its resistance to erosion, it does not control the pattern of streamflow. Rather, the pattern is determined chiefly by the direction of slope of the land.

Figure 3.13B illustrates a **rectangular** pattern, in which many right-angle bends can be seen. This pattern develops when the bedrock is crisscrossed by a series of joints and/or faults. Since these structures are eroded more easily than unbroken rock, their geometric pattern guides the directions of valleys.

When streams diverge from a central area like spokes from the hub of a wheel, the pattern is said to be **radial** (Figure 3.13C). This pattern typically develops on isolated volcanic cones and domal uplifts.

Figure 3.13D illustrates a **trellis** drainage pattern, a rectangular pattern in which tributary streams are nearly parallel to one another and have the appearance of a garden trellis. This pattern forms in areas underlain by alternating bands of resistant and less resistant rock and is particularly well displayed in the folded Appalachians, where both weak and strong strata outcrop in nearly parallel belts.

STAGES OF VALLEY DEVELOPMENT

Contrary to the popular belief of his day, James Hutton proposed that streams were responsible for cutting the valleys in which they flowed. Later geologic work conducted in stream valleys substantiated Hutton's proposal and further revealed that the development of stream valleys progresses in a somewhat orderly fashion. The evolution of a valley has been arbitrarily divided into three sequential stages: youth, maturity, and old age.

As long as the stream is downcutting to establish a graded condition with its base level, it is considered youthful. Rapids, an occasional waterfall, and a narrow V-shaped valley are all visible signs of the vigorous downcutting that is going on. Other features of a youthful stream include a steep gradient, little or no floodplain, and a rather straight course without meanders (Figure 3.14A).

When a stream reaches maturity, downward erosion diminishes and lateral erosion dominates. Thus the mature stream begins actively cutting its floodplain and meandering upon it (Figure 3.14B). During the mature stage cutoffs occur, producing oxbows, and a few streams may even produce low natural levees (Figure 3.14C). In contrast to the gradient of a youthful stream, the gradient of a mature stream is much lower and the profile is much smoother because all rapids and waterfalls have been eliminated.

A stream enters old age after it has cut its floodplain several times wider than its *meander belt,* which is the width of the meander (Figure 3.14D). When this stage is reached the stream is rarely near the valley walls; hence it ceases to enlarge the floodplain. Thus the primary work of a river in an old-age valley is the reworking of unconsolidated floodplain deposits. Because this task is easier than cutting bedrock, a stream in an old-age valley shifts more rapidly than a stream in a mature valley. For example, some meanders of the lower Mississippi move 20 meters (66 feet) a year, and its large floodplain is dotted with oxbow lakes and old cutoffs. Natural levees are also common features of old-age valleys and, when present, are accompanied by back swamps and yazoo tributaries.

Thus far we have assumed that the base level of a stream remains constant as a river progresses from youth to old age. On many occasions, however, the land is uplifted or the base level is lowered. The effect of uplifting on a youthful stream is to increase its gradient and accelerate its rate of downcutting. However, uplifting of a mature stream would cause it to abandon lateral erosion and revert to downcutting. Rivers of this type are said to be **rejuvenated,** and the meanders are known as **entrenched meanders** (Figure 3.15). Mature streams may eventually readjust to uplift by cutting a new floodplain at a level below the old one. The remnants of the old higher floodplain are often present in the form of flat surfaces called *terraces.*

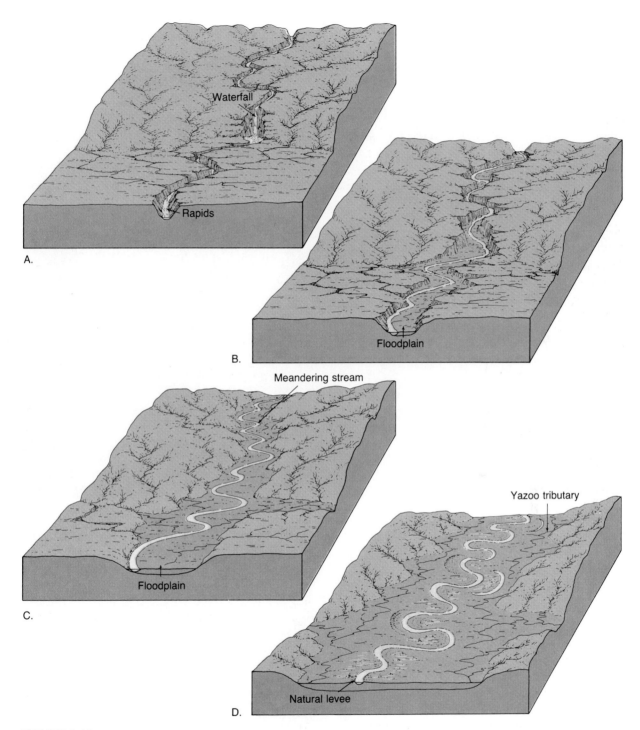

FIGURE 3.14
Stages of valley development. **A.** Youth. The youthful stage is characterized by downcutting and a V-shaped valley. **B.** and **C.** Maturity. Once a stream has sufficiently lowered its gradient, it begins to erode laterally, producing a wide valley. **D.** Old age. After the valley has been cut several times wider than the width of the meander belt, it has entered old age. (After Ward's Natural Science Establishment, Inc., Rochester, N.Y.)

FIGURE 3.15
Entrenched meanders of the San Juan River, Utah. (Photo by John S. Shelton)

Two additional points concerning valley development should be made. First, the time required for a stream to reach any given stage depends on several factors, including the erosive ability of the stream, the nature of the material through which the stream must cut, and the stream's height above base level. Consequently, a stream which starts out very near base level and only has to cut through unconsolidated sediments may reach maturity in a matter of a few hundred years. On the other hand, the Colorado River, where it is actively cutting the Grand Canyon, has retained its youthful nature for an estimated 10 million years. Second, individual portions of a stream reach each stage at different times. Often the lower reaches of a stream attain old age while the headwaters are still youthful in character.

CYCLE OF LANDSCAPE EVOLUTION

While streams are cutting their valleys they simultaneously sculpture the land. To describe this unending process, we need a starting point. For this reason only, we will assume the existence of a relatively flat upland area in a humid region. Until a well-established drainage system forms, lakes and ponds will occupy any depressions that exist. As streams form and begin to downcut and erode headward, they will drain the lakes. During the youthful stage the landscape retains its relatively flat surface, interrupted only by narrow stream valleys. As downcutting continues, relief increases, and the flat, youthful landscape is transformed into one consisting of the hills and valleys that characterize the mature stage. Eventually some of the streams will approach base level, and downcutting will give way to lateral erosion. As the cycle nears the old-age stage, the effects of overland flow and mass wasting, coupled with the lateral erosion by streams, will reduce the land to a **peneplain** ("near plain"), an undulating plain near base level. Although no peneplains are known to exist today, there is evidence that they formed in the past and have since been uplifted. Once a peneplain has formed, uplifting starts the cycle again. Most often, uplifting interrupts the cycle before it reaches old age.

WATER BENEATH THE SURFACE

Of all the world's water, only about six-tenths of one percent is found underground. Nevertheless, the amount of water stored in the rocks and sediments beneath the earth's surface is a vast and very significant natural resource. The U.S. Geological Survey estimates that the quantity of water in the upper 800 meters (0.5 mile) of the continental crust is about 3000 times greater than the volume of water in all rivers at any one time, and nearly 20 times greater than the combined volume in all lakes and rivers. In many parts of the world, wells and springs provide the water needs not only for great numbers of people, but for crops, livestock, and industry as well. In addition, subsurface water is important as an equalizer of streamflow and as an agent of erosion. It is the work of subsurface water that creates caverns and sinkholes.

DISTRIBUTION AND MOVEMENT OF GROUNDWATER

When rain falls, it may run off immediately, evaporate, be taken up by plants and transpired, or it may soak into the ground. This last path is the primary source of practically all subsurface water. The amount of water that takes each of these paths, however, varies considerably both in time and space. The steepness of the slope, the nature of the surface material, the intensity of the rainfall, and the type and amount of vegetation are all influential factors. Heavy rains that fall upon steep slopes underlain by impervious materials will obviously result in a high percentage of the water

running off. On the other hand, a gentle, steady rain that falls upon more gradual slopes composed of materials more easily penetrated by water would result in a much larger percentage of the water soaking into the ground.

Some of the water that soaks in does not travel far, because it is held by molecular attraction as a surface film on soil particles. A portion of this moisture evaporates back into the atmosphere, while much of the remainder serves as a source of water for use by plants between rains. Water that is not held near the surface penetrates downward until it reaches a zone where all of the open spaces in sediment and rock are completely filled with water. The water in this **zone of saturation** is called **groundwater.** The upper limit of this zone is known as the **water table.** The area above the water table where the soil, sediment, and rock are not saturated is known as the **zone of aeration** (Figure 3.16). The open spaces here are filled mainly with air.

The water table is rarely level as we might expect a table to be. Instead, its shape is usually a subdued replica of the surface topography, reaching its highest elevations beneath hills and decreasing in height toward valleys (Figure 3.16). Where a swamp is encountered, the water table is right at the surface, whereas lakes and streams generally occupy areas where the land surface is below the water table. Although many factors contribute to the irregular surface of the water table, the most important cause is that groundwater moves very slowly. For this reason, the water tends to "pile up" beneath hills. If rainfall were to cease completely, these water table "hills" would slowly subside and gradually approach the level of the valleys. However, new supplies of rainwater are usually added often enough to prevent this. Nevertheless, in times of extended drought, the water table may drop enough to dry up otherwise productive wells.

Depending upon the nature of the subsurface material, the flow of groundwater and the amount of water that can be stored vary greatly. The quantity of groundwater that can be stored depends upon the **porosity** of the material, that is, the percentage of the total volume of rock or sediment that consists of pore spaces. Although these openings often consist of spaces between particles of sediment or sedimentary rock, such features as vesicles (voids left by gases escaping from lava), joints and faults, and cavities formed by the solution of soluble rocks such as limestone, are also common.

Rock or sediment may be very porous and still not allow water to move through it. The **permeability** of a material, its ability to transmit a fluid, is also an important consideration. Groundwater moves by twisting and turning through small, interconnected openings. The smaller the pore spaces, the slower the water moves. If the spaces between particles are too small, the films of water clinging to the grains will come in contact or overlap. As a result, the force of molecular attraction which binds the water to the particles extends across the openings and the water is held quite firmly in place. Clay exemplifies this circumstance. Although the ability of clay to store water is often high, its pore spaces are so small that water is unable to move. Impermeable layers composed of materials such as clay that hinder or prevent water movement are termed **aquicludes.** On the other hand, larger particles, such as sand or gravel, have larger pore spaces. Therefore, the water in the centers of the openings is not bound to the particles by molecular attraction and can move with relative ease. Such permeable rock strata or sediment that transmit groundwater freely are called **aquifers.**

SPRINGS

Springs have aroused the curiosity and wonder of people for thousands of years. The fact that springs were (and to some people still are) rather mysterious phenomena is not difficult to understand, for here is water flowing freely from the ground in all kinds of weather in seemingly inexhaustible supply but with no obvious source. As a result, some interesting but false explanations for the source of springs were proposed that, at least to some extent, have managed to live on to the present. One such explanation is that springs draw their water from the ocean. How the salt is removed and the water elevated to the great heights it reaches in mountain springs, in defiance of gravity, remain unanswered questions. Another

explanation for springs, one supported by Aristotle, suggested that the water originated in cold subterranean caverns where water vapor condensed from air that had penetrated the earth. Today we know that the source of springs is water from the zone of saturation and that the ultimate source of this water is precipitation.

Whenever the water table intersects the ground surface, a natural flow of groundwater results, which we call a **spring.** There are many circumstances which create springs. One situation is illustrated in Figure 3.16. In some places the main zone of saturation is overlain by unsaturated material that contains an aquiclude, above which a localized zone of saturation has formed. If this **perched water table** intersects a hillside, a spring or series of springs are found.

WELLS

The most common device used by people for removing groundwater is the **well,** an opening bored into the zone of saturation (Figure 3.16). Wells serve as reservoirs into which groundwater moves and from which it can be pumped to the surface. Digging for water dates back many centuries and continues to be an important method of obtaining water. Today in many parts of the world, including the United States, water from wells irrigates more land than does water from streams.

The level of the water table may fluctuate considerably during the course of a year, dropping during dry seasons and rising following periods of rain. Therefore, to ensure a continuous supply of water, a well should penetrate many meters below the water table. When water is pumped from a well, it produces a conical depression in the water table that is known as a **cone of depression.** If pumping is heavy, the water table may not only be lowered immediately around the well but may also be lowered over a large area (Figure 3.17). This has been the case in portions of the western United States. Under these circumstances, it can be said that the groundwater is literally being "mined," for even if pumping were to cease immediately, it could take hundreds of years for the groundwater to be replenished. The following example illustrates this point:

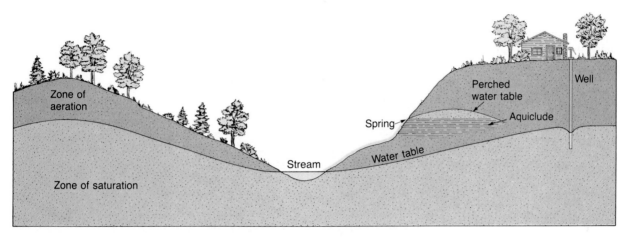

FIGURE 3.16
This diagram illustrates the relative positions of many features associated with subsurface water.

FIGURE 3.17
A cone of depression in the water table often forms around a pumping well. If heavy pumping lowers the water table, some wells may be left dry.

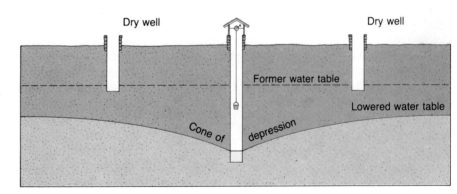

Consider, for example, a location in the dry southwestern United States where annual recharge to an aquifer is on the order of only two-tenths of an inch of water. In such areas it is not uncommon to pump two feet or more of water per year for irrigation or other uses. In this oversimplified example, if the entire aquifer were pumped at that rate, yearly pumpage would be equivalent to 120 years' recharge, and ten years of pumping would remove a 1200-year accumulation of water. New recharge during the pumping period would be negligible. Mechanical problems and economic factors prevent complete dewatering of an aquifer, but the example is valid in principle.*

---

*U.S. Geological Survey, *Water of the World,* 1968, p. 16.

ARTESIAN WELLS

The term **artesian** is applied to any situation in which groundwater rises in a well above the level where it was initially encountered. For such a situation to occur, two conditions must exist (Figure 3.18): (1) Water must be confined to an aquifer that is inclined so that one end is exposed at the surface, where it can receive water; and (2) Impermeable layers, both above and below the aquifer, must be present to prevent the water from escaping. When such a layer is tapped, the pressure created by the weight of the water above will cause the water to rise. If there were no friction the water in the well would rise to the level of the water at

FIGURE 3.18
Artesian systems occur when an inclined aquifer is surrounded by impermeable beds.

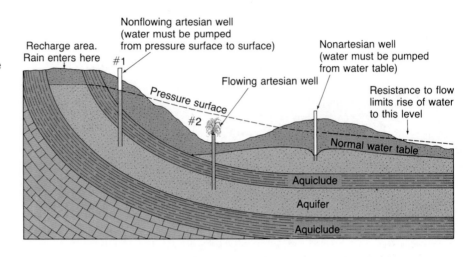

the top of the aquifer. However, friction reduces the height of this pressure surface. The greater the distance from the recharge area (area where water enters the inclined aquifer), the greater the friction and the less the rise of water. In Figure 3.18, Well 1 is a *nonflowing artesian well,* because at this location the pressure surface is below ground level. When the pressure surface is above the ground and a well is drilled into the aquifer, a *flowing artesian well* is created (Well 2, Figure 3.18).

Artesian systems act as conduits, transmitting water from remote areas of recharge great distances to the points of discharge. In this manner water which fell in central Wisconsin years ago is now taken from the ground and used by communities many kilometers away in Illinois. In South Dakota such a system has brought water from the Black Hills in the west, eastward across the state. On a different scale, city water systems may be considered as examples of artificial artesian systems. The water tower, into which water is pumped, may be considered the area of recharge, the pipes the confined aquifer, and the faucets in homes the flowing artesian wells.

ENVIRONMENTAL PROBLEMS ASSOCIATED WITH GROUNDWATER

As with many of our valuable natural resources, groundwater is being exploited at an ever-increas-

ing rate. Overuse threatens the groundwater supply in some areas. In other places, groundwater withdrawal has caused the ground and everything resting upon it to sink. Still other localities are concerned with the possible contamination of their groundwater supply.

As we shall see later in this chapter, surface subsidence can result from natural processes related to groundwater. However, the ground may also sink when water is pumped from wells faster than natural recharge processes can replace it. This effect is particularly pronounced in areas underlain by thick layers of unconsolidated sediments. As water is withdrawn, water pressure drops and the weight of the overburden is transferred to the sediment. The greater pressure packs the sediment grains tightly together and the ground subsides.

Many areas may be used to illustrate land subsidence resulting from the excessive pumping of groundwater from relatively loose sediment. A classic example in the United States occurred in the San Joaquin Valley of California. Here, in a region of extensive irrigation, the water table beneath the valley has gradually been drawn down by as much as 30 meters (100 feet). As a consequence, the land has subsided by 3 meters in some places. Similar effects have occurred in the Houston and Panhandle areas of Texas, as well as in Las Vegas and New Orleans. Damage to buildings, water and sewer lines, and roads can be extensive and costly. In addition, subsidence in coastal areas may require the construction of levees to keep out the encroaching sea.

The pollution of groundwater is a serious matter, particularly in areas where aquifers supply a large part of the water supply. A very common source of groundwater pollution is sewage which results from an ever-increasing number of septic tanks, as well as inadequate or broken sewer systems, and barnyard wastes.

If water contaminated with bacteria from sewage enters the groundwater system, it may become purified through natural processes. The harmful bacteria may be mechanically filtered out by the sediment through which the water percolates, destroyed by chemical oxidation, and/or assimilated by other organisms. In order for purification to occur, however, the aquifer must be of the correct composition. For example, extremely permeable aquifers such as highly fractured crystalline rock, coarse gravel, or cavernous limestone have such large openings that contaminated groundwater may travel long distances without being cleansed. In this case, the water flows too

rapidly and is not in contact with the surrounding material long enough for purification to occur. This is the problem at Well 1 in Figure 3.19A. On the other hand, when the aquifer is composed of sand or permeable sandstone, the water can sometimes be purified within distances as short as a few tens of meters. The openings between sand grains are large enough to permit water movement, yet the movement of the water is slow enough to allow ample time for its purification (Well 2, Figure 3.19B).

Sanitary landfills and garbage dumps are another source of pollutants that may endanger the groundwater supply of an area. As rainwater oozes through the refuse, it may dissolve a variety of organic and inorganic materials, some of which may be harmful. If water containing material leached from the landfill reaches the water table, it will mix with the groundwater and contaminate the supply. Since groundwater movement is usually slow, the polluted water may go undetected for a considerable time. When the problem is finally discovered, the volume of contaminated

FIGURE 3.19
A. Although the contaminated water has traveled more than 100 meters before reaching Well 1, the water moves too rapidly through the cavernous limestone to be purified. **B.** Since the aquifer is sandstone, the water is purified in a relatively short distance.

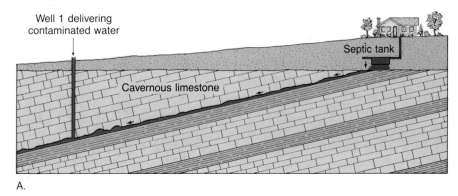

A.

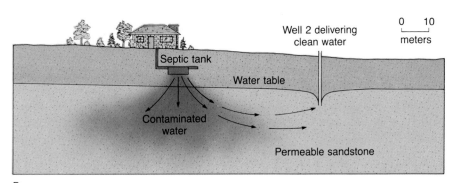

B.

water may already be very large. Thus, even if the source of pollution is eliminated immediately (which is most unlikely), the problem could linger for many years until the contaminated water has migrated from the area of use.

HOT SPRINGS AND GEYSERS

By definition, the water in **hot springs** is 6–9°C (10–15°F) warmer than the mean annual air temperature for the localities where they occur. In the United States alone, there are well over 1000 such springs.

Mineral explorations over the world have shown that temperatures in deep mines and oil wells usually rise with an increase in depth below the surface. Temperatures in such situations increase an average of about 2°C per 100 meters (1°F per 100 feet). Therefore when groundwater circulates at great depths, it becomes heated, and if it rises to the surface, the water may emerge as a hot spring. The water of some hot springs in the United States, particularly in the East, is heated in this manner. The great majority (over 95 percent) of the hot springs (and geysers) in the United States are found in the West. The reason for such a distribution is that the source of heat for most hot springs is cooling igneous rock, and it is in the West that igneous activity has been most recent.

Geysers are intermittent hot springs or fountains in which columns of water are ejected with great force at various intervals, often rising 30–60 meters (100–200 feet) (Figure 3.20). After the jet of water ceases, a column of steam rushes out, usually with a thundering roar. Perhaps the most famous geyser in the world is Old Faithful in Yellowstone National Park, which erupts about once each hour. Geysers are also found in other parts of the world, including New Zealand and Iceland, where the term *geyser,* meaning "spouter" or "gusher," was coined.

Geysers occur where extensive underground chambers exist within hot igneous rocks. As relatively cool groundwater enters the chambers, it is heated by the surrounding rock. At the bottom of the chamber, the water is under great pressure because of the weight of the overlying water. Consequently, a temperature above 100°C (212°F) is

FIGURE 3.20
Castle Geyser, Yellowstone National Park. (Photo by Stephen Trimble)

required before it will boil. For example, at the bottom of a 300-meter (1000-foot) chamber water must attain a temperature of nearly 230°C (450°F) before it will boil. The heating causes the water to expand, with the result that some flows out at the surface. This decreases the pressure, and the water quickly turns to steam and causes the geyser to erupt (Figure 3.21).

Groundwater from hot springs and geysers usually contains more material in solution than groundwater from other sources because hot water is a more effective dissolver than cold. When the water contains much dissolved silica, *geyserite* is deposited around the spring. *Travertine,* a form of calcite, is a characteristic deposit at hot springs in limestone regions. Some hot springs contain

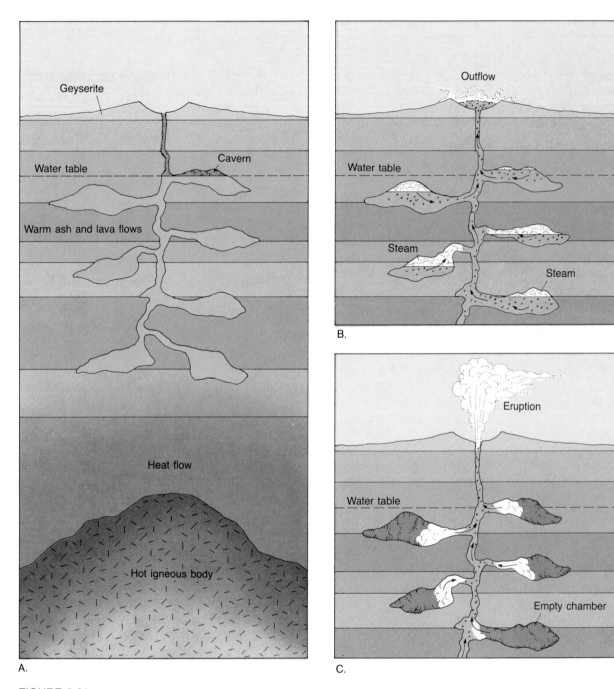

FIGURE 3.21

Idealized diagrams of a geyser. A geyser can form if the heat is not distributed by convection. **A.** In this figure, the water near the bottom is heated to near its boiling point. The boiling point is higher there than at the surface because the weight of the water above increases the pressure. **B.** The water higher in the geyser system is also heated; therefore, it expands and flows out at the top, reducing the pressure on the water at the bottom. **C.** At the reduced pressure on the bottom, boiling occurs. The bottom water flashes into steam, and the expanding steam causes an eruption.

sulfur. In addition to making the water taste bad, sulfur-rich springs emit an unpleasant odor. Undoubtedly Rotten Egg Spring, Nevada, is such a situation.

GEOTHERMAL ENERGY

Geothermal energy is produced by tapping naturally-occurring steam and hot water located beneath the surface in regions where subsurface temperatures are high due to relatively recent volcanic activity. Although the development of geothermal power plants has grown quite rapidly in recent years, the idea of using natural steam to generate electricity is not new. As early as 1904, natural steam vents at Larderello, Italy, were used to make power. By 1985, the U.S. Geological Survey reported that 188 separate power units in 17 countries were operating, with a combined capacity of almost 4800 megawatts (million watts). In addition to the United States, countries that lead in the use of geothermal energy to produce electricity are the Philippines, Mexico, Italy, and New Zealand.

The first commercial geothermal power plant in the United States was built in 1960 at The Geysers, north of San Francisco. Development at this location grew to more than 1500 megawatts by 1985, with increasing production planned for the future. In addition, geothermal development is occurring elsewhere in the western United States, including Nevada, Utah, and the Imperial Valley in southern California.

Geothermal energy is not used exclusively for generating electricity. In Iceland's capital, Reykjavik, steam and hot water are pumped into buildings throughout the city for space heating and used to warm greenhouses, where fruits and vegetables are grown all year. In the United States, localities in several western states use hot water from geothermal sources for space heating.

The most favorable geologic factors for a geothermal reservoir of commercial value include:

1   A potent source of heat, such as a large magma chamber. The chamber should be deep enough to ensure adequate pressure and a slow rate of cooling, and yet not be so deep that the natural circulation of water is inhibited. Magma chambers of this type are most likely to occur in regions of recent volcanic activity;

2   Large and porous reservoirs with channels connected to the heat source, near which water can circulate and then be stored in the reservoir;

3   Capping rocks of low permeability that inhibit the flow of water and heat to the surface. A deep, well-insulated reservoir is likely to contain much more stored energy than an uninsulated, but otherwise similar, reservoir.

As with other methods of power production, geothermal sources are not expected to provide a high percentage of the world's growing energy needs. Nevertheless, in regions where its potential can be developed, its use will no doubt continue to grow.

THE GEOLOGIC WORK OF GROUNDWATER

The primary erosional work carried out by groundwater is that of dissolving rock. Since soluble rocks, especially limestone, underlie millions of square kilometers of the earth's surface, it is here that groundwater carries on its rather unique and important role as an erosional agent. Although nearly insoluble in pure water, limestone is quite easily dissolved by water containing small quantities of carbonic acid. Most natural water contains this weak acid because rainwater readily dissolves carbon dioxide from the air and from decaying plants. Therefore, when groundwater comes in contact with limestone, the carbonic acid reacts with calcite in the rocks to form calcium bicarbonate, a soluble material that is then carried away in solution.

Among the most spectacular results of groundwater's erosional handiwork is the creation of limestone **caverns.** Most are relatively small, yet some have spectacular dimensions. In the United States, Carlsbad Caverns in southeastern New Mexico and Mammoth Cave in Kentucky are famous examples. One chamber in Carlsbad Caverns has an area equivalent to fourteen football fields and enough height to accommodate the U.S. Capitol Building. At Mammoth Cave, the total length of interconnected caverns extends for hundreds of kilometers.

Most caverns are believed to be created at or below the water table in the zone of saturation. Here the groundwater follows lines of weakness in the rock, such as joints and bedding planes. As time passes, the dissolving process slowly creates cavities and gradually enlarges them into caverns. Material that is dissolved by the groundwater is carried away and discharged into streams.

Certainly the features that arouse the greatest curiosity for most cavern visitors are the stone formations that often exhibit quite bizarre patterns and give some caverns a wonderland appearance. These features are created by the seemingly endless dripping of water over great spans of time. The calcite that is left behind produces the limestone we call travertine. These cave deposits, however, are also commonly called *dripstone,* an obvious reference to their mode of origin.

Although the formation of caverns takes place in the zone of saturation, the deposition of dripstone is not possible until the caverns are above the water table in the zone of aeration. This commonly occurs as nearby streams cut their valleys deeper, lowering the water table as the elevation of the river drops. As soon as the chamber is filled with air, the conditions are right for the decoration phase of cavern building to begin.

Of the various dripstone features found in caverns, perhaps the most familiar are **stalactites.** These icicle-like pendants hang from the ceiling of the cavern and form where water seeps through cracks above. When water reaches air in the cave, some of the dissolved carbon dioxide escapes from the drop and calcite begins to precipitate. Deposition occurs as a ring around the edge of the water drop. As drop after drop follows, each leaves an infinitesimal trace of calcite behind, and a hollow limestone tube is created. Water then moves through the tube, remains suspended momentarily at the end, contributes a tiny ring of calcite, and falls to the cavern floor. The stalactite just described is appropriately called a *soda straw* (Figure 3.22A). Often the hollow tube of the soda straw

A.

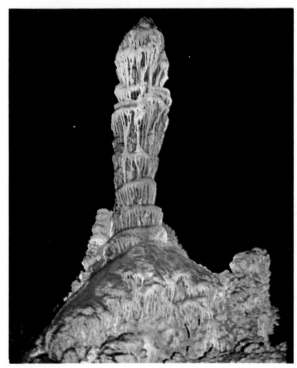

B.

FIGURE 3.22
A. A "live" solitary soda straw stalactite. (Photo by Clifford Stroud, National Park Service). B. Stalagmites grow upward from the floor of a cavern. (Photo by E. J. Tarbuck)

becomes plugged or its supply of water increases. In either case, the water is forced to flow, and hence deposit, along the outside of the tube. As deposition continues, the stalactite takes on the more common conical shape.

Formations that develop on the floor of a cavern and reach upward toward the ceiling are called **stalagmites** (Figure 3.22B). The water supplying the calcite for stalagmite growth falls from the ceiling and splatters over the surface. As a result, stalagmites do not have a central tube and are usually more massive in appearance and rounded on their upper ends than stalactites.

Many areas of the world have landscapes that to a large extent have been shaped by the dissolving power of groundwater. Such areas are said to exhibit **karst topography.** The term is derived from a plateau region located along the northeastern shore of the Adriatic Sea in the border area between Yugoslavia and Italy where such topography is strikingly developed. In the United States, karst landscapes occur in many areas, including portions of Kentucky, Tennessee, Alabama, southern Indiana, and central and northern Florida. Generally, arid and semiarid areas do not develop karst topography. When solution features exist in such regions, they are likely to be remnants of a time when more humid conditions prevailed.

Karst areas characteristically exhibit an irregular terrain punctuated with many depressions called **sinkholes,** or simply, **sinks.** In the limestone areas of Florida, Kentucky, and southern Indiana, there are literally tens of thousands of these depressions varying in depth from just a

meter or two to a maximum of more than 50 meters (Figure 3.23A).

Sinkholes commonly form in one of two ways. Some develop gradually over many years without any physical disturbance to the rock. In these situations, the limestone immediately below the soil is dissolved by downward-seeping rainwater that is freshly charged with carbon dioxide. These depressions are usually not deep and are characterized by relatively gentle slopes. By contrast, sinkholes can also form suddenly and without warning when the roof of a cavern collapses under its own weight. Typically the depressions created in this manner are steep-sided and deep. When they form in populous areas, they may represent a serious geologic hazard. The craterlike sinkhole in Figure 3.23B began forming in Winter Park, Florida, on May 8, 1981, just one day before this photograph was taken. Newspaper accounts were front page news and made this sinkhole one of the most publicized ever.

In addition to a surface pockmarked by sinkholes, karst regions characteristically show a striking lack of surface drainage. Following a rainfall, runoff is funneled below ground by way of sinks, where it then flows through caverns until it finally reaches the water table. When streams do exist at the surface, their paths are usually short. The names of such streams often give a clue to their fate. In the Mammoth Cave area of Kentucky, for example, there is Sinking Creek, Little Sinking Creek, and Sinking Branch. Other sinkholes become plugged with clay and debris to create small lakes or ponds.

FIGURE 3.23
A. This high altitude infrared image shows an area of karst topography in Florida. The numerous lakes occupy sinkholes. (Courtesy of USDA-ASCS) **B.** Aerial view of a large sinkhole that formed in Winter Park, Florida, in May, 1981. (Photo courtesy of George Remaine, *Orlando Sentinel Star*)

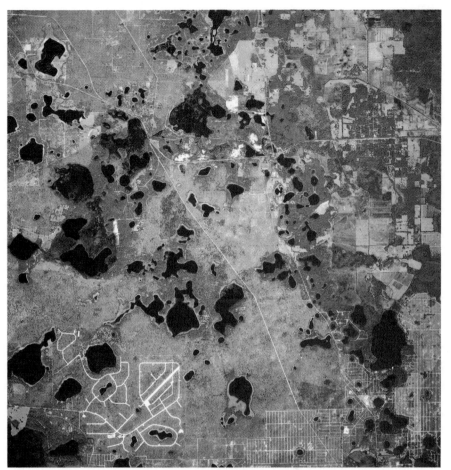

A.

B.

REVIEW QUESTIONS

1   Describe the movement of water through the hydrologic cycle. Is there more than one path which precipitation may take after it has fallen?

2   A stream starts out 2000 meters above sea level and travels 250 kilometers to the ocean. What is its average gradient in meters per kilometer?

3   Suppose that the stream mentioned in Question 2 developed extensive meanders so that its course was lengthened to 500 kilometers. Calculate its new gradient. How does meandering affect gradient?

4   Why is the Amazon considered the largest river on earth whereas the Nile is actually longer?

5   When the discharge of a stream increases, what happens to the stream's velocity?

6   Why does the downstream portion of a river have a gentle gradient when compared to the headwater region?

7   Define *base level*. Name the main river in your area. For what streams does it act as base level? What is the base level for the Mississippi River?

8   In what three ways does a stream transport its load?

9   If you collect a jar of water from a stream, what part of its load will settle to the bottom of the jar? What portion will remain in the water?

10 Differentiate between competency and capacity.

11 In what way is a delta similar to an alluvial fan? In what way are they different?

12 Why must the height of many artificial levees be increased periodically?

13 What is the purpose of artificial cutoffs?

14 What is a divide?

15 Why is it possible for a youthful valley to be older (in years) than a mature valley?

16 Do mature and old-age valleys make good political boundaries? Explain.

17 Define groundwater and relate it to the water table.

18 How do porosity and permeability differ?

19 Under what circumstances can a material have a high porosity but not be a good aquifer?

20 Why is the pumping of water in some areas of southwestern United States a serious problem?

21 What is meant by the term *artesian*? Under what circumstances do artesian wells form?

22 Briefly explain what happened in the San Joaquin Valley of California as the result of excessive groundwater withdrawal.

23 Which would be most effective in purifying polluted groundwater: an aquifer composed mainly of coarse gravel, sand, or cavernous limestone?

24 What is the source of heat for most hot springs and geysers? How is this reflected in the distribution of these features?

25 List two conditions required for the development of karst topography.

26 Differentiate between stalactites and stalagmites. How do these features form?

KEY TERMS

hydrologic cycle

infiltration

runoff

transpiration

gradient

discharge

base level

ultimate base level

temporary (local) base
level

graded stream

pothole

dissolved load

suspended load

bed load

saltation

competence

capacity

alluvium

delta

distributary

alluvial fan

natural levee

back swamp

yazoo tributary

rapids

waterfall

floodplain

meander

cutoff

oxbow lake

drainage basin

divide

dendritic pattern

rectangular pattern

radial pattern

trellis pattern

rejuvenation

entrenched meander

peneplain

zone of saturation

groundwater

water table

zone of aeration

porosity

permeability

aquiclude

aquifer

spring

perched water table

well

cone of depression

artesian

hot spring

geyser

geothermal energy

cavern

stalactite

stalagmite

karst topography

sinkhole (sink)

4

GLACIERS, DESERTS, AND WIND

Today glaciers cover nearly 10 percent of the earth's land surface; however, in the recent geologic past ice sheets were three times more extensive, covering vast areas with ice thousands of meters thick. Many present-day landscapes still bear the mark of these glaciers. The first part of this chapter examines glaciers and the erosional and depositional features they create. The second part is devoted to dry lands and the geologic work of wind. Since desert and near-desert conditions prevail over an area as large as that affected by the massive glaciers of the Ice Age, the nature of such landscapes is indeed worth investigating.

When alpine glaciers merge, their lateral moraines join to create a medial moraine. (Photo courtesy of the Glaciology Office, U.S. Geological Survey)

GLACIERS

A **glacier** is a thick mass of ice that originates on land from the compaction and recrystallization of snow and shows evidence of past or present movement. Since snow is the raw material that eventually produces glacial ice, glaciers must form in areas where more snow falls in winter than melts during the summer. A region that has such a net accumulation is termed a *snowfield* and its outer limits are defined by the *snowline*. The elevation of the snowline varies greatly. In the frigid polar realm, it may be sea level, while in tropical areas near the equator, the snowline exists only high in the mountains, often at elevations exceeding 4500 meters (15,000 feet). If the accumulation in the snowfield is great enough, the pressure of overlying layers transforms the snow below into glacial ice.

In Chapter 3 we learned that the earth's water is in constant motion. Time and again the same water is transferred from the oceans to the atmosphere, dropped upon the land, and carried by rivers and underground flow back to the sea. However, when precipitation falls at high elevations or high latitudes, the water may not immediately soak in or run off. Rather, it may become part of a large mass of moving ice, that is, a glacier. Although the ice will eventually melt and continue its path to the sea, it can be tied up in a glacier for many tens, hundreds, or even thousands of years. For example, data collected from the glacier covering Greenland show that some of this ice is more than 25,000 years old.

How much water is stored as glacial ice? Estimates by the U.S. Geological Survey indicate that only slightly more than two percent of the world's water supply is accounted for by glaciers. But this figure can be misleading when the actual amounts of water are considered.

Although glaciers are found in many parts of the world today, most are located in areas remote from populous regions. Literally thousands of relatively small glaciers exist in mountainous regions. Such glaciers are generally confined to mountain valleys and are most often termed **alpine glaciers.** The total volume of all alpine glaciers today is about 210,000 cubic kilometers (50,000 cubic miles), comparable to the combined volume of the world's large saline and freshwater lakes.

On a different scale, **continental glaciers** are massive accumulations of ice that are not confined to valleys but cover all of the land surface of extensive areas. Two such major accumulations exist on earth today and cover most of the island of Greenland and the continent of Antarctica. Their combined area represents almost 10 percent of the earth's land surface. The Greenland ice sheet covers 80 percent of this large island, occupying about 1.7 million square kilometers (667,000 square miles) and averaging nearly 1500 meters (5000 feet) thick. However, when compared to the glacier covering the continent of Antarctica, the Greenland ice sheet seems quite small. Eighty percent of the world's ice is represented by Antarctica's glacier, which covers an area almost one and one-half times that of the United States. If this ice melted, sea level would rise an estimated 60–70 meters (200–230 feet), inundating many densely populated coastal areas. The hydrologic importance of the continent and its ice can be illustrated another way. If Antarctica's ice sheet melted at a suitable rate it could feed (1) The Mississippi River for more than 50,000 years, (2) All the rivers in the United States for about 17,000 years, (3) The Amazon River for approximately 5000 years, or (4) All the rivers of the world for about 750 years.

GLACIAL MOVEMENT

The movement of glacial ice is generally referred to as *flow*. The fact that glacial movement is described in this way would seem to constitute a paradox—ice is solid, yet it is capable of flow. The way ice flows is complex but is believed to be of two basic types. One mechanism involves internal movement within the ice. Ice behaves as a brittle solid until the pressure or load upon it is equivalent to the weight of about 50 to 60 meters (165 to 200 feet) of ice. Once that load is surpassed, the ice will behave as a plastic material and flow continuously. A second and often equally important mechanism of glacial movement consists of the whole ice mass slipping along the ground. With the exception of some glaciers in polar regions

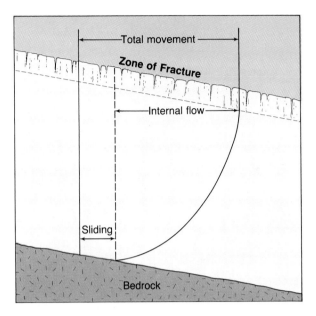

FIGURE 4.1
Glacial movement is divided into two components. Below approximately 50 meters, ice behaves plastically and flows. In addition, the entire ice mass may slide along the ground. The ice in the zone of fracture is carried along "piggyback" style. Notice that the rate of movement is slowest at the base of the glacier where frictional drag is greatest.

where the ice is probably frozen to the solid bedrock floor, the lowest portions of most glaciers are thought to move by this sliding process.

Figure 4.1 illustrates the effects of these two basic types of glacial motion. This vertical profile through a glacier also shows that all the ice does not flow forward at the same rate. Just as in streams, frictional drag with the bedrock floor results in the lower portions of the glacier moving more slowly.

The uppermost portion of a glacier is often quite appropriately referred to as the *zone of fracture.* Since there is not enough overlying ice to cause plastic flow, this upper part of the glacier consists of brittle ice. Consequently, the ice in this zone is carried along piggyback style by the ice below. When the glacier moves over irregular terrain, the zone of fracture is subjected to tension, with cracks called **crevasses** resulting (Figure 4.2). These gaping cracks, which often make travel across glaciers dangerous, may extend to depths

of 50 meters (165 feet). Beyond this depth, plastic flow seals them off.

Unlike streamflow, the movement of glaciers is not readily apparent to the casual observer. If we could watch a glacier in a mountain valley move, we would see that, like the water in a river, all of the ice in the valley does not move downstream at an equal rate. Just as friction with the bedrock floor slows the movement of the ice at the bottom of the glacier, the drag created by the valley walls leads to the flow being greatest in the center of the glacier.

FIGURE 4.2
Crevasses form in the brittle ice of the zone of fracture. They do not continue down into the zone of flow. (Courtesy of Ward's Natural Science Establishment, Inc., Rochester, N.Y.)

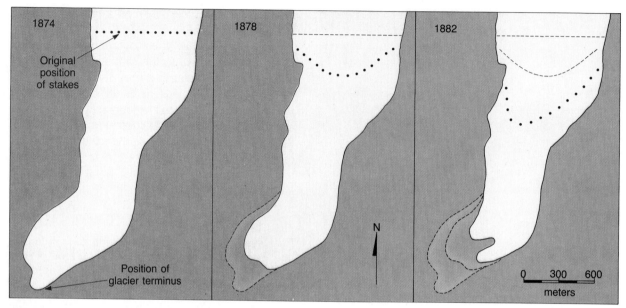

FIGURE 4.3

Ice movement and changes in the terminus at Rhone Glacier, Switzerland. In this classic study of an alpine glacier, the movement of stakes clearly showed that ice along the sides of the glacier moves slowest. Also notice that even though the ice front was retreating, the ice within the glacier was advancing.

More than 100 years ago the first measurements of glacial movement were made. In this experiment stakes were carefully placed in a straight line across the top of an alpine glacier. Periodically the positions of the stakes were noted, revealing the type of movement just described (Figure 4.3).

How rapidly does glacial ice move? Average velocities vary considerably from one glacier to another. Some move so slowly that trees and other vegetation may become well established in the debris that has accumulated on the glacier's surface, whereas others move at rates of up to several meters per day. The advance of some glaciers is characterized by periods of extremely rapid movement followed by periods during which movement is practically nonexistent. For example, Hassanabad Glacier in the Karakoram, a mountain range in Kashmir and northwestern India, advanced 10 kilometers in less than 3 months—a rate of almost 130 meters per day. The precise cause or causes of these sporadic, short-lived advances is not well understood. One proposal is that

the base of the glacier may have been frozen to the bedrock and then sudden melting released the ice. Another hypothesis suggests that a block of stagnant ice at the terminus of a glacier in an alpine valley may act as a dam until the buildup of pressure by the flowing ice behind it forces it to give way.

Glaciers are constantly gaining and losing ice. As we learned earlier, snow accumulation and ice formation occur above the snowline. Here the addition of snow thickens the glacier and promotes movement. Below the snowline, snow from the previous winter as well as some of the glacial ice melt. In addition, large pieces of ice may break off from the front of the glacier, a process called *calving,* sometimes creating icebergs where the glacier meets the sea. Whether the margin of a glacier is advancing, retreating, or remaining stationary depends upon the economy, or budget, of the glacier. That is, it depends upon the balance or lack of balance between accumulation on the one hand and wastage (also termed **ablation**) on the other. If ice accumulation exceeds ablation, the glacial front advances until the two factors balance. At this point, the terminus of the glacier is stationary. At a later time when ablation exceeds accumulation, the ice front will retreat until a balance is again reached. Whether the margins of a glacier are advancing, retreating, or stationary, the ice within the glacier continues to flow forward. In the case of a receding glacier, the ice simply does not flow forward rapidly enough to offset ablation. This point is illustrated rather well in Figure 4.3. While the line of stakes within Rhone Glacier continued to move downstream, the terminus of the glacier slowly retreated upstream.

GLACIAL EROSION

Glaciers are capable of carrying on great amounts of erosional work. For anyone who has observed the terminus of an alpine glacier, the evidence of its erosive force is plain. The observer can witness firsthand the melting ice unlocking rock material of all sizes. All signs lead to the conclusion that the ice has scraped, scoured, and torn rock debris from the floor and walls of the valley and carried them downslope. Indeed, as a transporter of sediment, ice has no equal, because once the debris is acquired by the ice, it will not settle out as will the load carried by a stream or by the wind. Consequently, glaciers can carry huge blocks that no other erosional agent could possibly budge. Although glaciers are of limited importance as an erosional agent today, many landscapes that were modified by the widespread glaciers of the recent ice age still reflect to a high degree the work of ice.

Glaciers primarily erode land in two ways. First, as a glacier flows over a fractured bedrock surface, it loosens and lifts blocks of rock, incorporates them into the ice, and carries them off. This process, known as **plucking,** occurs when meltwater penetrates the cracks and joints along the rock floor of the glacier and refreezes. As the water expands, it exerts tremendous leverage that pries the rock loose. In this manner sediment of all sizes becomes part of the glacier's load.

The second major erosional process is **abrasion.** As the ice with its load of rock fragments moves along, it acts as a giant rasp or file and grinds the surface below as well as the rocks within the ice. The pulverized rock produced by the glacial "grist mill" is appropriately called **rock flour.** So much rock flour may be produced that meltwater streams leaving a glacier often have the grayish appearance of skimmed milk—visible evidence of the grinding power of the ice. When the embedded material consists of large fragments, long scratches and grooves called **glacial striations** may be gouged out (Figure 4.4). These linear scratches on the bedrock surface provide

FIGURE 4.4
Results of glacial abrasion. Scratches and grooves in limestone, St. Johns Bay, Newfoundland. (Photo by Peter Kresan)

FIGURE 4.5
Prior to glaciation, a mountain valley is typically narrow and V-shaped. During glaciation, an alpine glacier widens, deepens, and straightens the valley, creating a U-shaped glacial trough. (Photo by Peter Kresan)

FIGURE 4.6
Bridalveil Falls in Yosemite National Park cascades from a hanging valley into the glacial trough below. (Courtesy of Ward's Natural Science Establishment, Inc., Rochester, N.Y.)

clues to the direction of glacial movement. By mapping the striations over large areas, glacial flow patterns can often be reconstructed. On the other hand, not all abrasive action produces striations. When the sediment consists primarily of fine silt-sized particles, the rock surfaces over which the glacier moves may become highly polished.

The erosional effects of alpine and continental glaciers are quite different from each other. A visitor to an alpine-glaciated region is likely to see sharp and very angular topography. The reason is that as alpine glaciers move downvalley, they tend to accentuate the irregularities of the mountain landscape by creating steeper canyon walls and making bold peaks even more jagged. By contrast, continental ice sheets generally override the terrain and hence tend to subdue rather than accentuate the irregularities they encounter.

Although the erosional accomplishments of continental glaciers can be tremendous, landforms carved by these huge ice masses usually do not inspire the same degree of wonderment and awe as do the erosional features created by alpine glaciers. In regions where the erosional effects of continental ice sheets are significant, glacially scoured surfaces and subdued terrain are the rule. By contrast, in mountainous areas, erosion by alpine glaciers yields many truly spectacular features. Much of the rugged mountain scenery so celebrated for its majestic beauty is the product of glacial erosion.

Prior to glaciation alpine valleys are characteristically V-shaped because streams are well above base level and are therefore downcutting. However, in mountainous regions that have been glaciated, the valleys are no longer narrow. As a glacier moves down a valley once occupied by a stream, the ice modifies it in three ways: The glacier widens, deepens, and straightens the valley, so that what was once a youthful V-shaped valley is transformed into a U-shaped **glacial trough** (Figure 4.5).

Since the magnitude of the glacial erosion depends upon the thickness of the ice, main or trunk glaciers cut their valleys deeper than their tributaries are able to do. After the glaciers have receded, the tributary valleys stand high above the main trough and are termed **hanging valleys.** Hanging valleys often produce spectacular cascading waterfalls, such as those in Yosemite National Park, California (Figure 4.6).

At the head of a glacial valley is a characteristic and often imposing feature associated with an alpine glacier—a **cirque**. As Figure 4.7 illustrates, these hollowed-out, bowl-shaped depressions have precipitous walls on three sides but are open on the downvalley side. The cirque represents the focal point of the glacier's source, that is, the area of snow accumulation and ice formation. Although the origin of cirques is still not totally clear, they are believed to begin as irregularities in the mountainside that are subsequently enlarged by frost wedging and plucking along the sides and bottom of the glacier. After the glacier has melted away, the cirque basin is usually occupied by a small lake.

Fiords are deep, often spectacular, steep-sided inlets of the sea that exist in many high-latitude areas of the world where mountains are adjacent

FIGURE 4.7
Aerial view of bowl-shaped depressions called cirques in the Uinta Range, Utah.
(Photo by John S. Shelton)

to the ocean (Figure 4.8). Norway, British Columbia, Greenland, New Zealand, Chile, and Alaska all have coastlines characterized by fiords. They represent glacial troughs that were partially submerged as the ice left the valley and sea level rose following the Ice Age. The depths of fiords are often dramatic, in some instances exceeding 1000 to 1500 meters (3300 to 5000 feet). The great depths of these flooded troughs is only partly explained by the post-Ice Age rise in sea level. Unlike the situation governing the downward erosional work of rivers, sea level does not act as base level for glaciers. As a consequence, glaciers are capable of eroding their beds far below the surface of the sea. For example, an alpine glacier 300 meters (1000 feet) thick can carve its valley floor more than 250 meters (800 feet) below sea level before downward erosion ceases and the ice begins to float.

The Alps, Northern Rockies, and many other mountain landscapes carved by alpine glaciers reveal more than glacial troughs and cirques. In addition, sinuous, sharp-edged ridges called **arêtes** and sharp, pyramid-like peaks called **horns** project above the surroundings. Both features can originate from the same basic process—the enlargement of cirques produced by plucking and frost action. A group of cirques around a single high mountain create the spires of rock called horns. As the cirques enlarge and converge, an isolated horn is produced. The most famous example is the Matterhorn in the Swiss Alps (Figure 4.9). Arêtes can form in a similar manner except that the cirques are not clustered around a point but rather exist on opposite sides of a divide. As the cirques grow, the divide separating them is reduced to a very narrow, knifelike partition. An arête, however, may also be created in another way. When two glaciers occupy parallel valleys, an arête can form when the divide separating the

FIGURE 4.8
Like other fiords, Muir Inlet, Alaska, is a drowned glacial trough. (Photo by Bruce F. Molnia, courtesy of Terraphotographics/BPS)

FIGURE 4.9
Horns are sharp, pyramid-like peaks that are fashioned by alpine glaciers. This example is the famous Matterhorn in the Swiss Alps. (Photo by E. J. Tarbuck)

moving tongues of ice is progressively narrowed as the glaciers scour and widen their valleys. The landforms carved by alpine glaciers are summarized in Figure 4.10 on page 104.

GLACIAL DEPOSITS

Glaciers are capable of acquiring and transporting a huge load of debris as they slowly yet steadily advance across the land. Ultimately these materials must be deposited when the ice melts. In regions where glacial sediment is deposited, the sediment can play a truly significant role in forming the physical landscape. For example, in many areas once covered by the continental ice sheets of the recent Ice Age, the bedrock is rarely exposed because glacial deposits that are tens or even hundreds of meters thick completely mantle the terrain. The general effect of these deposits is to reduce the local relief and thus level the topography. Indeed, much of the familiar country scenery—rocky pastures in New England, wheat fields in the Dakotas, rolling farmland in the Midwest—results directly from glacial deposition.

Long before the theory of an extensive Ice Age was proposed, much of the soil and rock debris covering portions of Europe was recognized as coming from elsewhere. At the time, these foreign materials were believed to have been "drifted" into their present positions by floating ice during an ancient flood. As a consequence, the term **drift** was applied to this sediment. Although rooted in a concept that was not correct, this term was so well established by the time the true glacial origin of the debris became widely recognized, that it remained in the glacial vocabulary. Today the word *drift* is an all-embracing term for sediments of glacial origin, no matter how, where, or in what form they were deposited.

Glacial deposits are of two types: (1) Those deposited directly by the glacier, which are known as **till;** and (2) Materials deposited by glacial meltwater, called **outwash.** Till is deposited as glacial ice melts and drops its load of rock fragments. Deposits of till are characteristically unsorted, a mixture of many sediment sizes (Figure 4.11). When boulders are found in till, they are called **glacial erratics,** indicating that they were derived from a source outside the area where they are found. In many areas erratics may be seen dotting pastures and farm fields, and sometimes these large rock fragments are cleared from the fields and piled into fences. By examining the erratics as well as the mineral composition of the remaining till, geologists are often able to trace the path taken by the glacier.

Outwash is sorted according to the size and weight of the fragments. Since ice is not capable of

FIGURE 4.10

Landforms created by alpine glaciers. **A.** A mountain mass not affected by glacial erosion. **B.** The mountain mass during the period of maximum glacial activity. **C.** The mountain mass shortly after the glaciers have melted from its valleys. [After William Morris Davis, "The Sculpture of Mountains by Glaciers," *Scottish Geographical Magazine* 22 (1906)]

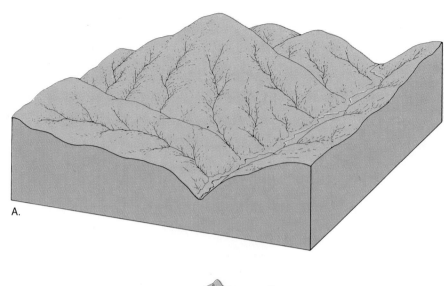

A.

B.

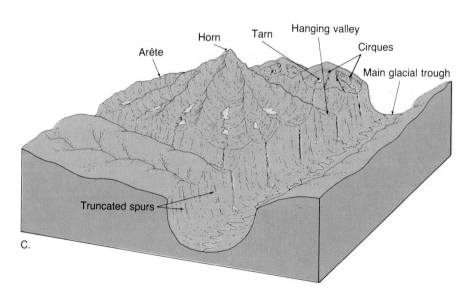

C.

FIGURE 4.11
Glacial till is an unsorted mixture of many sediment sizes. (Photo by E. J. Tarbuck)

such sorting activity, these sediments are not deposited directly by the glacier as till is, but rather they reflect the sorting action of the glacial meltwater that was responsible for dropping them. Accumulations of outwash often consist largely of sand and gravel, that is, bed load material, because the finer rock flour remains suspended and

is commonly carried far from the glacier by the meltwater streams. An indication that outwash consists primarily of sand and gravel can be seen in many areas where these deposits are actively mined as aggregate for road work and other construction projects.

MORAINES, OUTWASH PLAINS, AND KETTLE HOLES

Perhaps the most widespread features created by glacial deposition are *moraines,* which are simply layers or ridges of till. Several types of moraines are identified; some are common only to mountain valleys, and others are associated with areas affected by either continental or alpine glaciers. Lateral and medial moraines fall in the first category, while end moraines and ground moraines are in the second.

The sides of an alpine glacier accumulate large quantities of debris from the valley walls. When the glacier wastes away, these materials are left as ridges, called **lateral moraines,** along the sides of the valley (Figure 4.12). **Medial moraines** are

FIGURE 4.12
A well-developed lateral moraine deposited by the shrinking Athabaska Glacier in the Canadian Rockies. (Photo by James E. Patterson)

FIGURE 4.13
These ponds occupy depressions called kettles, which form when a block of ice that was buried in drift melts and leaves a pit. (Photo by Bruce F. Molnia, courtesy of Terraphotographics/BPS)

formed when two alpine glaciers coalesce to form a single ice stream. The till that was once carried along the edges of each glacier joins to form a single dark stripe of debris within the newly enlarged glacier. The creation of these dark stripes within the ice stream is one obvious proof that glacial ice moves, because the medial moraine could not form if the ice did not flow downvalley (see chapter-opening photo).

As the name implies, **end moraines** form at the terminus of a glacier. Here, while the ice front is stationary, the glacier continues to carry in and deposit large quantities of rock debris, creating a ridge of till tens to hundreds of meters high. The end moraine marking the farthest advance of the glacier is called the *terminal moraine,* and those moraines formed as the ice front periodically became stationary during retreat are termed *recessional moraines.* As the glacier recedes, a layer of till is laid down, forming a gently undulating surface of **ground moraine.** Ground moraine has a leveling effect, filling in low spots and clogging old stream channels, often leading to a disruption of drainage.

At the same time that an end moraine is forming, water from the melting glacier cascades over the till, sweeping some of it out in front of the growing ridge of unsorted debris. Meltwater generally emerges from the ice in rapidly moving streams that are often choked with suspended material and carry a substantial bed load as well. As the water leaves the glacier, it moves onto the relatively flat surface beyond and rapidly loses velocity. As a consequence, much of its bed load is dropped and the meltwater begins weaving a complex pattern of braided channels. In this way a broad, ramplike surface called an **outwash plain** is built adjacent to the downstream edge of most end moraines.

Often outwash plains are pockmarked with basins or depressions known as **kettles** (Figure 4.13). Kettles also occur in deposits of till. Kettles form when a block of stagnant ice becomes wholly or partially buried in drift and ultimately melts, leaving a pit in the glacial sediment. Although most kettles do not exceed two kilometers in diameter, some with diameters exceeding 10 kilometers (6 miles) occur in Minnesota. Likewise, the typical depth of most kettles is less than 10 meters (33 feet), although the vertical dimensions of some approach 50 meters. In many cases, water eventually fills the depression and forms a pond or lake.

OTHER DEPOSITIONAL FEATURES

Drumlins are streamlined asymmetrical hills composed of till (Figure 4.14). They range in height from 15 to 60 meters (50 to 200 feet) and average 0.4–0.8 kilometer (0.25–0.50 mile) in length. The steep side of the hill faces the direction from which the ice advanced, while the gentler slope points in the direction the ice moved. Drumlins are not found singly, but rather occur in clusters, sometimes called drumlin fields. Although drumlin formation is not well understood, their streamlined shape seems to indicate that they were molded in the zone of flow within an active glacier. Some geologists believe drumlin fields were created when a glacier advanced and reshaped an end moraine.

FIGURE 4.14
Drumlins, such as this one in upstate New York, are depositional features associated with continental glaciers. (Courtesy of Ward's Natural Science Establishment, Inc., Rochester, N.Y.)

In areas once occupied by continental ice sheets, sinuous ridges composed largely of sand and gravel may be found. These ridges, called **eskers,** are thought to be deposits made by streams flowing in tunnels beneath the ice, near the terminus of a glacier. They may be several meters high and extend for many kilometers. In some areas they are mined for sand and gravel, and for this reason, eskers are disappearing in some localities. **Kames** are steep-sided hills that, like eskers, are composed of sand and gravel. Kames are believed to have originated when sediment collected in openings in stagnant ice.

Figure 4.15, on page 108, depicts a hypothetical area during and following glaciation. It illustrates many of the landforms described in the preceding sections that may be found in regions affected by continental glaciers.

GLACIERS IN THE GEOLOGIC PAST

At various points in the preceding pages mention was made of the Ice Age, a time when ice sheets and alpine glaciers were far more extensive than they are today. As was noted earlier, there was a time when the most popular explanation for what

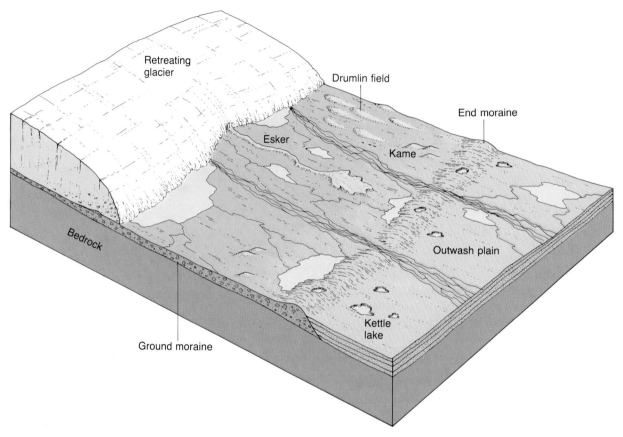

FIGURE 4.15
This hypothetical area illustrates many common depositional landforms.

we now know to be glacial deposits was that the materials had been drifted in by means of icebergs or perhaps simply swept across the landscape by a catastrophic flood. However, during the nineteenth century, field investigations by many scientists provided convincing proof that an extensive ice age was responsible for these deposits as well as for many other features.

By the beginning of the twentieth century, geologists had largely determined the areal extent of Ice Age glaciation. Further, during the course of their investigations they discovered that many glaciated regions had not one layer of drift, but several. Moreover, close examination of these older deposits showed well-developed zones of chemi-

cal weathering and soil formation as well as the remains of plants that require warm temperatures. The evidence was clear; there had not been just one glacial advance but several, each separated by long periods when climates were as warm or warmer than at present. The Ice Age had not simply been a time when the ice advanced over the land, lingered for a while, and then receded. Rather, the period was a very complex event characterized by a number of advances and withdrawals of glacial ice. In North America four major stages of glaciation have been identified (Figure 4.16). Each was named for the midwestern state where the deposits of that ice sheet are well exposed and/or were first studied. These are, in order of occurrence, the Nebraskan, Kansan, Illinoian, and Wisconsinan. Additional evidence from Europe, Alaska, and elsewhere further indicates that other glacial advances probably preceded the Nebraskan. During the glacial age, ice left its imprint on almost 30 percent of the earth's land area, including about 10 million square kilometers of North America, 5 million square kilometers of Europe, and 4 million square kilometers of Siberia (Figure 4.17, page 110). The amount of glacial ice in the Northern Hemisphere was roughly twice that of the Southern Hemisphere. The primary reason is that the southern polar ice could not spread far beyond the margins of Antarctica. By contrast, North America and Eurasia provided great expanses of land for the spread of ice sheets.

Today we know that the Ice Age began between two and three million years ago. This means that most of the major glacial stages occurred during a division of the geologic calendar called the **Pleistocene epoch.** Although the Pleistocene is commonly used as a synonym for the Ice Age, note that this epoch does not encompass all of the last glacial period. The Antarctic ice sheet, for example, is believed to have formed about 10 million years ago.

Glaciers have not been ever-present features throughout the earth's long history. In fact, for most of geologic time glaciers have been absent. Evidence does indicate that in addition to the Pleistocene epoch there were at least three other periods of glacial activity: 2 billion, 600 million, and 250 million years ago. However, the most recent period of glaciation is of greatest interest, because the features of many present-day landscapes are a reflection of the work of Pleistocene glaciers.

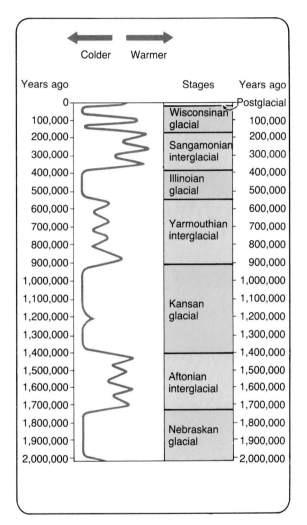

FIGURE 4.16
The Pleistocene epoch was marked by fluctuating climatic conditions that led to alternating glacial and interglacial periods. Four separate stages of glaciation are recognized for North America. (After D. B. Ericson and G. Wollin, "Pleistocene Climates and Chronology in Deep-Sea Sediments," *Science* 162 (1968): 1233. Copyright © 1968 by the American Association for the Advancement of Science)

SOME INDIRECT EFFECTS OF ICE AGE GLACIERS

In addition to the massive erosional and depositional work carried on by Pleistocene glaciers, the ice sheets had other, sometimes profound, effects upon the landscape. For example, as the ice advanced and retreated, animals and plants were forced to migrate. This led to stresses that some organisms could not tolerate. Furthermore, many present-day stream courses bear little

FIGURE 4.17
At their maximum extent,
Pleistocene glaciers covered
about 10 million square kilometers (4 million square
miles) of North America

resemblance to their preglacial routes. The Missouri River once flowed northward toward Hudson Bay, while the Mississippi River followed a path through central Illinois, and the head of the Ohio River reached only as far as Indiana. Other rivers that today carry only a trickle of water but nevertheless occupy broad channels are testimony to the fact that they once carried torrents of glacial meltwater.

In areas that were centers of ice accumulation, such as Scandinavia and the Canadian Shield, the land has been slowly rising for the past several thousand years. The land is rising because the added weight of the three-kilometer-thick mass of ice downwarped the earth's crust. Following the removal of this immense load, the crust has been adjusting by gradually rebounding upward ever since.

Certainly one of the most interesting and perhaps dramatic effects of the Ice Age was the fall and rise of sea level that accompanied the advance and retreat of the glaciers. Since we know that the snow from which glaciers are made ultimately comes from the evaporation of ocean water, the growth of ice sheets must have caused a worldwide drop in sea level. Indeed, estimates suggest that sea level was as much as 130 meters (425 feet) lower than today. Thus, some land that is presently flooded by the oceans was dry. The Atlantic Coast of the United States was located more than 100 kilometers (60 miles) to the east of New York City; France and Britain were joined where the English Channel is today; Alaska and Siberia were connected across the Bering Strait; and Southeast Asia was tied by dry land to the islands of Indonesia.

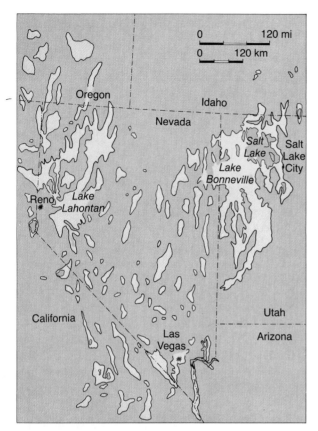

FIGURE 4.18
Pluvial lakes of the western United States. (After R. F. Flint, *Glacial and Quaternary Geology,* New York: John Wiley and Sons.)

While the formation and growth of ice sheets was an obvious response to significant changes in climate, the existence of the glaciers themselves triggered important climatic changes in the regions beyond their margins. In arid and semiarid areas on all continents, temperatures, and thus evaporation rates, were lower. At the same time, precipitation was moderate. This cooler, wetter climate resulted in the formation of many lakes called **pluvial lakes,** from the Latin term *pluvia* meaning *rain.* In North America, the greatest concentration of pluvial lakes occurred in the vast Basin and Range region of Nevada and Utah (Figure 4.18). Although most of the lakes completely disappeared, there are a few small remnants, the Great Salt Lake being the largest and best known.

CAUSES OF GLACIATION

A great deal is known about glaciers and glaciation. Much has been learned about glacier formation and movement, the extent of glaciers past and present, and the features created by glaciers, both erosional and depositional. However, scientists have not yet developed a completely satisfactory explanation for the causes of ice ages.

Any theory that attempts to explain the causes of glacial ages must successfully address two basic questions. First, what causes the onset of glacial conditions? For continental ice sheets to have formed, average temperatures must have been somewhat lower than at present and perhaps substantially lower than throughout much of geologic time. For that reason, a successful explanation would have to account for the gradual cooling that finally leads to glacial conditions. The second question is: What caused the alternation of glacial and interglacial stages that have been documented for the Pleistocene epoch? While the first question deals with long-term trends in temperature that occur on a scale of millions of years, this second question relates to much shorter-term changes.

Although the literature of science contains a vast array of hypotheses that attempt to explain the possible causes of glacial periods, we will discuss only a few major ideas in an effort to give the current thought on this problem.

Probably the most attractive theory for explaining why extensive glaciations have occurred only a few times in the geologic past comes from the theory of plate tectonics.* Not only does this theory provide geologists with explanations about many previously misunderstood processes and features, it also provides a possible explanation for some previously unexplainable climatic changes including the onset of glacial conditions. Since glaciers can form only on the continents, we know that landmasses must exist somewhere in the higher latitudes before an ice age can commence. Many believe that ice ages have only occurred when the earth's shifting crustal plates carried the continents from tropical latitudes to more poleward positions.

---

* A complete discussion of plate tectonics is presented in Chapter 6.

Glacial features in present-day Africa, Australia, South America, and India indicate that these regions experienced an ice age near the end of the Paleozoic era, about 250 million years ago. For many years this puzzled scientists. Was the climate in these relatively tropical latitudes once like it is today in Greenland and Antarctica? Why did glaciers not form in North America and Eurasia? Until the plate tectonics theory was formulated and proven, there was no reasonable explanation. Today scientists realize that the areas containing these ancient glacial features were joined together as a single super-continent called Pangaea that was located at high latitudes far to the south of their present positions. Later, this landmass broke apart and its pieces, each moving on a different plate, drifted toward their present locations (Figure 4.19). It is now believed that during the geologic past, plate movements accounted for many dramatic climatic changes as landmasses shifted in relation to one another and moved to different latitudinal positions. Changes in oceanic circulation also must have occurred, altering the transport of heat and moisture, and consequently, the climate as well. Since the rate of plate movement is very slow, on the order of a few centimeters per year, appreciable changes in the positions of the continents occur only over great spans of geologic time. Thus, climatic changes brought about by shifting plates are extremely gradual and happen on a scale of millions of years.

Since climatic changes brought about by moving plates are extremely gradual, the plate tectonics theory cannot explain the alternation of glacial and interglacial climates that occurred during the Pleistocene epoch. Therefore we must look to some other triggering mechanism that may cause climatic change on a scale of thousands rather than millions of years. Today many scientists believe or strongly suspect that the climatic oscillations that characterized the Pleistocene may be linked to variations in the earth's orbit. This hypothesis was first developed and strongly advocated by the Yugoslavian scientist Milutin Milankovitch and is based on the premise that variations in incoming solar radiation are a principal factor in controlling the earth's climate.

Milankovitch formulated a comprehensive mathematical model based on the following elements:

1 Variations in the shape (*eccentricity*) of the earth's orbit around the sun;
2 Changes in *obliquity;* that is, changes in the angle that the earth's axis makes with the plane of the earth's orbit; and
3 The wobbling of the earth's axis, called *precession.*

Using these factors Milankovitch calculated variations in the receipt of solar energy and the corresponding surface temperature of earth back into time in an attempt to correlate these changes with the climatic fluctuations of the Pleistocene. In explaining climatic changes that result from these three variables, it should be noted that they cause little or no variation in the total amount of solar energy reaching the ground. Instead, their impact is felt because they change the degree of contrast between seasons. Somewhat milder winters in the middle to high latitudes means greater snowfall totals while cooler summers would bring a reduction in snowmelt.

Over the years the astronomical theory of Milankovitch has been widely accepted, then largely rejected, and now, in light of recent investigations, has once again gained significant support. Among the studies that have added credibility and support to the astronomical theory is one in which deep-sea sediments containing certain climatically sensitive microorganisms were analyzed in order to establish a chronology of temperature changes going back nearly one-half million years.[*] This time scale of climatic change was then compared to astronomical calculations of eccentricity, obliquity, and precession to determine if a correlation did indeed exist. Although the study was very involved and mathematically complex, the conclusions were straightforward. The authors found that major variations in climate over the past several hundred thousand years were closely associated with changes in the geometry of the earth's orbit; that is, cycles of climatic

[*] J. D. Hays, John Imbrie, and N. J. Shackleton, "Variations in the Earth's Orbit: Pacemaker of the Ice Ages," *Science,* 194 (4270): 1121–32.

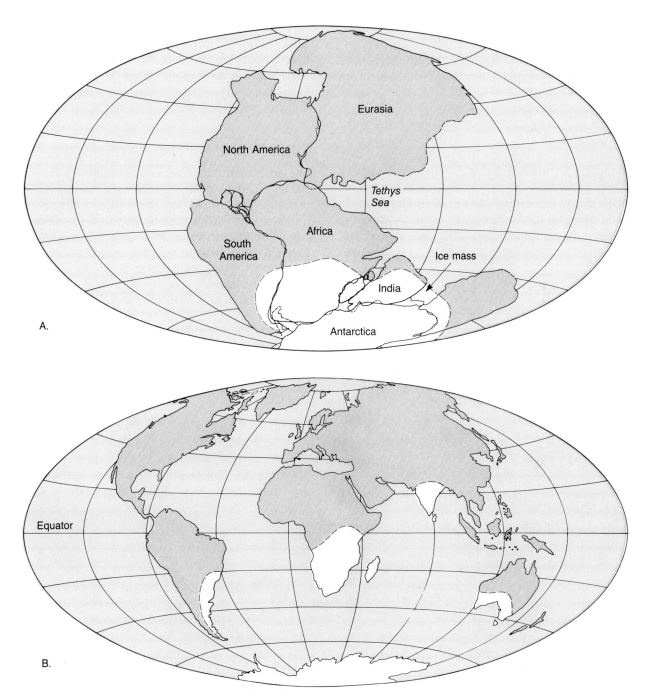

FIGURE 4.19
A. The supercontinent Pangaea showing the area covered by glacial ice 300 million years ago. **B.** The continents as they are today. The shading outlines areas where evidence of the old ice sheets exists. (After R. F. Flint and B. J. Skinner, *Physical Geology,* 2nd ed., p. 418, New York: Wiley, 1977)

change were shown to correspond closely with the periods of obliquity, precession, and orbital eccentricity. More specifically, they stated, "It is concluded that changes in the earth's orbital geometry are the fundamental cause of the succession of Quaternary ice ages."*

Let us briefly summarize the ideas that were just described. The theory of plate tectonics provides an explanation for the widely spaced and nonperiodic onset of glacial conditions at various times in the geologic past, whereas the astronomical theory proposed by Milankovitch and recently supported by the work of J. D. Hays and his colleagues furnishes an explanation for the alternating glacial and interglacial episodes of the Pleistocene.

In conclusion, it should be emphasized at this point that the ideas just discussed do not represent the only possible explanations for glacial ages. Although interesting and attractive, these theories are certainly not without critics nor are they the only proposals currently under study. Other factors may, and in fact probably do, enter into the picture.

DESERTS

The desert (arid) and steppe (semiarid) regions of the world encompass about 48 million square kilometers (18.7 million square miles), nearly one-third of the earth's land surface. These areas, which are characterized by scanty and unreliable precipitation, are shown in Figure 4.20. No other climatic group covers so large a land area.*

The angular hills, sheer canyon walls, and pebble or sand covered surface of the desert contrast

---

*Ibid., p. 1131. *Quaternary* refers to the period on the geologic time scale that encompasses the last few million years.

*A more complete discussion of dry climates is given in Appendix F.

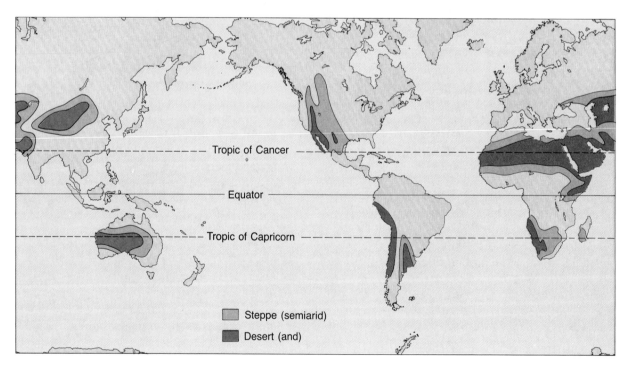

FIGURE 4.20
Arid and semiarid climates cover nearly one-third of the earth's land surface. No other climatic group covers so large a land area.

FIGURE 4.21
Desert stream courses are called washes. Most of the time washes are dry. Following a rain, they can be transformed into rushing torrents. (Photo by James E. Patterson)

sharply with the rounded hills and curving slopes of more humid areas. A desert landscape may at first seem to have been shaped by forces different from those operating in regions where water is more abundant. While the contrasts may be striking, they are not a reflection of different processes but merely the differing effects of the same processes operating under contrasting climatic conditions.

In humid regions, relatively fine textured soils mantle the surface to support an almost continuous cover of vegetation. Here the slopes and rock edges are rounded. Such a landscape reflects the greater importance of chemical weathering in a humid climate. Since most of the weathered debris in deserts consists of unaltered rock and mineral fragments, we can conclude that mechanical weathering processes are relatively more important in arid regions than in humid regions. In dry lands, however, rock weathering of any type is greatly reduced because of the lack of moisture and the scarcity of organic acids from decaying plants. Chemical weathering is not completely lacking in deserts. Over long periods, clays and thin soils do form and many iron-bearing silicate minerals oxidize to produce the rust-colored stain found tinting some desert landscapes.

Unlike the drainage in humid regions, stream courses in arid regions are seldom well integrated

with larger and larger tributaries feeding into main rivers. In fact, a characteristic of deserts is that most streams that originate in them are small and die out before reaching the sea. Because the water table is usually far below the surface, few desert streams can draw upon it. Without a steady supply of water, evaporation soon depletes the stream and the remaining water sinks into the ground. The few permanent streams that do cross arid regions, such as the Colorado and Nile rivers, originate outside the desert, often in mountains where water is more plentiful. Here the supply of water must be great to compensate for the losses that occur as the stream crosses the desert. For example, after the Nile leaves the rainy regions that are its source, it traverses almost 2000 kilometers (1250 miles) of the Sahara Desert without a single tributary.

Most of the time desert stream courses, called **washes** in the western United States, are dry (Figure 4.21). This is obvious even to the casual observer who, while traveling, may notice the number of bridges with no streams beneath them or the number of dips in the road where roads cross dry washes. However, when the rare heavy showers do come, so much rain falls in such a short time that it cannot all soak in. Since the vegetative cover is sparse, runoff is largely unhindered and therefore rapid, often creating flash floods along

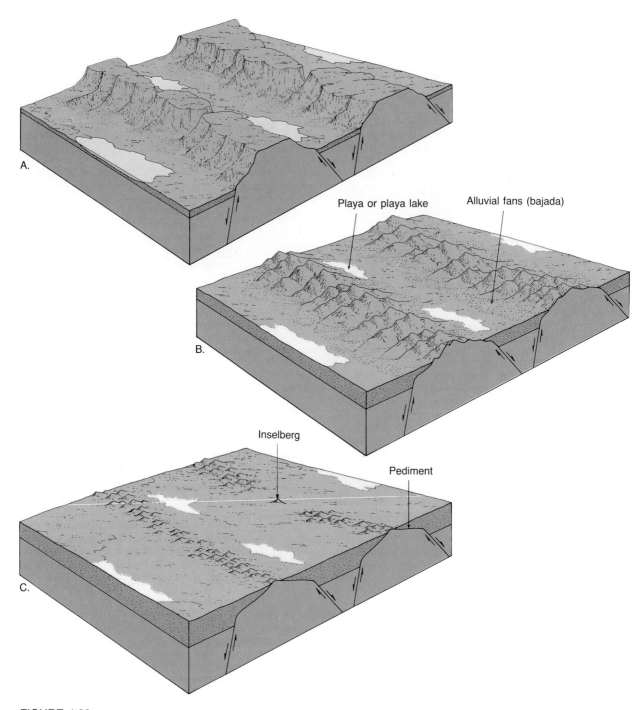

FIGURE 4.22
Stages of landscape evolution in a mountainous desert. As erosion of the mountains and deposition in the basins continue, relief diminishes. **A.** Early stage **B.** Middle stage. **C.** Late stage.

valley floors. Such floods are quite unlike floods in humid regions. A flood on a river like the Mississippi may take several days to reach its crest and then subside, whereas desert floods arrive suddenly and subside quickly. Because the loose material on the surface is not anchored by vegetation, the amount of erosional work that occurs during one of these short-lived events is substantial.

Even though it is an infrequent occurrence, running water is responsible for most of the erosional work in deserts. This is contrary to a commonly held belief that wind is the most important erosional agent sculpturing desert landscapes. Although wind is more significant in dry areas than elsewhere, most desert landforms are nevertheless carved by running water. The main role of wind, as we shall see later in this chapter, is in the transportation and deposition of sediment, creating and shaping the ridges and mounds we call dunes.

THE EVOLUTION OF A DESERT LANDSCAPE

Due to the fact that arid regions typically lack permanent streams, they are characterized as having interior drainage, that is, a discontinuous pattern of intermittent streams that do not flow out of the desert to the ocean. In the United States, the dry Basin and Range region provides an excellent example. The region includes southern Oregon, all of Nevada, western Utah, southeastern California, as well as southern Arizona and New Mexico.

The name Basin and Range is an apt description for this almost 800,000-square-kilometer region, since it is characterized by more than 200 relatively small mountain ranges which rise 900–1500 meters above the basins that separate them. In this region, as in others like it around the world, erosion is carried out for the most part without reference to the ocean (ultimate base level), because drainage is in the form of local interior systems. Even in areas where permanent streams flow to the ocean, few tributaries exist, and thus only a relatively narrow strip of land adjacent to the stream has sea level as the ultimate level of land reduction.

The block models in Figure 4.22 depict the stages of landscape evolution in a mountainous desert such as the Basin and Range region. During and following uplift of the mountains, running water begins carving the elevated mass and depositing large quantities of debris in the basin. In this early stage relief is greatest, for as erosion lowers the mountains and sediment fills the basins, elevation differences diminish.

When the occasional torrents of water produced by sporadic rains move down the mountain canyons, they are heavily loaded with sediment. Emerging from the confines of the canyon, the runoff spreads over the gentler slopes at the base of the mountains and quickly loses velocity. Consequently most of its load is dumped within a short distance. The result is a cone of debris known as an *alluvial fan* at the mouth of a canyon. (Figure 4.23). Since the coarsest material is

FIGURE 4.23
Alluvial fans develop where the gradient of a stream abruptly changes from steep to flat, such as at the base of a mountain.

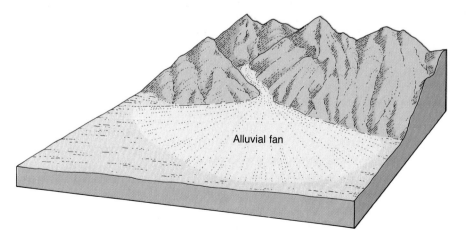

Alluvial fan

dropped first, the head of the fan is steepest. Moving down the fan, the size of the sediment and the steepness of the slope decrease and merge imperceptibly with the basin floor. Over the years, a fan enlarges, eventually coalescing with fans from adjacent canyons to produce an apron of sediment along the mountain front.

On the rare occasions when rainfall is abundant, streams may flow across the alluvial fans to the center of the basin, converting the basin floor into a shallow **playa lake.** Playa lakes are temporary features that last only a few days or at best a few weeks before evaporation and infiltration remove the water. The dry, flat lake bed that remains is termed a *playa*.

Playas are typically composed of fine silts and clays, and occasionally they become encrusted with salts precipitated during evaporation. These precipitated salts may be uncommon. A case in point is the sodium borate (better known as borax) mined from ancient playa lake deposits in Death Valley, California.

With time the mountain front is worn back and a sloping bedrock platform, called a **pediment,** is created adjacent to the steep mountain front. A pediment is an erosional surface, usually covered by a thin veneer of debris, that is formed by the action of running water. Just how the water carves the pediment, however, is unclear and still a matter for debate.

With the ongoing dissection of the mountain mass into an intricate series of valleys and sharp divides as well as the accompanying sedimentation, the local relief continues to diminish. After more time passes, the steady retreat of the mountain front enlarges the pediment. Eventually this pediment growth results in nearly the entire mountain mass being consumed. Thus, by the late stages of erosion, the mountain areas are reduced to a few large bedrock knobs projecting above the surrounding pediment and sediment-filled basin. These isolated erosional remnants on an old-age desert landscape are called **inselbergs,** a German word meaning "island mountains."

Each of the stages just described can be observed in the Basin and Range region. Recently uplifted mountains in an early stage of erosion are found in southern Oregon and northern Nevada.

Death Valley, California, and southern Nevada fit into the more advanced middle stage, while the late stage, with its inselbergs and extensive pediments, can be seen in southern Arizona.

WIND EROSION

Moving air, like moving water, is capable of picking up loose debris and moving it to another location. Although wind erosion is not restricted to arid and semiarid regions, it does its most effective work in these areas. In humid regions moisture binds particles together and vegetation anchors the soil so that wind erosion is negligible. For wind to be effective, dryness and scanty vegetation are important prerequisites. When such circumstances exist, wind may pick up, transport, and deposit great quantities of fine sediment. During the 1930s parts of the Great Plains experienced great dust storms. The plowing under of the natural vegetative cover for farming, followed by severe drought, made the land prime for wind erosion and led to the area being labeled the Dust Bowl.

Wind erosion differs from stream erosion in two significant ways. First, wind has a low density compared to water; thus it is not capable of picking up and transporting coarse materials. Second, because wind is not confined to channels, it can spread over large areas, as well as high into the atmosphere.

One way that winds erode is by **deflation,** the lifting and removal of loose material. Since the competence (ability to transport different size particles) of moving air is low, it can only suspend fine sediment such as clay and silt. Sand grains are rolled or skipped along the surface and comprise the bed load. Coarser particles are usually not transported by wind. Although the effects of deflation are sometimes difficult to notice because the entire surface is being lowered at the same time, they can be significant. In portions of the Dust Bowl, the land was lowered by as much as 1 meter.

The most noticeable result of deflation in some places are shallow depressions called **blowouts.** In the Great Plains region, from Texas north to Montana, thousands of blowouts can be seen.

FIGURE 4.24
Formation of desert pavement. As these cross-sections illustrate, coarse particles gradually become concentrated into a tightly packed layer as deflation lowers the surface by removing sand and silt.

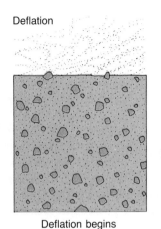

Deflation

Deflation begins

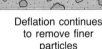

Deflation

Deflation continues to remove finer particles

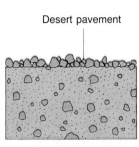

Desert pavement

Desert pavement established, deflation ends

They range in size from small dimples less than 1 meter deep and 3 meters wide to depressions that are over 45 meters deep and several kilometers across. In the past, several explanations for the depressions were put forth, including one which suggested they were sites of buffalo wallows. Today it is believed that the vast majority were created by deflation.

In portions of many deserts the surface is characterized by a layer of coarse pebbles and gravel. Such a layer, called **desert pavement,** is created as the wind removes fine material, leaving the coarse particles behind (Figure 4.24). Once desert pavement becomes established, a process which may take hundreds of years, the surface is effectively protected from further deflation.

Like glaciers and streams, wind erodes by abrasion. In dry regions as well as along some beaches, windblown sand cuts and polishes exposed rock surfaces. However, abrasion is often credited for accomplishments beyond its actual capabilities. Such features as balanced rocks that stand high atop narrow pedestals and intricate detailing on tall pinnacles are not the results of abrasion. Since sand seldom travels more than a meter above the surface, the wind's sandblasting effect is obviously quite limited in vertical extent. However, in areas prone to such activity, telephone poles have actually been cut through near their bases. For this reason, collars are often fitted on the poles to protect them from being "sawed" down.

WIND DEPOSITS

Although wind is relatively unimportant as a producer of erosional landforms, wind deposits are significant features in some regions. Accumulations of windblown sediment are particularly conspicuous landscape elements in the world's dry lands and along many sandy coasts. Wind deposits are of two distinctive types: (1) Extensive blankets of silt that once were carried in suspension; and (2) Mounds and ridges of sand from the wind's bed load.

LOESS

In some parts of the world the surface topography is mantled with deposits of windblown silt. Over periods of perhaps thousands of years dust storms deposited this material, which is called **loess.** As can be seen in Figure 4.25, when loess is breached by streams or road cuts it tends to maintain vertical cliffs and lacks any visible layers. The distribution of loess indicates that two primary sources for this sediment are deserts and glacial outwash deposits. The thickest and most extensive loess deposits occur in western and northern China, where accumulations of 30 meters are not uncommon and thicknesses of more than 100 meters have been measured. It is this fine, buff-colored sediment which gives the Yellow River (Hwang Ho) and the adjacent Yellow Sea their names. The

FIGURE 4.25
A vertical loess bluff near the Mississippi River in southern Illinois. (Photo by James E. Patterson)

source of China's 800,000 square kilometers of loess are the extensive desert basins of central Asia.

In the United States, deposits of loess are significant in many areas, including South Dakota, Nebraska, Iowa, Missouri, and Illinois as well as portions of the Columbia Plateau in the Pacific Northwest. The correlation between the distribution of loess and important farming regions in the Midwest and eastern Washington is not just a coincidence, because soils derived from this wind-deposited sediment are among the most fertile in the world. Unlike the deposits in China, the loess in the United States, as well as in Europe, is an indirect product of glaciation, for its source was deposits of outwash. During the retreat of the glacial ice, many river valleys were choked with sediment provided by meltwater. Strong westerly winds sweeping across the barren floodplains picked up the finer sediment and dropped it as a blanket on the east side of the valleys. Such an origin is confirmed by the fact that loess deposits are thickest and coarsest on the lee side of such

major glacier drainage outlets as the Mississippi and Illinois rivers and thin rapidly with increasing distance from the valleys. Furthermore, the angular, mechanically weathered particles composing the loess are essentially the same as the rock flour produced by the grinding action of glaciers.

SAND DEPOSITS

Like running water, wind releases its load of sediment when its velocity falls and the energy available for transport diminishes. Thus sand begins to accumulate wherever an obstruction across the path of the wind slows the movement of the air. Unlike deposits of loess, which form blanketlike layers over large areas, winds commonly deposit sand in mounds or ridges called *dunes.*

As moving air encounters an object, such as a clump of vegetation or a rock, the wind sweeps around and over it leaving a shadow of more slowly moving air behind the obstacle as well as a smaller zone of quieter air just in front of the obstacle. Some of the saltating sand grains moving with the wind come to rest in these "wind shadows." As the accumulation of sand continues, an increasingly efficient wind barrier forms to trap even more sand. If there is a sufficient supply of sand and the wind blows steadily long enough, the mound of sand grows into a dune.

The profile of some dunes shows an asymmetrical shape with the leeward slope being steep and the windward slope more gently inclined. Sand is rolled up the gentle slope on the windward side by the force of the wind. Just beyond the crest of the dune, the wind velocity is reduced and the sand accumulates. As more sand collects, the slope steepens and eventually some of it slides or slumps under the pull of gravity. In this way the leeward slope of the dune, called the **slip face,** maintains an angle of about 34 degrees. Continued sand accumulation coupled with periodic slides down the slip face result in the slow migration of the dune in the direction of air movement. As sand is deposited on the slip face, it forms layers inclined in the direction the wind is blowing. These sloping layers are called **cross beds.** When the dunes are eventually buried under layers of sediment and become part of the sedimentary

FIGURE 4.26
The cross-bedding in this sandstone indicates it was once a sand dune. (Photo by Stephen Trimble)

rock record, their asymmetrical shape is destroyed, but the cross beds remain as a testimony to their origin. Nowhere is cross-bedding more prominent than in the sandstone walls of Zion Canyon in Utah (Figure 4.26).

TYPES OF SAND DUNES

Although often complex, dunes are not just random heaps of sediment. Rather they are accumulations that usually assume surprisingly consistent patterns (Figure 4.27). Addressing this point, a leading early investigator of dunes, the British engineer R. A. Bagnold observed, "Instead of finding chaos and disorder, the observer never fails to be amazed at a simplicity of form, an exactitude of repetition, and a geometric order unknown in nature on a scale larger than that of crystalline structure. In places, vast accumulations of sand weighing millions of tons move inexorably, in regular formation, over the surface of the country retaining their shape. . . . "

Barchan Dunes Solitary sand dunes shaped like crescents and with their tips pointing downwind are called **barchan dunes** (Figure 4.27A). These dunes form where supplies of sand are limited and the surface is relatively flat, hard, and lacking vegetation. They migrate slowly with the wind at a rate of up to 15 meters per year. Their size is usually modest with the largest barchans reaching a height of about 30 meters while the maximum spread between their horns approaches 300 meters. When the wind direction is nearly constant, the crescent form of these dunes is nearly

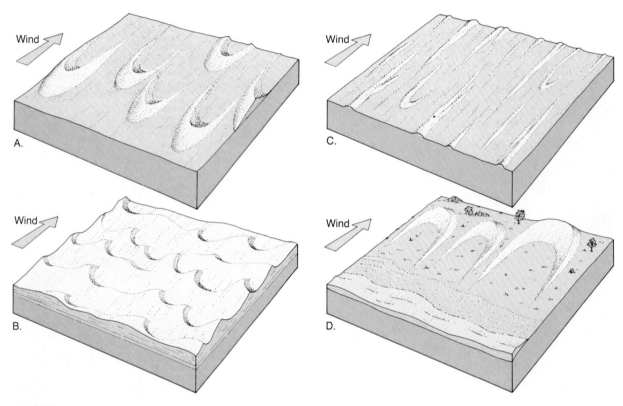

FIGURE 4.27
A. Barchan dunes. **B.** Transverse dunes. **C.** Longitudinal dunes. **D.** Parabolic dunes.

symmetrical. However, when the wind direction is not perfectly fixed, one tip becomes longer than the other.

Transverse Dunes   In regions where vegetation is sparse or absent and sand is very plentiful, the dunes form a series of long ridges that are separated by troughs and oriented at right angles to the prevailing wind. Because of this orientation, they are termed **transverse dunes** (Figure 4.28). Typically, many coastal dunes are of this type. In addition, they are common in arid regions where the extensive surface of wavy sand is sometimes called a sand sea.

Longitudinal Dunes   **Longitudinal dunes** are long ridges of sand that generally form parallel to the prevailing wind and where sand supplies are limited (Figure 4.27C). Apparently the prevailing wind direction must vary somewhat, but not by more than about 90 degrees. Although the smaller

types are only three or four meters high and several tens of meters long, in some large deserts longitudinal dunes can reach great size. For example, in portions of North Africa, Arabia, and central Australia, these dunes may approach a height of 100 meters and extend for distances of more than 100 kilometers.

Parabolic Dunes   Unlike the dunes that have been described thus far, **parabolic dunes** form where vegetation partly covers the sand (Figure 4.27D). The shape of these dunes resembles the shape of barchans except that their tips point into the wind rather than downwind. Parabolic dunes often form along coasts where there are strong onshore winds and abundant sand. If the sand's sparse vegetative cover is disturbed at some spot, deflation creates a blowout. Sand is then transported out of the depression and deposited as a curved rim which grows higher as deflation enlarges the blowout.

FIGURE 4.28
Transverse dunes in Great Sand Dunes National Monument, Colorado. (Photo by
Stephen Trimble)

REVIEW QUESTIONS

1 What is a glacier? Under what circumstances does glacial ice form?

2 Where are glaciers found today? What percentage of the earth's land area do they cover?

3 Describe glacial flow. At what rates do glaciers move? In an alpine glacier does all of the ice move at the same rate? Explain.

4 Why do crevasses form in the upper portion of a glacier but not below 50 meters?

5 Under what circumstances will the front of a glacier advance? Retreat? Remain stationary?

6 Describe the processes of glacial erosion.

7 How does a glaciated mountain valley differ from a mountain valley that was not glaciated?

8 List and describe the erosional features you might expect to see in an area where alpine glaciers exist or have recently existed.

9 What is glacial drift? What is the difference between till and outwash? What general effect do glacial deposits have on the landscape?

10 List the four basic moraine types. What do all moraines have in common? What is the significance of terminal and recessional moraines?

11 List and discuss depositional features other than moraines.

12 How does a kettle form?

13 How does the area covered by Pleistocene glaciers compare with the area presently covered by glacial ice?

14 How many glacial advances have been recognized for North America? List them in the order of their occurrence (Figure 4.16).

15 List three indirect effects of Ice Age glaciers.

16 How might plate tectonics help us understand the cause of ice ages? Can plate tectonics explain the alternation between glacial and interglacial climates during the Pleistocene?

17 How extensive are the desert and steppe regions of the earth?

18 What erosional agent does the greatest amount of work in deserts?

19 Why is sea level not a significant factor influencing stream erosion in desert regions?

20 Describe the features and characteristics associated with each of the stages in the evolution of a mountainous desert. Where in the United States can these stages be observed?

21 Why is wind erosion relatively more important in arid regions than in humid areas?

22 List two types of wind erosion and the features which may result from each.

23 Although sand dunes are the best-known wind deposits, accumulations of loess are very significant in some parts of the world. What is loess? Where are such deposits found? What are the origins of this sediment?

24 How do sand dunes migrate?

25 Identify each of the dunes described below.
   (a) Long ridges of sand oriented parallel to the prevailing wind.
   (b) Solitary, crescent-shaped dunes oriented with their tips pointing downwind.
   (c) Ridges of sand oriented at right angles to the prevailing wind.

KEY TERMS

glacier	drift	wash
alpine glacier	till	playa lake
continental glacier	outwash	pediment
crevasse	glacial erratic	inselberg
ablation	lateral moraine	deflation
plucking	medial moraine	blowout
abrasion	end moraine	desert pavement
rock flour	ground moraine	loess
glacial striation	outwash plain	slip face
glacial trough	kettle	cross beds
hanging valley	drumlin	barchan dune
cirque	esker	transverse dune
fiord	kame	longitudinal dune
arête	Pleistocene epoch	parabolic dune
horn	pluvial lake	

5

EARTHQUAKES
AND THE
EARTH'S INTERIOR

What is an earthquake? How does a seismograph record the location of a "quake"? Can earthquakes ever be controlled, or at least predicted? If we could view the earth's interior, what would it look like? In this chapter we shall try to answer these questions, as well as many others. The study of earthquakes is important, not only because of the devastating effect that some earthquakes have on us, but also because they furnish clues about the structure of the earth's interior.

Damage caused by the September 19, 1985 Mexican earthquake. (Photo by James L. Beck)

On September 19, 1985, a violent earthquake jolted southwestern Mexico, devastating sections of Mexico City and killing an estimated 7000 persons (see chapter-opening photo). After viewing the destruction from a helicopter, the United States Ambassador remarked that, "It looked as if a giant foot has stepped on the buildings." In less than two minutes the quake battered downtown Mexico City, causing an estimated 250 buildings to collapse and damaging over 7000 others. The following day a strong aftershock crumbled additional structures that had been weakened by the initial event.

This earthquake, which occurred during the morning rush hour, was centered on the Pacific Coast nearly 400 kilometers (250 miles) from Mexico City. The vibrations were felt as far north as Houston, Texas, where skyscrapters swayed. Ironically, coastal towns such as Ixtapa, located less than 80 kilometers from the epicenter, suffered less during the event than Mexico City. The fact that the vibrations were actually caused by two shocks that had an average magnitude of 8.1 on the Richter scale may have lengthened the time of intense shaking. In addition, the central district of Mexico City is built on the soft, moist sediments of an ancient lake bed. According to George Housner, a researcher at the California Institute of Technology, this earthquake caused the lake bed to vibrate "like a bowl of jelly when the seismic waves came through the ground beneath it."

Although the destruction to downtown Mexico City was staggering, the damage was very restricted. Out of a total of over 600,000 structures located in the city, only a few hundred buildings collapsed. Further, only one percent of the city was heavily damaged. Great earthquakes can be much worse.

This was not the first earthquake to batter Mexico City, and it will not be the last. It is estimated that nearly one million earthquakes occur worldwide each year. Fortunately, not many are as devastating as the 1985 Mexican earthquake. Generally only a few major earthquakes take place each year, but when they do, they are among the most destructive natural forces on earth. The shaking of the ground coupled with the liquefaction of some soils wreak havoc on manmade structures (Figure 5.1). In addition, when a quake occurs in a populated area, power and gas lines are often ruptured, causing numerous fires. In the 1906 San Francisco earthquake, much of the damage was caused by

FIGURE 5.1
This tilted building rests on unconsolidated sediment that imitated quicksand during the 1985 Mexican earthquake. (Photo by James L. Beck)

128

FIGURE 5.2
San Francisco in flames after the 1906 earthquake. (Reproduced from the collection of the Library of Congress)

fires which ran unchecked when broken water mains left firefighters with only trickles of water (Figure 5.2).

WHAT IS AN EARTHQUAKE?

An **earthquake** is the vibration of the earth produced by the rapid release of energy. This energy radiates in all directions from its source, the **focus,** in the form of waves analogous to those produced when a bell is struck, vibrating the air around it (Figure 5.3). During an earthquake and for many hours following, the earth could be described as "ringing like a bell." Even though the energy dissi-

pates rapidly with increasing distance from the focus, instruments located throughout the world record the event.

The tremendous energy released by atomic explosions or by volcanic eruptions can produce an earthquake, but these events are usually weak and infrequent. What mechanism does produce a destructive earthquake? Ample evidence exists that the earth is not a static planet. Numerous ancient wave-cut benches can be found many meters above the level of the highest tides, which indicates crustal uplifting of comparable magnitude. Other regions exhibit evidence of extensive subsidence. In addition to these vertical displacements, offsets in fence lines, roads, and other

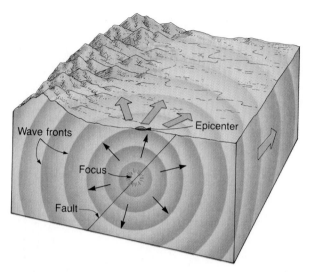

FIGURE 5.3
The focus of most earthquakes is located at depth. The surface location directly above it is called the epicenter.

structures indicate that horizontal movement is also prevalent. These movements are frequently associated with large fractures in the earth called **faults.** Most of the motion along faults can be satisfactorily explained by the plate tectonics theory. This theory proposes that large slabs of the earth are continually in motion. These mobile plates interact with neighboring plates, straining and deforming the rocks at their edges. It is along these plate boundaries that most earthquakes occur.

The actual mechanism of earthquake generation eluded geologists until H. F. Reid conducted a study following the great 1906 San Francisco earthquake. The earthquake was accompanied by displacements of several meters along the northern portion of the San Andreas fault, a 950-kilometer (600-mile) fracture which runs northward through southern California. Using land surveys conducted several years apart, Reid discovered that during the 50 years prior to the 1906 earthquake the land at distant points on both sides of the San Andreas fault showed a relative displacement of slightly more than 3 meters (10 feet). The mechanism for earthquake formation which Reid deduced from this information is illustrated in Fig-

ure 5.4. Tectonic forces ever so slowly deform the crustal rocks on both sides of the fault as illustrated by the bent features. Under these conditions, rocks are bending and storing elastic energy, much like a wooden stick would if bent. Eventually, the frictional resistance holding the rocks together is overcome. As slippage occurs at the weakest point (the focus), displacement will exert stress farther along the fault where additional slippage will occur until most of the built-up strain is released. This slippage allows the deformed rock to "snap back." The vibrations we know as an earthquake occur as the rock elastically returns to its original shape. The "springing back" of the rock was termed **elastic rebound** by Reid, since the rock behaves elastically, much like a stretched rubber band does when it is released.

The intense vibrations of the 1906 San Francisco earthquake lasted about 40 seconds. Although most of the displacement along the fracture occurred in this rather short period, additional movements and adjustments in the rocks occurred for several days following the main quake. The adjustments that follow a major earthquake often generate smaller earthquakes called **aftershocks.** Although these aftershocks are usually much weaker than the main earthquake, they can sometimes cause significant destruction to already badly weakened structures. In addition, small earthquakes called **foreshocks** often precede a major earthquake by days or in some cases by as much as several years. Monitoring of these foreshocks has been used as a means of predicting forthcoming major earthquakes. We will consider the topic of earthquake prediction in a later section of this chapter.

The tectonic forces responsible for the strain that was eventually released during the 1906 San Francisco earthquake are still active. Currently laser beams* are used to establish the relative motion between the opposite sides of this fault. These measurements have revealed a displacement of 2 centimeters per year. Although this rate of movement seems slow, it is fast enough to pro-

---

*Laser beams are used in very precise surveying instruments because of their incredibly accurate straight-line qualities.

FIGURE 5.4
Elastic rebound. As rock is deformed it bends, storing elastic energy. Once the rock is strained beyond its breaking point it ruptures, releasing the stored up energy in the form of earthquake waves.

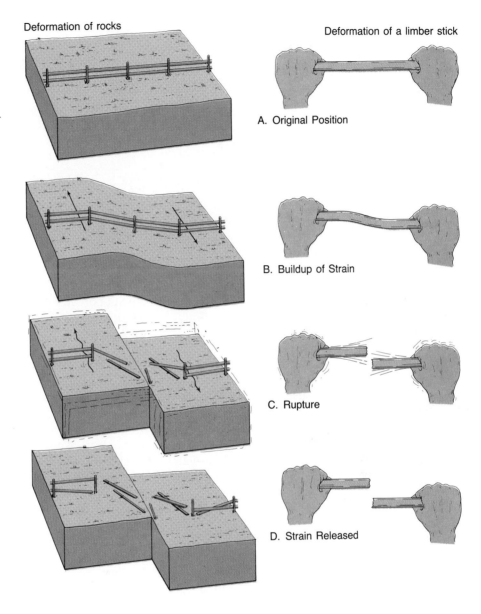

Deformation of rocks

Deformation of a limber stick

A. Original Position

B. Buildup of Strain

C. Rupture

D. Strain Released

duce substantial movement over millions of years of geologic time. In 30 million years such a rate of displacement is sufficient to slide the western portion of California northward so that Los Angeles would be adjacent to San Francisco. Once the strain along any segment of this active fault again reaches sufficient levels, another slippage with an accompanying earthquake can be expected. It is estimated that great earthquakes occur about every 50 to 200 years along plate boundaries such as the San Andreas fault. This repetitive process is often described as *stick-slip motion,* since elastic energy is stored over a period of time and then released through slippage.

Not all motion along the San Andreas fault is of the stick-slip type. Along certain portions of this

FIGURE 5.5
Scarp that resulted from vertical movement along a fault during the Madison Canyon earthquake in Montana. (Photo by Stephen Trimble)

fault the motion is a slow *creep.* Thus, while some sections of the fault are continually creeping, other "locked" segments are building up strain that could result in a major earthquake.

In addition, not all movement along faults is horizontal. Vertical displacement along faults, in which one side is lifted higher than the other, is also common. Figure 5.5 shows a scarp (cliff) produced by such vertical displacement. In the same manner, the 1964 Good Friday earthquake in Alaska produced a 15-meter vertical offset at one location. Further, many earthquakes occur at such great depths that no displacement is evident at the surface.

EARTHQUAKE WAVES

The study of earthquake waves, **seismology,** dates back to attempts by the Chinese almost 2000 years ago to determine the direction to the source of each earthquake. The principle used in modern **seismographs,** instruments which record earthquake waves, is rather simple. A weight is freely suspended from a support that is attached to bedrock (Figure 5.6). When waves from a distant earthquake reach the instrument, the inertia* of

---

*Simply stated, inertia refers to the fact that objects at rest tend to stay at rest and objects in motion tend to remain in motion unless acted upon by an outside force. You probably have experienced this phenomenon when you tried to quickly stop your automobile and your body continued to move forward.

the weight keeps it stationary, while the earth and the support vibrate. The movement of the earth in relation to the stationary weight is recorded on a rotating drum.

The principle of a seismograph can be demonstrated by attaching a heavy mass to a string and holding the other end of the string so that the mass is just off the floor. Rapid side-to-side movements of the string represent the vibrations of an earthquake. Notice that the mass remains relatively motionless. Any noticeable movement will be small and will represent the natural oscillation of your pendulum (this is similar to the oscillation of a clock pendulum). Modern seismographs use a damping mechanism to remove the effect of the natural oscillation of the suspended masses.

The records of seismographs, called **seismograms,** provide a great deal of information about the behavior of seismic waves. Simply stated, seismic waves are elastic energy which radiates outward in all directions from the focus. The propagation (transmission) of this energy can be compared to the shaking of gelatin in a bowl which results as some is spooned out. Whereas the gelatin will have one mode of vibration, seismograms reveal that two main groups of seismic waves are generated by the slippage of a rock mass. One of these wave types travels along the outer layer of the earth. These are called **surface waves.** Others travel through the earth's interior and are called **body waves.** Body waves are further divided into two types called **primary, or P, waves** and **secondary, or S, waves.**

These two wave forms are identified by their method of travel (propagation) through the earth. P waves push (compress) and pull (dilate) rocks in the direction the wave is traveling. This wave motion is the same as that generated by your vocal cords as they move air to and fro in order to transmit sound. S waves on the other hand "shake" the particles at right angles to their direction of travel. This can be illustrated by tying one end of a rope to a fence post and shaking the other end while holding the rope tense.

Notice in Figure 5.7, page 134, that the propagation of P waves involves changing the volume and shape of the intervening material, whereas S waves change only the shape. Solids, liquids, and gases all resist being compressed and will elastically spring back once the force is removed. For this reason, P waves can travel through all types of matter. On the other hand, S waves change only the shape of the medium through which they travel, and because fluids do not resist changes in shape, fluids will not transmit S waves.

The motion of surface waves is somewhat more complex. As surface waves travel along the ground, they cause the ground and anything resting upon it to move, much like ocean swells toss a ship. In addition to their up-and-down motion, surface waves have a side-to-side motion similar to an S wave oriented in a horizontal plane. This latter motion causes most of the structural damage to buildings and their foundations.

By observing a "typical" seismic record, as shown in Figure 5.8, one of the differences between these seismic waves becomes apparent: P waves arrive at the recording station before S waves, which themselves arrive before the surface waves. This is a consequence of their relative velocities. For purposes of illustration, the velocity of P waves through granite within the crust is about 6 kilometers per second, whereas S waves under the same conditions travel at 3.5 kilometers per second. Differences in density and elastic properties of the transmitting material greatly influence the velocities of these waves. However, in any solid material, P waves travel about 1.7 times faster than S waves, and surface waves can be expected to travel at 90 percent of the velocity of the S waves that are traveling in the layer directly below.

As we shall see, seismic waves allow us to determine the location and magnitude of earthquakes. In addition, seismic waves provide us with a tool for probing the earth's interior.

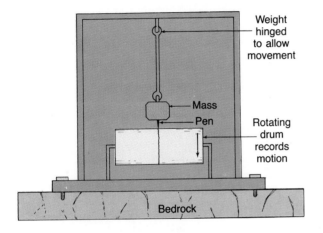

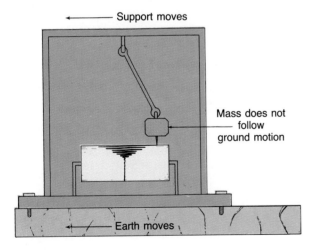

FIGURE 5.6

Principle of the seismograph. The inertia of the suspended mass tends to keep it motionless, while the recording drum, which is anchored to bedrock, vibrates in response to seismic waves. Thus, the stationary mass provides a reference point from which to measure the amount of displacement occurring as the seismic wave passes through the ground below.

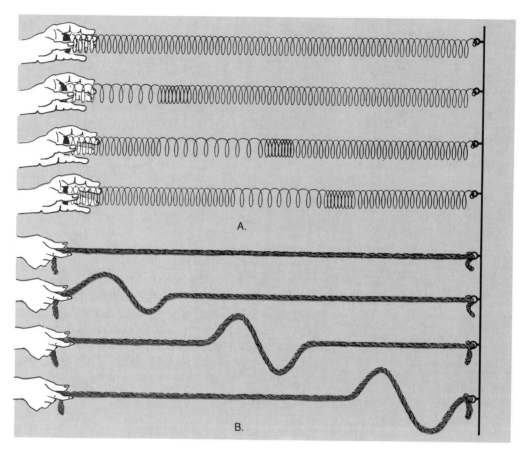

FIGURE 5.7
Types of seismic waves and their characteristic motion. **A.** P waves cause the
particles in the material to vibrate back and forth in the same direction as the
waves move. **B.** S waves cause particles to oscillate at right angles to the direction of wave motion.

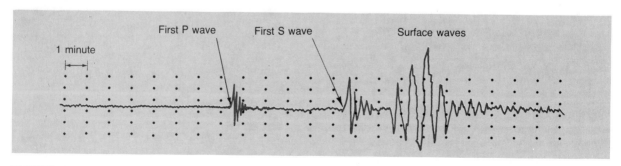

FIGURE 5.8
Typical seismic record. Note the time interval between the arrival of each wave
type.

LOCATION OF EARTHQUAKES

Recall that the focus is the place where the earthquake originates, usually below ground. The **epicenter** is the location on the surface directly above the focus (see Figure 5.3). The difference in velocities of P and S waves provides a method for determining the epicenter. The principle used is analogous to a race between two autos, one faster than the other. The greater the distance of the race, the greater will be the difference in the arrival times at the finish line. Therefore, the greater the interval between the arrival of the first P wave and the first S wave, the greater the distance to the earthquake source.

A system for locating earthquake epicenters was developed through the use of seismograms from earthquakes whose epicenters could be easily pinpointed from physical evidence. From these seismograms, travel-time graphs as shown in Figure 5.9 were constructed. The first travel-time graphs were greatly improved when seismograms became available from nuclear explosions, because the location and time of detonation were well established.

Using the sample seismogram in Figure 5.8 and the travel-time curves in Figure 5.9, we can determine the distance separating the recording station from the earthquake. The time interval between the arrival of the first P wave and the first S wave is determined, then the place on the travel-time graph which exhibits an equivalent time spread between the P and S wave curves is found. From this information we can determine that this earthquake occurred 3800 kilometers (2350 miles) from the recording instrument. Although the distance to an earthquake is established in this manner, its location could be in any direction from the observer. As shown in Figure 5.10, page 136, the precise location can be found only when the distance is known from three or more different seismic stations. By drawing circles representing the epicenter distance for each of these observatories, an accurate location is established.

About 95 percent of the energy released by earthquakes is concentrated in a few relatively narrow zones that wind around the globe (Figure 5.11, page 137). The greatest energy is released

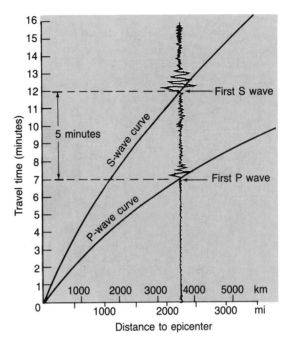

FIGURE 5.9
A travel-time graph is used to determine the distance to the epicenter. The difference in arrival times of the first P and S waves in the example is 5 minutes. Thus, the epicenter is roughly 3800 kilometers (2350 miles) away.

along a path near the outer edge of the Pacific Ocean known as the *circum-Pacific belt.* Included in this zone are regions of great seismic activity, such as Japan, the Philippines, Chile, and numerous volcanic island chains, as exemplified by the Aleutian Islands. Another major concentration of strong seismic activity runs through the mountainous regions that flank the Mediterranean Sea and continues through Iran and on past the Himalayan complex. Figure 5.11 indicates that yet another continuous belt extends for thousands of kilometers through the world's oceans. This zone coincides with the oceanic ridge system, an area of frequent but low-intensity seismic activity. By comparing this figure with Figure 6.6 on pages 162–63, we can see a close correlation between the location of earthquake epicenters and plate boundaries, a phenomenon which will be explored in the next chapter.

FIGURE 5.10
Earthquake epicenter is located using the distances obtained from three seismic stations.

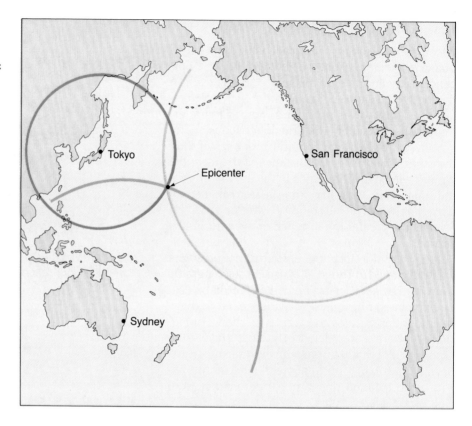

EARTHQUAKE INTENSITY AND MAGNITUDE

Early attempts to establish the intensity of earthquakes relied heavily on subjective descriptions. There was an obvious problem with this method—people's accounts varied widely, making an accurate classification of the quake's intensity difficult. Then in 1902 a fairly reliable scale based on the amount of damage caused to various types of structures was developed by Giuseppe Mercalli. A modified form of this tool is presently used by the U.S. Coast and Geodetic Survey (Table 5.1, page 138). However, the destruction wrought by earthquakes is not an adequate means for comparison. Many factors, including distance from the epicenter, nature of surface materials, and building design, cause variations in the amount of damage. Consequently, methods were devised to deter-

mine the total amount of energy released during an earthquake, a measurement referred to as **magnitude.**

Ideally, the magnitude of an earthquake would be determined from the amount of material which was displaced along a fault and the distance it was displaced. Even in an ideal setting such as that of the 1906 San Francisco earthquake, where the fault trace is visible and displacement can be measured from physical evidence, this method can only provide a crude estimate of the forces involved. In most earthquakes, the fault does not penetrate the surface, therefore the amount of displacement cannot be measured directly. In 1935, Charles Richter of the California Institute of Technology attempted to rank the earthquakes of southern California into groups of large, medium, and small magnitude. The system he developed determines earthquake magnitudes from the motions measured by seismic instruments.

Today a refined **Richter scale** is used world-wide to describe earthquake magnitude. Using Richter's scale, the magnitude is determined by measuring the amplitude of the largest wave recorded on the seismogram. Although seismographs greatly magnify the ground motion, large-magnitude earthquakes will cause the recording pen to be displaced farther than small-magnitude earthquakes. In order for seismic stations world-wide to obtain the same magnitude for a given earthquake, adjustments must be made for the weakening of seismic waves as they move from the focus and for the sensitivity of the recording instrument.

The largest earthquakes ever recorded have Richter magnitudes near 8.6. These great shocks released energy roughly equivalent to the detona-tion of one billion tons of TNT. Conversely, earth-quakes with a Richter magnitude of less than 2.5 are usually not felt by humans. Table 5.2, page 138, shows how earthquake magnitudes and their effects are related.

As we have seen, earthquakes vary enormously in strength; consequently, the wave amplitudes generated vary by factors of thousands of times as well. To accommodate this wide variation, Richter used a logarithmic scale to express magnitude. On this scale a tenfold increase in wave amplitude corresponds to an increase of one on the magnitude scale. Thus, the amplitude of the largest surface wave for a 5-magnitude earthquake is 10 times greater than the wave amplitude produced by an earthquake having a magnitude of 4. Further, each unit of magnitude increase on the

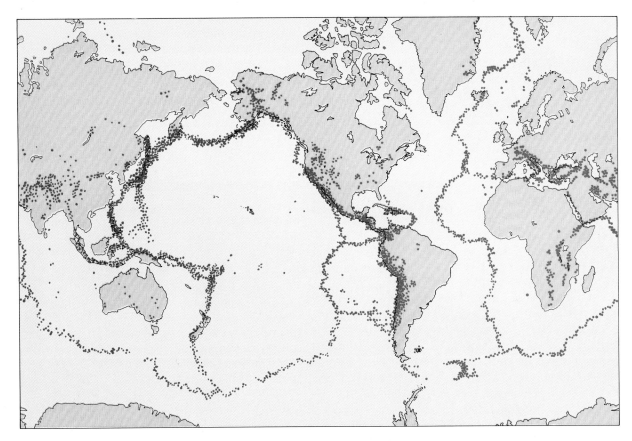

FIGURE 5.11
World distribution of eathquakes for a nine-year period. (Data from NOAA)

TABLE 5.1
Modified Mercalli intensity scale.

I	Not felt except by a very few under especially favorable circumstances.
II	Felt only by a few persons at rest, especially on upper floors of buildings.
III	Felt quite noticeably indoors, especially on upper floors of buildings, but many people do not recognize as an earthquake.
IV	During the day felt indoors by many, outdoors by few. Sensation like heavy truck striking building.
V	Felt by nearly everyone, many awakened. Disturbances of trees, poles, and other tall objects sometimes noticed.
VI	Felt by all; many frightened and run outdoors. Some heavy furniture moved; few instances of fallen plaster or damaged chimneys. Damage slight.
VII	Everybody runs outdoors. Damage negligible in buildings of good design and construction; slight to moderate in well-built ordinary structures; considerable in poorly built or badly designed structures.
VIII	Damage slight in specially designed structures; considerable in ordinary substantial buildings with partial collapse; great in poorly built structures. (Fall of chimneys, factory stacks, columns, monuments, and other vertically oriented features.)
IX	Damage considerable in specially designed structures. Buildings shifted off foundations. Ground cracked conspicuously.
X	Some well-built wooden structures destroyed. Most masonry and frame structures destroyed with foundations. Ground badly cracked.
XI	Few, if any, (masonry) structures remain standing. Bridges destroyed. Broad fissures in ground.
XII	Damage total. Waves seen on ground surfaces. Objects thrown upward into air.

SOURCE: U.S. Coast and Geodetic Survey.

TABLE 5.2
Earthquake magnitudes and expected world incidence.

Richter Magnitudes	Earthquake Effects	Estimated Number per Year
< 2.5	Generally not felt, but recorded.	900,000
2.5–5.4	Often felt, but only minor damage detected.	30,000
5.5–6.0	Slight damage to structures.	500
6.1–6.9	Can be destructive in populous regions.	100
7.0–7.9	Major earthquakes. Inflict serious damage.	20
≥ 8.0	Great earthquakes. Produce total destruction to communities near epicenter.	One every 5–10 years

SOURCE: *Earthquake Information Bulletin* and others.

Richter scale equates to roughly a 30-fold increase in the energy released. Thus, an earthquake with a magnitude of 6.5 releases 30 times more energy than one with a magnitude of 5.5, and roughly 900 times (30 × 30) more energy than a 4.5-magnitude quake. A major earthquake with a magnitude of 8.5 releases millions of times more energy than the smallest earthquakes felt by humans. This dispels the notion that a moderate earthquake decreases the chances for the occurrence of a major quake in the same region. Thousands of moderate tremors would be needed to release the amount of energy released by one "great" earthquake.

Some of the world's major earthquakes and their corresponding Richter magnitudes are listed in Table 5.3. Great earthquakes such as these can be expected in a tectonically active region every 50 to 200 years. The region of the San Andreas fault, along which slippage occurred to produce the 1906 San Francisco earthquake, has not generated a large earthquake in over 75 years—an alarming statistic to the residents of this region.

EARTHQUAKE DESTRUCTION: THE 1964 ALASKAN EARTHQUAKE

The most violent earthquake to jar North America this century—the Good Friday Alaskan Earthquake—occurred at 5:36 P.M. on March 27, 1964.

TABLE 5.3
Some notable worldwide and U.S. earthquakes.

Year	Location	Deaths (est.)	Magnitude	Comments
1290	Chihli (Hopei), China	100,000		
1556	Shensi, China	830,000		Possibly the greatest natural disaster.
1737	Calcutta, India	300,000		
1755	Lisbon, Portugal	70,000		Tsunami damage extensive.
*1811–12	New Madrid, Missouri	Few		Three major earthquakes.
*1886	Charleston, South Carolina	60		
*1906	San Francisco, California	700	8.25	Fires caused extensive damage.
1908	Messina, Italy	120,000	7.5	
1920	Kansu, China	180,000	8.5	
1923	Tokyo, Japan	150,000	8.2	Fire caused extensive destruction.
1960	Southern Chile	5700	8.5–8.7	Possibly the largest magnitude earthquake ever recorded.
*1964	Alaska	131	8.4–8.6	
1970	Peru	66,000	7.8	Great rockslide.
*1971	San Fernando, California	65	6.5	Damage exceeded one billion dollars.
1975	Liaoning Province, China	Few	7.5	First major earthquake to be predicted.
1976	Tangshan, China	650,000	7.6	Not predicted.
1985	Mexico City	7000	8.1	Major damage occurred 400 km from epicenter.

*U.S. earthquakes.
SOURCE: U.S. National Oceanic and Atmospheric Administration.

FIGURE 5.12
Region most affected by the Good Friday earthquake of 1964. Note the location of the epicenter (red dot). (After U.S. Geological Survey)

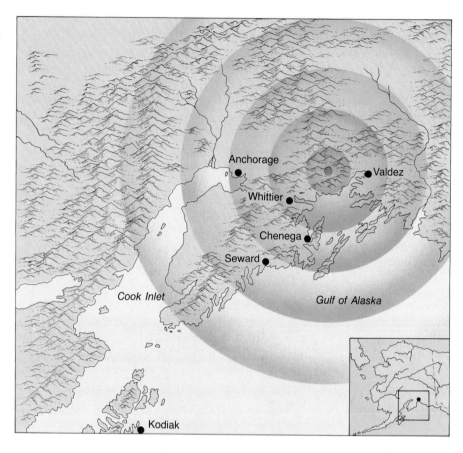

Felt throughout that state, the earthquake had a magnitude of 8.4–8.6 on the Richter scale and reportedly lasted 3 to 4 minutes. This brief event left 131 persons dead, thousands homeless, and the economy of the state badly disrupted. Had the schools and business districts been open, the toll surely would have been higher. The total financial loss has been estimated at over 300 million dollars. The location of the epicenter and the towns which were hardest hit by the quake are shown in Figure 5.12. Within 24 hours of the initial shock, 28 aftershocks were recorded, 10 of which exceeded a magnitude of 6 on the Richter scale.

Many factors determine the amount of destruction that accompanies an earthquake. The most obvious of these are the magnitude of the earthquake and the proximity of the quake to a populated area. Fortunately, most earthquakes are small and occur in remote regions of the earth.

However, about 20 major earthquakes are reported annually, one or two of which are catastrophic.

DESTRUCTION CAUSED BY SEISMIC VIBRATIONS

The 1964 Alaskan earthquake provided geologists with new insights into the role of ground shaking as a destructive force. As the energy released by an earthquake travels along the earth's surface, it causes the ground to vibrate in a complex manner by moving up and down as well as from side to side. The amount of structural damage attributable to the vibrations depends on several factors, including: (1) The intensity and duration of the vibrations; (2) The nature of the material upon which the structure rests; and (3) The design of the structure.

All of the multistory structures in Anchorage were damaged by the vibrations; the more flexible wood frame residential buildings fared best. However, many homes were destroyed when the ground failed. A striking example of how construction variations affect earthquake damage is shown in Figure 5.13. We can see in this photo that the steel-frame building on the left withstood the vibrations, whereas the relatively rigid concrete structure was badly damaged.

The greatest loss of life from an earthquake can be partially attributed to the type of structures that are inhabited. In 1556, in the Shensi region of China, an estimated 830,000 persons perished when an early morning earthquake struck the region. Many of these people lived in dwellings carved out of a compacted windblown sediment called loess (see Chapter 4). The walls of these structures failed, allowing the roofs to bury the inhabitants. The dense population and present-day construction practices in this region are such that a repetition of the 1556 event is possible.

FIGURE 5.13
Damage to the five-story J. C. Penney Co. building, Anchorage, Alaska. Very little structural damage was incurred by the adjacent building. (Courtesy of NOAA)

Most of the large structures in Anchorage were damaged even though they were built to conform to the earthquake provisions of the Uniform Building Code of California. Perhaps some of that destruction can be attributed to the unusually long duration of this earthquake, which was estimated at 3 to 4 minutes. Most earthquakes consist of tremors lasting from 20 seconds to one minute. The San Francisco earthquake of 1906 was felt for about 40 seconds.

Although the region within 20 to 50 kilometers of an epicenter will experience about the same degree of ground shaking, the destruction will vary considerably within this area. This difference is mainly attributable to the nature of the ground on which the structures are built. Soft sediments, for example, generally amplify the vibration more than solid bedrock. Thus, the buildings in Anchorage, which were situated on unconsolidated sediments, experienced heavy structural damage. By contrast, most of the town of Whittier, although located much nearer to the epicenter than Anchorage, rests on a firm foundation of granite and hence suffered much less damage from the seismic vibrations. However, Whittier was damaged by a seismic sea wave.

The 1985 Mexican earthquake gave seismologists and engineers a vivid reminder of what had been learned following the 1964 Alaskan earthquake. The Pacific coast of Mexico, where the earthquake was centered, experienced unusually mild tremors despite the strength of the quake. As expected, the seismic waves became progressively weaker with increasing distance from the epicenter. However, in the central section of Mexico City, nearly 400 kilometers (250 miles) from the source, the vibrations intensified to five times that experienced in outlying districts. Much of this amplified ground motion can be attributed to soft sediments, remnants of an ancient lake bed, that underlie portions of the city.

To understand what happened in Mexico City, recall that as seismic waves pass through the earth they cause the intervening material to vibrate much as a tuning fork vibrates when struck. Although most objects can be forced to vibrate over a wide range of frequencies, each has a natural period of vibration that is preferred. Various earth materials, like different-length tuning forks, also have different natural periods of vibration.*

Ground motion is amplified when the natural period of ground vibration matches that of the seismic waves. A common example of this phenomenon occurs when a parent pushes a child on a swing. As the parent periodically pushes the child in rhythm with the frequency of the swing, the child moves back and forth in a greater and greater arc. By chance, the column of sediment beneath Mexico City had a natural period of vibration of about two seconds, matching that of the strongest seismic waves. Thus, when the seismic waves began shaking the soft sediments, a resonance developed which greatly increased the amplitude of the vibrations. This amplified motion began throwing the ground back and forth about 40 centimeters (16 inches) every two seconds for a period of nearly two minutes. Such shaking was too intense for many of the poorly designed buildings in the city. In addition, buildings of intermediate height (5 to 15 stories) swayed back and forth with a period of about two seconds. Thus, resonance also developed between these buildings and the ground, and most of the building failures were in this height range (Figure 5.14).

TSUNAMI

Most deaths associated with the 1964 Alaskan quake were caused by **seismic sea waves,** or **tsunami.**** These destructive waves have popularly been called "tidal waves." However, this name is not accurate since these waves are not generated by the tidal effect of the moon or sun.

Most tsunamis result from vertical displacement of the ocean floor during an earthquake as illustrated in Figure 5.15 on page 144. Once formed, a tsunami resembles the ripples formed

---

* To demonstrate the natural period of vibration of an object, hold a ruler over the edge of a desk so that most of it is not supported by the desk. Start it vibrating and notice the sound it makes. By changing the length of the unsupported portion of the ruler, the natural period of vibration will change accordingly, as reflected by the change in sound.

** Seismic sea waves were given the name *tsunami* by the Japanese, who have suffered a great deal from them. The term *tsunami* is now used worldwide.

FIGURE 5.14
During the 1985 Mexican earthquake, multistory buildings swayed back and forth
as much as one meter. Many, including the hotel shown here, collapsed or were
seriously damaged. (Photo by James L. Beck)

when a pebble is dropped into a pond. In contrast
to ripples, tsunamis advance at speeds between
500 and 800 kilometers (300 and 500 miles) per
hour. Despite this striking characteristic, a tsunami in the open ocean can pass undetected because its height is usually less than one meter and
the distance between wave crests ranges from 100
to 700 kilometers. However, upon entering shallower coastal waters, these destructive waves are
slowed and the water begins to pile up to heights
that occasionally exceed 30 meters (100 feet). As a
tsunami approaches shore, it appears as a rapid

rise in sea level with a turbulent and chaotic surface.

Usually the first warning of a tsunami is a rather
rapid withdrawal of water from beaches (Figure
5.15B). Residents of coastal areas have learned to
heed this warning and move to higher ground.
About 5 to 30 minutes later the retreat of water is
followed by a surge capable of extending hundreds of meters inland. In a successive fashion,
each surge is followed by rapid oceanward retreat
of the water. These waves, separated by intervals
of between 10 and 60 minutes, are able to traverse

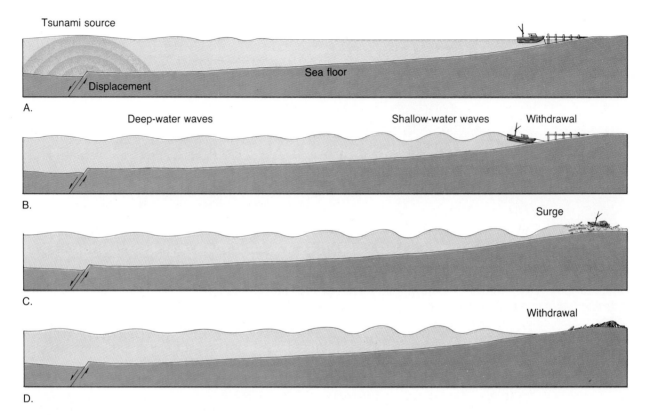

FIGURE 5.15
Schematic drawing of a tsunami generated by displacement of the ocean floor.
The size and spacing of the swells are not to scale.

large stretches of the ocean before their energy is totally dissipated.

The tsunami generated in the 1964 Alaskan earthquake inflicted heavy damage to communities in the vicinity of the Gulf of Alaska, completely destroying the town of Chenega. The town of Seward was also heavily damaged as most of its port facilities were demolished by this seismic sea wave (Figure 5.16). The deaths of 107 persons have been attributed to this tsunami. By contrast, only 9 persons died in Anchorage as a direct result of the vibrations. Tsunami damage following the Alaskan earthquake extended along much of the west coast of North America, and in spite of a one-hour warning, 12 persons perished in Crescent City, California, where all of the deaths and most of the destruction were caused by the fifth wave.

FIGURE 5.16
The effects of a tsunami at Seward, Alaska. (Courtesy of NOAA)

The first wave crested about 4 meters (13 feet) above low tide and was followed by three progressively smaller waves. Believing that the tsunami had ceased, people returned to the shore, only to be met by the fifth and most devastating wave, which, superimposed upon high tide, crested about 6 meters higher than the level of low tide.

FIRE

Fire was only a minor consequence in the 1964 Alaskan earthquake, but often it is the most destructive result. The 1906 earthquake centered near the city of San Francisco reminds us of the formidable threat of fire. The central city contained mostly large, older wooden structures and brick buildings. Although many of the unreinforced brick buildings were extensively damaged, the greatest destruction was caused by innumerable fires which started when gas and electrical lines were severed. The fires raged out of control for three days and devastated over 500 blocks of the city (see Figure 5.2). The problem was compounded by the initial ground shaking which broke the city's water lines into hundreds of unconnected pieces.

The fire was finally contained when buildings were dynamited along a wide boulevard to provide a fire break. Although only a few deaths were attributed to the fires, that is not always the case. An earthquake which rocked Japan in 1923 triggered an estimated 250 fires, which devastated the city of Yokohama and destroyed more than half the homes in Tokyo. Over 100,000 deaths were attributed to the fires, which were driven by unusually high winds.

LANDSLIDES AND GROUND SUBSIDENCE

In the 1964 Alaskan earthquake, it was not ground vibrations directly, but landslides and ground subsidence triggered by the vibrations that probably caused the greatest damage to structures. At Valdez and Seward the violent shaking caused deltaic materials to liquefy; the subsequent slumping carried both waterfronts away. Because of the

threat of recurrence, the entire town of Valdez was relocated about 7 kilometers away in a region of stable ground. The destruction at Valdez was compounded by the tragic loss of 31 lives. While waiting for an incoming vessel, the 31 persons and the dock slid into the sea.

Most of the damage in the city of Anchorage was also attributed to landslides caused by the shaking and lurching ground. Many homes were destroyed in Turnagain Heights when a layer of clay lost its strength and over 200 acres of land slid toward the ocean. The destruction was so complete that this area was bulldozed over and made into a park, which was appropriately named "Earthquake Park." Downtown Anchorage was equally disrupted as blocks of earth broke loose and sections of the main business district dropped by as much as 3 meters (10 feet).

EARTHQUAKE PREDICTION AND CONTROL

The vibrations that shook the San Fernando, California, area on the morning of February 9, 1971, inflicted 65 deaths and almost one billion dollars in damages (Figure 5.17)—all from an earthquake that lasted only 60 seconds and had a moderate rating of 6.5 on the Richter scale. Fortunately, because of the early hour, freeways, businesses, and schools were sparsely occupied, reducing the possible toll. Also, had the Lower Van Norman Lake Dam, badly damaged during the earthquake, actually broken, 80,000 additional lives might have been lost, which would have made it the most catastrophic event ever in the United States. This quake in populous southern California reemphasized the need for reliable methods of earthquake prediction and control.

Substantial research programs are underway in Japan, the United States, China, and the Soviet Union—countries where the risks of earthquakes are high. These projects strive to identify those geologic phenomena that precede major earthquakes. In California, for example, uplift or subsidence of the land, changes in the movements along a fault zone, and a period of seismic

FIGURE 5.17
Collapsed overpass to
Golden State Freeway, San
Fernando earthquake, 1971.
(Courtesy of James L. Ruhle)

quiescence often followed by renewed activity have been found to precede some earthquakes. It therefore seems reasonable that earthquakes may be predicted by continually monitoring ground tilt, fault movement, and seismic activity. Perhaps the most ambitious earthquake prediction experiment of this type is underway along a segment of the San Andreas fault near the town of Parkfield in central California. Here, earthquakes of moderate intensity have occurred on a regular basis about once every 22 years since 1857. The most recent rupture was a 5.6-magnitude quake that occurred in 1966. With the next event expected about 1988, the U.S. Geological Survey has established an elaborate monitoring network. Included are creepmeters, tiltmeters, and bore hole strain meters that are used to measure the accumulation and release of strain. Moreover, 70 seismometers of various designs have been installed to record foreshocks as well as the main event. Finally, a network of distance-measuring devices that employ lasers measures movement across the fault. The object is to identify shifts that may precede a sizeable rupture.

Although no reliable method of short-range prediction has yet been devised, a few successful predictions have been made. In 1966, an earthquake in Tashkent, U.S.S.R., was apparently predicted by monitoring the radon level in nearby wells. Radon is an inert gas generated by the radioactive decay of radium, a small amount of which is found in certain rocks. Normally this gas remains locked within the rock, but during the buildup of stress, newly formed cracks allow for its release.

Perhaps the most notable earthquake prediction to date foretold the 1975 earthquake in the Lainoning Province of China. Here, for the first and only time, seismologists succeeded in forecasting a large earthquake that was about to destroy a major city. By evacuating an estimated 3 million residents from unreinforced masonry structures, tens of thousands of lives were spared. Western observers confirmed Chinese reports that almost 90 percent of the structures in the city of Haicheng were heavily damaged. Although this event was predicted months earlier, swarms of small earthquakes which preceded this earthquake undoubtedly aided the prediction and also prompted the citizenry to heed the warning.

Unfortunately, the Chinese were unable to pinpoint the date of the great Tangshan earthquake of 1976. Their long-range warning of a forthcoming earthquake was not sufficiently precise to prevent the loss of an estimated 240,000 lives. The Chinese have also issued false alarms. In a province near

Hong Kong, people left their dwellings for over a month, but no earthquake followed. The debate that would precede an order to evacuate a large city in the United States, such as Los Angeles or San Francisco, would be considerable. The cost of evacuating millions of people, to say nothing of lost work time, and the innumerable other problems associated with an evacuation would have to be weighed against the earthquake's probability.

What constitutes success or failure for earthquake predictions? Most experts agree that a successful earthquake prediction must specify the geographic area involved, the expected magnitude, and the probable time of occurrence. As you might expect, researchers have found that establishing a time for the event is the most elusive factor. This is particularly true if the forecasts attempt to establish the time of a particular quake within the relatively narrow period of a few days, or at most a few weeks. On the other hand, substantial progress has been made in producing long-term earthquake predictions.

Long-term forecasts are based on the premise that earthquakes are cyclic. In other words, strain is continually building along active faults until rupture occurs. Immediately thereafter, the accumulation of strain begins anew. With this concept in mind, a group of seismologists plotted the distribution of rupture zones associated with great earthquakes that have occurred in the seismically active regions of the Pacific basin. The maps revealed that the individual rupture zones tended to abut without overlapping, thereby tracing out a

plate boundary. Recall that most earthquakes are generated by the relative motions of large crustal blocks along plate boundaries. Because the plates are in constant motion, these researchers predicted that over a long period of time (one or two centuries), major earthquakes would occur along the entire plate boundary. When historical records were studied, it was found that some segments of these seismic belts had not produced a great earthquake in more than a century. These quiet zones, called *seismic gaps,* were identified as probable sites for major earthquakes in the next few decades (Figure 5.18). In the 20 years since the original studies were conducted, some of these gaps have ruptured. Included in this group is the zone that produced the earthquake which devastated portions of Mexico City in September, 1985. Hence, some long-range predictions have already been fulfilled.

The actual control of earthquakes is another matter altogether. The discovery that humans have inadvertently triggered earthquakes has given earth scientists some encouragement. The most convincing evidence that people can initiate earthquakes came between 1962 and 1966 when studies of the seismic activity at the Rocky Mountain Arsenal near Denver were conducted. For a period of 80 years prior to 1962, the U.S. Coast and Geodetic Survey reported no significant earthquake activity in the Denver region. In 1962 the arsenal began disposing wastes from its chemical warfare production into a well over 3600 meters deep. During the period of fluid waste injection,

FIGURE 5.18
The distribution of rupture areas of large, shallow earthquakes from 1930 to 1979 along the southwestern coast of Alaska and the Aleutian Islands. The three seismic gaps denote the most likely locations for the next large earthquakes along this plate boundary. (After J. C. Savage et. al., U.S. Geological Survey)

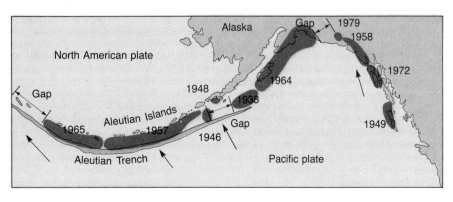

from April, 1962 to September, 1965, about 700 microearthquakes were reported, 75 intense enough to be felt. The injection of water under pressure is believed to have "lubricated" the fault, which had been building up strain over the years. This lubricating effect is not one of making rocks along the fault zone slippery. Rather the water exerts an outward force which is directed perpendicular to the fault plane. The induced outward force opposes the natural inward force caused by the weight of rock piled above. When injection was halted for about a year, a marked drop in seismic activity was also detected. When pumping resumed, the frequency of tremors increased markedly.

Other earthquakes caused by human activity have occurred in regions adjacent to large reservoirs such as Lake Mead on the Arizona-Nevada border. Ever since Lake Mead was filled in 1936, hundreds of small tremors have been recorded. They are thought to have been caused by the added weight of the lake, and perhaps aided by the "lubricating" effect of water seeping into the rock below. Another large reservoir in India is believed responsible for triggering a disastrous earthquake in which 200 persons were killed. Underground nuclear explosions have also initiated numerous small aftershocks, although none has been as great as the explosion itself.

The hope of many scientists is that we may someday reduce the threat of earthquakes by triggering numerous small earthquakes using fluid injections or nuclear explosions. Such methods would slowly and continually release the elastic strain that might otherwise build up and be released as a high-magnitude earthquake. Recall, however, that many thousands of minor tremors are required to equal the energy released by one strong earthquake. This fact coupled with the inaccessibility of many fault zones make the possibility of earthquake control not quite as feasible as it might first appear. A favorable condition for earthquake control in California is the shallow depth of earthquake foci, making drilling operations possible. Many tests will have to be made in remote regions before we dare risk such a venture along a fault system like the San Andreas, at least in those portions located in populous regions.

THE EARTH'S INTERIOR

The earth's interior lies just below us, yet its accessibility to direct observation is very limited. Most of our knowledge of the earth's interior comes from the study of P and S waves that travel through the earth and emerge at some distant point. Simply stated, the technique involves accurately measuring the time required for seismic waves to travel from the focus of an earthquake or nuclear explosion to a seismographic station. Since the time required for P and S waves to travel through the earth depends upon the properties of the rock materials encountered, seismologists search for variations in travel times that cannot be accounted for simply by differences in the distances traveled. These variations correspond to changes in rock properties. Based upon these seismological data the earth has been divided into four major layers: (1) The **crust,** a very thin outer layer; (2) The **mantle,** a rocky layer located below the crust and having a thickness of 2885 kilometers (1789 miles); (3) The **outer core,** a layer about 2270 kilometers (1407 miles) thick which exhibits the characteristics of a mobile liquid; and (4) The **inner core,** a solid metallic sphere about 1216 kilometers (754 miles) in radius (Figure 5.19).

In 1909, a pioneering Yugoslavian seismologist, Andrija Mohorovičić, presented the first convincing evidence for layering within the earth. By studying seismic records, he found that the velocity of waves increases abruptly below a depth of 50 kilometers. This boundary that he discovered separates the crust from the underlying mantle and is known as the **Mohorovičić discontinuity** in his honor (Figure 5.19). For reasons that are obvious, the name for this boundary was quickly shortened to **Moho.**

A few years later another major boundary was discovered by the German seismologist Beno Gutenberg. This discovery was based primarily on the observation that P waves diminish and eventually die out completely about 105 degrees from an earthquake. Then, about 140 degrees away, the P waves reappear, but about two minutes later than would be expected based on the distance traveled. This belt where direct seismic waves are absent is about 35 degrees wide and has been named the

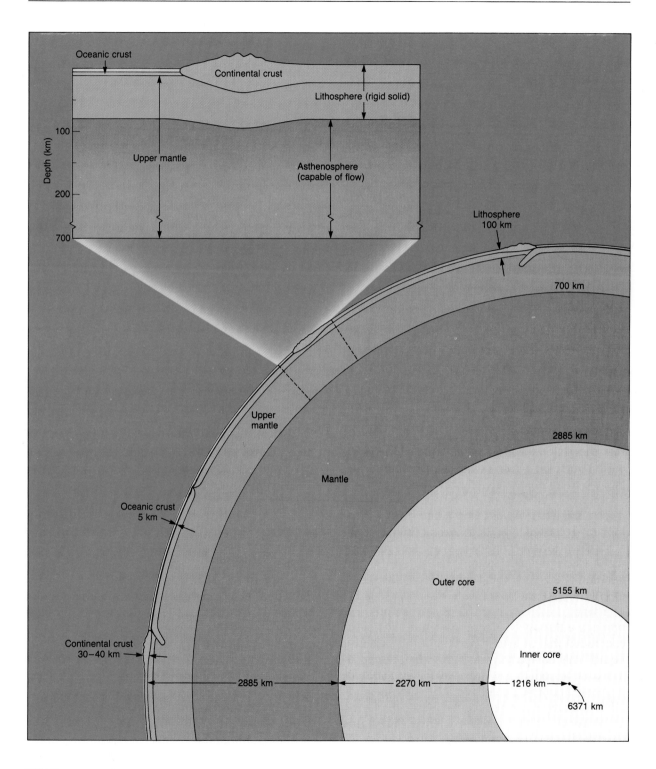

FIGURE 5.19
Cross-sectional view of the earth showing internal structure.

FIGURE 5.20
The abrupt change in physical properties at the mantle-
core boundary causes the
wave paths to bend sharply.
This abrupt change in wave
direction results in a shadow
zone for P waves between
about 105 and 140 degrees.

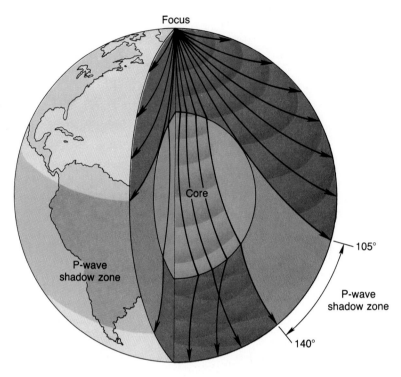

shadow zone* (Figure 5.20). Gutenberg realized that the shadow zone could be explained if the earth contained a core composed of material un-like the overlying mantle and had a radius of 3420 kilometers. The core must somehow hinder the transmission of P waves in a manner similar to the light rays blocked by an opaque object which casts a shadow. However, rather than actually stopping the P waves, the shadow zone is produced by the bending of P waves which enter the core as shown in Figure 5.20.

It was further learned that S waves could not propagate through the core; therefore, geologists concluded that at least a portion of this region is liquid. This conclusion was further supported by the observation that P-wave velocities suddenly decrease about 40 percent as they enter the core. Since melting would reduce the elasticity of rock, all evidence points to the existence of a liquid layer below the rocky mantle.

In 1936, the last major subdivision of the earth's interior was predicted by the discovery of seismic waves believed to be reflected from a boundary within the core. Hence, a core within a core was discovered. The actual size of the inner core was not accurately calculated until the early 1960s when underground nuclear tests were conducted in Nevada. Because the precise location and time of the explosions were known, echoes from seis-mic waves which bounced off the inner core provided an accurate means of determining its size. From these data and subsequent studies, the inner core was found to have a radius of about 1216 kilometers. Further, P waves passing through the inner core have appreciably faster travel times than those penetrating the outer core exclusively. The apparent increase in elasticity of the inner core material is considered evidence for the solid nature of the earth's innermost region.

A most important zone exists within the upper mantle and deserves special mention. This region, called the **asthenosphere,** is located between the depths of 70 kilometers and 700 kilometers (Fig-

---

* As more sensitive instruments were developed, weak and de-layed P waves that enter this zone via reflection were detected.

ure 5.19). In the asthenosphere the velocity of S waves decreases, indicating to seismologists that this zone consists partly of melted rock (approximately 10 percent). Situated above the asthenosphere is the outer solid portion of the earth, called the **lithosphere,** which includes part of the upper mantle and the crust. It is believed that the "plastic" material of the asthenosphere moves and carries along the rigid lithosphere. Thus, the discovery of the asthenosphere was an important contribution to the theory which proposes that the continents "drift" about. This is the topic of the next chapter. It is also from the asthenosphere that some, but not all, molten material for volcanic activity is thought to originate.

COMPOSITION OF THE EARTH

The crust of the earth varies in thickness, being greater than 70 kilometers in some mountainous regions and less than 5 kilometers in some oceanic regions (see Figure 5.19). Early seismic data indicated that the continental crust, which is mostly made of felsic (granitic) rocks, is quite different in composition from the oceanic crust. Until recently, however, scientists had only seismic evidence from which to determine the composition of oceanic crust, which lies beneath 3 kilometers of water as well as hundreds of meters of sediment. With the development of the deep-sea drilling ship *Glomar Challenger* (see Figure 10.9A), recovery of ocean floor samples became possible. The samples were of basaltic composition—indeed different from the rocks which compose the continents.

Our knowledge of the compositions of the mantle and core is much more speculative. However, we do have some clues. Recall that some of the lava that reaches the earth's surface originates in the partially melted asthenosphere located within the mantle. In the laboratory, experiments have shown that partial melting of a rock called perido-

tite results in a melt that has a basaltic composition similar to those associated with the volcanic activity of oceanic islands. Ultramafic rocks such as peridotite are thought to make up the mantle and provide the lava for oceanic eruptions.

Surprisingly, meteorites—"shooting stars"—which fall to the earth from space, are considered evidence of the earth's inner composition. Since meteorites are part of the solar system, they are assumed to be representative samples. Their composition ranges from metallic types made of iron and nickel to stony meteorites composed of ultramafic rock similar to peridotite. Because the earth's crust contains a much smaller percentage of iron than meteorites do, geologists believe that the heavy minerals sank during the early history of the earth. By the same token, the lighter minerals may have floated to the top, creating the crust. Thus, the core of the earth is thought to be mainly iron and nickel, similar to metallic meteorites, while the surrounding mantle is believed to be composed of ultramafic rocks, similar to stony meteorites.

The concept of a molten iron outer core is further supported by the existence of the earth's magnetic field, which acts as a large bar magnet. The most widely accepted mechanism explaining the magnetic field requires that the earth's core be made of a material which conducts electricity, such as iron, and which is mobile enough that circulation can occur. Both of these conditions are met by the model of the earth's core that was established on the basis of seismic data.

Not only does an iron core explain the earth's magnetic field, it also explains the high density of the inner earth, about 13.5 times that of water. Even under the extreme pressure at those depths, average crustal rocks with densities 2.8 times that of water would not have the density calculated for the core. But iron, which is 3 times more dense than crustal rocks, has the required density.

REVIEW QUESTIONS

1 What is an earthquake? Under what circumstances do earthquakes occur?

2 How are faults, foci, and epicenters related?

3 Who was first to explain the actual mechanism by which earthquakes are generated?

4 Explain what is meant by elastic rebound.

5 Faults which are experiencing no active creep may be considered "safe." Rebut or defend this statement.

6 Describe the principle of a seismograph.

7 Using Figure 5.9, determine the distance between an earthquake and a seismic station if the first S wave arrives 3 minutes after the first P wave.

8 List the major differences between P and S waves.

9 An earthquake measuring 7 on the Richter scale releases about ___ times more energy than an earthquake with a magnitude of 6.

10 List three factors that affect the amount of destruction caused by seismic vibrations.

11 In addition to the destruction created directly by seismic vibrations, list three other types of destruction associated with earthquakes.

12 Distinguish between the Mercalli scale and the Richter scale.

13 What is a tsunami? How is one generated?

14 Cite some reasons why an earthquake with a moderate magnitude might cause more extensive damage than a quake with a high magnitude.

15 How might earthquakes be controlled in the future?

16 What evidence do we have that the earth's outer core is molten?

17 Contrast the physical makeup of the asthenosphere and the lithosphere.

18 Earthquakes occur only in the rigid lithosphere, not in the plastic asthenosphere. Using the elastic rebound idea, explain this phenomenon.

19 Why are meteorites considered important clues to the composition of the earth's interior?

20 Describe the chemical (mineral) makeup of the following:
   (a) Continental crust.
   (b) Oceanic crust.
   (c) Mantle.
   (d) Core.

KEY TERMS

earthquake

focus

fault

elastic rebound

aftershock

foreshock

seismology

seismograph

seismogram

surface wave

body wave

primary (P) wave

secondary (S) wave

epicenter

magnitude

Richter scale

seismic sea wave
  (tsunami)

crust

mantle

outer core

inner core

Mohorovičić
  discontinuity
  (Moho)

shadow zone

asthenosphere

lithosphere

6

PLATE TECTONICS

Will California eventually "slide" into the ocean, as some predict? Have continents really "drifted" apart over the centuries? Answers to these questions and many others which have intrigued geologists for decades are now being provided by a new and exciting theory on large-scale movements taking place within the earth. This theory, called plate tectonics, represents the real frontier of the earth sciences, and its implications are so far-reaching that it can be considered the framework from which most other geologic processes should be viewed.

The Sinai Peninsula is bordered by rifts believed to be caused by sea-floor spreading. (Courtesy of NASA)

Early in this century geologic thought was dominated by a belief in the geographic permanency of the ocean basins and continents. During the last few decades, however, vast accumulations of new data have dramatically changed our ideas about the nature and workings of the earth. Earth scientists now realize that the positions of landmasses are not fixed. Rather, the continents gradually migrate over the surface of the globe. The splitting of continental blocks has resulted in the formation of new ocean basins, while older segments of the sea floor are continually being recycled in areas where we find deep-ocean trenches. Further, because of this movement, once-disjointed segments of continental material have collided and formed the earth's great mountain ranges. In short, a revolutionary new model of the earth's tectonic* processes has emerged in marked contrast to what was accepted just a few decades ago.

This profound reversal of scientific opinion has been appropriately described as a scientific revolution. Like other scientific revolutions, an appreciable length of time elapsed between the idea's inception and its general acceptance. The revolution began in the early part of the twentieth century as a relatively straightforward proposal that the continents drift about the face of the earth. After many years of heated debate, the idea of drifting continents was rejected by the vast majority of earth scientists as being improbable. However, during the 1950s and 1960s new evidence began to rekindle interest in this abandoned proposal. By 1968 these new developments led to the unfolding of a far more encompassing theory than continental drift—a theory known as plate tectonics.

CONTINENTAL DRIFT: AN IDEA BEFORE ITS TIME

The idea that continents, particularly South America and Africa, fit together like pieces of a jigsaw puzzle originated with improved world maps. However, little significance was given this idea

until 1915, when Alfred Wegener,* a German climatologist and geophysicist, published an expanded version of a 1912 lecture in his book *The Origin of Continents and Oceans.* In this monograph, Wegener set forth the basic outline of his radical hypothesis of **continental drift.** One of his major tenets suggested that a supercontinent he called **Pangaea** (meaning "all land") once existed (Figure 6.1). He further hypothesized that about 200 million years ago this supercontinent began breaking into smaller continents, which then "drifted" to their present positions. Wegener and others who advocated this position collected substantial evidence to support these claims. The fit of South America and Africa, ancient climatic similarities, fossil evidence, and rock structures all seemed to support the idea that these now-separate landmasses were once joined.

FIT OF THE CONTINENTS

Like a few others before him, Wegener first suspected that the continents might have been joined when he noticed the remarkable similarity between the coastlines on opposite sides of the South Atlantic. However, his use of present-day shorelines to make a fit of the continents was challenged immediately by other earth scientists. These opponents correctly argued that shorelines are continually modified by erosional processes and even if continental displacement had taken place, a good fit today would be unlikely. Wegener appeared to be aware of this problem, and, in fact, his original jigsaw fit of the continents was only very crude.

A much better approximation of the outer boundary of the continents is the seaward margin of the continental shelf. Today the continental shelf's edge lies several hundred meters below sea level. In the early 1960s, Sir Edward Bullard and two associates produced a map with the aid of computers that attempted to fit the continents at a depth of 900 meters. The remarkable fit that was obtained is shown in Figure 6.2 on page 158. Al-

---

*Tectonics refers to the deformation of the earth's crust and results in the formation of structural features such as mountains.

*Wegener's ideas were actually preceded by those of an American geologist, F. B. Taylor, who in 1910 published a paper on continental drift. Taylor's paper provided little supporting evidence for continental drift, which may have been the reason that it had a relatively small impact on the geologic community.

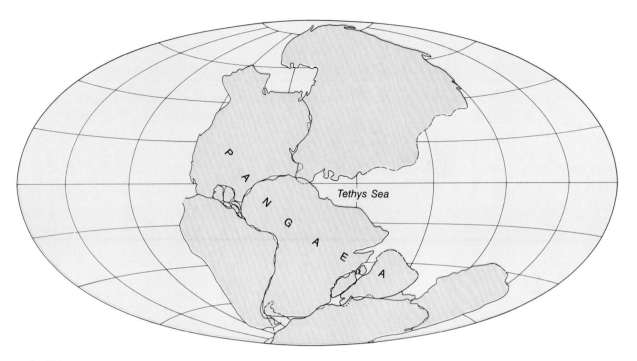

FIGURE 6.1
Reconstruction of Pangaea as it is thought to have appeared 200 million years ago. (After R. S. Dietz and J. C. Holden. *Journal of Geophysical Research* 75: 4943. Copyright by American Geophysical Union)

though the continents overlap in a few places, these are regions where streams have deposited large quantities of sediment, thus enlarging the continents. The overall fit obtained by Bullard and his associates was better than even the supporters of the continental drift theory suspected it would be.

FOSSIL EVIDENCE

Although Wegener was intrigued by the remarkable similarities of the shorelines on opposite sides of the Atlantic, he at first thought the idea of a mobile earth improbable. Not until he came across an article citing fossil evidence for the existence of a land bridge connecting South America and Africa did he begin to take his own idea seriously. Through a search of the literature Wegener learned that most paleontologists were in agreement that some type of land connection was needed to explain the existence of identical fossils on the widely separated landmasses.

To add credibility to his argument for the existence of the supercontinent of Pangaea, Wegener

cited documented cases of several fossil organisms that had been found on different landmasses but which could not have crossed the vast oceans presently separating the continents. Of particular interest were organisms that were restricted in geographical distribution but which nevertheless appeared in two or more areas that are presently separated by major barriers. The classic example is *Mesosaurus,* a presumably aquatic, snaggle-toothed reptile whose fossil remains are known to be limited to eastern South America and southern Africa (Figure 6.3). If *Mesosaurus* had been able to swim well enough to cross a vast ocean, its remains should be widely distributed. Since this was not the case, Wegener argued that South America and Africa must have been joined.

Wegener also cited the distribution of the fossil fern *Glossopteris* as evidence for Pangaea's existence, because this plant was known to be widely dispersed among Africa, Australia, India, and South America during the late Paleozoic era. Later, fossil remains of *Glossopteris* were

FIGURE 6.2
The best fit of South America and Africa along the continental slope at a depth of 500 fathoms (about 900 meters). The area where continental blocks overlap appears in brown. (After A. G. Smith. "Continental Drift." In *Understanding the Earth,* edited by I. G. Gass. Courtesy of Artemis Press)

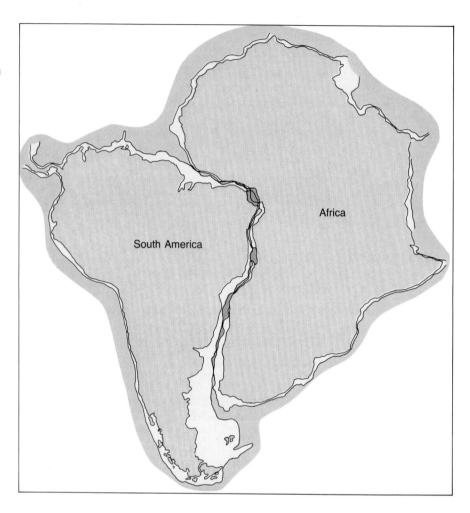

discovered in Antarctica as well. Wegener knew that these seed ferns and associated flora grew only in a subpolar climate; therefore, he concluded that these landmasses must have been joined, since they presently include climatic regions that are too diverse to support such flora. For Wegener, fossils proved without question that a supercontinent had existed.

How could these fossil flora and fauna be so similar in places separated by thousands of kilometers of open ocean? The idea of land bridges (isthmian links) was the most widely accepted solution to the problem of migration (Figure 6.4). We know for example that during the recent glacial period the lowering of sea level allowed animals to cross the narrow Bering Straits between Asia and North America. Was it possible then that land bridges once connected Africa and South Amer-

ica? We are now quite certain that land bridges of this magnitude did not exist, for their remnants should still lie below sea level, but are nowhere to be found.

ROCK TYPE AND STRUCTURAL SIMILARITIES

Anyone who has worked a picture puzzle knows that in addition to the pieces fitting together, the picture must be continuous as well. The "picture" that must match in the "Continental Drift Puzzle" is represented by the rock types and mountain belts found on the continents. If the continents were once together, the rocks found in a particular region on one continent should closely match in age and type with those found in adjacent positions on the matching continent.

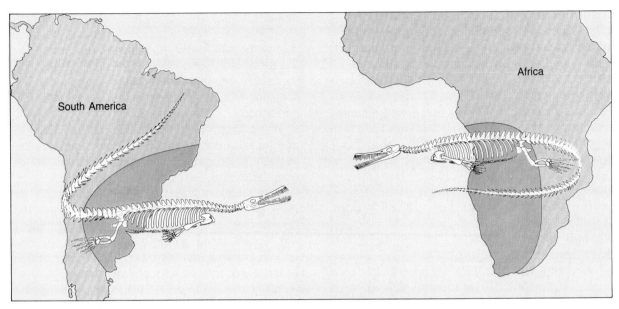

FIGURE 6.3
Fossils of *Mesosaurus* have been found on both sides of the South Atlantic and nowhere else in the world. Fossil remains of this and other organisms found on the continents of Africa and South America appear to link these landmasses during the late Paleozoic and early Mesozoic eras.

FIGURE 6.4
These sketches by John Holden lightheartedly illustrate various explanations for the occurrence of similar species on landmasses that are presently separated by vast oceans. (Reprinted with permission of John Holden)

Such evidence has been found in the form of several mountain belts which appear to terminate at one coastline only to reappear on a landmass across the ocean. For instance, the mountain belt that includes the Appalachians trends northeastward through the eastern United States and disappears off the coast of Newfoundland. Mountains of comparable age and structure are found in Greenland and Northern Europe. When these landmasses are reassembled, the mountain chains form a nearly continuous belt.

Wegener was very satisfied that the similarities in rock structure on both sides of the Atlantic linked these landmasses. In fact, he was too zealous with this evidence and incorrectly suggested that glacial moraines in North America matched up with those of Northern Europe. In his own words, "It is just as if we were to refit the torn pieces of a newspaper by matching their edges and then check whether the lines of print run smoothly across. If they do, there is nothing left but to conclude that the pieces were in fact joined in this way."*

PALEOCLIMATIC EVIDENCE

Since Alfred Wegener was a climatologist by training, he was keenly interested in obtaining paleoclimatic (ancient climatic) data in support of continental drift. His efforts were rewarded when he found evidence for apparently dramatic climatic changes. For instance, glacial deposits indicate that near the end of the Paleozoic era (between 220 and 300 million years ago), ice sheets covered extensive areas of the Southern Hemisphere. Layers of glacial till were found in southern Africa and South America, as well as in India and Australia. Below these beds of glacial debris lay striated and grooved bedrock. In some locations the striations and grooves indicated the ice had moved from the sea onto land (Figure 6.5). Much of the land area containing evidence of this late Paleozoic glaciation presently lies within 30 degrees of the equator in a subtropical or tropical climate.

Could the earth have gone through a period sufficiently cold to have generated extensive conti-

nental glaciers in what is presently a tropical region? Wegener rejected this explanation because during the late Paleozoic, large swamps existed in the Northern Hemisphere. These swamps with their lush vegetation eventually became the major coal fields of the eastern United States, Europe, and Siberia.

Fossils from these coal fields indicate that the tree ferns which produced the coal deposits were indigenous to the tropics, because their fronds were large and their trunks lacked growth rings. Wegener believed that a better explanation for these paleoclimatic regimes is provided when the landmasses are fitted together as a supercontinent with South Africa centered over the South Pole. This would account for the conditions necessary to generate extensive expanses of glacial ice over much of the Southern Hemisphere. At the same time, this geography would place the northern landmasses nearer the tropics and account for their vast coal deposits. Wegener was so convinced that his explanation was correct that he wrote, "This evidence is so compelling that by comparison all other criteria must take a back seat."*

How does a glacier develop in hot, arid Australia? How do land animals migrate across wide expanses of open water? As compelling as this evidence may have been, fifty years passed before most of the scientific community would accept it and the logical conclusions to which it led.

THE GREAT DEBATE

Wegener's proposal did not attract much open criticism until 1924 when his book was translated into English. From this time on, until his death in 1930, his drift hypothesis encountered a great deal of hostile criticism. To quote the respected American geologist R. T. Chamberlin, "Wegener's hypothesis in general is of the foot-loose type, in that it takes considerable liberty with our globe, and is less bound by restrictions or tied down by awkward, ugly facts than most of its rival theories. Its appeal seems to lie in the fact that it plays a game in which there are few restrictive rules and no sharply drawn code of conduct."

---

* Alfred Wegener, *The Origin of Continents and Oceans.* Translated from the 4th revised German edition of 1929 by J. Birman. (London: Methuen, 1966)

---

* Ibid.

One of the main objections to Wegener's hypothesis stemmed from his inability to provide a mechanism for continental drift. Wegener proposed two possible energy sources for drift. One of these, the tidal influence of the moon, was presumed by Wegener to be strong enough to give the continents a westward motion. However, the prominent physicist Harold Jeffreys quickly countered with the argument that tidal friction of the magnitude needed to displace the continents would bring the earth's rotation to a halt in a matter of a few years. Further, Wegener proposed that the larger and sturdier continents broke through the oceanic crust, much like ice breakers cut

FIGURE 6.5
A. Direction of ice movement in the southern supercontinent called Gondwanaland by the founders of the continental drift concept. **B.** Glacial striations in the bedrock of Hallet Cove, South Australia, indicate direction of ice movement. (Photo by W. B. Hamilton, U.S. Geological Survey)

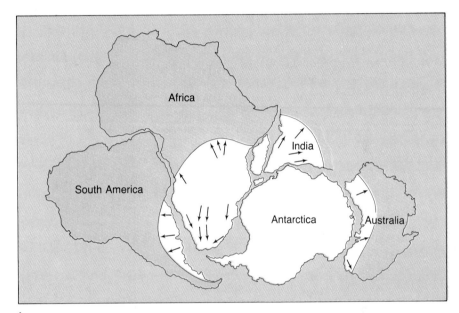

A.

B.

through ice. However, no evidence existed to suggest that the ocean floor was weak enough to permit passage of the continents without themselves being appreciably deformed in the process. By 1929 criticisms of Wegener's ideas were pouring in from all areas of the scientific community. Despite these attacks, Wegener wrote the fourth and final edition of his book, maintaining his basic hypothesis and adding supporting evidence.

Although most of Wegener's contemporaries opposed his views, even to the point of openly ridiculing them, a few considered his ideas plausible. For these few geologists who continued the search, the concept of continents in motion evidently provided enough excitement to hold their interest. Others undoubtedly viewed continental drift as a solution to previously unexplainable observations.

PLATE TECTONICS: A MODERN VERSION OF AN OLD IDEA

During the years that followed Wegener's proposal, great strides in technology permitted mapping of the ocean floor, and extensive data on seismic activity and the earth's magnetic field became available. By 1968 these developments led to the unfolding of a far more encompassing theory than continental drift, known as **plate tectonics.** The implications of plate tectonics are so far-reaching that this theory can be considered the framework from which to view most other geologic processes. Since this concept is relatively new, it most surely will be modified as additional information becomes available; however, the main tenets appear to be sound and are presented here in their current state of refinement.

The theory of plate tectonics holds that the outer, rigid lithosphere consists of about twenty rigid segments called **plates** (Figure 6.6). Of these, the largest is the Pacific plate, which is located mostly within the ocean proper, except for a small sliver of North America that includes southwestern California and the Baja Peninsula. Notice from Figure 6.6 that all of the other large plates contain both continental and oceanic crust—a major departure from the continental drift theory, which proposed that the continents moved through, not with, the ocean floor. Most smaller plates, on the

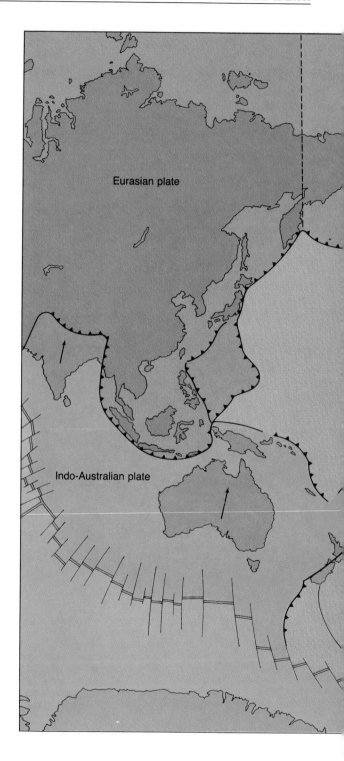

FIGURE 6.6
Mosaic of rigid plates that constitute the earth's outer shell.

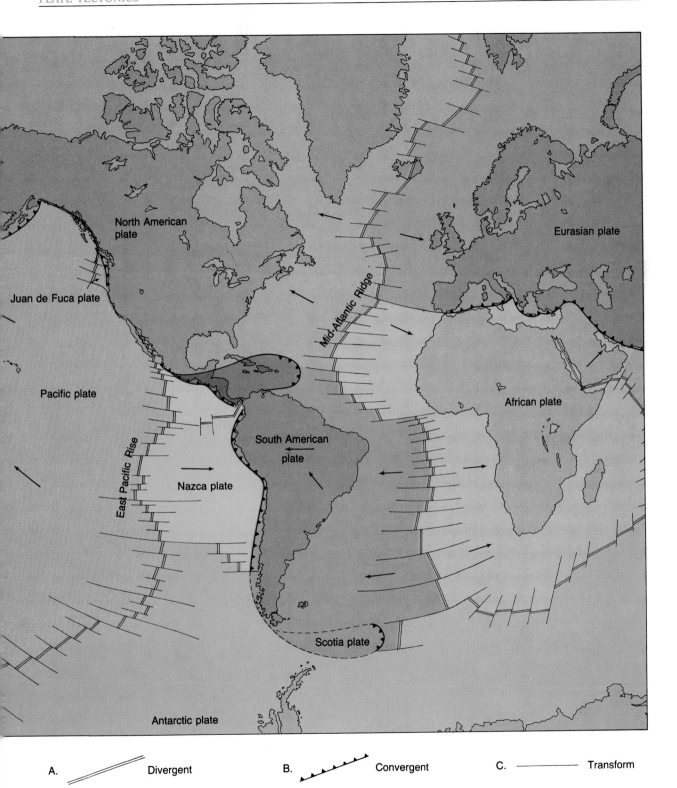

A. ⟍⟍ Divergent   B. ◣◣◣◣ Convergent   C. ——— Transform

other hand, consist exclusively of oceanic material, as for example, the Nazca plate located off the west coast of South America. Although not clearly defined in Figure 6.6, one small plate that roughly coincides with Turkey is located exclusively within a continent.

The lithosphere overlies a zone of much weaker and hotter material known as the asthenosphere. Hence, the lithospheric plates form a rigid outer shell supported from below by the more "plastic" material of the asthenosphere. A relationship appears to exist between the thickness of the lithospheric plates and the nature of the crustal material that caps them. Plates are thinnest in the oceans, where their thickness varies from 50 to

100 kilometers. By contrast, continental blocks are 100 kilometers thick, and in some regions may exceed 150 kilometers in thickness.

One of the main tenets of the plate tectonics theory is that each plate moves as a distinct unit in relation to other plates. The mobile behavior of the rock within the asthenosphere is believed to allow this motion in the earth's rigid outer shell. As the plates move, the distance between two cities on the same plate, New York and Denver, for example, remains constant, while the distance between New York and London, which are located on different plates, is continually changing. Since each plate moves as a distinct unit, all major interactions between plates occur along plate bounda-

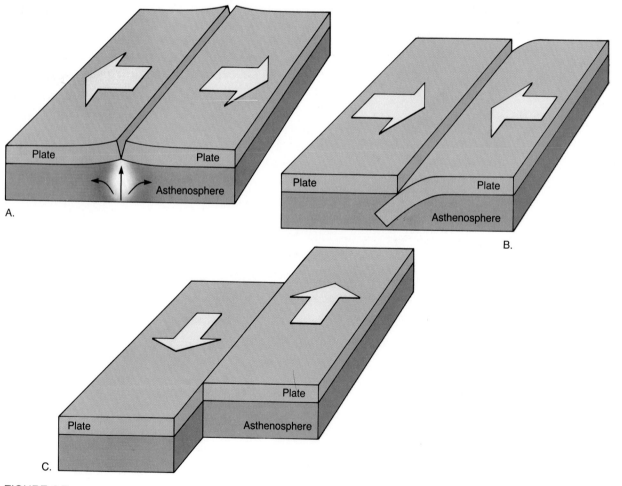

FIGURE 6.7
Schematic of plate boundaries. **A.** Divergent boundary. **B.** Convergent boundary.
C. Transform boundary.

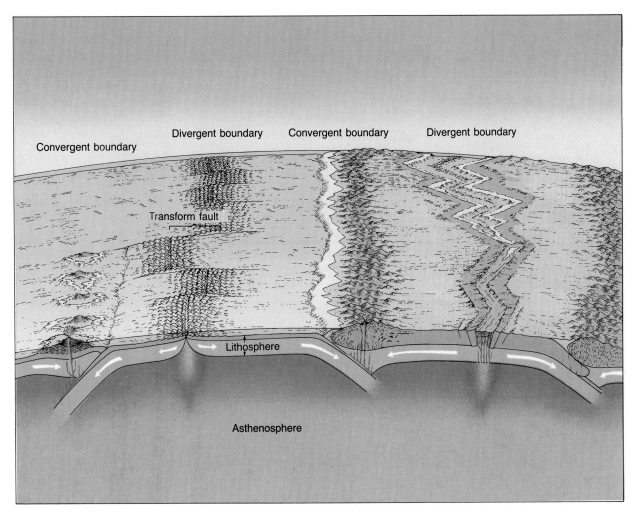

FIGURE 6.8
View of the earth showing the relationship between various types of boundaries.

ries. Thus, most of the earth's seismic activity, volcanism, and mountain building occur along these dynamic margins.

PLATE BOUNDARIES

For some time now, tectonic activity has been known to be restricted to narrow zones, such as the so-called *Ring of Fire* that encircles the Pacific. Thus, the first approximations of plate margins relied on the distribution of earthquake and volcanic activity. Later work indicated the existence of three distinct types of plate boundaries, each differentiated by the movement it exhibits (Figure 6.7). These are:

1 **Divergent boundaries**—where plates move apart, resulting in upwelling of material from the mantle to create new sea floor.
2 **Convergent boundaries**—where plates move together, causing one of the slabs of lithosphere to be consumed into the mantle as it descends beneath an overriding plate.
3 **Transform boundaries**—where plates slide past each other without creating or destroying lithosphere.

Each plate is bounded by a combination of these zones. Movement along any boundary requires that adjustments be made at the others (Figure 6.8).

DIVERGENT BOUNDARIES

Divergent boundaries, where plate spreading occurs, are situated at the crests of oceanic ridges. Here, as the plates move away from the ridge axis, the gaps created are immediately filled with molten rock that oozes up from the hot astheno-sphere. This material cools slowly to produce new slivers of sea floor. In a continuous manner successive separations and injections of magma add new oceanic crust (lithosphere) between the diverging plates. This mechanism, which has produced the floor of the Atlantic Ocean during the

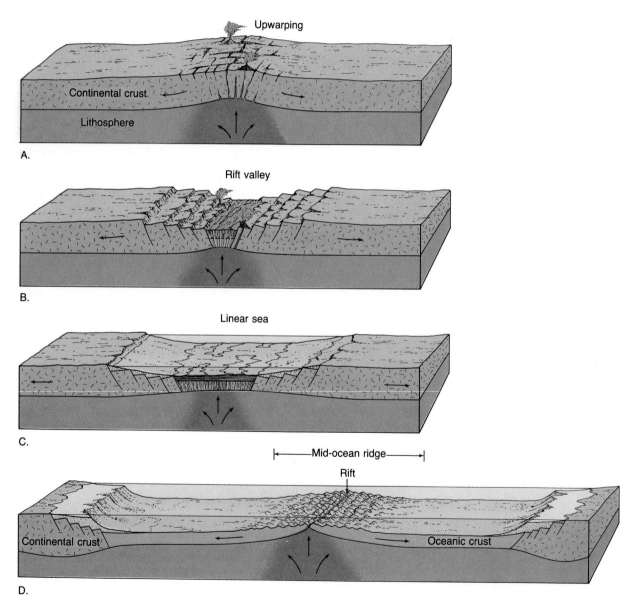

FIGURE 6.9
A. Rising magma upwarps the crust, causing numerous cracks in the rigid lithosphere. **B.** As the crust is pulled apart, large slabs of rock sink, generating a rift zone. **C.** Further spreading generates a narrow sea. **D.** Eventually, an expansive ocean basin and ridge system are created.

past 165 million years, is called **sea-floor spreading.** The typical rate of spreading at these ridges is estimated to be between 2 and 10 centimeters per year, and averages about 6 centimeters (2 inches) per year. Since new rock is added equally to the trailing edges of both diverging plates, the rate of ocean floor growth is twice the value of the spreading rate. That fact notwithstanding, these seemingly slow rates are rapid enough to have opened and reclosed the Atlantic Ocean more than ten times during the 5-billion-year history of our planet, although this probably did not happen.

Not all spreading centers are as old as the Mid-Atlantic Ridge and not all are found in the middle of large oceans. The Red Sea is believed to be the site of a recently formed divergent boundary. Here the Arabian Peninsula separated from Africa and began to move toward the northeast. Consequently, the Red Sea is providing oceanographers with a view of how the Atlantic Ocean may have looked in its infancy. Another result of sea-floor spreading in the recent geologic past has been the formation of the Gulf of California.

When a spreading center develops within a continent, the landmass may split into smaller segments as Wegener had proposed for the breakup of Pangaea. The fragmentation of a continent is thought to be initiated by an upward movement of hot rock from below. The effect of this activity is to upwarp the crust directly above the hot rising plume. Crustal stretching associated with the doming generates numerous tensional cracks as shown in Figure 6.9A. Then, as the hot plume spreads laterally from the region of upwelling, the broken lithosphere is pulled apart. Gradually the broken slabs slide downward into the gaps created by the diverging plates (Figure 6.9B). The large downfaulted valleys generated by this process are called **rifts,** or **rift valleys.** The Great Rift Valley of East Africa is an excellent example of such a feature (Figure 6.10, page 168). If the spreading process continues in East Africa, the rift valley will lengthen and deepen, eventually extending out into the ocean. At this point the valley will become a narrow linear sea with an outlet to the ocean similar to the Red Sea today (Figure 6.9C). The zone of rifting will remain the site of igneous activity, continually generating new sea floor in an ever-expanding ocean basin (Figure 6.9D).

Our knowledge of oceanic ridge systems, the sites of sea-floor spreading, comes from soundings taken of the ocean floor, core samples obtained from deep-sea drilling, visual inspection using submersibles, and even first-hand inspection of slices of ocean floor which have been shoved up onto dry land. Ocean ridge systems are characterized by an elevated position and numerous volcanic structures which have grown on the newly formed crust. Because of its accessibility, the Mid-Atlantic Ridge has been studied more thoroughly than other ridge systems. The Mid-Atlantic Ridge is a gigantic submerged mountain range standing 2500–3000 meters (8200–10,000 feet) above the adjacent deep-ocean basins. It extends southward from the Arctic Ocean to beyond the southern tip of Africa. In a few places, such as Iceland, the Mid-Atlantic Ridge has actually grown above sea level. Throughout most of its length, however, this divergent boundary lies 2500 meters below sea level.

Although volcanic structures do contribute to the height of a ridge, the warm, buoyant nature of the intruding magma is the primary reason for its elevated position. As the newly formed lithosphere travels away from the spreading center, it gradually cools and contracts. This thermal contraction accounts in part for the greater ocean depths that exist away from the ridge. Almost 100 million years must pass before cooling and contraction cease completely. By this time rock that was once a part of the majestic ocean mountain system becomes part of the deep-ocean basin.

CONVERGENT BOUNDARIES

At spreading centers new lithosphere is continually being generated; however, since the total surface area of the earth remains constant, lithosphere must also be destroyed. The zones of plate convergence are the sites of this destruction. When two plates collide, the leading edge of one is bent downward, allowing it to descend beneath the other. Upon entering the hot asthenosphere, the plunging plate begins to warm and loses its rigidity. Generally the descending plate is relatively cold and approaches 100 kilometers in thickness. Thus, depending upon its angle of descent, it may reach a depth of 700 kilometers

FIGURE 6.10
East African rift valleys and
associated features.

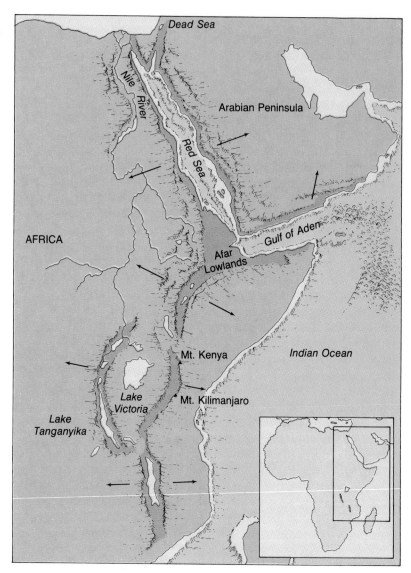

before its leading edge is completely assimilated within the material of the upper mantle.

Although all convergent zones are basically similar, the nature of plate collisions is influenced by the type of crustal material involved. Collisions can occur between two oceanic plates, one oceanic and one continental plate, or two continental plates, as shown in Figure 6.11. Whenever the leading edge of a plate capped with continental crust converges with oceanic crust, the less dense continental material apparently remains "floating," while the more dense oceanic slab sinks into the asthenosphere. The region where an oceanic

plate descends into the asthenosphere because of convergence is called a **subduction zone.** As the oceanic plate slides beneath the overriding plate, the oceanic plate bends, thereby producing a **deep-ocean trench** adjacent to the zone of subduction (Figure 6.11A). Trenches formed in this manner may be thousands of kilometers long and 8 to 11 kilometers deep.

Oceanic-Continental Convergence During a collision between an oceanic slab and a continental block, the oceanic crust is bent, permitting its descent into the asthenosphere (Figure 6.11A). The

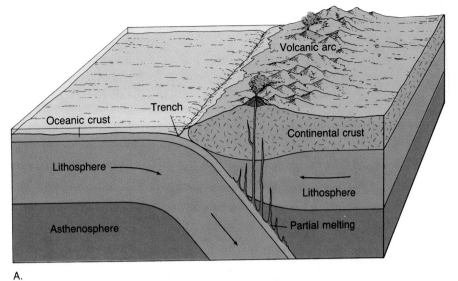

A.

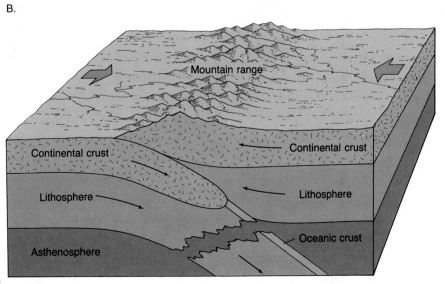

B.

C.

angle at which the plate descends is usually 45 degrees, or greater. Upon entering the hot asthenosphere, the downward moving plate and the water-soaked sediments carried upon it begin to melt. The newly formed magma created in this manner is less dense than the surrounding mantle rocks and, consequently, when sufficient quantities have gathered, the molten rock will slowly rise. Most of the rising magma will be emplaced in the 'overlying continental crust where it will cool and crystallize at a depth of several kilometers. The remaining magma will eventually migrate to the surface where it can give rise to numerous and occasionally explosive volcanic eruptions. The volcanic Andes Mountains are believed to have been produced by such activity when the Nazca plate melted as it plunged beneath the continent of South America (Figure 6.6). The frequent earthquakes that occur within the Andes testify to the activity beyond our view.

Mountains such as the Andes that are believed to be produced in part by volcanic activity associated with the subduction of oceanic lithosphere are called **volcanic arcs.** Two of these volcanic arcs are in the western United States. One, the Cascade Range, is composed of several well-known volcanic mountains, including Mounts Rainier, Shasta, and St. Helens. The second is the Sierra Nevada, in which Yosemite National Park is located. The Sierra Nevada system is the older of the two and has been inactive for several million years as evidenced by the absence of volcanic cones. Here erosion has stripped away most of the obvious traces of volcanic activity and left exposed the large, crystallized magma chambers that once fed lofty volcanoes. As the recent eruptions of Mount St. Helens testify, the Cascade Range is still quite active. The magma here arises from the melting of a small remaining segment of the Juan de Fuca plate.

Oceanic-Oceanic Convergence   When two oceanic slabs converge, one descends beneath the other, initiating volcanic activity in a manner similar to that which occurs at an oceanic-continental convergent boundary. However, in this case, the volcanoes form on the ocean floor rather than on the continents (Figure 6.11B). If this volcanic activity is sustained, dry land will eventually emerge from the ocean depths. In the early stages of development, this newly formed land consists of a chain of small volcanic islands called an **island arc.** The Aleutian, Mariana, and Tonga islands exemplify such features. Island arcs such as these are generally located a few hundred kilometers from an ocean trench where active subduction of the lithosphere is occurring. Adjacent to the island arcs just mentioned are the Aleutian trench, Mariana trench, and the Tonga trench, respectively.

Over an extended period, numerous episodes of volcanic activity build large volcanic piles on the ocean floor. This volcanic activity, plus the buoyancy of the intrusive igneous rock emplaced within the crust below, gradually increase the size and elevation of the developing arc. This growth, in turn, increases the amount of eroded sediments added to the sea floor. Some of these sediments reach the trench and are deformed and metamorphosed by the compressional forces exerted by the two converging plates. The result of these diverse activities is the development of a mature island arc composed of a complex system of volcanic rocks, folded and metamorphosed sedimentary rocks, and intrusive igneous rocks. Examples of mature island arc systems are the Alaskan Peninsula, the Philippines, and Japan.

Continental-Continental Convergence   When two plates carrying continental crust collide, a continental collision occurs (Figure 6.11C). This is thought to take place because of the light composition, and thus buoyant nature, of continental rocks. Such a collision is believed to have happened when the once-separated continent of India "rammed" into Asia and produced the Himalayas, perhaps the most spectacular mountain range on earth (Figure 6.12). During this collision, the continental crust buckled, fractured, and was generally shortened. In addition to the Himalayas, several other complex mountain systems, including the Alps, Appalachians, and Urals, are thought to have formed in this manner.

Prior to a continental collision, the landmasses involved are separated by the oceanic crust formed during an earlier episode of sea-floor spreading. As the continental blocks converge, the intervening sea floor is subducted beneath one of the plates. The partial melting of the descending oceanic slab and the sediments carried with it generate a volcanic arc. Depending on the location of the subduction zone, the volcanic arc could develop on either of the converging landmasses, or if the subduction zone developed at an appreciable distance into the ocean, an island arc would form. In any case, erosion of the newly formed volcanic arc would add large quantities of sediment to the already sediment-laden continental margins. Eventually, as the intervening sea floor was

consumed, these continental masses would collide, thereby squeezing, folding, and generally deforming the sediments as if they were placed in a gigantic vise. The result would be the formation of a new mountain range composed of deformed sedimentary rocks and fragments of the volcanic arc.

TRANSFORM BOUNDARIES

The third type of plate boundary is the transform fault, where plates slide past one another without

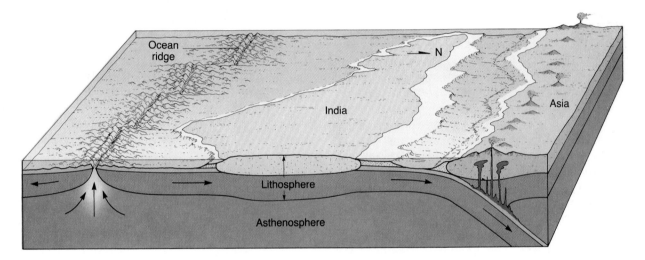

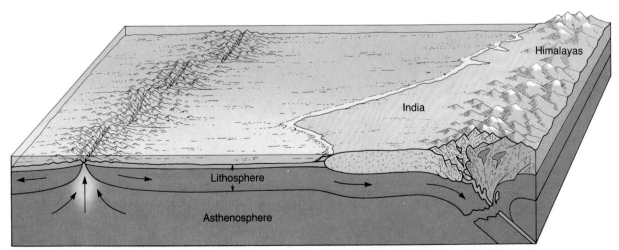

FIGURE 6.12
The collision of India and Asia about 45 million years ago produced the majestic Himalayas.

the production of crust, as occurs along oceanic ridges, or without the destruction of crust, as occurs at subduction zones. Transform faults roughly parallel the direction of plate movement and were first identified where they join segments of the oceanic ridge system (see Figure 6.6).

The nature of transform faults was proposed in 1965 by J. Tuzo Wilson of the University of Toronto. Wilson suggested that these large fractures connected convergent and divergent plate boundaries into a continuous network that divides the earth's outer shell into several rigid plates. In this role, transform faults provide the means by which the oceanic crust created at the ridge crests can be transported to its site of destruction: the deep-ocean trenches. Figure 6.13 illustrates this activity. Notice that the Juan de Fuca plate moves in a southeasterly direction, eventually being subducted under the west coast of the United States. The southern end of this relatively small plate is bounded by the Mendocino escarpment. This transform fault boundary connects an active spreading center to a subduction zone. Therefore, the fault facilitates the movement of the crustal material created at the ridge crest to its destination beneath the North American continent.

Wilson called these special faults transform faults because the relative motion of the plates can be changed, or transformed, along them. As we saw in the preceding example, divergence occurring at a spreading center can be transformed into convergence at a subduction zone. Since transform faults connect convergent and divergent boundaries in various combinations, other changes in relative plate motion are possible along transform faults.

Most transform faults are located in oceanic crust, but a few, including the San Andreas fault in California, are situated within continents (Figure 6.13). Along the San Andreas fault, the Pacific plate is moving toward the northwest, past the North American plate. If this movement continues for millions of years, that part of California west of the fault zone, including the Baja Peninsula, will become an island off the west coast of the United States and Canada, and could eventually reach Alaska. However, the more immediate concerns

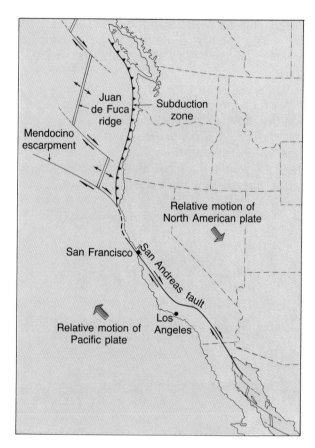

FIGURE 6.13
The role of transform faults in permitting relative motion between adjacent plates. The Mendocino escarpment permits sea floor generated at the Juan de Fuca ridge to move southeastward past the Pacific plate.

are the earthquakes triggered by movements along this fault system.

TESTING THE MODEL

As the plate tectonics theory was being developed, an avalanche of data was compiled to test this revolutionary idea. The model just presented will surely be modified to fit this wealth of data; however, the basic premises appear able to withstand the test of time.

Although most geologists have accepted this theory energetically, there remains an ever-diminishing number who reject it in part or in total. Some of the evidence supporting continental drift and sea-floor spreading has already been presented in this chapter. In addition, some of the evidence which was instrumental in solidifying the support for this new concept follows. It should be pointed out that some of this evidence was not new, rather it was a new interpretation of old data that swayed the tide of opinion.

PLATE TECTONICS AND PALEOMAGNETISM

Probably the most persuasive evidence to the geologic community for the acceptance of the plate tectonics theory comes from the study of the earth's magnetic field. Anyone who has used a compass to find direction knows that the earth's magnetic field has a north pole and a south pole. These magnetic poles align closely, but not exactly, with the respective geographic poles. In many respects the earth's magnetic field is very much like that produced by a simple bar magnet. Invisible lines of force pass through the earth and extend from one pole to the other. A compass needle, itself a small magnet free to move about, becomes aligned with these lines of force and thus points toward the magnetic poles.

The technique used to study ancient magnetic fields relies on the fact that certain rocks contain minerals which serve as fossil compasses. These iron-rich minerals such as magnetite are abundant, for example, in lava flows of basaltic composition. When heated above a certain temperature called the **Curie point,** these magnetic minerals lose their magnetism. However, when these iron-rich grains cool below their Curie point (about 580°C) they become magnetized in the direction parallel to the existing magnetic field. Once the minerals solidify, the magnetism they possess will remain "frozen" in this position. In this regard, they behave much like a compass needle inasmuch as they "point" toward the existing magnetic poles. If the rock is moved or the magnetic pole changes position, the rock magnetism will, in most instances, retain its original alignment. Rocks formed thousands or millions of years ago

thus "remember" the location of the magnetic poles at the time of their formation and are said to possess fossil magnetism, or **paleomagnetism.**

Polar Wandering   A study of lava flows conducted in Europe in the 1950s led to an interesting discovery. The magnetic alignment in the iron-rich minerals in lava flows of different ages was found to vary widely. A plot of the position of the magnetic north pole through time revealed that during the past 500 million years the position of the pole had gradually wandered from a location near Hawaii northward through eastern Siberia and finally to its present location (Figure 6.14). This was clear evidence that either the magnetic poles had migrated through time, an idea known as **polar wandering,** or the continents had drifted.

Although the magnetic poles are known to move, studies of the magnetic field indicate that the average positions of the magnetic poles correspond closely to the positions of the geographic poles. This is consistent with our knowledge of the

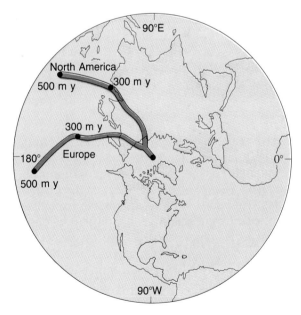

FIGURE 6.14
Simplified apparent polar wandering paths for North America and Europe. If these landmasses are brought together to close the North Atlantic, the paths roughly coincide.

FIGURE 6.15
New sea floor records the
polarity of the magnetic field
at the time it formed. Hence
it behaves much like a tape
recorder, as it continually
keeps count of changes in
the earth's magnetic field.

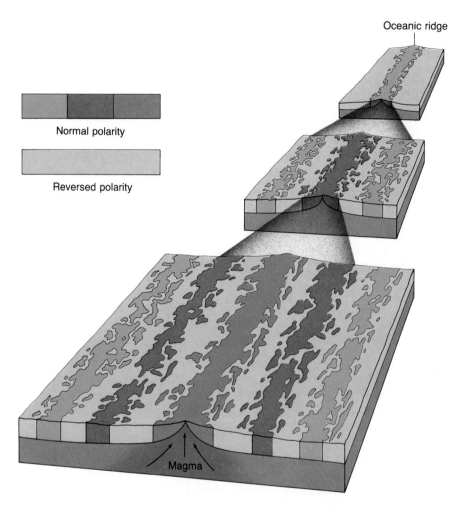

Normal polarity

Reversed polarity

Oceanic ridge

Magma

earth's magnetic field, which is generated in part
by the rotation of the earth about its axis. If the
geographic poles do not wander appreciably,
which we believe is true, neither can the magnetic
poles. Therefore, a more acceptable explanation
for the apparent polar wandering is provided by
the plate tectonics theory. If the magnetic poles
remain stationary, their apparent movement can
be produced by moving the continents.

Further evidence for plate tectonics came a few
years later when a polar wandering curve was constructed for North America (Figure 6.14). To nearly
everyone's surprise the curves for North America
and Europe had similar paths, except that they
were separated by about 30 degrees of longitude.
When these rocks solidified, could there have

been two magnetic north poles which migrated
parallel to each other? This is very unlikely. The
differences in these migration paths, however, can
be reconciled if the two presently separated continents are placed next to one another, as we now
believe they were prior to opening of the Atlantic
Ocean.

Magnetic Reversals Another discovery in the
field of paleomagnetism came when geophysicists
learned that the earth's magnetic field periodically
reverses polarity; that is, the north magnetic pole
becomes the south magnetic pole and vice versa.
The cause of these reversals is apparently linked
to the fact that the earth's magnetic field changes

intensity. Recent calculations, for example, indicate that the magnetic field has weakened about 5 percent over the past century. If this trend continues for the next thousand years or so, we might expect the earth's magnetic field to become very weak or nonexistent. During periods when the earth's magnetic field is very weak, some external influence such as sunspot activity could possibly contribute to a reversal of polarity. After a reversal has taken place, the field would rebuild itself with opposite polarity. A rock solidifying during one of the periods of reverse polarity will be magnetized with the opposite polarity of those rocks being formed today. When rocks exhibit the same magnetism as the present magnetic field, they are said to possess **normal polarity,** while those rocks exhibiting the opposite magnetism are said to have **reverse polarity.** Using the potassium-argon method of radiometric dating, the polarity of the earth's magnetic field has been reconstructed for a period of several million years.

A significant relationship between magnetic reversals and the sea-floor spreading hypothesis was developed from data obtained when very sensitive instruments called *magnetometers* were towed by research vessels across a segment of the ocean floor located off the west coast of the United States. Here workers from the Scripps Institution of Oceanography discovered alternating strips of high- and low-intensity magnetism that trended in roughly a north-south direction. This relatively simple pattern of magnetic variation defied explanation until 1963, when Fred Vine and D. H. Matthews tied the discovery of the high- and low-intensity strips to the concept of sea-floor spreading. Vine and Matthews suggested that the strips of high-intensity magnetism are regions where the paleomagnetism of the ocean crust is of the normal type (Figure 6.15). Consequently, these positively magnetized rocks enhance the existing magnetic field. Conversely, the low-intensity strips represent regions where the ocean crust is polarized in the reverse direction and, therefore, weaken the existing magnetic field. But how do parallel strips of normally and reversely magnetized rock become distributed across the ocean floor?

Vine and Matthews reasoned that as new basalt was added to the ocean floor at the oceanic ridges, it would be magnetized according to the existing magnetic field. Since new rock is added in approximately equal amounts to the trailing edges of both plates, we should expect strips of equal size and polarity to parallel both sides of the ocean ridges as shown in Figure 6.15. This explanation of the alternating strips of normal and reverse polarity, which lay as mirror images across the ocean ridges, was the strongest evidence so far presented in support of the concept of sea-floor spreading.

PLATE TECTONICS AND EARTHQUAKES

By 1968, the basic outline of plate tectonics was firmly established. In this same year, three Lamont-Doherty Observatory seismologists, B. Isacks, J. Oliver, and L. R. Sykes, published papers demonstrating how much more successful the new plate tectonics model was than older models in accounting for the global distribution of earthquakes (Figure 6.16). In particular these seismologists were able to explain the distribution of earthquakes that are associated with deep-ocean trenches. For example, examine the distribution of earthquakes in the vicinity of the Japan trench just to the east of Japan (Figure 6.16). Here most shallow-focus earthquakes occur within, or adjacent to, the trench, whereas intermediate- and deep-focus earthquakes occur toward the mainland. A similar distribution pattern exists along the western margin of South America where the Nazca plate is being subducted beneath the continent.

In the plate tectonics model, deep-ocean trenches are produced where slabs of oceanic lithosphere plunge into the mantle (Figure 6.17, page 177). Very shallow focus earthquakes are generated at the outer edge of the trenches where the plate is bent to permit its descent. Shallow-focus earthquakes are also produced as the descending plate scrapes beneath the overriding slab. As the slab continues to descend into the asthenosphere, deeper-focus earthquakes are generated. The

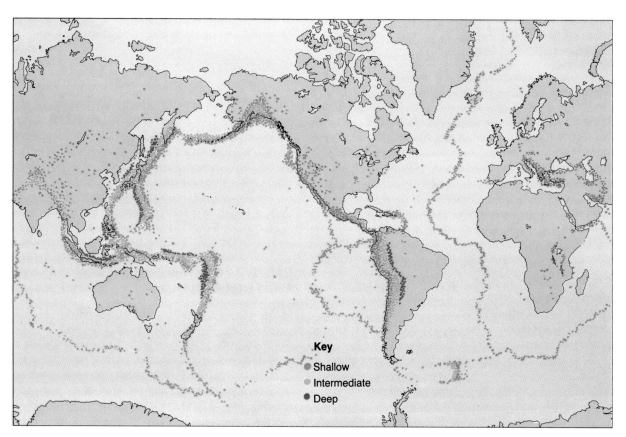

FIGURE 6.16
Distribution of shallow-, intermediate-, and deep-focus earthquakes. (Data from NOAA)

deep-focus earthquakes appear to originate within the descending plate and are caused by resistance to the plate's downward movement. Since the earthquakes occur within the rigid lithosphere rather than within the mobile asthenosphere, they provide a method for tracking the plate's descent into the mantle. Very few earthquakes have been recorded below 700 kilometers (435 miles), possibly because the lithosphere has been completely assimilated into the mantle by the time it reaches this depth.

The plate tectonics model also explains why deep-focus earthquakes are confined to areas adjacent to oceanic trenches (subduction zones), while earthquakes which occur along divergent boundaries and transform fault boundaries have only shallow foci. Recall that earthquakes result

from the rapid release of strain that can only build up in the rigid material of the lithosphere. Since subduction zones are the only regions where rigid material is forced to great depths, these should be the only sites of deep-focus earthquakes. Indeed, the absence of deep-focus earthquakes along divergent boundaries and transform faults also substantiates the plate tectonics theory.

EVIDENCE FROM OCEAN DRILLING

Some of the most convincing evidence confirming the plate tectonics theory has come from drilling directly into ocean floor sediment. From 1968 until 1983, the source of this important data was the Deep Sea Drilling Project, an international program sponsored by several major oceanographic

FIGURE 6.17
Relationship between the
descending plate and depth
of earthquake foci.

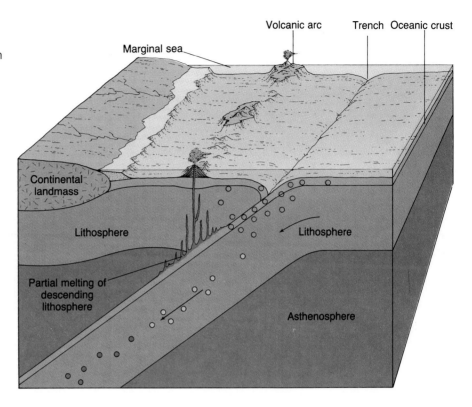

institutions and the National Science Foundation. The primary goal was to gather firsthand information about the age of ocean basins and processes of ocean basin formation. Researchers felt that the predictions concerning sea-floor spreading that were based on paleomagnetic data could best be confirmed by the direct sampling of sediments from the floor of the deep-ocean basins. To accomplish this, a new drilling ship, the *Glomar Challenger,* was built. The *Glomar Challenger* represented a significant technological breakthrough, because this ship was capable of lowering drill pipe thousands of meters to the ocean floor and then drilling hundreds of meters into the sediments and underlying basaltic crust.

Operations began in August, 1968, and shortly thereafter important evidence was gathered in the South Atlantic. At several sites holes were drilled through the entire thickness of sediments to the basaltic rock below. An important objective was to gather samples of sediment from just above the igneous crust as a means of dating the sea floor at

each site.* Since sedimentation begins immediately after the oceanic crust forms, fossils found in the oldest sediments (that is, those resting directly above the basalt) can be used to date the ocean floor at that site. When the oldest sediment from each drill site was plotted against its distance from the ridge crest, it was revealed that the age of the sediment increased with increasing distance from the ridge. This finding was in agreement with the sea-floor spreading hypothesis, which predicted that the youngest oceanic crust is to be found at the ridge crest and that the oldest oceanic crust flanks the continental margins. Further, the rate of sea-floor spreading determined from the ages of sediments was identical to the rate previously estimated from magnetic evidence. Subsequent drilling in the Pacific Ocean verified these findings. These excellent correlations were a striking confirmation of sea-floor spreading.

---

*Radiometric dates of the ocean crust itself are unreliable because of the alteration of basalt by seawater.

FIGURE 6.18
The *JOIDES Resolution,* the drilling ship of the Ocean Drilling Program. This modern drilling ship has replaced the *Glomar Challenger* in the important work of sampling the floors of the world's oceans. (Photo courtesy of Ocean Drilling Program)

The data from the Deep Sea Drilling Project also reinforced the idea that the ocean basins are geologically youthful because no sediment with an age in excess of 160 million years was found. By comparison, some continental crust has been dated at more than 3.8 billion years.

The thickness of ocean-floor sediments provided additional verification of sea-floor spreading. Drill cores from the *Glomar Challenger* revealed that sediments on the ridge crest are almost entirely absent and that the sediment thickens with increasing distance from the ridge. Since the ridge crest is younger than the areas farther from it, this pattern of sediment distribution is exactly what was predicted by the plate tectonics theory.

During its 15 years of operation the *Glomar Challenger* drilled 1092 holes and obtained more than 96 kilometers (60 miles) of invaluable core samples. Although the Deep Sea Drilling Project ended and the *Glomar Challenger* was retired, the important work of sampling the floors of the world's ocean basins continues. The Ocean Drill-

ing Program has succeeded the Deep Sea Drilling Project and, like its predecessor, it is a major international program. A more technologically advanced drilling ship, the *JOIDES Resolution,* now continues the work of the *Glomar Challenger.** The *JOIDES Resolution* can drill in water depths as great as 8100 meters (27,000 feet) and contains onboard laboratories equipped with the largest and most varied array of seagoing scientific research equipment in the world (Figure 6.18). The work of the Ocean Drilling Program is to continue until at least 1995.

HOT SPOTS

Mapping of seamounts in the Pacific revealed a chain of volcanic structures extending from the Hawaiian Islands to Midway Island and then continuing northward toward the Aleutian trench. Potassium-argon dating of numerous volcanoes in this chain revealed an increase in age with an in-

---

*JOIDES is an acronym for Joint Oceanographic Institutions for Deep Earth Sampling.

crease in distance from Hawaii. Suiko Seamount, which is located near the Aleutian trench, is 65 million years old, Midway Island is 27 million years old, and the island of Hawaii rose from the sea less than 1 million years ago (Figure 6.19).

Researchers have proposed that a **hot spot** exists within the mantle and emits magma onto the overlying sea floor. Presumably, as the Pacific plate moved over the hot spot, successive volcanic structures emerged. The age of each volcano indi-

cates the time when it was situated over the relatively stationary hot spot. Kauai is the oldest of the large islands in the Hawaiian chain. Five million years ago, when it was positioned over the hot spot, Kauai was the only Hawaiian Island in existence (Figure 6.19). Visible evidence of the age of Kauai can be seen by examining the extinct volcanoes which have been eroded into jagged peaks and vast canyons. By contrast, the south slopes of the island of Hawaii consist of fresh lava flows and

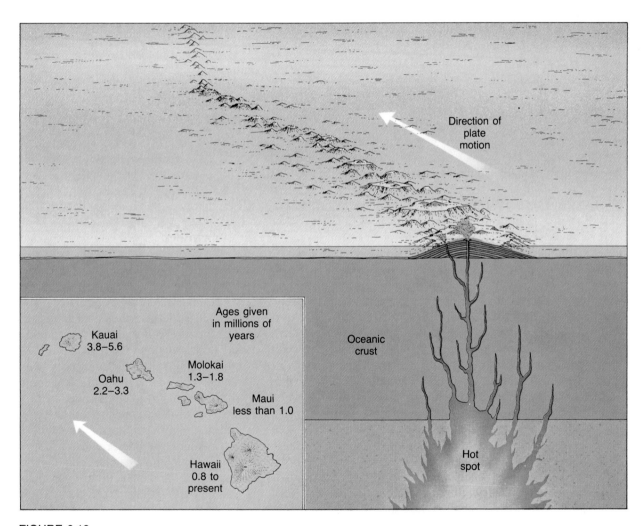

FIGURE 6.19
The chain of islands and seamounts that extend from Hawaii to the Aleutian trench results from the movement of the Pacific plate over an apparently stationary hot spot. Radiometric dating of the Hawaiian Islands shows that the volcanic activity decreases in age toward the island of Hawaii.

two of Hawaii's volcanoes, Mauna Loa and Kilauea, remain active. Recent evidence indicates that a new volcanic pile is forming on the ocean floor just off the coast of Hawaii. Geologically speaking, it should not be long before another tropical island will be added to the Hawaiian chain.

Two island groups parallel the Hawaiian Island-Emperor Seamount chain. One chain consists of the Tuamotu and Line islands, and the other includes the Austral, Gilbert, and Marshall islands. In each case, the most recent volcanic activity has occurred at the southeastern end of the chain, and the islands get progressively older to the northwest. Thus, like the Hawaiian Island-Emperor Seamount chain, these volcanic structures apparently formed by the same motion of the Pacific plate over fixed hot spots. This evidence not only supports the fact that the plates do indeed move relative to the earth's interior, but also the hot spot "tracks" provide a frame of reference for tracing the direction of plate motion. Notice, for example, in Figure 6.19 that the Hawaiian Island-Emperor Seamount chain bends. This particular bend in the trace occurred about 40 million years ago when the motion of the Pacific plate changed from nearly due north to a northwesterly path.

Although the existence of hot spots is well documented, their exact nature or role in plate tectonics is not altogether clear. Apparently hot spots are unusually warm regions found deep within the earth's mantle. Here the high temperatures produce a rising plume of molten rock which frequently initiates volcanism at the surface. Most evidence indicates that hot spots remain relatively stationary; however, some appear to have exhibited movement. Of the 50 to 120 hot spots believed to exist, about 20 are located near divergent plate boundaries, whereas the others are not associated with plate boundaries. A hot spot beneath Iceland is thought to be responsible for the unusually large accumulation of lava found in that portion of the Mid-Atlantic Ridge. Another hot spot is believed to be located beneath Yellowstone National Park and may be responsible for the large outpourings of lava and volcanic ash in this area. If the Yellowstone region was indeed modified by hot spot volcanism, there is good reason to expect additional activity in the future.

PANGAEA: BEFORE AND AFTER

Robert Dietz and John Holden have rather precisely projected the gross details of the migrations of individual continents over the past 500 million years. By extrapolating plate motion back in time using such evidence as the orientation of volcanic structures left behind on moving plates, the distribution and movements of transform faults, and paleomagnetism, Dietz and Holden were able to reconstruct Pangaea (see Figure 6.1). The use of radiometric dating helped them establish the time frame for the formation and eventual breakup of Pangaea, and the relatively stationary positions of hot spots through time helped to fix the locations of the continents.

BREAKUP OF PANGAEA

The fragmentation of Pangaea began about 200 million years ago. Figure 6.20 illustrates the breakup and subsequent paths taken by the landmasses involved. As we can see in Figure 6.20A, two major rifts initiated the breakup. The rift zone between North America and Africa generated numerous outpourings of Triassic-age basalts which are presently visible along the eastern seaboard of the United States. Radiometric dating of these basalts indicates that rifting occurred between 200 and 165 million years ago. This date can be used as the birth date of this section of the North Atlantic. The rift that formed in the southern landmass of Gondwanaland developed a "Y"-shaped fracture which sent India on a northward journey and simultaneously separated South America-Africa from Australia-Antarctica.

Figure 6.20B illustrates the position of the continents 135 million years ago, about the time Africa and South America began splitting apart to form the South Atlantic. India can be seen halfway into its journey to Asia, and the southern portion of the North Atlantic has widened considerably. By the end of the Cretaceous period, about 65 million years ago, Madagascar had separated from Africa, and the South Atlantic had emerged as a full-fledged ocean (Figure 6.20C, page 182).

The current map (Figure 6.20D) shows India in contact with Asia, an event that occurred about 45 million years ago and created the highest

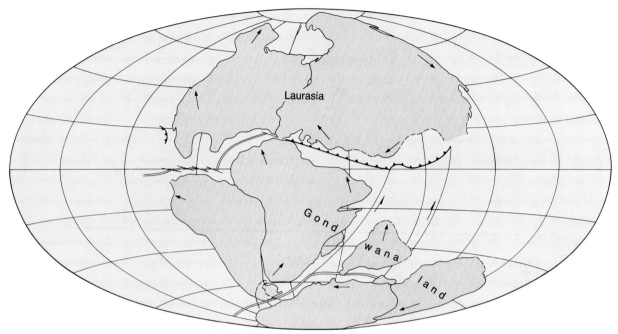

A. 180 Million Years Ago (Triassic Period)

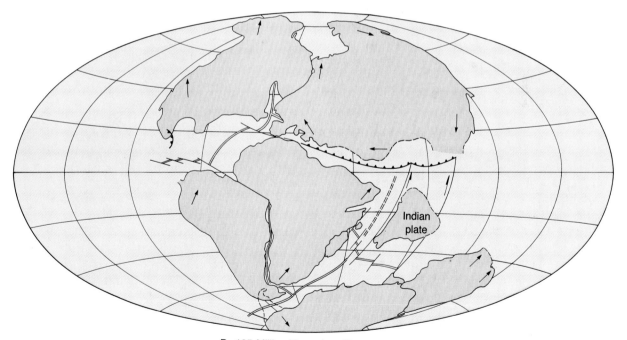

B. 135 Million Years Ago (Jurassic Period)

FIGURE 6.20A., B.
Several views of the breakup of Pangaea over a period of 200 million years according to Dietz and Holden. (Robert S. Dietz and John C. Holden, *Journal of Geophysical Research* 75:4939–56. 1970. Copyright by American Geophysical Union)

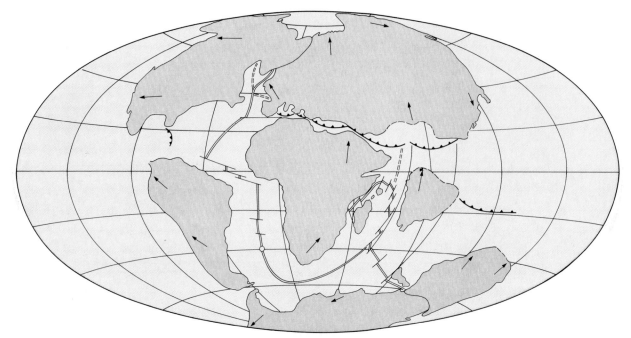

C. 65 Million Years Ago (Cretaceous Period)

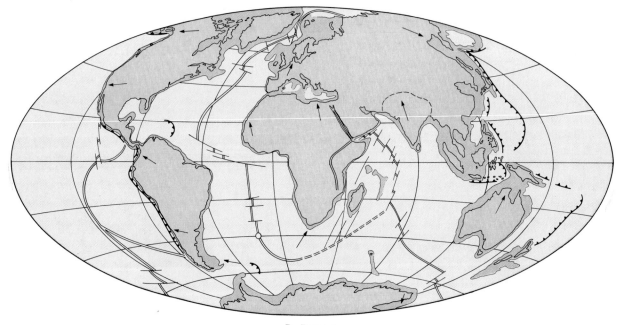

D. Present

FIGURE 6.20C., D.

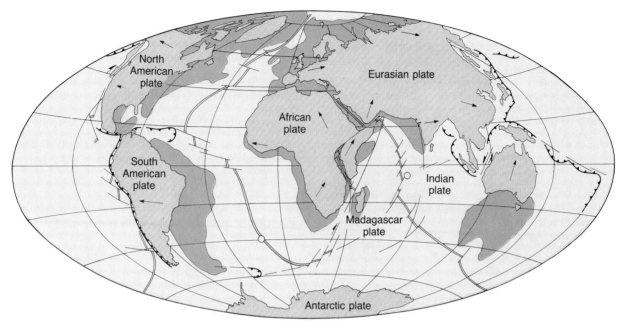

FIGURE 6.21
The world as it may look 50 million years from now. (From "The Breakup of Pangaea," Robert S. Dietz and John C. Holden. Copyright 1970 by Scientific American, Inc. All rights reserved)

mountains on earth, the Himalayas, along with the Tibetan Highlands. By comparing Figures 6.20C and 6.20D, we can see that the separation of Greenland from Eurasia was a recent event in geologic history. Also notice the recent formation of the Baja Peninsula and the Gulf of California. This event is thought to have occurred less than 10 million years ago.

BEFORE PANGAEA

Prior to the formation of Pangaea, the landmasses had probably gone through several episodes of fragmentation similar to what we see happening today. Also like today, these ancient continents moved away from each other only to collide again at some other location. During the period between 500 and 225 million years ago, the fragments of an earlier dispersal began collecting to form the continent of Pangaea. Evidence of these earlier continental collisions include the Ural Mountains of the Soviet Union and the Appalachians of North America.

A LOOK INTO THE FUTURE

After Dietz and Holden drew together the events that over the past 500 million years resulted in the current configuration of the continents, they went one step further and extrapolated plate motion into the future. Figure 6.21 illustrates what they envision the earth's landmasses will look like 50 million years from now. Important changes are seen in Africa where a new sea emerges as East Africa parts company with the mainland. In North America we see that the Baja Peninsula and the portion of southern California that lies west of the San Andreas fault have slid past the North American plate. If this northward migration takes place as predicted, Los Angeles and San Francisco will pass each other.

These projections into the future, although interesting, must be viewed with caution since many assumptions must be correct for these events to unfold as just described. Nevertheless, similar types of changes in the shapes and positions of the continents will undoubtedly occur for many millions of years to come.

THE DRIVING MECHANISM

The plate tectonics theory describes plate motion and the effects of this motion. Therefore, acceptance does not rely on a knowledge of the force or forces moving the plates. This is fortunate, since none of the driving mechanisms yet proposed can account for all of the major facets of plate motion. Nevertheless, the unequal distribution of heat within the earth is accepted by most as the underlying cause of plate movement. The unequal distribution of heat, in turn, is thought by many geologists to generate large convection cells within the mantle (Figure 6.22A,B). The warm, less dense material of the lower mantle rises very slowly in the regions of oceanic ridges. As the material spreads laterally it cools, becomes more dense, and begins to sink back into the mantle, only to be reheated. Note that rocks need not be molten to flow. Just as a hot, solid metal can be pounded into different shapes, so too can rock move when subjected to heat and stress over an extended period. Measurements indicate a higher rate of heat flow at the oceanic ridges than in other oceanic regions—a good indication that some type of thermal convection cells might exist. However, many of the details of their motions remain unclear. How many cells exist? At what depth do they originate? What is the structure of these thermal cells?

Although unequal distribution of heat is generally accepted as the underlying driving force of plates, many geologists are not convinced that large convection cells can exist in the mantle. Thus, many other mechanisms have been suggested. One mechanism relies on the fact that the cold oceanic slab has a greater density than the asthenosphere supporting it from below. This being the case, it has been proposed that as a plate begins to descend, the heavy, sinking slab might pull the trailing lithosphere along. This hypothesis is similar to another model which suggests that the elevated position of an ocean ridge could cause the lithosphere to slide under the influence of gravity (Figure 6.22C). These push-pull models are themselves a type of convection current. As the sinking plate enters the mantle, material is forced aside which then migrates toward the ridge systems. The cell is completed as molten rock moves up from the asthenosphere to fill the gap in the diverging oceanic plates.

Some oceans, notably the Atlantic, lack subduction zones; thus, the slab-pull mechanism cannot adequately explain the spreading occurring at these ridges. Other ridge systems are rather subdued, which would reduce the effectiveness of the alternate slab-push model. Perhaps the slab-push and slab-pull phenomena are active in different ridge systems and occasionally even work in tandem.

Another version of the thermal convection model suggests that relatively narrow, hot plumes (hot spots) of mantle rock generate plate motion (Figure 6.22D). These hot plumes are thought to originate near the mantle-core boundary. Upon reaching the lithosphere, these plumes would spread laterally and carry the plates away from the zone of upwelling. The hot plumes usually reveal themselves as volcanic structures growing up from the ocean floor in such places as Iceland. About twenty hot spots have been identified along ridge systems, where they may contribute to plate divergence. Recall, however, that some hot plumes, for example, the one which generated the Hawaiian Islands, are not located in ridge areas. Therefore, we must conclude that this model is not without its shortcomings. Perhaps a combination of all these phenomena generates the plate motion observed.

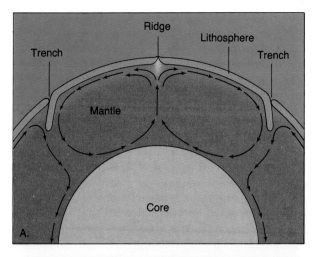
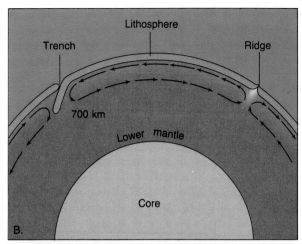
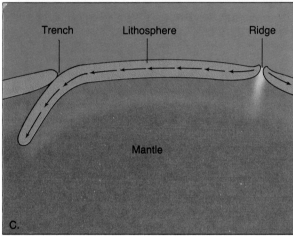
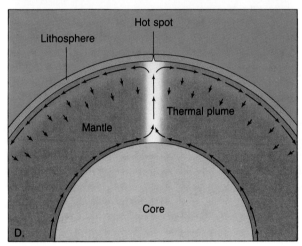

FIGURE 6.22

Proposed models of the driving force for plate tectonics. **A.** Large convection cells in the mantle carry the lithosphere in a conveyor belt fashion. **B.** Convection cells confined to the mobile upper mantle are responsible for plate motion. **C.** Push-pull models are also a type of convection. Here the cold, sinking oceanic slab pulls the trailing sea floor along, while gravity-sliding of material down the elevated ridge crest pushes the slab. **D.** The hot plume model suggests that all upward movement is confined to a few narrow plumes, while downward flow occurs slowly throughout the remaining mantle. (Parts A, B, and D after "The Earth's Mantle," by Peter J. Wyllie, © 1975, by Scientific American, Inc. All rights reserved)

REVIEW QUESTIONS

1   Who is credited with developing the continental drift hypothesis?

2   What was probably the first evidence that led some to suspect the continents were once connected?

3   What was Pangaea?

4   List the evidence that Wegener and his supporters gathered to substantiate the continental drift hypothesis.

5   Explain why the discovery of the fossil remains of *Mesosaurus* in both South America and Africa, but nowhere else, supports the continental drift hypothesis.

6   Early in this century, what was the prevailing view of how land animals migrated across vast expanses of ocean?

7   How did Wegener account for the existence of glaciers in the southern landmasses, while at the same time areas in North America, Europe, and Siberia supported lush tropical swamps?

8   On what basis were plate boundaries first established?

9   What are the three major types of plate boundaries? Describe the relative plate motion at each of these boundaries.

10   What is sea-floor spreading? Where is active sea-floor spreading occurring today?

11   What is a subduction zone? With what type of plate boundary is it associated?

12   Where is lithosphere being destroyed? Why must the production and destruction of lithosphere be going on at approximately the same rate?

13   Why is the oceanic portion of a lithospheric plate destroyed, whereas the continental portion is not?

14   Relate the formation of the Andes Mountains to the movement of plates.

15   Discuss the formation of an island arc.

16   In what ways may the origin of the Japanese Islands be considered similar to the formation of the Andes Mountains? How do they differ?

17   Briefly describe how the Himalaya Mountains formed.

18   Differentiate between transform faults and the other two types of plate boundaries.

19   Some predict that California will sink into the ocean. Is this idea consistent with the theory of plate tectonics?

20   Define the term *paleomagnetism.*

21   How does the continental drift hypothesis account for polar wandering?

22   Describe the distribution of earthquake epicenters and foci depths as they relate to oceanic trench systems.

23   What is the age of the oldest sediments recovered by deep-ocean drilling? How do the ages of these sediments compare to the ages of the oldest continental rocks?

24   How does sediment thickness vary with increasing distance from the crests of mid-ocean ridges?

25   How do hot spots and the plate tectonics theory account for the fact that the Hawaiian Islands vary in age?

26   With what type of plate boundary are the following places or features associated (be as specific as possible): Himalayas, Aleutian Islands, Red Sea, Andes Mountains, San Andreas fault, Iceland, Japan, Mount St. Helens?

KEY TERMS

continental drift	sea-floor spreading	paleomagnetism
Pangaea	rift (rift valley)	polar wandering
plate tectonics	subduction zone	normal polarity
plates	deep-ocean trench	reverse polarity
divergent boundary	volcanic arc	hot spot
convergent boundary	island arc	
transform boundary	Curie point	

7

IGNEOUS ACTIVITY

The significance of igneous activity may not be obvious at first glance. However, since volcanoes extrude molten rock that formed at great depth, they provide the only windows we have for direct observation of processes which occur many kilometers below the earth's surface. Furthermore, the gases emitted during volcanic eruptions are thought to be the material from which the atmosphere and oceans evolved. Either of these facts is reason enough for igneous activity to warrant our attention. All volcanic eruptions are spectacular, but why are some destructive and others quiescent? Is the entire island of Hawaii a volcano as high as Mount Everest resting on the ocean floor? Do the large volcanoes of Washington and Oregon present a threat to human lives? This chapter considers these and other questions as we explore the formation and movement of magma.

Mount Shasta, California, one of the most picturesque volcanoes of the Cascade Range. (Photo by John S. Shelton)

At 8:32 A.M. on Sunday, May 18, 1980, the largest volcanic eruption to occur in North America in recent times transformed a picturesque volcano into a decapitated remnant (Figure 7.1). On this date in southwestern Washington state, Mount St. Helens erupted with a force hundreds of times greater than that of the atomic bombs dropped on Japan during World War II. The blast blew out the entire north flank of the volcano, leaving a gaping hole. A once prominent volcano that had grown to more than 2900 meters (9500 feet) had in one brief moment been lowered by about 410 meters (1350 feet).

The early morning blast totally devastated a wide swath of timber-rich land on the north side of the mountain (Figure 7.2, page 192). Trees within a 400-square-kilometer area lay intertwined and flattened, stripped of their branches and appearing from the air like toothpicks strewn about. The immense force caused trees as far away as 25 kilometers to topple. The gases and ash unleashed from the volcano had temperatures that probably exceeded 800°C (1470°F)! As many as 59 persons were killed by the eruption. Some died from the intense heat and the suffocating cloud of ash and gases. Others perished as they were hurled from the mountain by the force of the blast. Still others were trapped by debris-laden mudflows.

The blast and accompanying mudflows carried ash, trees, and water-saturated rock debris 29 kilometers down the Toutle River. The river quickly became a mud-filled torrent and reached depths

FIGURE 7.1
Before and after photographs show the transformation of Mount St. Helens caused by the May 18, 1980 eruption. ("Before" photo by Stephen Trimble, "after" photo by Jim Hughes, courtesy USDA Forest Service)

of 60 meters in some places. Further, a debris dam was deposited at the outlet of Spirit Lake, causing its level to rise by more than 30 meters (100 feet). For several days the threat of pent-up waters breaching the dam posed another potential hazard.

The eruption of May 18th ejected an estimated three to four cubic kilometers of ash and rock debris. By comparison, this is roughly equal to the quantity of ash that buried the city of Pompeii during the historic eruption of Mount Vesuvius in 79 A.D.

Following the devastating explosion, Mount St. Helens continued to emit great quantities of hot gases and ash. Only minutes after the eruption began, a dark plume rose from the volcano. The force of the blast was so strong that some ash was propelled high into the stratosphere, more than 18,000 meters (60,000 feet) above the ground. During the next hours and days, this very fine grained material was carried around the earth by strong upper air winds. Measurable deposits were reported as far away as Oklahoma and Minnesota. Meanwhile, ash fallout in the immediate vicinity accumulated to depths exceeding 2 meters, and the air over Yakima, Washington, 130 kilometers to the east, was so filled with ash that residents experienced midnight-like darkness at noon. Crop damage from the volcanic fallout was reported as far away as central Montana.

The events leading to the May 18th eruption began about two months earlier, on March 20th,

FIGURE 7.2
Forest lands and battered
van northwest of Mount St.
Helens destroyed by lateral
blast of May 18, 1980.
(Photo courtesy of USDA
Forest Service)

as a series of minor earth tremors centered beneath the awakening mountain. The first volcanic
activity took place on March 27th, when a small
amount of ash and steam rose from the summit.
Over the next several weeks, sporadic eruptions of
varied intensity occurred.

Prior to the main eruption, the primary concern
had been the potential hazard of mudflows. These
moving lobes of saturated debris were created
when ice and snow were melted by heat from the
magma within the volcano. The only sign of a potentially hazardous eruption was a bulge on the
volcano's north flank. Careful monitoring of this
dome-shaped structure indicated a very slow but
steady growth rate of a few meters per day. Geologists monitoring the activity suggested that if the
growth rate of the bulge changed appreciably, an
eruption might quickly follow. Unfortunately, no
such variation was detected prior to the explosion.
In fact, the seismic activity decreased during the
two days preceding the huge blast.

"Vancouver, Vancouver, this is it!" was the only
warning to precede the unleashing of tremendous
quantities of pent-up gases. The trigger was an
earth tremor with a magnitude of 5.1 on the

Richter scale. The vibrations sent the north slope
of the cone plummeting into the Toutle River, effectively removing the overburden which had
trapped the magma below. With the pressure reduced, the water-rich magma is thought to have
ruptured like an overheated steam boiler. Since
the eruption originated in the vicinity of the bulge,
which was several hundred meters below the
summit, the main impact of the eruption was directed laterally rather than vertically. Had the full
force of the eruption been upward, far less destruction would have occurred.

Mount St. Helens is only one of the 15 large
volcanoes and innumerable smaller ones extending from British Columbia to northern California.
Eight of the largest cones have been active in the
past few hundred years, while the last eruptive
phase of Mount St. Helens came to an end in 1857.
Of the remaining seven "active" volcanoes, Mount
Baker, Mount Shasta (see chapter-opening photo),
Lassen Peak, and Mount Rainier are believed most
likely to erupt again. It is hoped that the eruptions
of Mount St. Helens will provide geologists with
enough data to more effectively evaluate the potential hazards of future volcanic eruptions.

THE NATURE OF VOLCANIC ACTIVITY

Volcanic activity is commonly perceived as a process that produces a picturesque, cone-shaped structure which periodically erupts in a violent manner. However, although some eruptions may be cataclysmic, many are relatively quiescent. The primary factors which determine the nature of volcanic eruptions include the magma's composition, its temperature, and the amount of dissolved gases it contains. These factors affect the magma's mobility, or **viscosity.** The more viscous the material, the greater its resistance to flow. For example, molasses is more viscous than water. The effect of temperature on viscosity is easily seen. Just as heating molasses makes it more fluid, the mobility of lava is also influenced by temperature changes. As a lava flow cools and begins to congeal, its mobility decreases and eventually the flowing halts.

The chemical composition of magmas was discussed in Chapter 1 with the classification of igneous rocks. One major difference between various igneous rocks and therefore between the magmas from which they originate is their silica (SiO_2) content (Table 7.1). Magmas that produce basaltic rocks contain about 50 percent silica, whereas rocks of granitic composition (granite and its extrusive equivalent, rhyolite) contain over 70 percent silica. The intermediate rock types, andesite and diorite, contain around 60 percent silica. It is important to note that a magma's viscosity is directly related to its silica content. In general, the higher the percentage of silica in magma, the greater its viscosity. It is believed that the flow of magma is impeded because the silica molecules link into long chains even before crystallization begins. Consequently, because of their lower silica content, basaltic lavas tend to be quite fluid, whereas lavas of granitic composition are very viscous and incapable of flow over appreciable distances even at relatively high temperatures.

The gas content of a magma also affects its mobility. Dissolved gases tend to increase the fluidity of magma. Of far greater consequence is the fact that escaping gases provide enough force to propel molten rock from a volcanic vent. As magma moves into a near-surface environment, such as within a volcano, the confining pressure in the uppermost portion of the magma body is greatly reduced. This reduced confining pressure allows the gases, which had been dissolved when they were at greater depths, to be released suddenly. At temperatures of 1000°C (1830°F) and low, near-surface pressures, these gases will expand to occupy hundreds of times their original volume. Very fluid basaltic magmas allow the expanding gases to migrate upward and escape from the vent with relative ease. As they escape, the gases will often carry incandescent lava hundreds of meters into the air, producing lava fountains. Although spectacular, such fountains are not generally associated with major explosive events of the type which cause great loss of life and property. Rather, eruptions of fluid basaltic lavas, such as those that occur in Hawaii, are relatively quiescent. At the other extreme, highly viscous magmas impede the upward migration of gases. As a consequence, gases collect as bubbles and pockets

TABLE 7.1
Variations in properties among magmas of differing compositions.

Property	Basaltic	Andesitic	Granitic
Silica content	Least (about 50%)	Intermediate (about 60%)	Most (about 70%)
Viscosity	Least	Intermediate	Highest
Tendency to form lavas	Highest	Intermediate	Least
Tendency to form pyroclastics	Least	Intermediate	Highest
Melting point	Highest	Intermediate	Lowest

that increase in size and pressure until they explosively eject the semimolten rock from the volcano.

To summarize, we have seen that the quantity of dissolved gases, as well as the ease with which the gases can escape, largely determine the nature of a volcanic eruption. We can now understand why the volcanic eruptions on Hawaii are relatively quiet, whereas the volcanoes bordering the Pacific are explosive and pose the greatest threat to people, because these latter volcanoes generally contain great quantities of gas and emit viscous lavas.

MATERIALS EXTRUDED DURING AN ERUPTION

Many people believe that lava is the primary material extruded from a volcano. This is not always the case. Explosive eruptions which eject huge quantities of broken rock, lava bombs, and fine ash and dust, occur just as frequently. Moreover, all volcanic eruptions emit large amounts of gas. In this section we will examine each of these materials associated with a volcanic eruption.

LAVA FLOWS

Due to their low silica content, basaltic lavas are usually very fluid and flow in thin, broad sheets or tongues. On the island of Hawaii such lavas have been clocked at speeds of 30 kilometers (20 miles) per hour on steep slopes. These velocities are rare however, and flow rates of 10 to 300 meters per hour are more common. In contrast, the movement of silica-rich lava is often too slow to be perceptible.

When fluid basaltic lavas of the Hawaiian type congeal, they often form a relatively smooth skin that sometimes wrinkles as the still-molten subsurface lava continues to advance (Figure 7.3A). These are known as **pahoehoe flows** and resemble the twisted braids in ropes. Another common type of basaltic lava has a surface of rough, jagged blocks with dangerously sharp edges and spiny projections (Figure 7.3B). The name **aa** (pronounced "ah ah") is given to these flows. Active aa

flows are relatively cool and thick and, depending upon the slope, advance at rates from 5 to 50 meters per hour. Further, escaping gases fragment the cool surface and produce numerous voids and sharp spines in the congealing lava. As the molten

A.

B.

FIGURE 7.3
A. Typical pahoehoe (ropy) lava flow, Kilauea, Hawaii. (Photo by D. W. Peterson, U.S. Geological Survey)
B. Typical slow-moving aa flow. (Photo by J. D. Griggs, U.S. Geological Survey)

FIGURE 7.4
Parícutin Volcano in eruption at night, 1943. (Photo by K. Segerstrom, U.S. Geological Survey)

interior advances, the outer crust is broken further, giving the flow the appearance of an advancing mass of lava rubble.

GASES

Magmas contain varied amounts of dissolved gases held in the molten rock by confining pressure, just as carbon dioxide is held in soft drinks. As with soft drinks, as soon as the pressure is reduced, the gases begin to escape. Obtaining samples from an erupting volcano is very difficult and dangerous, so geologists usually only estimate the amount of gas originally contained within the magma.

The gaseous portion of most magmas is believed to compose from 1 to 5 percent of the total weight, and most of this is in the form of water vapor. Although the percentage may be small, the actual quantity of emitted gas can exceed thousands of tons per day. The composition of the gases is also of interest to scientists, since much evidence points to these as the source of the oceans and the earth's atmosphere. Analysis of samples taken during Hawaiian eruptions indicated that the gases emitted there consist of about 70 percent water vapor, 15 percent carbon dioxide, 5 percent each of nitrogen and sulfur compounds, and lesser amounts of chlorine, hydrogen, and argon. Sulfur compounds are easily recognized by their pungent odor and because they readily form sulfuric acid, which when inhaled produces a burning sensation.

PYROCLASTIC MATERIALS

When basaltic lava is extruded, the dissolved gases escape quite freely and continually. As stated earlier, these gases often carry incandescent blobs of lava to great heights, thereby producing spectacular lava fountains (see Figure 7.8). Some ejected material may land near the vent and produce a cone structure, while smaller particles will be carried great distances by the wind. The gases in highly viscous magmas, on the other hand, are less able to escape and may build up an internal pressure capable of producing a violent eruption. Upon release, these superheated gases expand a thousandfold as they blow pulverized rock and lava from the vent. The particles produced by these processes are called **pyroclastics.** These ejected lava fragments range in size from very fine dust and sand-sized volcanic ash, to large volcanic bombs and blocks (Figure 7.4, page 195).

The fine *ash* particles are produced when the extruded lava contains so many gas bubbles that it resembles the froth flowing from a newly opened bottle of champagne. As the hot gases expand explosively, the lava is disseminated into very fine fragments. When the hot ash falls, the glassy shards often fuse to form *welded tuff.* Sheets of this material, as well as ash deposits that consolidate later, cover vast portions of the western United States. In some instances the froth-like lava is ejected in larger pieces called *pumice.* This material has so many voids that it is often light enough to float in water.

Pyroclastics the size of walnuts, called *lapilli* ("little stones"), and pea-sized particles called *cinders* are also very common. Cinders contain numerous voids and form when ejected lava blobs are pulverized by escaping gases. Particles larger than lapilli are called *blocks* when they are made of hardened lava and *bombs* when they are ejected as incandescent lava. Since volcanic bombs are semimolten upon ejection, they often take on a streamlined shape as shown in Figure 7.5.

VOLCANOES AND VOLCANIC ERUPTIONS

Successive eruptions from a central vent result in a mountainous accumulation of material known as a **volcano.** Located at the summit of many volcanoes is a steep-walled depression called a **crater,** which is connected to a magma chamber via a pipelike conduit, or vent. Some volcanoes have unusually large summit depressions that exceed one kilometer in diameter and are known as **calderas.** When fluid lava leaves a conduit, it is often stored in the crater or caldera until it overflows. On the other hand, lava that is very viscous forms a plug in the pipe which rises slowly or is blown out, often enlarging the crater. However, lava does not always issue from a central crater. Sometimes it is easier for the magma or escaping gases to push through fissures on the volcano's flanks. Continued activity from a flank eruption may build a so-called *parasitic cone.* Mount Etna

FIGURE 7.5
Volcanic bombs. Ejected lava fragments take on a streamlined shape as they sail through the air.

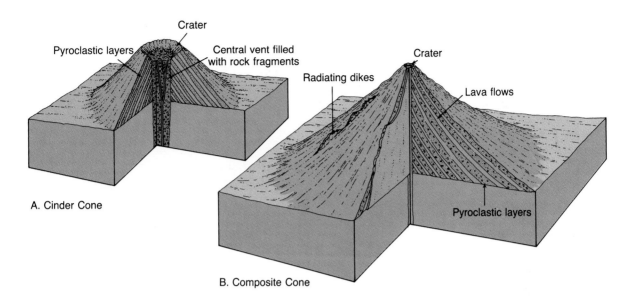

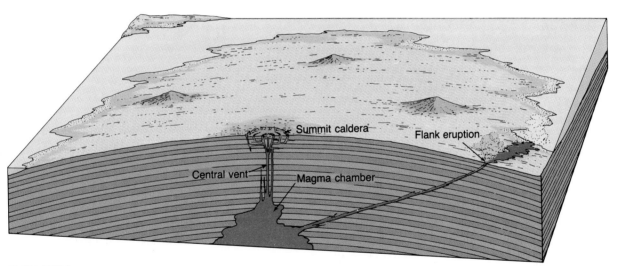

C. Shield Volcano

FIGURE 7.6
The three basic types of volcanic structures. **A.** Cinder cone. **B.** Composite cone.
C. Shield volcano.

in Italy, for example, has more than 200 secondary vents. Some of these secondary vents extrude only gases and are appropriately called *fumaroles*.

The eruptive history of each volcano is unique; consequently, all volcanoes are somewhat different in form and size. Nevertheless, volcanologists have recognized that volcanoes exhibiting somewhat similar eruptive styles can be grouped. Based on their "typical" eruptive patterns and characteristic form, three groups of volcanoes are generally recognized: shield volcanoes, cinder cones, and composite cones (Figure 7.6).

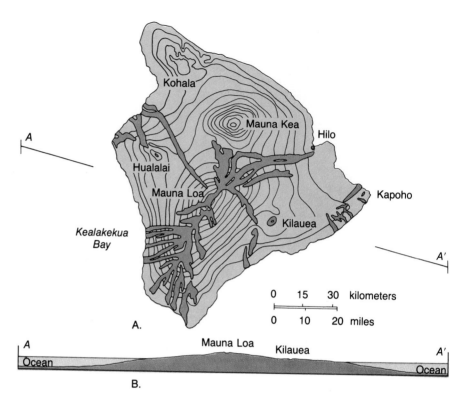

FIGURE 7.7
Map of the island of Hawaii. **A.** Five volcanoes collectively make up the island. Contour interval is 300 meters (1000 feet). **B.** Illustration of the very gentle slope which is characteristic of a shield volcano (no vertical exaggeration). (After H. T. Stearns and G. A. MacDonald, U.S. Geological Survey)

SHIELD VOLCANOES

When fluid lava is extruded, the volcano takes the shape of a broad, slightly domed structure called a **shield volcano** (Figure 7.6C). Shield volcanoes are built primarily of basaltic lava flows and contain only a small percentage of pyroclastic material. Typically they have a slope of a few degrees at their flanks and generally do not exceed 15 degrees near their summit, as exemplified by the volcanoes of the Hawaiian Islands. Mauna Loa, probably the largest volcano on earth, is one of the five shield volcanoes that together make up the island of Hawaii (Figure 7.7). Its base rests on the ocean floor 5000 meters below sea level, while its summit reaches a height of 4170 meters (13,680 feet) above the water. Nearly one million years and numerous eruptive cycles were required to build this truly gigantic pile of volcanic rock. Many other volcanic structures, including Midway Island and the Galapagos Islands, have been built in a similar manner from the ocean's depths.

Perhaps the most active and intensively studied shield volcano is Kilauea, located on the island of Hawaii on the southeastern flank of the larger volcano Mauna Loa. Kilauea has erupted more than 50 times in recorded history and is still active today. A testimony to the relatively quiescent nature of these eruptions is the fact that the U.S. Geological Survey's Hawaiian Volcano Observatory is situated on the very rim of the summit caldera. Several months before each eruptive phase, the summit inflates as magma rises from its source 60 to 100 kilometers below the surface. This molten rock gradually works its way upward and accumulates in smaller reservoirs 3 to 5 kilometers below the summit. For up to 24 hours in advance of each eruption, swarms of earth tremors warn of the impending activity. The 1983 eruption occurred along a 6.5-kilometer (4-mile) fissure located east of the summit caldera (Figure 7.8). Here fountains of fluid lava with heights approaching 100 meters (330 feet) fed a lava flow that extended for 6 kilometers and eventually inundated a few dwellings in the sparsely populated settlement of Royal Gardens. This eruptive phase, like most others, was accompanied by a gradual subsidence of the summit area, nearly equal in volume to the extruded lava.

CINDER CONES

As the name suggests, **cinder cones** are built from ejected lava fragments. Because unconsolidated pyroclastic material maintains a high angle of repose (between 30 and 40 degrees), volcanoes of this type have very steep slopes. Cinder cones are rather small, usually less than 300 meters (1000 feet) high, and often form as parasitic cones on or near larger volcanoes. In addition, they frequently occur in groups, where the cones apparently represent the last phase of activity in a region of older basaltic flows. This may result because the contributing magma has cooled and become more viscous.

One of the very few volcanoes whose formation has been observed by geologists from beginning to end is a cinder cone called Parícutin. This volcano's history serves to illustrate the formation and structure of a relatively large cinder cone.

In 1943, about 200 miles west of Mexico City, the volcano Parícutin was born (see Figure 7.4).

FIGURE 7.8
Lava fountain along the east rift zone of Kilauea, 1983. (Photo by J. D. Griggs, U.S. Geological Survey)

The eruption site was a cornfield owned by Dionisio Pulido, who with his wife, Paula, witnessed the event as they were preparing the field for planting. For two weeks prior to the first eruption, numerous earth tremors caused apprehension in the village of Parícutin about 3.5 kilometers away. Then around 4:00 P.M. on February 20th, smoke with a sulfurous odor began billowing from a small hole that had been in the cornfield for as long as Dionisio could remember. During the night hot, glowing rock fragments thrown into the air from the hole produced a spectacular fireworks display. By the next day the cone had grown to a height of 40 meters and by the fifth day it was over 100 meters high. At this time explosive eruptions were throwing hot fragments 1000 meters (3300 feet) above the crater rim. The larger fragments fell near the crater, some remaining incandescent as they rolled down the slope. These fragments built an aesthetically pleasing cone, while finer ash fell over a much larger area, burning and eventually covering the village of Parícutin. Within two years the cone had grown to 400 meters (1310 feet) high and would rise only a few tens of meters more.

The first lava flow came from a fissure that had opened just north of the cone; but after a few months of activity, flows began to emerge from the base of the cone itself. In June of 1944, a clinkery flow 10 meters thick moved over the village of San Juan Parangaricutiro, leaving only the church steeple exposed (Figure 7.9). After nine years the activity ceased almost as quickly as it began. Now Parícutin is just another one of the numerous cinder cones dotting the landscape in this region of Mexico. Like the others, it will probably not erupt again.

COMPOSITE CONES

The earth's most picturesque volcanoes are **composite cones,** or **stratovolcanoes.** Just as shield volcanoes owe their shape to the fluid nature of the extruded lavas, so too do composite cones reflect the nature of the erupted material. Composite cones are produced when relatively viscous lavas of andesitic composition are extruded. A composite cone may extrude viscous lava for long periods. Then suddenly the eruptive style changes and the volcano violently ejects pyroclastic material. Most of the ejected pyroclastic material falls near the summit, building a steep-sided mound of cinders. In time this debris will be covered by lava. Occasionally both activities occur simultaneously, and the resulting structure consists of alternating layers of lava and pyroclastics. Two of the most perfect cones, Mount Mayon in the Philippines and Fujiyama in Japan, exhibit the classic form of the composite cone with its steep summit area and rather gently sloping flanks.

Although composite cones are the most picturesque, they also represent the most violent type of volcanic activity. Their eruption can be unexpected and devastating as was the 79 A.D. eruption of the Italian volcano we now call Vesuvius. Prior to this eruption, Vesuvius was dormant for centuries. Although minor earthquakes probably warned of the events to follow, Vesuvius was covered with a heavy coat of vegetation and hardly looked threatening. On August 24th, however, the tranquillity ended, and in the next three days the city of Pompeii (near Naples) and more than 2000 of its 20,000 residents were buried. They remained so for nearly seventeen centuries, until the city was rediscovered and excavated.

FIGURE 7.9
The village of San Juan Parangaricutiro engulfed by lava from Parícutin, shown in the background. Only the church towers remain. (Photo by Tad Nichols)

The destruction of Pompeii was truly catastrophic, yet eruptions of a more devastating nature occur when hot gases infused with incandescent ash are ejected, producing a fiery cloud called **nuée ardente.** Also referred to as *glowing avalanches,* these flows, which are black in daylight and glow red at night, move down steep volcanic slopes at speeds exceeding 150 kilometers per hour (Figure 7.10). These glowing avalanches are very dense, but their weight is supported by the expanding gases emitted from the hot lava particles. In this way, the material, which can be composed of rather large lava fragments, flows downslope in an almost frictionless environment cushioned by the expanding gases.

In 1902, a nuée ardente from Mount Pelée, a small volcano on the Caribbean island of Martinique, destroyed the port town of St. Pierre. The destruction was instantaneous and so devastating that almost all of St. Pierre's 28,000 inhabitants were killed. Reportedly only a prisoner protected in a dungeon, a shoemaker, and a few people on ships in the harbor were spared (Figure 7.11).

FIGURE 7.10
Nuée ardente races down the slope of Mount St. Helens on August 7, 1980 at speeds in excess of 100 kilometers (60 miles) per hour. (Photo by Peter W. Lipman, U.S. Geological Survey)

FIGURE 7.11
St. Pierre as it appeared shortly after the eruption of Mount Pelée, 1902. (Reproduced from the collection of the Library of Congress)

FIGURE 7.12
A house damaged by debris from lahars that flowed along the Toutle River, west-northwest of Mount St. Helens. The end section of the house was torn free and lodged against trees. (Photo by D. R. Crandell, U.S. Geological Survey)

When we compare the destruction of St. Pierre with that of Pompeii, several differences are noticed. Pompeii was totally buried by an event lasting three days, whereas St. Pierre was destroyed in a brief instant and its remains were only mantled by a thin layer of volcanic debris. Also, the structures of Pompeii remained intact except for roofs that collapsed under the weight of the ash. In St. Pierre masonry walls nearly one meter thick were knocked over like dominoes; large trees were uprooted and cannons were torn from their mounts.

In addition to their violent eruptions, volcanoes are potential hazards in other ways. In particular, large composite cones often generate a type of mudflow known as a *lahar.* These destructive events occur when volcanic ash and debris become saturated with water and flow down steep volcanic slopes, generally following stream valleys. Some lahars are produced when rainfall saturates volcanic deposits, whereas others are triggered as large volumes of ice and snow melt during an eruption.

When Mount St. Helens erupted in May, 1980, several lahars formed. The flows and accompanying floods raced down the valleys of the north and south forks of the Toutle River at speeds in excess of 30 kilometers per hour. Water levels reached four meters above flood stage, and nearly all the homes and bridges along the river were destroyed or severely damaged (Figure 7.12). Fortunately, the affected area was not densely populated. This was not the case in November, 1985, when Nevado del Ruiz, a volcano in the Andes, erupted and generated a lahar that killed nearly 20,000 people (see the section entitled "Mudflow" in Chapter 2).

VOLCANIC PIPES AND NECKS

Most volcanoes are fed by conduits called **pipes** or **vents,** which are believed to be connected to a magma source emplaced near the surface. By contrast, some pipes are thought to extend in a tubelike fashion directly into the asthenosphere, a distance of 200 kilometers (135 miles). Consequently, the materials in these vents are considered to be samples of the asthenosphere that have undergone very little alteration during their ascent. Geologists consider pipes, therefore, to be "windows" into the the earth since they allow us to view rocks found at greater depths.

Occasionally a volcanic pipe will reach the surface, but the eruptive phase will cease before any lava is extruded. The upper portions of these pipes often contain a jumble of lava fragments and fragments that were torn from the walls of the vent by violently escaping gases. The best known of these structures are the diamond-bearing pipes of South Africa. Here the rocks filling the pipes are thought to have originated at depths of about 200 kilometers, where pressures are great enough to generate diamonds and other high-pressure minerals. In these instances the diamonds crystallized at depth and were carried upward by the still-molten portion of the magma.

Volcanoes, like all land areas, are continually being lowered by the forces of weathering and erosion. Cinder cones are easily eroded, because they are composed of unconsolidated materials. However, all volcanic structures will eventually be worn away. As erosion progresses, the rock occupying the vent is often more resistant and may remain standing above the terrain long after most of the cone has vanished. Shiprock, New Mexico, is thought to be such a feature, called a **volcanic neck** (see Figure 7.17). This structure, higher than many skyscrapers, is but one of many that protrude conspicuously from the red desert landscapes of the southwestern United States.

FORMATION OF CALDERAS

Earlier it was pointed out that some volcanoes have unusually large craters known as calderas. Most calderas are thought to form when the summit of a volcano collapses into the partially emptied magma chamber below (Figure 7.13). Crater Lake in Oregon, 8–10 kilometers wide and 1300 meters deep, is located in such a depression (Figure 7.14). The creation of Crater Lake began about 7000 years ago when the volcano, later to be named Mount Mazama, put forth a violent ash eruption much like that of Vesuvius. However, this ancient eruption was on a much larger scale, extruding an estimated 50–70 cubic kilometers of volcanic material. With the loss of support, 1500 meters of this once prominent 3600-meter cone

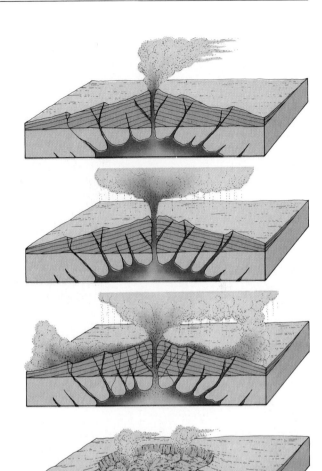

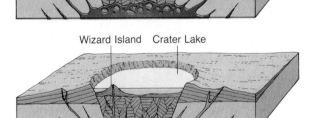

Wizard Island   Crater Lake

FIGURE 7.13
Sequence of events that formed Crater Lake, Oregon. About 7000 years ago, the summit of former Mount Mazama collapsed following a violent eruption which partly emptied the magma chamber. Subsequent eruptions produced the cinder cone called Wizard Island. Rainfall and groundwater contributed to form the lake. (After H. Williams. *The Ancient Volcanoes of Oregon,* p. 47. Courtesy of the University of Oregon)

FIGURE 7.14
Crater Lake occupies a caldera about 10 kilometers (6 miles) in diameter. (Courtesy of the National Park Service)

collapsed. After the collapse, rainwater filled the caldera. Later activity built a small cinder cone called Wizard Island, which today provides a mute reminder of past activity.

Calderas of varying sizes are known to exist, the largest more than 20 kilometers (12 miles) across. Some, such as the calderas of Mauna Loa and Kilauea, clearly result from subsidence that occurred as supporting magma was diverted to a flank eruption. Others were formed by an explosive event that blasted the upper portion of the volcano away, as was the case at Mount St. Helens in 1980.

FISSURE ERUPTIONS

Although volcanic eruptions from a central vent are the most familiar, by far the largest amounts of volcanic material are extruded from cracks in the crust called **fissures.** Rather than building a cone, these long, narrow cracks distribute volcanic materials over a wide area. An extensive region in the northwestern United States known as the Columbia Plateau was formed in this manner. Here numerous fissure eruptions extruded very fluid basaltic lava. Successive flows, some 50 meters thick, buried the old landscape as they built a lava plain, which in some places is nearly a mile thick (Figure 7.15). The fluidity is evident, since some lava remained molten long enough to flow 150 kilometers from its source. The term **flood basalts** appropriately describes these waterlike flows.

When silica-rich magma is extruded from fissures, **pyroclastic flows** consisting largely of ash and pumice fragments usually result. When these pyroclastic materials are ejected, they move away from the vent at high speeds and may blanket extensive areas before coming to rest. Once deposited, the pyroclastic materials closely resemble lava flows.

Extensive pyroclastic flow deposits are found in many parts of the world and are most often associated with large calderas. Perhaps the best-known region of pyroclastic flows is the Yellowstone Plateau in northwestern Wyoming. Here a large magma body, rich in silica, still exists a few kilo-

FIGURE 7.15
Basalt flows of the Columbia Plateau. (Photo by E. T. Jones, U.S. Geological Survey)

meters below the surface. Several times over the past two million years, fracturing of the rocks overlying the magma chamber has resulted in huge eruptions accompanied by the formation of calderas. In the northwestern portion of Yellowstone National Park, 27 fossil forests have been discovered, one resting upon another. During periods of inactivity, a forest developed upon the newly-formed volcanic surface, only to be covered by ash from the next eruptive phase. Fortunately, no eruption of this type has occurred in modern times.

VOLCANOES AND CLIMATE

The idea that explosive volcanic eruptions change the earth's climate was first proposed many years ago and is still regarded as an explanation for some aspects of climatic variability. Explosive eruptions emit huge quantities of gases and fine-grained debris into the atmosphere. The greatest eruptions are sufficiently powerful to inject material high into the stratosphere, where it spreads around the globe and remains for many months or even years. The basic premise is that this suspended volcanic material (most importantly, droplets of sulfuric acid) will filter out a portion of the incoming solar radiation which, in turn, will lower air temperatures.

Perhaps the most notable cool period linked to a volcanic event is the "year without a summer" that followed the 1815 eruption of Mount Tambora in Indonesia. In many Northern Hemisphere locations, including New England, the abnormally cold spring and summer of 1816 were believed to be caused by the cloud of volcanic debris ejected from Tambora.

When Mount St. Helens erupted on May 18, 1980, there was almost immediate speculation about the possible effects of this event on our climate. Can an eruption such as this change our climate? Although spectacular, a single explosive volcanic eruption of the magnitude of Mount St. Helens occurs somewhere in the world every 2 to 3 years. Studies of these events indicate that a very slight cooling of the lower atmosphere does occur. However, it is believed that the cooling is so slight, less than one-tenth of one degree Celsius, as to be inconsequential.

On April 4, 1982, El Chichón, a little-known volcano on Mexico's Yucatan peninsula, erupted. The cloud of debris and sulfur gases lofted into the atmosphere was huge, probably 20 times greater than the cloud from Mount St. Helens. Following the blast, scientists predicted a gradual lowering of temperatures in the Northern Hemisphere, perhaps as great as 0.3–0.5°C. Such a change is large enough to be distinguishable from normal temperature fluctuations but is probably too small to affect our life-styles. Nevertheless, many scientists agree that such a hemispheric cooling could alter the general pattern of atmospheric circulation for a limited period. Such a change, in turn, could have an effect on the weather in some regions. However, the prediction of specific regional effects still presents a considerable challenge to atmospheric scientists.

INTRUSIVE IGNEOUS ACTIVITY

Volcanic eruptions can be among the most violent and spectacular events in nature and are therefore worthy of detailed study. Yet most magma is believed to be emplaced at depth. An understanding of intrusive igneous activity is therefore as important to geologists as the study of volcanic events. Figure 7.16 shows several types of intrusive igneous bodies that form when magma crystallizes within the earth's crust. Notice that some of these structures have a tabular shape, while others are quite massive. Also observe that some of these bodies cut across existing structures, such as layers of sedimentary rocks, while others form when magma is injected between sedimentary layers. Due to these differences, intrusive igneous bodies are generally classified according to their shape as either *tabular* or *massive,* and by their orientation with respect to the country (host) rock. Intrusive igneous bodies are said to be *discordant* if they cut across existing sedimentary beds and *concordant* if they form parallel to the existing sedimentary beds.

Intrusive igneous bodies occur in a great variety of sizes and shapes. **Dikes** are discordant masses produced when magma is injected into fractures that cut across rock layers. Once crystallized, these tabular structures have thicknesses

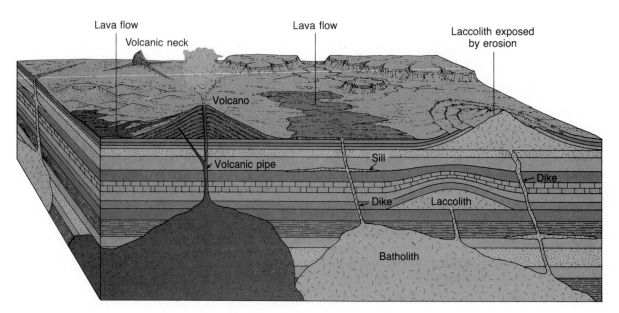

FIGURE 7.16
Block diagram illustrating intrusive and extrusive igneous structures.

FIGURE 7.17
Shiprock, New Mexico, is the remnant neck of a volcano which erosion has almost completely removed. The tabular shaped structure in the foreground is a dike. (Photo by Stephen Trimble)

ranging from less than one centimeter to more than one kilometer. The largest have lengths of one hundred kilometers or more. Dikes are often oriented vertically and represent pathways followed by molten rock which fed ancient lava flows. Frequently dikes are more resistant to weathering than the surrounding rock. When exposed, these dikes have the appearance of a wall as shown in Figure 7.17.

Sills are tabular bodies formed when magma is injected along sedimentary bedding surfaces (Figure 7.18). Horizontal sills are the most common although all orientations, even vertical, are known to exist where the strata have been tilted. Because of their relatively uniform thickness and large extent, sills are believed to form from very fluid magma. As we may expect, sills are most often composed of basaltic magma, which is typically quite fluid. The emplacement of a sill requires that the overlying sedimentary rock be lifted to a height equal to the thickness of the sill. Conse-

quently, sills form only at rather shallow depths where the pressure exerted by the weight of overlying strata is relatively low.

FIGURE 7.18
Taylor Glacier region, Antarctica. The dark horizontal band is a sill intruded into flat-lying sandstone. (Photo by W. B. Hamilton, U.S. Geological Survey)

FIGURE 7.19
Yosemite National Park, located in the Sierra Nevada batholith. This batholith extends approximately 400 kilometers and is thought to have formed over a 130-million-year period. (Photograph used by permission of Dennis Tasa)

Laccoliths are similar to sills because they form when magma is intruded between sedimentary layers in a near-surface environment. However, unlike sills, the magma that generates laccoliths is believed to be quite viscous. This thick, nonfluid magma collects as a lens-shaped mass that arches the overlying strata upward. Consequently, a laccolith can be detected because of the dome it creates at the surface even before the overlying rock is stripped away by erosional forces.

By far the largest intrusive igneous bodies are **batholiths.** The Idaho batholith, for example, encompasses an area of more than 40,000 square kilometers. These massive, discordant bodies are usually composed of rock types having mineral compositions near the granitic end of the igneous rock spectrum. Small batholiths can be relatively simple structures composed almost entirely of one rock type. However, studies of large batholiths have shown that they resulted from several distinct events that occurred over millions of years (Figure 7.19). Batholiths frequently compose the cores of mountain systems. Here uplift and erosion have removed the surrounding rock to expose the resistant igneous body. Some of the highest mountain peaks, such as Mount Whitney in the Sierra Nevada, are carved from such a granitic mass. Large expanses of granitic rock are also exposed in the stable interiors of the continents, such as the Canadian Shield of North America. These relatively flat outcrops are believed to be the remnants of ancient mountains that erosion has long since leveled.

IGNEOUS ACTIVITY AND PLATE TECTONICS

The origin of magma has been a controversial topic in geology almost from the very beginning of the science. How do magmas of different compositions form? Why do volcanoes located in the deep-ocean basins primarily extrude basaltic lava, whereas those adjacent to oceanic trenches extrude mainly andesitic lava? Why does an area of igneous activity, commonly called the Ring of Fire, surround the Pacific Ocean? New insights gained from the theory of plate tectonics are providing answers to these questions. We will first examine the origin of magma and then look at the global distribution of volcanic activity as viewed from the model provided by plate tectonics.

ORIGIN OF MAGMA

We know that magma can be produced when rock is heated to its melting point. In a surface environment, rocks of granitic composition begin to melt at temperatures near 750°C (1400°F), whereas basaltic rocks must reach temperatures above 1000°C (1850°F) before melting will begin. One important difference exists between the melting of a substance that consists of a single compound, such as ice, and the melting of igneous rocks, which are mixtures of several minerals. Whereas ice melts at 0°C, most igneous rocks melt over a temperature range of a few hundred degrees. As a rock is heated, the first liquid to form will contain a higher percentage of the low-melting-point minerals than the original rock. If melting continues, the composition of the melt will steadily approach the overall composition of the rock from which it is derived. Most often, however, melting is not complete. This process, known as **partial melting,** produces most, if not all, magma.

A significant result of partial melting is the production of a melt with a higher silica content than the parent rock. Recall that basaltic rocks have a relatively low silica content and that granitic rocks have a much higher silica content. Consequently, magmas generated by partial melting are nearer the granitic end of the compositional spectrum than the parent material from which they formed. As we shall see, this idea will help us to understand the global distribution of the various types of volcanic activity.

What is the heat source to melt rock? One source is the heat liberated during the decay of radioactive elements that are thought to be concentrated in the upper mantle and crust. Workers in underground mines have long recognized that temperatures increase with depth.

If temperature were the sole factor to determine whether or not a rock melts, the earth would be a molten ball covered with only a thin, solid outer shell. However, pressure also increases with depth. Since rock expands when heated, extra heat is needed to melt buried rocks in order to overcome the effect of confining pressure. In general, an increase in the confining pressure causes an increase in the rock's melting point.

In nature, deeply buried rocks melt for one of two reasons. First, rocks melt when they are heated to their melting points. Second, without increasing temperature, a reduction in the confining pressure can lower the melting temperature sufficiently to trigger melting. Both processes are thought to play significant roles in magma formation.

DISTRIBUTION OF IGNEOUS ACTIVITY

Most of the more than 600 active volcanoes that have been identified are located in the vicinity of convergent plate margins. Further, extensive volcanic activity occurs out of view along spreading centers of the oceanic ridge system. In this section we will examine three zones of volcanic activity and relate them to global tectonic activity. These active areas are found along the oceanic ridges, adjacent to ocean trenches, and within the plates themselves (Figure 7.20).

Spreading Center Volcanism The greatest volume of volcanic rock is produced along the oceanic ridge system, where sea-floor spreading is active (Figure 7.20). As the rigid lithosphere pulls apart, the pressure on the underlying rocks is lessened. This reduced pressure, in turn, lowers the melting point of the mantle rocks. Partial melting of these rocks (primarily peridotite) generates large quantities of basaltic magma that moves upward to fill the newly formed cracks.

Some of the molten basalt reaches the ocean floor, where it produces extensive lava flows or occasionally grows into a volcanic pile. Sometimes this activity produces a volcanic cone that rises above sea level as the island of Surtsey did in 1963 (Figure 7.21). Numerous submerged volcanic cones also dot the flanks of the ridge system and the adjacent deep-ocean floor. Many of these

formed along the ridge crests and were moved away as new oceanic crust was created by the process of sea-floor spreading.

Subduction Zone Volcanism Rocks with andesitic to granitic composition are confined to the continents and to volcanic island chains, such as the Aleutians, which lie along oceanic margins. Very little granitic rock is found as part of the volcanoes in the deep-ocean basins. Further, most active volcanoes that extrude andesitic magma are found on continental areas or island arcs located adjacent to deep-ocean trenches. Recall that ocean trenches are sites where slabs of oceanic crust are bent and move downward into the upper mantle.

When cold oceanic lithosphere reaches depths of about 125 kilometers (80 miles), melting is be-

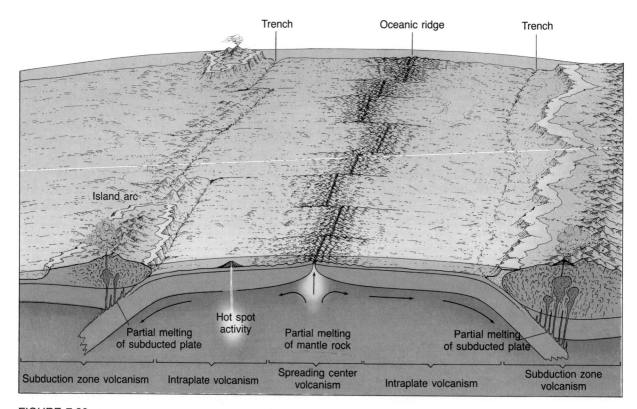

FIGURE 7.20
Three zones of volcanism. Two of these zones are plate boundaries, and the third includes areas within the plates themselves.

FIGURE 7.21
Surtsey emerged from the
ocean just south of Iceland
in 1963. (Courtesy of Sólar-
filma)

lieved to take place. The partial melting of these wet, sediment-laden basalts yields a magma of andesitic composition. After a sufficient quantity has melted, this magma buoys upward, because it is less dense than the surrounding rock. The volcanoes of the Andes Mountains, from which andesite obtains its name, are examples of this mechanism at work.

The *Ring of Fire* is associated with subduction and melting of the Pacific plate. The volcanoes in this very active zone primarily extrude magma having an intermediate silica content. The volcanoes of the Cascade Range in the northwestern United States, including Mounts St. Helens, Rainier, and Shasta, are all of this type.

Intraplate Volcanism The processes that actually trigger volcanic activity within a rigid plate are difficult to establish. Activity such as in the Yellowstone region and other nearby areas produced rhyolitic lava, pumice, and ash flows, while extensive basaltic flows cover vast portions of our Northwest. Yet these rocks of greatly varying compositions actually overlie one another in several locations.

Since basaltic extrusions occur on the continents as well as within the ocean basins, the partial melting of upper mantle rocks is the most probable source for this activity. One proposal is that a small percentage of the rocks of the asthenosphere exists in the molten state. From these molten pockets, called *hot spots,* plumes of magma are thought to migrate upward where they often penetrate to the surface. Hot spots are believed to be located beneath Hawaii and Iceland, and may have formerly existed beneath the Columbia Plateau.

Generally lavas and ash of granitic composition are extruded from vents located landward of the continental margins. This suggests that remelting of the continental crust may be one mechanism for the formation of these silica-rich magmas. But what mechanism causes large quantities of continental material to melt? One proposal is that a thick segment of continental crust occasionally becomes situated over a plume of rising magma, that is, a hot spot. Rather than producing vast outpourings of basaltic lava as occurs at oceanic sites such as Hawaii, magma from the rising plume is emplaced at depth. Here the incorporation and melting of surrounding country rock results in the formation of a secondary, silica-rich magma, which slowly migrates upward. Continued hot spot activity supplies heat to the rising mass, thereby aiding its ascent. The activity in the Yellowstone region may have resulted from just this type of activity.

Although the theory of plate tectonics has answered many of the questions which have plagued volcanologists for decades, many new questions have arisen; for example, Why does sea-floor spreading occur in some areas and not others? How do hot spots originate? These are just two of the many unanswered questions.

REVIEW QUESTIONS

1 Name four volcanoes in the western United States that geologists believe will erupt again.

2 What is the difference between magma and lava?

3 What three factors determine the nature of a volcanic eruption? What role does each play?

4 Why is a volcano fed by highly viscous magma likely to be a greater threat than a volcano supplied with very fluid magma?

5 Describe pahoehoe and aa lava.

6 List the main gases released during a volcanic eruption.

7 Analysis of samples taken during Hawaiian eruptions on the island of Hawaii indicate that _____ was the most abundant gas released.

8 Describe each type of pyroclastic material.

9 Compare and contrast the main types of volcanoes (size, shape, eruptive style, and so forth).

10 Name one example of each of the three types of volcanoes.

11 Compare the formation of Hawaii with that of Parícutin.

12 How is a caldera different from a crater?

13 Describe the formation of Crater Lake.

14 What is Shiprock, New Mexico, and how did it form?

15 Describe each of the four intrusive features discussed in the text.

16 Why might a laccolith be detected at the earth's surface before being exposed by erosion?

17 What is the largest of all intrusive igneous bodies? Is it tabular or massive? Concordant or discordant (see Figure 7.16)?

18 Explain how most magma is thought to originate.

19 Volcanic islands in the deep ocean are composed primarily of what igneous rock type?

20 Explain why andesitic and granitic rocks are confined to the continents and oceanic margins but are absent from deep-ocean basins.

KEY TERMS

viscosity

pahoehoe flow

aa flow

pyroclastics

volcano

crater

caldera

shield volcano

cinder cone

composite cone
(stratovolcano)

nuée ardente

pipe

vent

volcanic neck

fissure

flood basalt

pyroclastic flows

dike

sill

laccolith

batholith

partial melting

8

MOUNTAIN BUILDING

Mountains provide some of the most spectacular scenery on our planet. This splendor has been captured by poets, painters, and songwriters alike. Geologists believe that at some time all continental regions were mountainous masses and have concluded that the continents grow by the addition of mountains to their flanks. Consequently, by unraveling the secrets of mountain formation, geologists will have taken a major step in understanding the evolution of the earth. If continents do indeed grow by adding mountains to their flanks, how do geologists explain the existence of mountains (the Urals, for example) that are located in the interior of a landmass? In order to answer this and other related questions, this chapter attempts to piece together the sequence of events which is believed to generate these lofty structures.

Mount Timpanogos in the Wasatch Range, Utah.
(Photo by Stephen Trimble)

Mountains often are spectacular features which rise several hundred meters or more above the surrounding terrain. Some occur as single isolated masses; the volcanic cone Kilimanjaro, for example, stands almost 6000 meters (20,000 feet) above sea level overlooking the grasslands of East Africa. Others make up a portion of an extensive mountainous chain, such as the American Cordillera, which reaches from the tip of South America northward through Alaska. Chains such as the Himalayas are youthful, gigantic mountains that are still rising, while others are very old and nearly worn down, as exemplified by the Appalachian Mountains in the eastern United States.

The name for the processes which collectively produce a mountain system is **orogenesis,** from the Greek *oros* ("mountain") and *genesis* ("to come into being"). Mountain systems show evidence of enormous forces which have folded, faulted, and generally deformed large sections of the earth's crust. Although the processes of folding and faulting have contributed to the majestic appearance of mountains, much of the credit for their beauty must be given to the work of running water and glacial ice, which sculpture these uplifted masses in an unending effort to lower them to sea level.

The first encompassing explanation of orogenesis came about two decades ago as part of the plate tectonics theory. As noted before, the idea of plates colliding has opened many new and exciting avenues to geologists. Before examining mountain building according to the plate tectonics model however, it will be advantageous first to view the processes of crustal uplifting and rock deformation.

CRUSTAL UPLIFT

The fossilized shells of marine invertebrates are often found in mountain regions, an indication that the sedimentary rock composing the mountain was once below sea level. This is rather convincing evidence that some dramatic changes occurred between the time these animals died and when their fossilized remains were discovered. Evidence for crustal uplift such as this is common in the geologic record and is even present in the historical record. For example, Figure 8.1 depicts three columns remaining from a Roman temple. The columns have clam borings to a height of about 6 meters (20 feet), indicating that the land upon which the temple was built submerged and was later partially uplifted. These elevated clam borings might also be explained by a recent change in sea level; however, a similar change in sea level is not recorded at any other location for that same time period. Further evidence for crustal uplift can be found along the coastline of the western United States. When a coastal area remains undisturbed for an extended period, wave action cuts a gently sloping platform. In

Clam borings

FIGURE 8.1
Remaining columns of the ancient Roman temple of Serapis, Pozzuoli, Italy, in 1836. Clam borings 6 meters (20 feet) above sea level indicate former submergence. (After Charles Lyell, *Principles of Geology,* 10th ed. 1867)

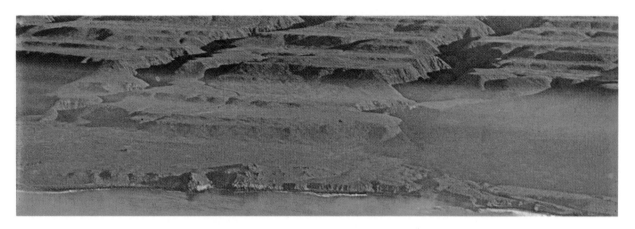

FIGURE 8.2
Wave-cut terraces on San Clemente Island south of Los Angeles, California.
Once at sea level, the highest terraces are now about 400 meters (1320 feet)
above it. (Photo by John S. Shelton)

parts of California, ancient wave-cut platforms can now be found as terraces, hundreds of meters above sea level (Figure 8.2). Each terrace represents a period when that area was at sea level. Unfortunately, the reasons for uplift are not always as easy to determine as the evidence for the movements.

We know that the force of gravity must play an important role in determining the elevation of the land. In particular, the less dense lithosphere is believed to float on top of the denser and more easily deformed rocks of the asthenosphere. The concept of a floating lithosphere in gravitational balance is called **isostasy.** Perhaps the easiest way to envision isostatic balance is to compare the lithosphere to floating logs. Imagine two logs, one much thicker than the other, floating in water. The larger log will float higher in the water than the smaller log. In the same manner, mountainous regions are believed to represent unusually thick sections of the earth's crust, whereas areas of low elevation do not have such crustal thickness. Mountains, like thick logs, not only stand high above the surface, but also extend farther into the supporting material below (Figure 8.3). This fact has been confirmed by seismic and gravitational data.

Carrying this idea one step further, the lithosphere beneath the oceans must be thinner than that of the continents because its elevation is lower. Although this is true, oceanic rocks also have a greater density than continental rocks, another factor contributing to their lower position.

If the concept of isostasy is correct, we should expect that when weight is added to the crust, the crust will respond by subsiding, and that when weight is removed there will be uplifting. (Visualize what happens to a ship as cargo is being loaded and unloaded.) Evidence for this type of movement exists, strongly supporting the theory of *isostatic adjustment.* For example, when Hoover Dam was built in the 1930s, the impounded waters of Lake Mead, and to a lesser degree the millions of tons of sediment collected by it, caused regional subsidence and a marked increase in seismic activity. Another classic example is provided by nature. When continental glaciers occupied portions of North America during the Pleistocene epoch, the added weight of the 3-kilometer-thick masses of ice caused downwarping of the earth's crust. In the 8000 to 10,000 years since the last ice sheets melted, uplifting of as much as 330 meters has occurred in the Hudson Bay region, where the thickest ice had accumulated.

As the foregoing examples illustrate, isostatic adjustment can account for considerable crustal movement. Thus, we can now understand why, as erosion lowers the summits of mountains, the crust will rise in response to the reduced load. The processes of uplifting and erosion will continue until the deeply buried portions of the mountains have reached the same height as the surrounding crust (Figure 8.3). In addition, as the mountains wear down, the weight of the eroded sediment deposited on the adjacent continental margin will cause it to subside.

To summarize, mountains are unusually thick portions of the earth's crust that remain elevated above their surroundings because of isostasy. As erosion removes material, isostatic adjustment gradually raises the mountains in response. Eventually the deepest portions of the mountains are brought up to the shallower depths of the surrounding crust. The question still to be answered is: how do these thick sections of the earth's crust come into existence?

ROCK DEFORMATION

When rocks are subjected to stresses greater than their own strength, they begin to deform, usually

FIGURE 8.3
This sequence illustrates how the combined effect of erosion and isostatic adjustment results in a thinning of the crust in mountainous regions.

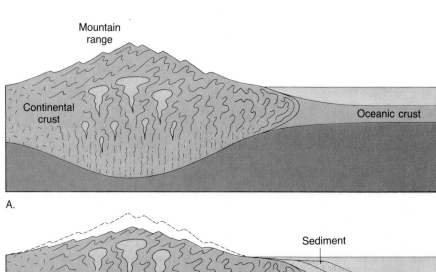

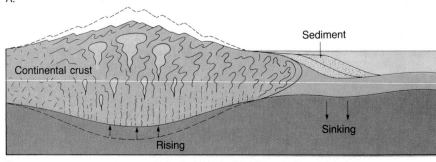

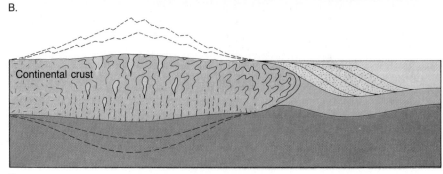

by folding or fracturing. It is easy to visualize how individual rocks break, but how are large rock units bent into intricate folds without being appreciably broken during the process? In an attempt to answer this, geologists turned to the laboratory and subjected rocks to stresses while simulating those conditions believed to exist at various depths within the crust. Although all rock types respond somewhat differently to stresses, the general characteristics of rock deformation were determined from these experiments. Geologists discovered that when stress is applied slowly and under low pressure, rocks first respond by deforming elastically. Changes resulting from *elastic deformation* are reversible; that is, like a rubber band, the rock will return to nearly its original size and shape when the stress is removed. However, once the elastic limit is surpassed, rocks either rupture or deform plastically. *Plastic deformation* results in permanent changes; that is, the size and shape of a rock unit are altered through folding and flowing. Laboratory experiments confirmed that at high temperatures and pressures, most rocks deform plastically once their elastic limit is surpassed. Rocks tested under surface conditions also deform elastically, but once they exceed their elastic limit, most behave like a brittle solid and rupture. Recall that the energy for most earthquakes comes from stored elastic energy that is released as rock ruptures and snaps back to its original shape.

One factor that researchers cannot duplicate in the laboratory is geologic time. We know that if stress is applied quickly, as with a hammer, rocks tend to fracture. On the other hand, these same materials may deform plastically if stress is applied over an extended period. For example, marble benches have been known to sag under their own weight over a period of a hundred years or so. In nature, small forces applied over long time periods surely play an important role in the deformation of rock strata.

FOLDING

During mountain building, flat-lying sedimentary and volcanic rocks are often bent into a series of broad folds, much like those that would form if you were to hold the ends of a sheet of paper and then push them together. Such folding shortens and thickens the crust. Experimental evidence shows that when sedimentary rocks are buried deep within the crust, where confining pressures are great, strata can be deformed into very tight folds stacked one atop the other, without appreciable fracturing. Figure 8.4 illustrates some common folded structures. Linear upfolded forms are commonly called **anticlines,** whereas downfolded

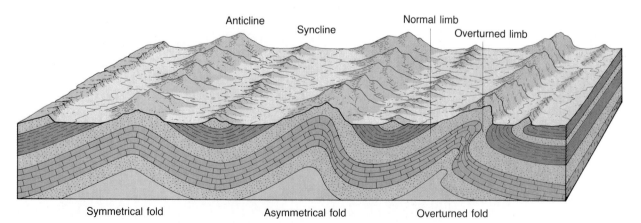

FIGURE 8.4
Block diagram of principal types of folded strata.

FIGURE 8.5
Sheep Mountain, a doubly
plunging anticline. Note that
erosion has cut the flanking
sedimentary beds into low
ridges that make a "V"
pointing in the direction of
plunge. (Photo by John S.
Shelton)

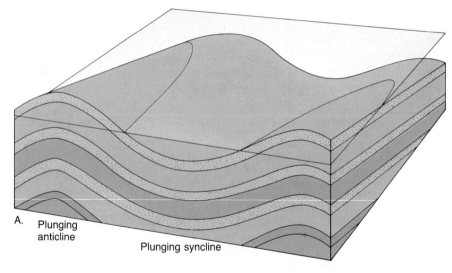

A. Plunging
anticline

Plunging syncline

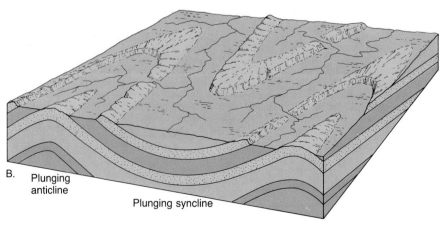

B. Plunging
anticline

Plunging syncline

FIGURE 8.6
Plunging folds. A. Idealized
view. B. View after extensive
erosion.

structures are typically referred to as **synclines.** Depending upon their orientation, anticlines and synclines are said to be *symmetrical, asymmetrical,* or *overturned* if one limb has been tilted beyond the vertical.

Folds do not continue on forever; rather, their ends die out much like the wrinkles in cloth. Some folds are said to be *plunging,* since the axis of the fold is plunging into the ground (Figure 8.5). Figure 8.6 shows some examples of plunging folds and the pattern produced when erosion removes the upper layers of these structures and exposes their interiors. Note that the outcrop pattern of an anticline points in the direction it is plunging, while the opposite is true for synclines. A good example of the kind of topography that results when erosional forces attack folded sedimentary strata is found in the Valley and Ridge Province of the Appalachians. Here resistant sandstone beds remain as imposing ridges separated by valleys cut into more easily eroded shale or limestone beds (see Figure 8.19).

Although most folds are caused by compressional stresses that squeeze and crumble strata, some folds are a consequence of vertical displacement. When upwarping produces a circular or somewhat elongated structure, the feature is called a **dome.** Downwarped structures having a similar shape are termed **basins.** The Black Hills of western South Dakota is one such domal structure in which erosion has stripped away the upwarped sedimentary beds, exposing older igneous and metamorphic rocks in the center. Several basins also exist in the United States. The basins of Michigan and Illinois have very gently sloping beds similar to saucers. Because large basins contain sedimentary beds sloping at such low angles, they are usually identified by the age of the rocks composing them. The youngest rocks are found near the center and the oldest rocks are at the flanks. This is just the opposite order of a domal structure such as the Black Hills, where the oldest rocks form the core.

FAULTING

As indicated in our discussion of earthquakes, faults are fractures in the earth's crust along which

FIGURE 8.7
Faulting caused the vertical displacement of these sedimentary beds in southern Nevada. (Photo by E. J. Tarbuck)

appreciable movement has taken place. Faults are categorized on the basis of the relative movement between the blocks on both sides of the fault plane. The movement can be horizontal, vertical, or oblique.

Faults having primarily vertical movement are called **dip-slip faults,** since the displacement is in the direction of the inclination, or dip, of the fault plane (Figure 8.7). Dip-slip faults are classified as **normal faults** when the rock above the fault plane moves *down* relative to the rock below (Figure 8.8). **Reverse faults** are created when the rock

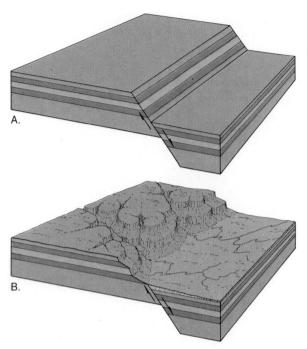

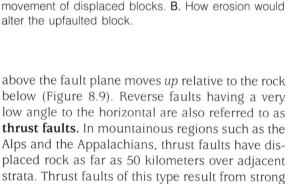

FIGURE 8.8
Block diagrams of a normal fault. **A.** The relative movement of displaced blocks. **B.** How erosion would alter the upfaulted block.

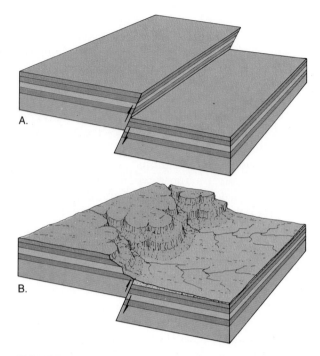

FIGURE 8.9
Block diagrams of a reverse fault. **A.** The relative movement of displaced blocks. **B.** How erosion would alter the upfaulted block.

above the fault plane moves *up* relative to the rock below (Figure 8.9). Reverse faults having a very low angle to the horizontal are also referred to as **thrust faults.** In mountainous regions such as the Alps and the Appalachians, thrust faults have displaced rock as far as 50 kilometers over adjacent strata. Thrust faults of this type result from strong compressional stresses.

Faults in which the dominant displacement is along the trend, or strike, of the fault are called **strike-slip faults.** Many large strike-slip faults are associated with plate boundaries and are called transform faults. Transform faults have nearly vertical dips and serve to connect large structures such as segments of an oceanic ridge. The San Andreas fault in California is a well-known transform fault in which the displacement has been on the order of several hundred kilometers. When faults have both vertical and horizontal movement, they are called **oblique-slip faults.**

Fault motion provides the geologist with a method of determining the nature of the forces at work within the earth. Normal faults indicate the existence of *tensional stresses* that pull the crust apart. This "pulling apart" can be accomplished either by uplifting that causes the surface to stretch and break, or by horizontal forces that actually rip the crust apart. Normal faulting is known to occur at spreading centers, where plate divergence is prevalent. Here a central block called a **graben** is bounded by normal faults and drops as the plates separate (Figure 8.10). These grabens produce an elongated valley bounded by upfaulted structures called **horsts.** The Great Rift Valley of East Africa consists of several large grabens, above which tilted horsts produce a linear mountainous topography. This valley, nearly 6000 kilometers (3700 miles) long, contains the excavation sites of some of the earliest human fossils. Other rift valleys include the Rhine Valley in

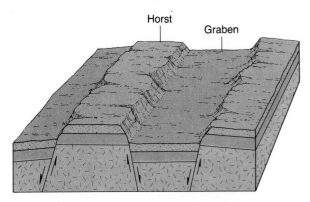

FIGURE 8.10
Diagrammatic sketch of downfaulted block (graben) and upfaulted block (horst).

Germany and the valley of the Dead Sea in the Middle East.

Since the blocks involved in reverse and thrust faulting are displaced toward one another, geologists conclude that *compressional forces* are at work. The primary regions of this activity are thought to be the convergent zones, where plates are colliding. Compressional forces generally produce folds as well as faults, and result in a general thickening and shortening of the material involved.

MOUNTAIN TYPES

Even though no two mountain ranges are exactly alike, they can be classified according to their most dominant characteristics. Using this approach, four main categories of mountains result: (1) Folded mountains (complex mountains); (2) Volcanic mountains; (3) Fault-block mountains; and (4) Upwarped mountains (Figure 8.11, pages 224–25). Mountain ranges of the same type are commonly found in close proximity forming a mountain system. For example, nearly the entire state of Nevada is composed of numerous elongated fault-block mountains that are separated by faulted basins. Further, within any mountainous belt, such as that portion of the American Cordillera in the western United States, mountain ranges representing each of these groups can be found. In addition to these basic varieties, some regions

have mountainous topography that was produced without appreciable crustal deformation. For example, plateaus (areas of high-standing rocks that are essentially horizontal) can be deeply dissected into rugged terrains. Although these highlands have the topographical expression of mountains, they lack the structure associated with orogenesis.

In the following sections we will examine three of the four basic mountain types. Volcanic mountains are treated in detail in Chapter 7.

FOLDED MOUNTAINS

The largest and most complex mountain systems are the so-called **folded mountains** or **complex mountains.** Although folding is often more conspicuous, faulting, metamorphism, and igneous activity are always present in varying degrees (Figure 8.11A). All major mountain belts, including the Alps, Urals, Himalayas, and Appalachians, are of this type. Since folded mountains represent the world's major mountain systems, the process of mountain building is usually described in terms of their formation. Thus a separate section on orogenesis is devoted to the evolution of these often majestic and always complex mountain systems.

FAULT-BLOCK MOUNTAINS

Fault-block mountains are generated when large rock units are displaced, often accompanied by tilting, along high-angle normal faults (Figure 8.11C). Recall that tensional stresses, which stretch crustal material, produce normal faults and the displacement along them.

Excellent examples of fault-block mountains are found in the Basin and Range Province, a region that encompasses Nevada and portions of Utah, New Mexico, Arizona, and California. Here the crust has literally been broken into hundreds of pieces, giving rise to nearly parallel mountain ranges that average about 80 kilometers in length and rise precipitously above the adjacent sediment-laden basins.

The most widely accepted proposal for fault-block mountain formation in the Basin and Range Province comes from the plate tectonics theory. According to this model, the nature of the

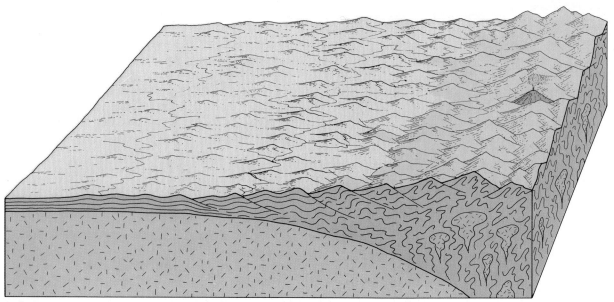

A. Folded (complex)

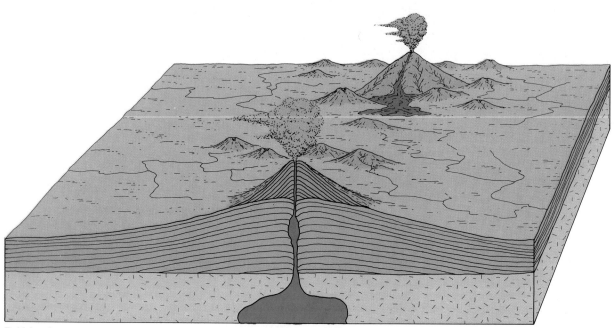

B. Volcanic

FIGURE 8.11
Classification of mountains.

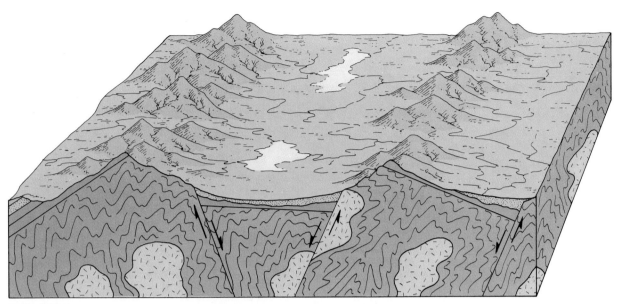

C. Fault-block

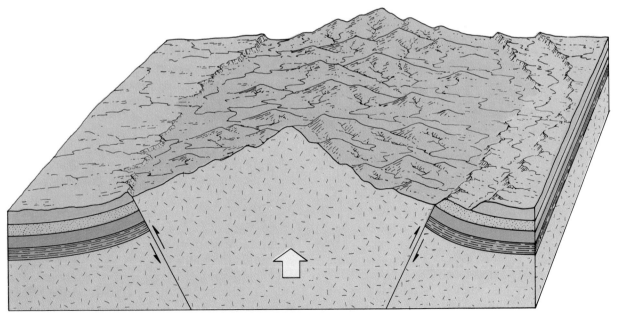

D. Upwarped

FIGURE 8.12
Formation of the fault-block mountains of the Basin and Range Province. **A.** Tensional forces extended and faulted the rocks, giving rise to a period of volcanism. **B.** Continued extension displaced rock units along numerous high-angle normal faults.

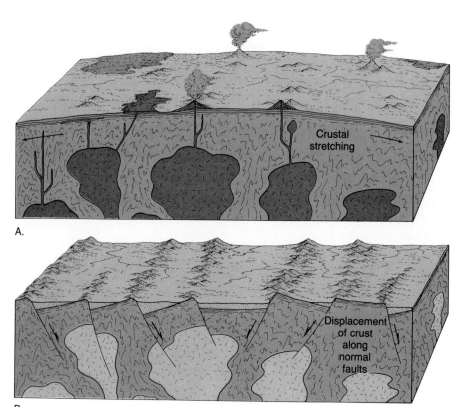

boundary between the Pacific plate and the North American plate changed about 30 million years ago. The relative movement along this boundary changed, and the forces acting upon the region became tensional rather than compressional. The resultant tensional forces extended (stretched) and faulted the Basin and Range rock (Figure 8.12A). This deformation, in turn, gave rise to a period of volcanism. Although the igneous activity has since waned, continued extension has displaced large rock units along high-angle normal faults (Figure 8.12B).

Other fault-block mountains in the United States are the Teton Range of Wyoming and the Sierra Nevada of California. Both are faulted along their eastern flanks, which were uplifted as the blocks tilted downward to the west. Looking west from Jackson Hole, Wyoming, and Owens Valley, California, respectively, the eastern fronts of these ranges rise over 2 kilometers, making them two of the most precipitous mountain fronts in the United States.

UPWARPED MOUNTAINS

Upwarped mountains are produced in association with a broad arching of the crust, or in some instances, because of vertical displacement along high-angle faults. Some, such as the Black Hills in western South Dakota and the Adirondack Mountains in upstate New York, consist of older igneous and metamorphic bedrock that was once eroded flat and subsequently mantled with sediment. As the regions were upwarped, erosion removed the veneer of sedimentary strata, leaving a core of igneous and metamorphic rocks standing above the surrounding terrain (Figure 8.13).

Other examples of upwarped mountains are found in the middle and southern Rockies from Montana south through Colorado and into New Mexico. In this group are the Front Range of Colorado, the Sangre de Cristo of New Mexico and Colorado, and the Bighorns of Wyoming. These mountains are structurally much different from the northern Rockies, which include the Canadian

Rockies and those portions of the Rockies found in Idaho, western Wyoming, and western Montana. Whereas the latter ranges are composed of thick sequences of sedimentary rocks that were de- formed by folding and low-angle thrust faulting, the portion of the southern Rockies bordering the Great Plains was pushed almost vertically up- ward along high-angle faults. In general these

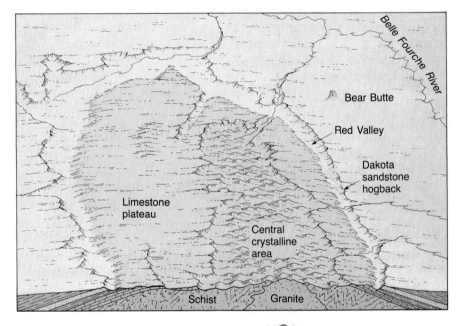

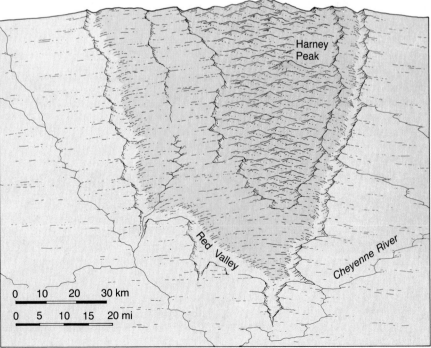

FIGURE 8.13
The Black Hills of South Dakota, an example of an upwarped mountain system in which the resistant igne- ous and metamorphic cen- tral core has been exposed by erosion. (After Arthur N. Strahler. *Introduction to Physical Geography,* 3rd ed. New York: John Wiley & Sons, 1973. Reprinted by permission)

mountains consist of older basement rocks covered by relatively thin layers of younger strata. However, since the time of deformation much of this mantle of sedimentary rocks has been eroded from the highest portions of the uplifted blocks, exposing the igneous and metamorphic cores. In many areas, remnants of these sedimentary layers are visible flanking the crystalline cores of the mountain ranges. They are often easy to identify because the upturned strata form prominent angular ridges called **hogbacks** (Figure 8.14). Examples of exposed Precambrian cores include a number of granitic outcrops that project as steep summits, such as Pikes Peak and Longs Peak in Colorado's Front Range. It has been suggested that the upwarping of the eastern Rockies was in response to compressional stresses generated when the North American plate overrode a portion of the adjacent oceanic plate.

MOUNTAIN BUILDING

Mountain building has operated during the recent geologic past in several locations around the world. These relatively young mountainous belts include the American Cordillera, which runs along the western margin of the Americas from Cape Horn to Alaska; the Alpine-Himalayan chain, which extends from the Mediterranean through Iran to northern India and into Indochina; and the mountainous terrains of the western Pacific, which include mature island arcs such as Japan, the Philippines, and Sumatra. Most of these young mountain belts have come into existence within the last 100 million years. Some, including the Himalayas, began their growth as recently as 45 million years ago.

In addition to these recently formed complex (folded) mountains, several chains of much older mountains exist on the earth as well. Although these structures are deeply eroded and topographically less prominent, they clearly possess the same structural features found in younger mountains. Typical of this older group are the Appalachians in the eastern United States and the Urals in the Soviet Union.

Although complex mountains differ from one another in particular details, all possess the same basic structures. Mountain belts generally consist of roughly parallel ridges of folded and faulted sedimentary and volcanic rocks, portions of which

FIGURE 8.14
A view looking north along the east flank of the Front Range of the Colorado Rockies. The upturned remnants of sedimentary strata (center) once covered the igneous and metamorphic terrain which lies to the west. (Photo by T. S. Lovering, U.S. Geological Survey)

FIGURE 8.15
Highly deformed sedimentary strata in the Rocky Mountains of British Columbia.
(Photo by John Montagne)

have been strongly metamorphosed and intruded by somewhat younger igneous bodies. In most cases the sedimentary rocks formed from enormous accumulations of deep-water marine sediments that occasionally exceeded 15,000 meters in thickness, as well as from thinner shallow-water deposits. Moreover, these deformed sedimentary rocks are for the most part older than the mountain building event. This fact indicates that a long quiescent period of deposition was followed by an episode of deformation (Figure 8.15).

In order to unravel the events that produce mountains, many studies are conducted in regions that exhibit ancient mountain structures as well as at sites where orogenesis is thought to be in progress. Of particular interest are active subduction zones, where plates are converging. Here partial melting of a subducted plate and possibly frictional heating of mantle rocks generate a supply of magma that migrates upward. Thus, at most

modern-day subduction zones, volcanic arcs are forming. This situation is typified by Alaska's Aleutian Islands and by the chain of volcanoes that is found in the Andes of South America. Although the development of a volcanic arc does result in the formation of mountainous topography, this activity is viewed as but one of the phases in the development of a complex mountain system.

At sites where oceanic crust is being subducted, continental blocks are also being rafted toward one another. Recent studies indicate that the most important cause of orogenesis is the collision of two or more of these crustal fragments. Collisions can occur between a continental block and a variety of landmasses, including archipelagos such as the Aleutian Islands, or small crustal fragments similar in size to Madagascar, or even other continental sized blocks. We shall consider these sites of mountain building in the following sections.

FIGURE 8.16

Subduction-type orogenesis along an active continental margin. **A.** Passive plate margin. **B.** Plate convergence generates a subduction zone. **C.** Partial melting of subducted plate generates the volcanic arc. **D.** Continued growth of the complex mountain system through deformation of the shallow- and deep-water sediments.

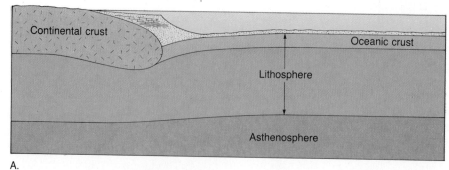

A.

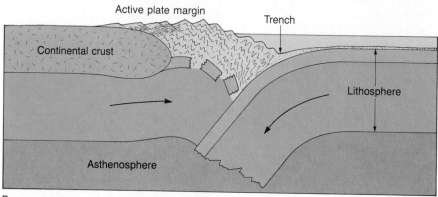

B.

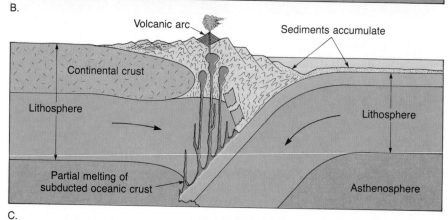

C.

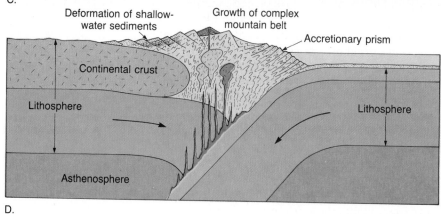

D.

OROGENESIS AT SUBDUCTION ZONES

Mountain building along continental margins involves the convergence of an oceanic plate and a plate whose leading edge contains continental material. However, the first stage in the development of a complex mountain system is thought to occur prior to the formation of the subduction zone. During this period the continental margin is passive; that is, it is not a plate boundary but a part of the same plate as the adjoining oceanic crust. The east coast of the United States is a present-day example of a passive continental margin. Here, as at other passive continental margins surrounding the Atlantic, sediment deposited on the continental shelf is producing a wedge of shallow-water sandstones, limestones, and shales (Figure 8.16A). Beyond the continental shelf, turbidity currents (dense slurries of mud and water) are depositing deep-water sediment upon the ocean floor.

At some point the continental margin becomes active; a subduction zone forms and the deformation process is initiated (Figure 8.16B). However, the cause of this event is unknown. A good place to examine an active continental margin is along the west coast of South America. Here the Nazca plate is being subducted beneath the South American plate (see Figure 6.6). This subduction zone probably formed in conjunction with the breakup of the supercontinent Pangaea. As South America separated from Africa and migrated westward, the oceanic crust adjacent to the west coast of South America was bent and thrust under the continental plate.

Subduction and partial melting of the oceanic plate initiate yet another stage—the development of a volcanic arc (Figure 8.16C). Since the break between the oceanic and continental lithosphere generally forms seaward of the coastline, a volcanic island arc often forms a few hundred kilometers out to sea. This occurred during the development of the Andean arc, where the first volcanic structures closely resembled the present-day volcanic island arcs of the western Pacific. As the descending plate plunged farther beneath South America, igneous activity in the Andes migrated inland from the initial volcanoes, which have been mostly eroded away. Remnants of the original volcanic arc are the exposed batholiths and metamorphosed terrains composing the crystalline Andes that flank the west coast of South America.

During the development of the volcanic arc, sediment derived from the land as well as that scraped from the subducting plate is plastered against the landward side of the trench. This chaotic accumulation of metamorphosed rocks and scraps of oceanic crust is called an **accretionary prism** (Figure 8.16D).

One of the best examples of a volcanic arc-accretionary prism is found in the western United States and includes the Sierra Nevada and the Coast Range of California (Figure 8.17). These

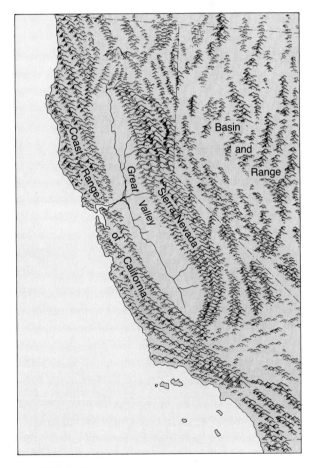

FIGURE 8.17
Map of the California Coast Range and the Sierra Nevada.

parallel mountainous belts formed as a portion of the Pacific basin subducted under the western edge of the North American plate. The Sierra Nevada batholith is a remnant of a portion of the volcanic arc that was produced by several surges of magma over a period of tens of millions of years. Subsequent uplifting and erosion have removed most evidence of past volcanic activity and exposed a core of crystalline rocks. In the trench region, sediments scraped from the subducting plate and those provided by the eroding volcanic arc were intensely folded and faulted into the accretionary prism that presently constitutes the Franciscan Formation of the Coast Range of California. Uplifting of the Coast Range took place quite recently, as evidenced by the unconsolidated sediments that still mantle portions of these highlands.

OROGENESIS AND CONTINENTAL COLLISIONS

Up to this point we have discussed the formation of orogenic belts where the leading edge of only one of the two converging plates contained continental crust. However, both of the colliding plates may be carrying continental crust. Because continental lithosphere is evidently too buoyant to undergo any appreciable amount of subduction, a collision between the continental fragments eventually results (Figure 8.18). An example of such a collision occurred about 45 million years ago when India collided with Asia. India, thought to have been previously a part of Antarctica, was rafted nearly 5000 kilometers due north before the collision occurred. The result was the formation of the spectacular Himalaya Mountains and the Tibetan Highlands. Although most of the oceanic crust which separated these landmasses prior to the collision was subducted, some was caught up in the squeeze along with sediment that lay offshore and can now be found elevated high above sea level. After such a collision, the subducted oceanic plate is believed to decouple from the rigid continental plate and continue its downward path.

The spreading center that separated India from Antarctica and moved it northward is still active; hence, India continues to be thrust into Asia at

an estimated rate of a few centimeters per year. However, numerous earthquakes recorded off the southern coast of India indicate that a new subduction zone may be in the making. If formed, it would provide a disposal site for the floor of the Indian Ocean, which is continually being produced at a spreading center located to the southwest. Should this occur, India's northward journey would come to an end and the growth of the Himalayas would cease.

A similar but much older collision is believed to have taken place when the European continent collided with the Asian continent to produce the Ural Mountains, which extend in a north-south direction through the Soviet Union. Prior to the discovery of plate tectonics, geologists had difficulty explaining the existence of mountain ranges such as the Urals which are located deep within the continental interiors. How could thousands of meters of marine sediment be deposited and then become highly deformed while situated in the middle of a large landmass?

Other mountain ranges showing evidence of continental collisions are the Alps and the Appalachians. The Appalachians are thought to have resulted from a collision between North America, Europe, and northern Africa. Although they have since separated, these landmasses were juxtaposed as part of the supercontinent Pangaea less than 200 million years ago. Detailed studies in the southern Appalachians indicate that the formation of this mountain belt was more complex than once thought. Rather than forming during a single continental collision, the Appalachians resulted from several distinct episodes of mountain building that occurred over a period of nearly 300 million years. As the two continental blocks began to converge, a subduction zone formed seaward of the ancient coastline of North America. The igneous activity associated with the subducting plate gave rise to a volcanic island arc, perhaps similar to those volcanic arcs presently rimming the western Pacific (Figure 8.18A). Then, about 380 million years ago, the ocean basin behind the volcanic arc began to close.

The final orogeny occurred about 250–300 million years ago when Africa and Europe collided with North America. This event is thought to have displaced the earlier-accreted volcanic arc farther

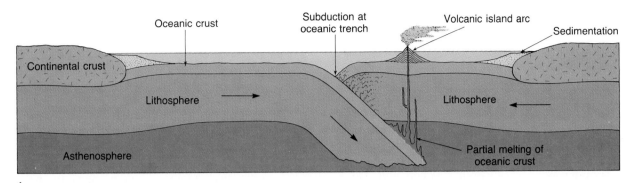

Oceanic crust   Subduction at oceanic trench   Volcanic island arc   Sedimentation

Continental crust

Lithosphere   Lithosphere

Asthenosphere   Partial melting of oceanic crust

A.

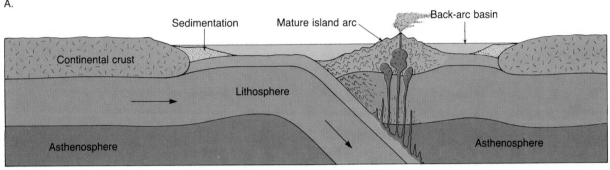

Sedimentation   Mature island arc   Back-arc basin

Continental crust

Lithosphere

Asthenosphere   Asthenosphere

B.

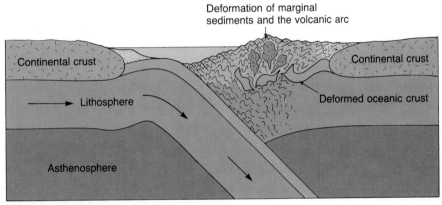

Deformation of marginal sediments and the volcanic arc

Continental crust   Continental crust

Lithosphere

                  Deformed oceanic crust

Asthenosphere

C.

FIGURE 8.18
Orogenesis and continental collisions. **A.** Converging plates generate a subduction zone and initiate island arc volcanism. **B.** Sediments scraped from the subducting plate and igneous activity add to the size of the volcanic arc. **C.** Closing of the back-arc basin deforms the entrapped marginal sediments and volcanic arc. **D.** A continental collision closes the ocean basin, resulting in further deformation and metamorphism of the sediments and volcanic rocks.

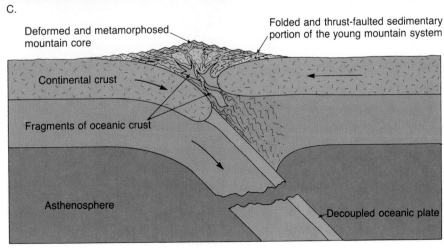

Deformed and metamorphosed mountain core   Folded and thrust-faulted sedimentary portion of the young mountain system

Continental crust

Fragments of oceanic crust

Asthenosphere   Decoupled oceanic plate

D.

inland along low-angle thrust faults. At some locations the total displacement may have exceeded 250 kilometers (155 miles). This landward displacement further deformed the shallow-water sediments which had flanked North America. Today these folded and faulted sandstones, limestones, and shales compose the essentially unmetamorphosed rocks of the Valley and Ridge Province (Figure 8.19).

In summary, the orogenesis of a complex mountain chain, as typified by the Appalachians, is thought to occur as follows:

1 A long period of uninterrupted deposition produces a thick wedge of sediments along a passive continental margin, thereby increasing the size of the continental block.

2 For reasons not yet understood, the ocean basin then begins to close and the continents start to converge.

3 Plate convergence results in subduction of the intervening oceanic slab and initiates an extended period of igneous activity. This activity results in the formation of a volcanic arc often located a few hundred kilometers seaward of the ancient coastline (Figure 8.18A).

4 Debris eroded from the volcanic arc and the mainland, plus sediment scraped from the descending plate, add to the wedge of sediment along the continental margin (Figure 8.18B).

5 Further convergence causes the narrow sea behind the volcanic arc to close. This orogenic event deforms and metamorphoses the back-arc sediments and associated volcanic debris as well as the volcanic arc itself (Figure 8.18C).

6 Eventually the continents collide. This event and the associated igneous activity further deform and metamorphose the entrapped sediments and volcanic arc to produce the crystalline core of the young mountain belt. As this deformed terrain is thrust landward, the shallow-water deposits that once formed the continental shelf are folded and displaced inland along low-angle thrust faults (Figure 8.18D).

7 Finally, a change in the plate boundary ends mountain belt growth. Only then does erosion become the dominant process in altering the landscape.

This sequence of events is thought to have been duplicated many times in geologic time. However,

FIGURE 8.19
Valley and Ridge Province.
(Courtesy of NASA)

the rate of deformation and the geologic and climatic settings varied in each instance. Thus, the formation of each mountain chain must be regarded as a unique event.

OROGENESIS AND CONTINENTAL ACCRETION

When originally formulated, the plate tectonics theory suggested two mechanisms for orogenesis. First, continental collisions were proposed to explain the formation of such mountainous terrains as the Alps, Himalayas, and Urals. Second, as typified by the Andes, orogenesis associated with the subduction of oceanic lithosphere was thought to be the underlying tectonic process for many circum-Pacific mountain chains. Recent investigations, however, indicate yet another mechanism of orogenesis. This new proposal suggests that relatively small crustal fragments collide and merge with continental margins and that through this process of collision and accretion, many of the mountainous regions rimming the Pacific have been generated.

What is the nature of the small crustal fragments and where do they come from? Researchers believe that prior to their accretion to a continental block, some of the fragments may have been microcontinents similar in nature to the present-day island of Madagascar. Many others may have been located below sea level and are represented today by submerged platforms rising high above the floor of the western Pacific (see Figure 10.10). Over one hundred of these so-called oceanic plateaus are known to exist. It is believed that these plateaus originated as submerged continental fragments, extinct volcanic island arcs, or as submerged volcanic chains associated with hot spot activity.

The widely accepted view today is that as oceanic plates move, they carry the embedded oceanic plateaus or microcontinents to a subduction zone. Here the upper portions of these thickened zones are peeled from the descending plate and thrust in relatively thin sheets upon the adjacent continental block. This newly added material increases the width of the continent and may later

be overridden and displaced farther inland by colliding with other fragments.

Geologists refer to these accreted crustal blocks as terranes. Simply, the term **terrane** designates any crustal fragment whose geologic history is distinct from the adjoining terranes. (Do not confuse the term *terrane* with the word *terrain,* which indicates the lay of the land.) Terranes come in a variety of shapes and sizes; some are no larger than the volcanic islands on the floor of the ocean. Others, such as the one composing the entire Indian subcontinent, are quite large.

The idea that orogenesis occurs in association with the accretion of small crustal fragments to a continental mass arose principally from studies conducted in the northern portion of the North American Cordillera. Here it was learned that some terranes, principally those in the orogenic belts of Alaska and British Columbia, contain fossil and magnetic evidence to indicate these strata originated nearer the equator. Further, these exotic terranes were found to consist of rock sequences vastly different from those of adjacent terranes.

It is now believed that the exotic terranes found in the North American Cordillera were once scattered throughout the eastern Pacific much as we find oceanic plateaus distributed in the western Pacific. Over the last 200 million years, these fragments migrated toward and collided with the west coast of North America. Apparently, this activity resulted in the piecemeal addition of fragments to the entire Pacific Coast from the Baja Peninsula to northern Alaska. If this is true, such collisions are responsible for the orogenesis of much of the North American Cordillera. In a like manner, many of the ocean plateaus today will eventually be accreted to active continental margins, thus resulting in the formation of new orogenic belts.

THE ORIGIN AND EVOLUTION OF CONTINENTAL CRUST

In the preceding section, we learned that the theory of plate tectonics provides a model from which to examine the formation of complex mountainous belts. But what roles have plate tectonics and mountain building played in events leading to the origin and evolution of continents? At this time no single answer has met with overwhelming acceptance. The lack of agreement among geologists can in part be attributed to the complex nature and antiquity of most continental material, which makes deciphering its history very difficult. Nevertheless, during the last two decades, great strides have been made in unraveling the secrets held by the rocks composing stable continental interiors.

One view, which has gained support in recent years, contends that the continents have grown larger through geologic time by the gradual accretion of material derived from the upper mantle. A main tenet of this hypothesis is that the primitive crust was of an oceanic type and the continents were small or possibly nonexistent. Further, this proposal suggests that the formation of continental material takes place in two distinct phases as shown in Figure 8.20. The first step occurs in the upper mantle directly beneath the oceanic ridges.

Here partial melting of the rock peridotite yields basaltic magma that rises to form oceanic crust. The ocean floor rocks are higher in silica, potassium, and sodium, and lower in iron and magnesium than the rocks of the upper mantle from which they were derived. As new ocean floor is generated at the ridge crests, older oceanic crust is being destroyed at the oceanic trenches. In trench regions, the subducted oceanic crust is heated sufficiently to cause partial melting. This gives rise to relatively light, silica-rich rocks, which are then emplaced in volcanic arcs. The subducted oceanic crust, depleted of its lighter constituents, continues to sink and is no longer involved in the process of generating crustal rocks.

According to this view, the earliest continental rocks came into existence at a few isolated island arcs. Once formed, these island arcs coalesced to form larger continental masses, while deforming the volcanic and sedimentary rocks which were deposited in the intervening oceans. Eventually this process generated masses of continental crust having the size and thickness of modern conti-

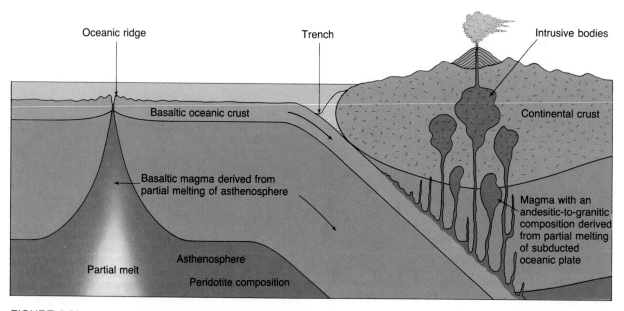

FIGURE 8.20
The two-stage process for transforming material from the asthenosphere into continental crust. Once continental crust is generated, its low density apparently keeps it afloat indefinitely.

nents. In this way, the process of mountain building not only restructures continental rocks, but also generates new continental materials.

If the continents do in fact grow by accretion of material at their flanks, then the continents have grown larger at the expense of oceanic crust. This view assumes the buoyancy and indestructibility of continental crust. Even sediment eroded from a continent and carried to great depths on an adjacent subducting plate eventually melts and returns to the continents. Although crustal rock apparently remains afloat indefinitely, some continents are occasionally fragmented and carried along in a conveyor belt fashion until they collide with other landmasses. Australia, which separated from Antarctica, is presently being rafted northward and will probably join Asia in much the same manner as India did about 45 million years ago. According to this view, fragmentation and the for-

mation of new crustal rocks that accompanied the reshuffling of these fragments are responsible for the present volume, structure, and configuration of continents.

A word of caution is in order. The views set forth in this section to explain the origin and evolution of the continents are still somewhat speculative. Plate tectonics appears to be the major force in crustal evolution over the last 600 million years. However, during the earth's early history, heat released by the decay of uranium, thorium, and potassium must have been at least twice as great as it is today. Was plate tectonics active early in the earth's history, only at a different rate, or were there much different processes in operation? Was the primitive crust composed primarily of continental rocks, or was it of the oceanic type? These are some of the questions that still require definitive answers.

REVIEW QUESTIONS

1   State the three lines of evidence which support the concept of crustal uplift. Can you think of another?

2   What happens to a floating object when weight is added? Subtracted? How do these principles apply to changes in the elevations of mountains? What term is applied to the adjustment that causes crustal uplift of this type?

3   List two lines of evidence which support the idea of a floating crust.

4   What conditions favor rock deformation by folding? By faulting?

5   Using the concept of elastic rebound, explain why earthquakes cannot occur at great depths.

6   Compare the movement of normal and reverse faults. What type of force produces each?

7   At which of the three types of plate boundaries does normal faulting predominate? Reverse faulting? Strike-slip faulting?

8   Describe a horst and a graben. Explain how a graben valley forms and name one.

9   Compare and contrast anticlines and synclines. Domes and basins. Anticlines and domes.

10   Although we classify many mountains as folded, why might this description be misleading?

11   What type of faults are associated with fault-block mountains?

12   During the formation of fault-block mountains, are the forces acting upon the region compressional or tensional?

13 Which of the four types of mountain ranges is exemplified by the following?
   (a) Black Hills
   (b) Basin and Range
   (c) Adirondacks
   (d) Cascades
   (e) Appalachians
   (f) Tetons
   (g) Bighorns
   (h) Himalayas
   (i) Front Range

14 Name the site where sediments are deposited and have a good chance of being squeezed into a mountain range.

15 Which type of plate boundary is most directly associated with mountain building?

16 Describe an accretionary prism and explain its formation.

17 Would the discovery of a sliver of oceanic crust in the interior of a continent tend to support or refute the theory of plate tectonics? Why?

18 Why might it have been difficult for geologists to conclude that the Appalachian Mountains were formed by plate collision if examples like the Himalayas did not exist?

19 How does the plate tectonics theory explain the existence of fossil marine life on top of the Ural Mountains?

20 In your own words, briefly enumerate the steps involved in the formation of a mountain system according to the plate tectonics model.

KEY TERMS

orogenesis	normal fault	folded mountains (complex mountains)
isostasy	reverse fault	
anticline	thrust fault	fault-block mountains
syncline	strike-slip fault	upwarped mountains
dome	oblique-slip fault	hogback
basin	graben	accretionary prism
dip-slip fault	horst	terrane

9

GEOLOGIC TIME AND EARTH HISTORY

In the late eighteenth century James Hutton recognized that the earth is very old. But how old? For many years there was no reliable method to determine the age of the earth or the times of various events in the geologic past. Rather, a geologic time scale was developed that relied on relative dating principles. What are these principles? What part do fossils play? With the discovery of radioactivity and the development and refinement of radiometric dating techniques, geologists now can assign fairly accurate dates to many of the events in earth history. What is radioactivity? Why is it a good "clock" for dating the geologic past? In this chapter we shall attempt to answer the questions raised here, as well as others.

The strata exposed in the Grand Canyon contain clues to millions of years of earth history. (Photo by E. J. Tarbuck)

In 1869 John Wesley Powell, who was later to head the U.S. Geological Survey, led a pioneering expedition down the Colorado River and through the Grand Canyon. Writing about the strata that were exposed by the downcutting of the river, Powell said, " . . . the canyons of this region would be a Book of Revelations in the rock-leaved Bible of geology." Powell was undoubtedly impressed with the countless millions of years of earth history exposed along the walls of the Grand Canyon (see chapter-opening photo). Interpreting earth history is a prime goal of the science of geology. Like a modern-day sleuth, the geologist must interpret clues found preserved in the rocks. By studying rocks, especially sedimentary rocks, and the features they contain, geologists can often unravel the complexities of the past.

Events by themselves, however, have little meaning until they are put into a time perspective. Studying history, whether it be the Civil War or the Age of Dinosaurs, requires a calendar. Among the major contributions that geology has made to the knowledge of humankind is the geologic time scale and the concept that earth history is exceedingly long. Over many years geologists have devised a time scale of earth history where geologic events can be put in their proper place. Recognizing that earth history has spanned an immense amount of time, geologists worked at finding out just how old the earth is.

SOME HISTORICAL NOTES ABOUT GEOLOGY

The nature of our earth—its materials and processes—has been a focus of study for centuries. Writings about such topics as fossils, gems, earthquakes, and volcanoes date back to the Greeks, more than 2300 years ago. Certainly the most influential of the Greek philosophers was Aristotle. Because Aristotle was a philosopher, his explanations were not always based on observations and experiments but often were arbitrary pronouncements. He believed that rocks were created under the "influence" of the stars and that earthquakes occurred when air crowded into the ground was heated by central fires and escaped explosively. When confronted with a fossil fish, he explained that, "a great many fishes live in the earth motionless and are found when excavations are made."

Although Aristotle's explanations may have been adequate for his day, they unfortunately continued to be expounded for many centuries, thus thwarting the acceptance of ideas that were more closely in accord with observations. Frank D. Adams states in *The Birth and Development of the Geological Sciences* (New York: Dover, 1938) that, "throughout the Middle Ages Aristotle was regarded as the head and chief of all philosophers; one whose opinion on any subject was authoritative and final."

CATASTROPHISM

During the seventeenth and eighteenth centuries the doctrine of **catastrophism** strongly influenced the formulation of explanations about the dynamics of the earth. Briefly stated, catastrophists believed that the earth's landscape had been developed primarily by great catastrophes. Features such as mountains and canyons, which today we know take great periods of time to form, were explained as having been produced by sudden and often worldwide disasters produced by unknowable causes that no longer operate. This philosophy was an attempt to fit the rate of earth processes to the then-current ideas on the age of the earth. In the mid-seventeenth century, James Ussher, Anglican Archbishop of Armagh, Primate of all Ireland, published a work that had immediate and profound influence. A respected scholar of the Bible, Ussher constructed a chronology of human and earth history in which he determined that the earth was only a few thousands of years old, having been created in 4004 B.C. Ussher's treatise earned widespread acceptance among scientific and religious leaders alike, and his chronology was soon printed in the margins of the Bible itself.

The relationship between catastrophism and the age of the earth has been summarized nicely as follows:

> That the earth had been through tremendous adventures and had seen mighty changes during its obscure past was plainly evident to every inquiring eye; but to concentrate these changes into a few brief millenniums required a tailor-made philosophy, a philosophy whose basis was sudden and violent change.*

*H. E. Brown, V. E. Monnett, and J. W. Stovall. *Introduction to Geology* (New York: Blaisdell, 1958).

THE BIRTH OF MODERN GEOLOGY

The late eighteenth century is generally regarded as the beginning of modern geology, for it was during this time that James Hutton, a Scottish physician and gentleman farmer, published his *Theory of the Earth* in which he put forth a principle that came to be known as the doctrine of **uniformitarianism** (Figure 9.1). Uniformitarianism is a fundamental concept in modern geology. It simply states that the physical, chemical, and biological laws that operate today have also operated in the geologic past. That is to say that the forces and processes that we observe presently shaping our planet have been at work for a very long time. Thus, to understand ancient rocks, we must first understand present-day processes and their results. This idea is commonly stated by saying "the present is the key to the past."

Prior to Hutton's *Theory of the Earth,* no one had effectively demonstrated that geology had to deal with extremely long periods of time. However,

FIGURE 9.1
James Hutton, the 18th century Scottish geologist who is often called the "father of modern geology." (Photo courtesy of the British Museum)

Hutton persuasively argued that processes which appear weak and slow-acting could, over long spans of time, produce effects that were just as great as those resulting from sudden catastrophic events. Unlike his predecessors, Hutton cited verifiable observations to support his ideas.

Since Hutton's literary style was cumbersome and difficult, his work was not widely read nor easily understood. It is the English geologist Charles Lyell who is given the most credit for advancing the basic principles of modern geology. Between 1830 and 1872 Lyell produced eleven editions of his great work, *Principles of Geology.* As was customary, Lyell's book had a rather lengthy subtitle that outlined the main theme of the work: *Being an Attempt to Explain the Former Changes of the Earth's Surface, by Reference to Causes now in Operation.* In the text, he painstakingly illustrated the concept of the uniformity of nature through time. He was able to show more convincingly than his predecessors that those geologic processes observed today can be assumed to have operated in the past. Although the doctrine of uniformitarianism did not originate with Lyell, he is the person who was most successful in interpreting and publicizing it for society at large.

Despite its importance in modern geology, the doctrine of uniformitarianism should not be taken too literally. To say that geologic processes in the past were the same as those occurring today is not to suggest that they always operated at precisely the same rate. Although the processes have remained essentially the same, their rates have undoubtedly varied during geologic time.

The acceptance of the concept of uniformitarianism, however, meant the acceptance of a very long history for the earth, for although processes vary in their intensity, they still take a very long time to create or destroy major landscape features.

For example, rocks containing fossils of organisms that lived in the sea more than 15 million years ago are now part of mountains that stand 3000 meters (9800 feet) above sea level. This means that the mountains were uplifted 3000 meters in about 15 million years, which works out to a rate of only 0.2 millimeter per year! Rates of erosion are equally slow. Estimates indicate that the North American continent is being lowered at a rate of just 3 centimeters per 1000 years. Thus, as you can see, it takes tens of millions of years for nature to build mountains and wear them down again. But even these time spans are relatively short on the time scale of earth history, for the rock record contains evidence that shows the earth has experienced many cycles of mountain building and erosion. Concerning the everchanging nature of the earth through great expanses of geologic time, Hutton stated: "We find no vestige of a beginning, no prospect of an end." A quote from William L. Stokes sums up the significance of Hutton's basic concept:

> In the sense that uniformitarianism implies the operation of timeless, changeless laws or principles, we can say that nothing in our incomplete but extensive knowledge disagrees with it.*

It is important to remember that although many features of our physical landscape may seem to be unchanging in terms of the tens of years we might observe them, they are nevertheless changing, but on time scales of hundreds, thousands, or even many millions of years.

RADIOACTIVITY AND RADIOMETRIC DATING

Current methods of radiometric dating put the age of the earth between 4.6 and 4.8 billion years. However, this great age for the earth is a relatively recent discovery. Although James Hutton and others who accepted the principle of uniformitarianism believed the earth was very old, they had no way of knowing its exact age. Solutions to this dating problem were sought, and several methods were subsequently devised. However, none of these early attempts proved reliable. Reasonably accurate determinations of the earth's age as well as dates for other events in the geologic past had to await the development of radiometric dating.

In Chapter 1 we learned that an atom is composed of electrons, protons, and neutrons. Electrons have a negative charge and protons have a positive charge. Since a neutron is actually a pro-

ton and electron combined, it has no charge. Protons and neutrons are found in the center, or nucleus, of the atom, and electrons spin around the nucleus in definite paths, called orbits. Practically all (99.9 percent) of an atom's mass is found in the nucleus, indicating electrons have practically no mass at all. By adding together the number of protons and neutrons in the nucleus, the mass number of the atom is determined. The atomic number (the atom's identifying number) is equal to the number of protons. Every element has a different number of protons in the nucleus, and thus a different atomic number. Many elements have atoms with different numbers of neutrons in the nucleus. Called isotopes, these atoms have different mass numbers, but have the same atomic number.

The forces binding protons and neutrons together in the nucleus are very strong; however, the nature of these forces is still poorly understood. In some isotopes the nuclei are unstable; that is, the forces binding protons and neutrons together are not strong enough. As a result, the nuclei spontaneously break apart (decay), a process called **radioactivity.** What happens when unstable nuclei break apart? Three common types of radioactive decay are illustrated in Figure 9.2 and are summarized as follows:

1 Alpha particles (α particles) may be emitted from the nucleus. An alpha particle is composed of 2 protons and 2 neutrons. Consequently, the emission of an alpha particle means the mass number of the isotope is reduced by 4 and the atomic number is decreased by 2.

2 When a beta particle (β particle), or electron, is given off from a nucleus, the mass number remains unchanged, because electrons have practically no mass. However, since the electron must have come from a neutron (remember, a neutron is a combination of a proton and an electron), the nucleus contains one more proton than before. Therefore, the atomic number increases by 1.

3 Sometimes an electron is captured by the nucleus. The electron combines with a proton and forms a neutron. As in the last ex-

*William L. Stokes. *Essentials of Earth History* (Englewood Cliffs, New Jersey: Prentice-Hall, 1966), p. 34.

FIGURE 9.2

Common types of radioactive decay. Notice that in each case the number of protons (atomic number) in the nucleus changes, thus producing a different element.

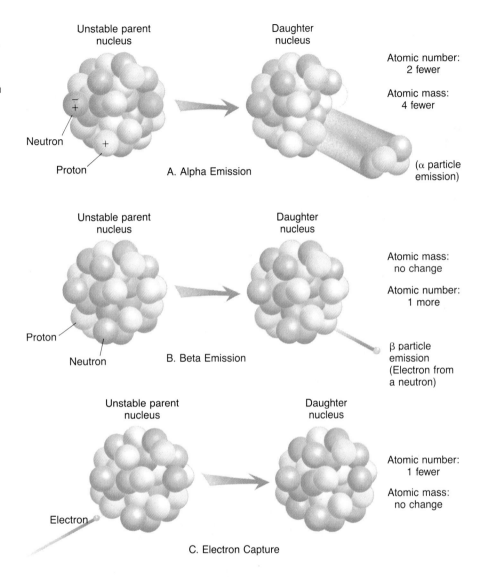

Unstable parent nucleus

Daughter nucleus

Atomic number: 2 fewer

Atomic mass: 4 fewer

Neutron

Proton

A. Alpha Emission

(α particle emission)

Unstable parent nucleus

Daughter nucleus

Atomic mass: no change

Atomic number: 1 more

Proton

Neutron

B. Beta Emission

β particle emission (Electron from a neutron)

Unstable parent nucleus

Daughter nucleus

Atomic number: 1 fewer

Atomic mass: no change

Electron

C. Electron Capture

ample, the mass number remains unchanged. However, since the nucleus now contains one less proton, the atomic number decreases by 1.

The radioactive isotope is referred to as the *parent,* and the isotopes resulting from the decay of the parent are termed the *daughter products.* Figure 9.3 provides an example of radioactive decay. Here it can be seen that when the radioactive parent, uranium-238 (atomic number 92, mass number 238), decays, it emits 8 alpha particles and 6 beta particles before becoming the stable daughter product lead-206 (atomic number 82, mass number 206).

Certainly among the most important results of the discovery of radioactivity is that it provided a reliable means of calculating the ages of rocks and minerals which contain particular radioactive isotopes, a procedure referred to as **radiometric dating.** Why is radiometric dating reliable? The answer lies in the fact that the rates of decay for many isotopes have been precisely measured and do not vary, at least under the physical conditions that exist in the outer layers of the earth. Therefore, each radioactive isotope used for dating has been decaying at a fixed rate since the formation of the rocks in which it occurs and the products of decay have been accumulating at a corresponding rate. For example, when uranium is incorporated into a mineral that crystallizes from magma, there is no lead (the stable daughter product) from previous decay. As the uranium in this newly formed mineral disintegrates, atoms of the daughter product are trapped and measurable amounts of lead eventually accumulate.

The time required for one-half of the nuclei in a sample to decay, called its **half-life,** is a common way of expressing the rate of radioactive disintegration. If the half-life is known, then the ratio be-

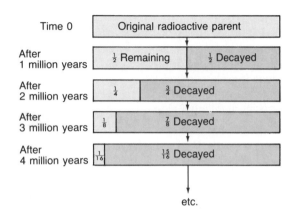

FIGURE 9.4
Decay of a hypothetical radioactive isotope with a half-life of one million years.

tween the amount of daughter product and the amount of parent will indicate how long the radioactive "clock" has been operating. Figure 9.4 illustrates the principle of radiometric dating using a hypothetical radioactive parent that decays directly into the stable daughter product. Its half-life is 1 million years. In this example, when the quantities of parent and daughter are equal (ratio 1:1), we know that one half-life has transpired and that the specimen is 1 million years old. When the ratio of parent to daughter reaches 1:15, we know the sample is 4 million years old.

Notice that the percentage of radioactive atoms that decay during one half-life is always the same. However, the actual number of atoms that decay with the passing of each half-life continually decreases. Thus, as the percentage of radioactive parent atoms declines, the proportion of stable daughter atoms rises, with the increase in daughter atoms just matching the drop in parent atoms (Figure 9.5). This fact is the key to radiometric dating.

Of the many radioactive isotopes that exist in nature, five have proven particularly important in providing radiometric ages for ancient rocks. Table 9.1 lists these most frequently used isotopes. Rubidium-87 and the two isotopes of uranium are used only for dating rocks that are millions of years old, but potassium-40 is more versatile. Although the half-life of potassium-40 is 1.3 billion years, recent analytical techniques have

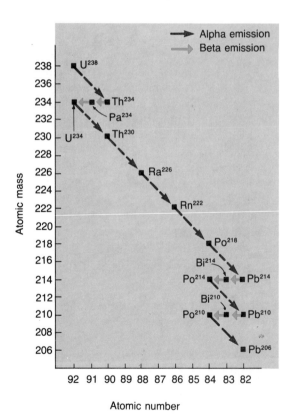

FIGURE 9.3
The most common isotope of uranium (U_{238}^{92}) is an example of a radioactive decay series. Before the stable end product (Pb_{206}^{82}) is reached, many different isotopes are produced as intermediate steps.

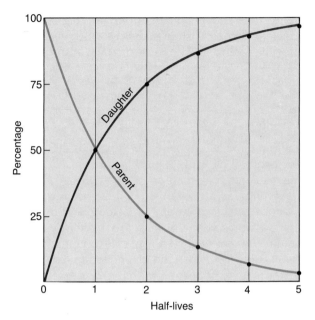

FIGURE 9.5

Decay of a radioactive isotope and growth of the daughter product. The parent/daughter ratio changes continually with time. As the proportion of parent decreases, the proportion of daughter rises, with the increase in daughter atoms just matching the decline in parent atoms.

made possible the detection of tiny amounts of its stable daughter product, argon-40, in some rocks that are younger than 100,000 years. Another important reason for its frequent use is that potassium is an abundant constituent of many common minerals, particularly micas and feldspars. The potassium-argon clock begins when potassium-bearing minerals crystallize from a magma or form within a metamorphic rock. At this point the new minerals will contain K^{40} but will be free of

Ar^{40}, because this element is an inert gas that does not chemically combine with other elements. As time passes, the K^{40} steadily decays by electron capture. The Ar^{40} produced by this process remains trapped within the mineral's crystal lattice. Since there was no Ar^{40} present when the mineral formed, all of the daughter atoms trapped in the mineral must have come from the decay of K^{40}. To determine a sample's age, the K^{40}/Ar^{40} ratio must be measured precisely and the known half-life for K^{40} applied.

It is important to realize that an accurate radiometric date can be obtained only if the mineral remained a closed system during the entire period since its formation; that is, a correct date is not possible unless there was neither the addition nor loss of parent or daughter isotopes. This is not always the case. In fact, an important limitation of the potassium-argon method arises from the fact that argon is a gas and may leak from the minerals in which it forms. Indeed, losses can be significant if the rock is subjected to relatively high temperatures. Of course, a reduction in the amount of Ar^{40} leads to an underestimation of the rock's age. Sometimes temperatures are high enough for a sufficiently long period that all argon escapes. When this happens, the potassium-argon clock is reset and dating the sample will give only the time of thermal resetting, not the true age of the rock. For other radiometric clocks, a loss of daughter atoms can occur if the rock has been subjected to weathering or leaching. To avoid such a problem, one simple safeguard is to use only fresh, unweathered material and not samples that may have been chemically altered.

If parent/daughter ratios are not always reliable, how can meaningful radiometric dates be

TABLE 9.1

Radioactive isotopes frequently used in radiometric dating.

Radioactive Parent	Stable Daughter Product	Currently Accepted Half-life Values
Uranium-238	Lead-206	4.5 billion years
Uranium-235	Lead-207	713 million years
Thorium-232	Lead-208	14.1 billion years
Rubidium-87	Strontium-87	47.0 billion years
Potassium-40	Argon-40	1.3 billion years

obtained? One common precaution against unknown errors is the use of cross checks. Often this simply involves subjecting a sample of two different radiometric methods. If the two dates agree, the likelihood is high that the date is reliable. If, on the other hand, there is an appreciable difference between the two dates, other cross checks must be employed to determine which, if either, is correct.

To date very recent events, carbon-14 (also called *radiocarbon*), the radioactive isotope of carbon, is used. Since the half-life of carbon-14 is only 5730 years, it can be used for dating events from the historic past as well as those from recent geologic history. Until the late 1970s radiocarbon was useful in dating events only as far back as 40,000–50,000 years. However, as was the case with potassium-40, the development of more sophisticated analytical techniques has increased the usefulness of this "clock." In some cases carbon-14 can now be used to date events as far back as 75,000 years. This is a significant accomplishment because geologists can now date many ice-age phenomena that previously could not be dated accurately.

Carbon-14 is continuously produced in the upper atmosphere as a consequence of cosmic ray bombardment, in which cosmic rays (high-energy nuclear particles) shatter the nuclei of gases, releasing neutrons. The neutrons are absorbed by nitrogen (atomic number 7, mass number 14), causing its nucleus to emit a proton. As a result, the atomic number decreases by 1 (to 6), and a new element, carbon-14, is created (Figure 9.6A). This isotope of carbon is quickly incorporated into carbon dioxide, circulates in the atmosphere, and is absorbed by living matter. As a result, all organisms contain a small amount of carbon-14.

As long as an organism is alive, the decaying radiocarbon is continually replaced, and the proportions of carbon-14 and carbon-12 (the stable and most common isotope of carbon) remain constant. However, when the plant or animal dies, the amount of carbon-14 gradually decreases as it decays to nitrogen-14 by beta emission (Figure 9.6B). By comparing the proportions of carbon-14 and carbon-12 in a sample, radiocarbon dates can be determined. Although carbon-14 is only useful in dating the last small fraction of geologic time, it has become a very valuable tool for anthropologists, archeologists, and historians, as well as for geologists who study very recent earth history. In fact, the development of radiocarbon dating was

FIGURE 9.6
A. Production and **B.** decay of carbon-14.

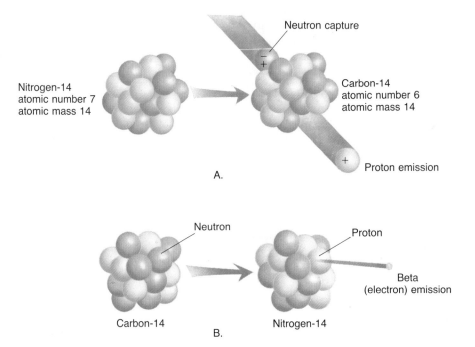

considered so important that the chemist who discovered this application, Willard F. Libby, received a Nobel Prize.

Bear in mind that although the basic principle of radiometric dating is rather simple, the actual procedure is quite complex, for the analysis which determines the quantities of parent and daughter that are present must be painstakingly precise. In addition, some radioactive materials do not decay directly into the stable daughter product as was the case with our hypothetical example, a fact which may further complicate the analysis. In the case of uranium-238, there are thirteen intermediate unstable daughter products formed before the fourteenth and last daughter product, the stable isotope lead-206, is produced (see Figure 9.3).

Radiometric dating methods have produced literally thousands of dates for events in earth history. Rocks from several localities have been dated at more than 3 billion years, and geologists realize that still-older rocks exist. For example, a granite from South Africa has been dated at 3.2 billion years and contains inclusions of quartzite. Quartzite is a metamorphic rock which originally was the sedimentary rock sandstone. Since sandstone is the product of the lithification of sediments produced by the weathering of pre-existing rocks, we have a positive indication that older rocks existed.

Radiometric dating has vindicated the ideas of Hutton, Darwin, and others who over 150 years ago assumed that geologic time must be immense. Indeed, it has proven that there has been enough time for the slow processes we observe to have accomplished tremendous tasks.

THE MAGNITUDE OF GEOLOGIC TIME

The magnitude of geologic time is difficult to grasp, because we must learn to think in spans of time that far exceed our common experience. Earth features, which seem to be everlasting and unchanging to us and in fact to generations of people, are indeed slowly changing. Over millions of years, mountains rise and are eroded to hills, and rivers excavate deep canyons. How long is 5 billion years? If you were to begin counting to 5 billion at the rate of one number per second and

continued 24 hours a day, 7 days a week, and never stopped, it would take about two lifetimes (150 years) to reach 5 billion! Don L. Eicher gives us another basis for comparison:

> Compress for example, the entire 4.5 billion years of geologic time into a single year. On that scale, the oldest rocks we know date from about mid-March. Living things first appeared in the sea in May. Land plants and animals emerged in late November and the widespread swamps that formed the Pennsylvanian coal deposits flourished for about four days in early December. Dinosaurs became dominant in mid-December, but disappeared on the 26th, at about the time the Rocky Mountains were first uplifted. Manlike creatures appeared sometime during the evening of December 31st, and the most recent continental ice sheets began to recede from the Great Lakes area and from northern Europe about 1 minute and 15 seconds before midnight on the 31st. Rome ruled the Western world for 5 seconds from 11:59:45 to 11:59:50. Columbus discovered America 3 seconds before midnight, and the science of geology was born with the writings of James Hutton just slightly more than one second before the end of our eventful year of years.*

RELATIVE DATING

Radiometric dating results in specific dates for rock units which represent various events in the earth's distant past. We can now state with some confidence that a particular geologic event took place a certain number of years ago. Such dates are referred to as **absolute dates,** for they pinpoint the time in history when something occurred. Prior to the discovery of radioactivity and the development of the technology of radiometric dating, geologists had no reliable method of absolute dating and had to depend solely on relative dating. **Relative dating** means that rocks are placed in their proper sequence or order. Relative dating will not tell us how long ago something took place, only that it followed one event and preceded another. The relative dating techniques which were developed are still widely used. Absolute dating methods did not replace these techniques; they

---

*Don L. Eicher. *Geologic Time* (Englewood Cliffs, New Jersey: Prentice-Hall, 1968), p. 19.

FIGURE 9.7
Applying the law of superposition to this cross section, the shale bed is oldest and the sandstone is youngest.

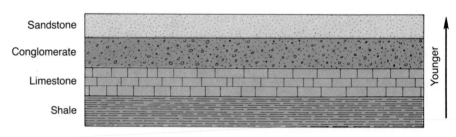

Sandstone

Conglomerate

Limestone

Shale

Younger

simply supplemented them. To establish a relative time scale, a few simple principles or rules had to be discovered and applied. Although they may seem rather obvious to us today, their discovery was a very important scientific achievement.

Nicolaus Steno, a physician in Florence, Italy, is credited with being the first to recognize a sequence of historical events in an outcrop of sedimentary rock layers. Working in the mountains of western Italy, Steno applied a very simple rule that has come to be the most basic principle of relative dating—the **law of superposition.** The law simply states that in an undeformed sequence of sedimentary rocks, each bed is older than the one above it and younger than the one below (Figure 9.7). Although it may seem obvious that a layer

could not be deposited with nothing beneath it for support, it was not until 1669 that Steno clearly stated the principle. This rule also applies to other surface-deposited materials such as lava flows and beds of ash from volcanic eruptions.

Steno is also credited with recognizing the importance of another basic principle, called the **principle of original horizontality.** Simply stated, most layers of sediment are deposited in a nearly horizontal position. Thus, if we observe rock layers that are folded or inclined at a steep angle, they must have been moved into that position by crustal disturbances sometime after their deposition.

When igneous intrusions or faults cut through other rocks, they are assumed to be younger than the structures they cut. For example, when two dikes intersect, the older one must have been opened up in order to allow the younger one to cut through it. The younger dike would be continuous, while the older dike would be interrupted at the point of their intersection. Figure 9.8 illustrates this principle of **cross-cutting.**

Sometimes inclusions can aid the relative dating process. **Inclusions** are pieces of one rock unit that are contained within another. The basic principle is logical and straightforward. The rock mass adjacent to the one containing the inclusions must have been there first in order to provide the rock fragments. Therefore, the rock mass containing inclusions is the younger of the two. Figure 9.9 provides an example. Here the inclusions of granite in the adjacent sedimentary layer indicate that the sedimentary layer was deposited on top of an eroded mass of granite rather than being intruded from below by a younger granite.

Layers of rock are said to be **conformable** when they are found to have been deposited essentially without interruption. Although particular sites

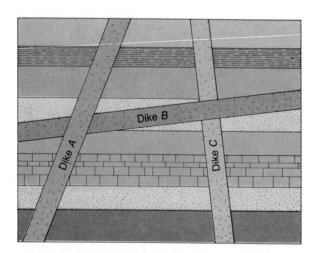

FIGURE 9.8
Cross-cutting relationships. All of the dikes are younger than the rock into which they were intruded. Since Dike *B* cuts through Dike *C* and Dike *A* cuts through Dike *B*, the order of intrusion from oldest to youngest is Dike *C*, Dike *B*, Dike *A*.

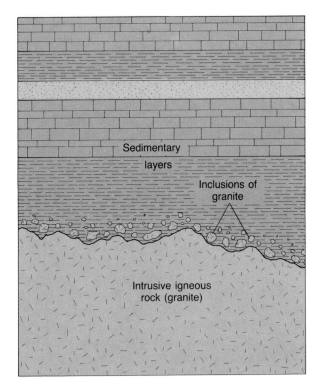

FIGURE 9.9
Since pieces of granite are contained within the over-lying sedimentary bed, we know the granite must be older. When older intrusive igneous rocks are overlain by younger sedimentary layers, a type of unconformity termed a nonconformity is said to exist.

may exhibit conformable beds representing significant spans of geologic time, there is no place on earth that contains a full set of conformable strata. Throughout earth history, the deposition of sediment has been interrupted over and over again. All such breaks in the rock record are termed unconformities. An **unconformity** represents a long period of time during which deposition ceased, erosion removed previously formed rocks, and then deposition resumed. In each case uplift and erosion are followed by subsidence and renewed sedimentation. Unconformities are important features because they represent significant geologic events in local earth history. Moreover their recognition helps us identify what intervals of time are not represented by strata.

Perhaps the most easily recognized type of unconformity consists of tilted or folded sedimentary

rocks that are overlain by younger, more flat-lying strata. These are **angular unconformities** and indicate that during the pause in deposition, a period of deformation (folding or tilting) as well as erosion occurred (Figure 9.10). More common, but usually far less conspicuous, are unconformi-

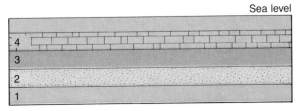

A. Deposition

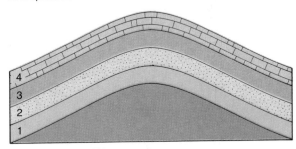

B. Folding and uplifting

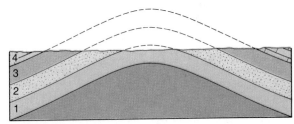

C. Erosion

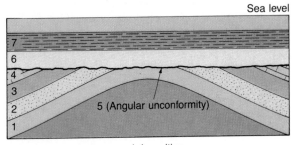

D. Subsidence and renewed deposition

FIGURE 9.10
Formation of an angular unconformity. An angular unconformity represents a period during which deformation and erosion occurred.

ties in which the strata on either side are essentially parallel. When this is the case, the term **disconformity** is applied. Many disconformities are difficult to identify because the rocks above and below are similar and there is little evidence of erosion. Such a break often resembles an ordinary bedding plane. Other disconformities are easier to identify because the ancient erosion surface is cut deeply into the older rocks below.

Nonconformities are the third basic type of unconformity. Here the break separates older metamorphic or intrusive igneous rocks from younger sedimentary strata (see Figure 9.9). Just as angular unconformities and disconformities imply crustal movements, so too do nonconformities. Intrusive igneous masses and metamorphic rocks originate far below the surface. Thus, for a nonconformity to develop, there must be a period of uplift and the erosion of overlying rocks. Once exposed at the surface, the igneous or metamorphic rocks are subjected to weathering and erosion prior to subsidence and the renewal of sedimentation.

By applying the principles of relative dating to the hypothetical geologic cross section shown in Figure 9.11, the rocks and the events in earth history they represent can be placed into their proper sequence. The following statements summarize the logic used to interpret the cross section:

1 Applying the law of superposition, beds *A*, *B*, *C*, and *E* were deposited, in that order. Since bed *D* is a sill (a concordant igneous intrusion), it is younger than the rock into which it was intruded. Further evidence that the sill is younger than beds *C* and *E* are the inclusions in the sill of fragments from these beds. If the igneous mass contains pieces of surrounding rock, the surrounding rock must have been there first.

2 Following the intrusion of the sill *(D)*, the intrusion of the dike *(F)* occurred. Since the dike cuts through beds *A* through *E*, it must be younger than all of them.

3 Next, the rocks were tilted and then eroded. We know the tilting happened first because the upturned ends of the strata have been eroded. The tilting and erosion, followed by further deposition, produced an angular unconformity.

4 Beds *G*, *H*, *I*, *J*, and *K* were deposited in that order, again using the law of superposition. Although the lava flow (bed *H*) is not a sedimentary rock layer, it is a surface-deposited layer, and thus superposition must be applied.

5 Finally, the irregular surface and the stream valley indicate that another gap in the rock record is being produced by erosion.

FIGURE 9.11
Geologic cross section of a hypothetical region.

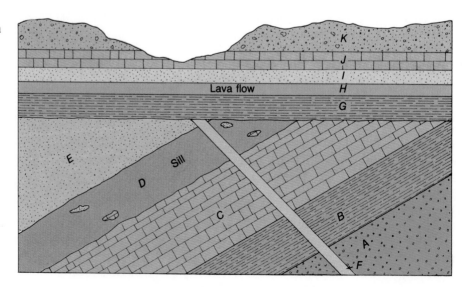

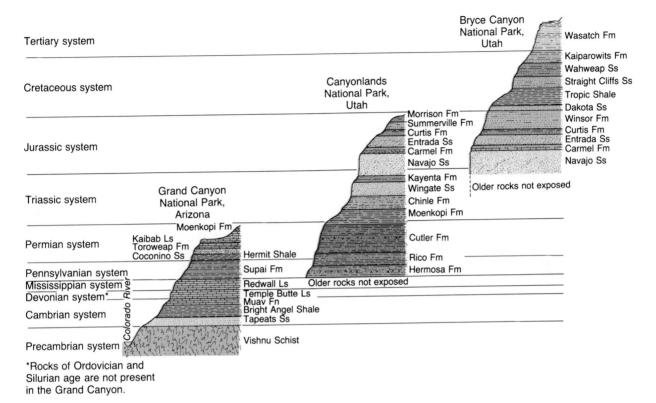

FIGURE 9.12

Correlation of strata at three locations on the Colorado Plateau reveals the extent of sedimentary rocks in the region. (After U.S. Geological Survey)

In the foregoing example our goal was to establish a relative time scale for the rocks and events in the area of the cross section. Remember, we do not have any idea how many years of earth history are represented, nor do we know how this area compares to any other.

CORRELATION

In order to develop a geologic time scale that is applicable to the whole earth, rocks of similar age in different regions must be matched up. Such a task is referred to as **correlation.** Within a limited area there are several methods of correlating the rocks of one locality with those of another. A bed or series of beds may be traced simply by walking along the outcropping edges. This may not be possible, however, when the continuity of the bed is interrupted. Correlation over short distances is often achieved by noting the place of a bed in a sequence of strata, or a bed may be identified in another location if it is composed of very distinctive or uncommon minerals. By correlating the rocks from one place to another, a more comprehensive view of the geologic history of a region is possible (Figure 9.12).

Most geologic studies involve rather small areas. Although they are usually important in their own right, their full value is realized only when the rocks are correlated with those of other regions. Although the methods just described may be sufficient to trace a rock formation over relatively short distances, they are not adequate for matching rocks at great distance. When correlation between widely separated areas or between continents is the objective, the geologist must rely upon fossils.

FIGURE 9.13
Petrified wood in Petrified Forest National Park, Arizona. (Photo by Steven Trimble)

FOSSILS

Paleontology, the branch of geology devoted to the study of ancient life, is based on the study of fossils. Originally the term **fossil** referred to any curious object dug from the ground, but it is now used to mean the remains or traces of organisms preserved from the geologic past. Although geologists studying past life must have a firm background in the biological sciences, they are often quick to point out the differences between the two fields: "If the remains stink they belong to zoology, but if not, to paleontology."

Fossils are of many types. The remains of relatively recent organisms may not have been altered at all. Such objects as teeth, bones, and shells are common examples. Far less common are entire animals, flesh included, that have been preserved because of rather unusual circumstances. Remains of prehistoric elephants called mammoths that were frozen in the Arctic tundra of Siberia and Alaska are examples, as are the mummified remains of sloths preserved in a dry cave in Nevada.

Given enough time, the remains of an organism are likely to be modified. Often fossils become *pet-rified* (literally, "turned into stone") meaning that the small internal cavities and pores of the original structure are filled with precipitated mineral matter (Figure 9.13). In other instances *replacement* may occur. Here the cell walls and other solid material are removed and replaced with mineral matter. Sometimes the microscopic details of the replaced structure are faithfully retained.

Molds and casts constitute another common class of fossils. When a shell or other structure is buried in sediment and then dissolved by underground water, a *mold* is created. The mold faithfully reflects only the shape and surface markings of the organism, but does not reveal any information concerning its internal structure. If these hollow spaces are subsequently filled with mineral matter, *casts* are created (Figure 9.14).

A type of fossilization called *carbonization* is particularly effective in preserving leaves and delicate animal forms. It occurs when fine sediment encases the remains of an organism. As time passes, pressure squeezes out the liquid and gaseous components and leaves behind a thin residue of carbon (Figure 9.15). Black shales deposited as organic-rich mud in oxygen-poor

FIGURE 9.14
Natural casts of shelled invertebrates. (Photo by E. J. Tarbuck)

FIGURE 9.15
A fossil bee preserved as a thin carbon film. (Photo courtesy of the National Park Service)

environments often contain abundant carbonized remains. If the film of carbon is lost from a fossil preserved in fine-grained sediment, a replica of the surface, called an *impression,* may still show considerable detail (Figure 9.16).

Delicate organisms, such as insects, are difficult to preserve and consequently are quite rare in the fossil record. Not only must they be protected from decay, they must not be subjected to any pressure that would crush them. One way in which some insects have been preserved is in *amber,* the hardened resin of ancient trees. The fly in Figure 9.17 was preserved after being trapped

FIGURE 9.16
Impressions are common fossils and often show considerable detail. (Photo by E. J. Tarbuck)

FIGURE 9.17
Insect in amber. (Courtesy of Ward's Natural Science Establishment, Inc., Rochester, N.Y.)

in a drop of sticky resin. Resin sealed off the insect from the atmosphere and protected the remains from damage by water and air. As the resin hardened, a protective, pressure-resistant case was formed.

In addition to the fossils already mentioned, there are numerous other types, many of them only traces of prehistoric life. Examples of such indirect evidence include:

1 Tracks—footprints made by animals in soft sediment that was later lithified.
2 Burrows—tubes in sediment, wood, or rock made by an animal. These holes may later become filled with mineral matter and preserved. Some of the oldest-known fossils are believed to be worm burrows.
3 Coprolites—fossil dung and stomach contents that can provide useful information pertaining to food habits of organisms.
4 Gastroliths—highly polished stomach stones that were used in the grinding of food by some extinct reptiles.

CONDITIONS FAVORING PRESERVATION

Only a tiny fraction of the organisms that have lived during the geologic past have been preserved as fossils. Normally the remains of an animal or plant are totally destroyed. Under what circumstances are they preserved? Two special conditions appear to be necessary: rapid burial and the possession of hard parts.

Usually when an organism perishes, its soft parts are quickly eaten by scavengers or decomposed by bacteria. Occasionally, however, the remains are buried by sediment. When this occurs the remains are removed from the environment where destructive processes operate. Rapid burial therefore is an important condition favoring preservation.

In addition, animals and plants have a much better chance of being preserved as part of the fossil record if they have hard parts. Although traces and imprints of soft-bodied animals such as jellyfish, worms, and insects exist, they are rare. Flesh usually decays so rapidly that preservation is exceedingly remote. Hard parts like shells, bones, and teeth predominate in the record of past life.

Since preservation is contingent on special conditions, the record of life in the geologic past is biased. The fossil record of those organisms with hard parts that lived in areas of sedimentation is quite abundant. However, we only get an occasional glimpse of the vast array of other life forms that did not meet the special conditions favoring preservation.

FOSSILS AND CORRELATION

The existence of fossils had been known for centuries, yet it was not until the late 1700s and early 1800s that their significance as geologic tools was made evident. During this period an English engineer and canal builder, William Smith, discovered that each rock formation in the canals contained fossils unlike those in the beds either above or below. Further, he noted that sedimentary strata in widely separated areas could be identified by their distinctive fossil content. Based upon Smith's classic observations and the findings of many geologists who followed, one of the most important and basic principles in historical geology was formulated: Fossil organisms succeed one another in a definite and determinable order, and therefore any time period can be recognized by its fossil content. This has come to be known as the **principle of faunal succession.** In other words, when fossils are arranged according to their age by using the law of superposition on the rocks in which they are found, they do not present a random or haphazard picture. To the contrary, fossils show progressive changes from simple to complex and reveal the advancement of life through time. For example, an Age of Trilobites is recognized quite early in the fossil record. Then, in succession, paleontologists recognize an Age of Fishes, an Age of Coal Swamps, an Age of Reptiles, and an Age of Mammals. These "ages" pertain to groups that were especially plentiful and characteristic during particular time periods. Within each of the "ages" there are many subdivisions based, for example, on certain species of trilobites, and certain types of fish, reptiles, and so on. This same succession of dominant organisms, never out of order, is found on every major landmass.

Since fossils were found to be time indicators, they became the most useful means of correlating

FIGURE 9.18
Overlapping ranges of fossils help date rocks more exactly than using a single fossil.

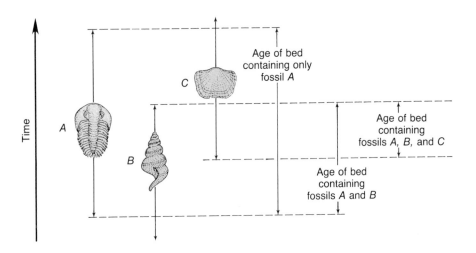

rocks of similar age in different regions. Geologists pay particular attention to certain fossils called **index fossils.** Due to the fact that these fossils are widespread geographically and are limited to a short span of geologic time, their presence provides an important method of matching rocks of the same age. Rock formations, however, do not always contain a specific index fossil. In such situations, groups of fossils are used to establish the age of the bed. Figure 9.18 illustrates how a group of fossils can be used to date rocks more precisely than could be accomplished by the use of only one of the fossils.

In addition to being important and often essential tools for correlation, fossils are important environmental indicators. Although much can be deduced about past environments by studying the nature and characteristics of sedimentary rocks, a close examination of the fossils present can usually provide a great deal more information. For example, when the remains of certain clam shells are found in limestone, the geologist feels it is quite reasonable to assume that the region was once covered by a shallow sea. Also, by using what we know of living organisms, we can conclude that fossil animals with thick shells capable of withstanding pounding and surging waves inhabited shorelines. On the other hand, animals with thin, delicate shells probably indicate deep, calm offshore waters. Hence by taking a closer look at the types of fossils, the approximate shoreline may be identified. Further, fossils can be used to indicate the former temperature of the water. Certain kinds of present-day corals must live in warm and shallow tropical seas like those around Florida and the Bahamas. When similar types of coral are found in ancient limestones, they give a good estimate of the marine environment that must have existed when they were alive. The preceding are just a few brief examples of how fossils can help unravel the complex story of earth history.

THE GEOLOGIC TIME SCALE

The whole of geologic history has been divided into units of varying magnitude, which together comprise the time scale of earth history (Figure 9.19, page 258). The major units of the calendar were delineated during the nineteenth century, principally by workers in western Europe and Great Britain. Since absolute dating was not a reality during this time, the entire calendar was created using methods of relative dating. Absolute dates have only recently been added.

By examining Figure 9.19, you can see that the largest of the subdivisions of the geologic time scale are called **eras.** Three eras are currently recognized: the **Paleozoic** ("ancient life"), the **Mesozoic** ("middle life"), and the **Cenozoic** ("recent life"). As the names imply, the eras are bounded by quite profound worldwide changes in life forms. Each era is subdivided into time units known as **periods.** The Paleozoic has seven, the Mesozoic three, and the Cenozoic two. Since we

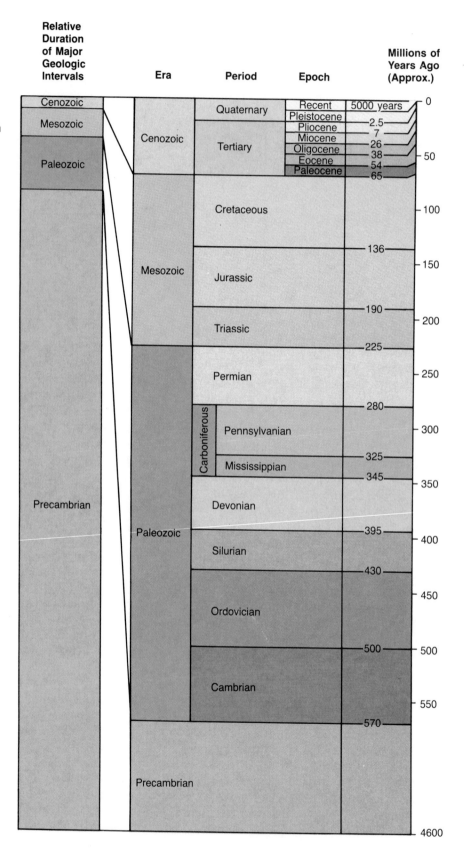

FIGURE 9.19
The geologic time scale. Absolute dates were added quite recently, long after the time scale had been established using relative dating techniques. The Precambrian accounts for more than 85 percent of geologic time.

Relative Duration of Major Geologic Intervals

Era	Period	Epoch	Millions of Years Ago (Approx.)
Cenozoic	Quaternary	Recent	5000 years — 0
		Pleistocene	2.5
	Tertiary	Pliocene	7
		Miocene	26
		Oligocene	38
		Eocene	54 — 50
		Paleocene	65
Mesozoic	Cretaceous		— 100
			136 — 150
	Jurassic		
			190 — 200
	Triassic		225
Paleozoic	Permian		— 250
			280
	Carboniferous	Pennsylvanian	— 300
			325
		Mississippian	345 — 350
	Devonian		395 — 400
	Silurian		430
	Ordovician		— 450
			500 — 500
	Cambrian		— 550
			570
Precambrian			4600

are currently living in the Cenozoic era, there may be more periods yet to come. Each period is characterized by a somewhat less profound change in life forms as compared with the eras. Finally, each of the twelve periods is divided into still smaller units called **epochs.** Except for the seven epochs which have been named for the periods of the Cenozoic era, those of other periods are not commonly referred to by specific names. Rather, the terms *early, middle,* and *late* are generally applied to the epochs of these earlier periods.

Notice that the detail of the geologic time scale does not begin until about 570 million years ago, the date for the beginning of the first period of the Paleozoic era: the Cambrian period. The more than 4 billion years prior to the Cambrian is simply referred to as the **Precambrian.** In terms of life forms, the Precambrian is often given another name, the *Cryptozoic eon.* The term *Cryptozoic* is derived from the Greek words for *hidden* and *life,* and refers to the obscure fossil record of this time span. By contrast, the time span beginning with the Cambrian period and extending to the present is referred to as the *Phanerozoic eon.* This term is also derived from the Greek and means *visible life.* Thus, the Phanerozoic eon represents the age of abundant life. A brief overview of the major changes in life forms that occurred during this span can be found near the end of this chapter.

Why is the huge expanse of Precambrian time not divided into numerous eras, periods, and epochs? The reason is that Precambrian history is not known in great enough detail. The quantity of information geologists have deciphered about the earth's past is somewhat analogous to the detail of human history. The farther back we go, the less that we know. Certainly more data and information exist about the past ten years than for the first decade of the twentieth century; the events of the nineteenth century have been documented much better than the events of the first century A.D.; and so on. So it is with earth history. The more recent past has the freshest, least disturbed, and most observable record. The farther back in time the geologist goes, the more fragmented the record and clues become. There are other reasons to explain our lack of a detailed time scale for this vast segment of earth history:

1   The first abundant fossil evidence does not appear in the geologic record until the beginning of the Cambrian period. Prior to the Cambrian, very simple life forms such as algae, bacteria, fungi, and worms predominated. All of these organisms lack hard parts, an important prerequisite for fossilization. For this reason, there is only a meager Precambrian fossil record. Many exposures of Precambrian rocks have been studied in some detail, but correlation is often very difficult when fossils are lacking. Consequently, there is no general agreement on the subdivisions of the Precambrian portion to the time scale.

2   Because Precambrian rocks are very old, many have been subjected to a great variety of changes. Significant portions of the Precambrian rock record are composed of highly distorted metamorphic rocks. This makes the interpretation of past environments very difficult, for many of the clues found in sedimentary rocks have been destroyed.

With the development of radiometric dating methods, a partial solution to the troublesome task of dating and correlating Precambrian rocks now exists. Untangling the complex Precambrian record, however, is still many years away.

DIFFICULTIES IN DATING THE GEOLOGIC TIME SCALE

Although reasonably accurate absolute dates have been worked out for the periods of the geologic time scale (see Figure 9.19), the task is not without its difficulties. The primary difficulty in assigning absolute dates to units of time is the fact that not all rocks can be dated radiometrically. Recall that for a radiometric date to be useful, all minerals in the rock must have formed at approximately the same time. For this reason, radioactive isotopes can be used to determine when minerals in an igneous rock crystallized and when pressure and heat created new minerals in a metamorphic rock. However, samples of sedimentary rock can only rarely be dated directly by radiometric

FIGURE 9.20
Absolute dates for sedimentary layers are usually determined by examining their relationship to igneous rocks. (After U.S. Geological Survey)

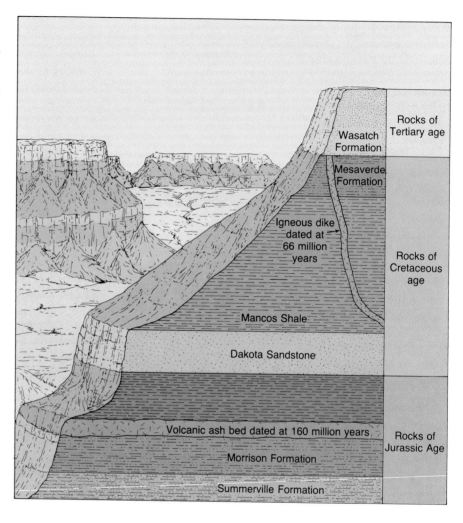

means. Although a detrital sedimentary rock may include particles that contain radioactive isotopes, the rock's age cannot be accurately determined because the grains composing the rock are not the same age as the rock in which they occur. Rather, the sediments have been weathered from rocks of diverse ages. Radiometric dates obtained from metamorphic rocks may also be difficult to interpret, since the age of a particular mineral in a metamorphic rock does not necessarily represent the time when the rock initially formed. Instead, the date may indicate any one of a number of subsequent metamorphic phases.

If samples of sedimentary rocks rarely yield reliable radiometric ages, how can absolute dates be assigned to sedimentary layers? Usually the geolo-

gist must relate them to igneous masses, as in Figure 9.20. In this example, the age of the volcanic ash bed within the Morrison Formation and the dike cutting the Mancos Shale and Mesaverde Formation are known. The sedimentary beds below the ash are obviously older than the ash, and all the layers above the ash are younger. The dike is younger than the Mancos Shale and the Mesaverde Formation but older than the Wasatch Formation because the dike does not intrude the Tertiary rocks. From this kind of evidence, geologists estimate that a part of the Morrison Formation was deposited about 160 million years ago as indicated by the ash bed. Further, they conclude that the Tertiary period began after the intrusion of the dike, 66 million years ago. This is one example of the literally thousands that illustrates how dated materials are used to bracket the various episodes in earth history within specific time periods and illustrates the necessity of combining laboratory dating methods with field observations of rocks.

LIFE OF THE GEOLOGIC PAST

A century ago the earliest fossils known dated from the beginning of the Cambrian period. An important unanswered problem facing science at that time was the abrupt appearance in the geologic record of complex organisms. Today we know that life existed during the Precambrian, but the forms were simple and soft bodied. Although fossils from that time are rare, well-preserved remains of minute organisms such as bacteria and blue-green algae have been discovered that extend the record of life back beyond 3.1 billion years. The fossil record of animals dates from the late Precambrian and includes distinctive trails and burrows believed to have been created by elongate wormlike animals. Impressions showing complete shapes have also been discovered. Although most, if not all, Precambrian animals were soft bodied and lacked shells, the fossil record is sufficient to demonstrate the existence of a diverse and complex group of multicelled organisms as the Precambrian ended. Thus, the stage was ready for the appearance of even more complex organisms with preservable hard parts as the Cambrian period began.

PALEOZOIC LIFE

The geologic time scale shows that the last 750 million years of earth history are divided into three eras: Paleozoic, Mesozoic, and Cenozoic. The Paleozoic era, which encompasses more than 345 million years, is by far the longest. Its beginning is marked by the appearance of life forms that possessed hard parts. Hard parts greatly enhanced an organism's chance of being preserved as part of the fossil record. Therefore, our knowledge of the diversification of life improves greatly from the Paleozoic onward (Figure 9.21, page 262). This abundant fossil evidence has allowed geologists to construct a far more detailed time scale for the last one-eighth of geologic time than for the preceding seven-eighths. Moreover, since every organism is associated with a particular environment, the greatly improved fossil record provided vast and invaluable information for deciphering past geologic environments.

Life in early Paleozoic time was restricted to the seas and consisted of several invertebrate groups (Figure 9.22, page 263). Included were trilobites, graptolites, brachiopods, bryozoans, and mollusks. The Cambrian period was the golden age of trilobites. There were more than 600 genera of these mud-burrowing scavengers. By Ordovician times, a group of small marine invertebrates called brachiopods outnumbered the trilobites. Brachiopods are among the most common and abundant Paleozoic fossils. Although the adults lived attached to the sea floor, the young larvae were free-swimming. This mobility accounts for the group's wide geographic distribution. The Ordovician also marked the appearance of abundant cephalopods (Figure 9.22, page 263). These highly developed mollusks were very mobile and became the major predators of their time. Cephalopods, whose descendants include the modern squid, octopus, and nautilus, were the first truly large organisms on earth. Whereas the largest trilobites seldom exceeded 30 centimeters (12 inches) in length and the biggest brachiopods were no more than about 20 centimeters (8 inches) across, one species of cephalopod reached a length of nearly 10 meters (30 feet).

During most of the late Paleozoic, numerous groups of organisms diversified greatly. As the Silurian gave way to the Devonian, some 400 million years ago, plants that had adapted to survival at the water's edge began to move inland. These earliest land plants were leafless, vertical spikes about the size of a person's index finger. However, by the end of the Devonian, the fossil record indicates the existence of forests with trees tens of meters high.

In the oceans, armor-plated fishes that had evolved during the Ordovician made major adaptive radiations. Their armor plates became thin, light-weight scales, which increased their speed and mobility. Other fishes evolved during the Devonian, including primitive sharks and boney fishes, the groups to which virtually all modern fishes belong. The Devonian period was a time when fishes evolved rapidly; thus, it is often called the "age of fishes."

By late Devonian time, two groups of boney fishes, the lung fish and the lobe-finned fish, adapted to land environments. Like their modern relatives, these fishes had primitive lungs that supplemented the exchange of gases through the gills. Lobe-finned fish are believed to have occupied tidal flats or small ponds. Moreover, in times of drought they may have been able to use their boney fins to "walk" from dried-up pools in search of other ponds. Through time, the lobe-finned fish

began to rely more on their lungs and less on their gills. By late Devonian time, true air-breathing amphibians with fish-like heads and tails began to invade the land.

Although modern amphibians, which include frogs, toads, and salamanders, are small and occupy rather limited biological niches, the conditions during the remainder of the Paleozoic were ideal for these newcomers to the land. Plants and insects, which were their main diet, were very abundant and large. Having only minimal competition from other land dwellers, the amphibians rapidly diversified. Some groups took on roles and

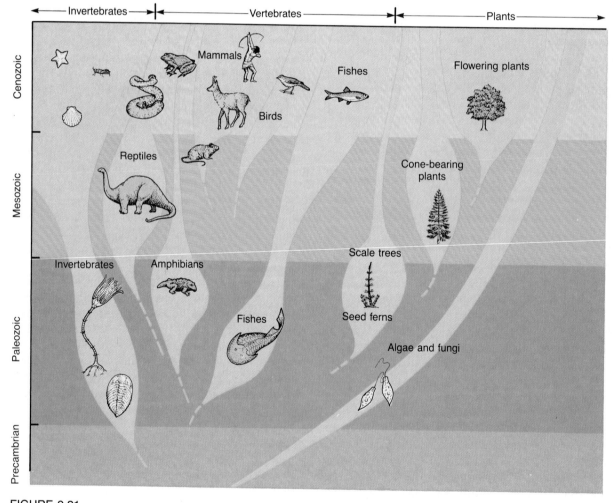

FIGURE 9.21
This chart indicates the times of appearance and relative abundance of major groups of organisms.

FIGURE 9.22
The shallow waters of an Ordovician inland sea contained an abundance of marine invertebrates. Shown here are straight-shelled cephalopods, trilobites, brachiopods, snails, and corals. (Courtesy of the Field Museum of Natural History, Chicago)

forms that were more similar to modern reptiles, such as crocodiles, than to modern amphibians.

By the Pennsylvanian period, large tropical swamps extended across North America, Europe, and Siberia (Figure 9.23, page 264). Trees grew to heights approaching 30 meters (100 feet) with trunks over one meter across. Common trees of the time were the scale trees *Lepidodendron* and *Sigillaria,* the seed ferns, and the scouring rushes called *Calamites*. The coal deposits that fueled the industrial revolution originated in these vast swamps. Further, it was in the lush coal-swamp environment of the late Paleozoic that the amphibians radiated quickly into a variety of species.

The Paleozoic ended with the Permian period, a time when the earth's major landmasses joined to form the supercontinent Pangaea. The redistribution of land and water that resulted from the creation of Pangaea, as well as the changes in elevations of the landmasses, brought about pronounced changes in world climates. Much of the

northern continent was elevated above sea level and the climate trended toward a period of greater aridity. This climatic change caused the great scale trees of the coal swamps to become nearly extinct. By the close of the Permian, 75 percent of the amphibian families had disappeared, and plants declined both in number and variety. Although the amphibians declined, their descendants became the most successful and advanced animals on earth. Marine life was not spared during this time of mass extinctions. Many marine invertebrates that had been dominant during the Paleozoic, including all of the remaining trilobites, as well as some types of corals and brachiopods, failed to adapt to the widespread environmental changes at the close of the Permian. These animals, like many others, became extinct.

MESOZOIC LIFE

Spanning nearly 160 million years, the Mesozoic era is divided into three periods: the Triassic,

FIGURE 9.23
Restoration of a Pennsylvanian coal swamp. Shown are scale trees (left), seed
ferns (lower left), and scouring rushes (right). Also note the large dragonfly.
(Courtesy of the Field Museum of Natural History, Chicago)

FIGURE 9.24
A composite Mesozoic landscape showing large carnivorous and herbivorous
dinosaurs. (Courtesy of the Peabody Museum of Natural History, Yale University)

Jurassic, and Cretaceous. This episode of geologic history was a time when organisms that had survived the great Permian extinction began to diversify in spectacular ways. On land, dinosaurs became dominant and remained unchallenged for over 100 million years. For this reason the Mesozoic era is often referred to as the "age of dinosaurs."

Early geologists recognized a profound difference between the kinds of fossils in Permian strata and those discovered in younger Triassic rocks. Clearly one-half of the fossil groups in late Paleozoic rocks were missing in rocks of Mesozoic age. On this basis, it was decided to separate the Paleozoic and Mesozoic at the Permian-Triassic boundary.

The life forms at the dawn of the Mesozoic era were those that had survived the great Permian extinctions. These survivors began to diversify to fill the voids left at the close of the Paleozoic. On land, conditions favored those groups that were best able to adapt to drier climates. Among members of the plant kingdom, the gymnosperms were one such group. Unlike the first plants to invade land, the seed-bearing gymnosperms were not dependent on free-standing water for fertilization. Consequently, these plants were not restricted to life near the water's edge.

The gymnosperms quickly replaced the scale trees as the dominant trees of the Mesozoic. Ancient gymnosperms included the cycads, conifers, and ginkgoes. The cycads had large, palmlike leaves and resembled a large pineapple plant, whereas the ginkgoes had fan-shaped leaves much like their modern relatives. The largest plants in the Mesozoic forests were the conifers, whose modern descendants include the pines, firs, and junipers. The best-known fossil occurrence of these ancient trees is found in northern Arizona's Petrified Forest National Park (see Figure 9.13). Here, huge petrified logs lie exposed at the surface, having been weathered from rocks of the Triassic Chinle Formation.

Another group that readily adapted to the drier Mesozoic environment was the reptiles, the first true terrestrial life forms. Unlike the amphibians, reptiles have shell-covered eggs that can be laid on land. The elimination of a water-dwelling stage, as, for example, the tadpole stage in the develop-

ment of frogs, was an important evolutionary step. Of interest is the fact that the watery fluid within the reptilian egg closely resembles seawater in chemical composition. Because the reptile embryo develops in this watery environment, the shelled egg has been characterized as a "private aquarium" in which the embryos of these land vertebrates spend their water-dwelling stage.

With the perfection of the shelled egg, reptiles quickly became the dominant land animals. They continued this dominance for more than 160 million years. The largest, most awe-inspiring of the Mesozoic reptiles were the dinosaurs (Figure 9.24). Some of the huge dinosaurs such as *Tyrannosaurus* were carnivorous, whereas others such as the ponderous *Brontosaurus* were herbivorous. The extremely long neck of *Brontosaurus* was thought to be an adaptation for feeding on tall conifer trees. However, not all dinosaurs were large. In fact, certain forms closely resembled modern fleet-footed lizards. Further, evidence indicates that some dinosaurs, unlike their present-day relatives, were probably warm blooded.

The reptiles made one of the most spectacular adaptive radiations in all of earth history (Figure 9.25). One group, the pterosaurs, took to the air. These "dragons of the sky" possessed huge membranous wings that allowed them to exhibit rudimentary flight. Another group of reptiles, examplified by the fossil *Archaeopteryx*, led to more successful flyers—the birds. Whereas some reptiles took to the skies, others returned to the sea. Included in this latter group were the fish-eating reptiles: the plesiosaurs and ichthyosaurs. Although these reptiles became proficient swimmers, they retained their reptilian teeth and breathed by means of lungs.

At the close of the Mesozoic many reptile groups became extinct, as had other dominant life forms before them. Of the large number of Mesozoic reptiles, only a few survived to recent times. Among these are the turtles, snakes, crocodiles, and lizards. The huge land-dwelling dinosaurs, the marine plesiosaurs, and the flying pterosaurs all became extinct (Figure 9.25). The decline of the reptiles provided vacancies in habitats for the mammals. Although small and inconspicuous during the Mesozoic, the mammals rose to dominance during the Cenozoic.

65 m.y. ago

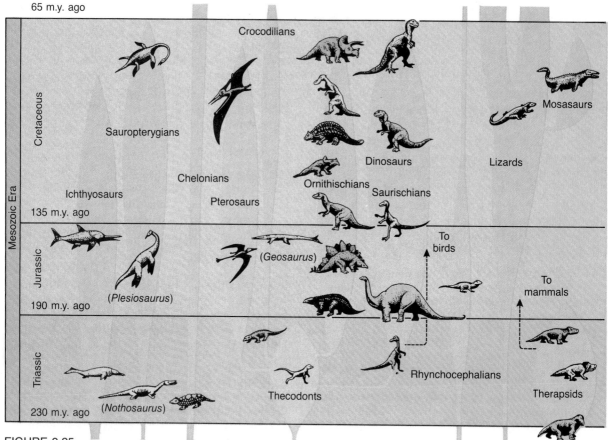

FIGURE 9.25
Adaptive radiation of reptiles during the Mesozoic era—"the golden age of reptiles." Dinosaurs, pterosaurs (flying reptiles), and large marine reptiles all became extinct by the end of the Mesozoic. (Drawing by Robert W. Tope. From Burchfiel et al. *Physical Geology*. Columbus, Ohio: Merrill, 1982)

CENOZOIC LIFE

The Cenozoic era, or "era of recent life," encompasses the past 65 million years of earth history. During this span the physical landscapes and life forms of our modern world came to be. The record for life during the Cenozoic era is more familiar than the records of preceding eras. Mammals replaced the reptiles that had dominated the land during Mesozoic time. Moreover, as the Cenozoic era opened, angiosperms (flowering plants with covered seeds) had replaced gymnosperms as the dominant land plants. Marine invertebrates also took on a modern look. Although mollusks, especially pelecypods (clams, oysters) and gastropods (snails), are the most familiar and abundant mac-

roscopic* fossils, microscopic foraminifera became especially important. Today foraminifers are among the most intensely studied of all fossils, because their widespread occurrence makes them invaluable in correlating Tertiary sediments. These strata are very important to the modern world, for they yield more oil than rocks of any other age.

The Cenozoic is often called the "age of mammals," because these animals dominated land life. It could also be called the "age of flowering plants," for the angiosperms enjoyed a similar status in the plant world. Due to advances in seed

---

Macroscopic means large enough to be examined with the unaided eye; that is, without instrumentation.

fertilization and dispersal, angiosperms experienced a period of rapid development and expansion as the Mesozoic drew to a close. Thus, as the Cenozoic era began, angiosperms were already the dominant land plants. Development of the flowering plants, in turn, strongly influenced the evolution of both birds and mammals. Birds that feed on seeds and fruits, for example, evolved rapidly during the Cenozoic in close association with the flowering plants. During the middle Tertiary, grasses developed rapidly and spread over the plains. This fostered the emergence of herbivorous (plant-eating) mammals, which were mainly grazers. The development and spread of grazing mammals in turn established the setting for the evolution of the carnivorous mammals that preyed upon them.

An important evolutionary event was the appearance of primitive mammals in late Triassic time, about the same time as the dinosaurs emerged. Yet throughout the period of dinosaur dominance, mammals remained in the background as small and inconspicuous animals. By the close of the Mesozoic era, dinosaurs and other reptiles no longer dominated the land. It was only after these large reptiles became extinct that mammals came into their own as the dominant land animals. This transition is a major example in the fossil record of the replacement of one large group by another.

Mammals are distinct from reptiles in several respects. Among these differences, mammalian young are born live and mammals maintain a constant body temperature; that is, they are "warm blooded."* This latter adaption allowed mammals to lead more active and diversified lives than reptiles, because they could survive in cold regions and search for food during any season or time of day. Other mammalian adaptations included the development of insulating body hair and a more efficient heart and lungs.

Following the reptilian extinctions at the close of the Mesozoic, two groups of mammals, the marsupials and the placentals, evolved and expanded to dominate the Cenozoic. The groups differ principally in their modes of reproduction. Young marsupials are born live but at a very early stage of development. After birth, the tiny and immature young crawl into the mother's external stomach pouch to complete their development. Placental mammals, on the other hand, develop within the mother's body for a much longer period, so that birth occurs after the young are relatively mature and independent.

Today marsupials are found primarily in Australia, where they went through a separate evolutionary expansion during the Cenozoic, largely isolated from placental mammals. In South America both primitive marsupials and placentals reached the continent before the landmass became completely isolated. Evolution and specialization of both groups continued undisturbed for approximately 40 million years, until the close of the Pliocene epoch when the Central American land bridge was established. Then an invasion of advanced carnivores from North America brought the extinction of many hoofed mammals that had persisted for millions of years. The marsupials, except for opossums, also could not compete and became extinct. Both Australia and South America provide excellent examples of how isolation caused by the separation of continents increased the diversity of animals in the world.

As we have seen, during the Cenozoic era mammals diversified quite rapidly. One tendency was for members of some groups to become very large. For example, by the Oligocene epoch a hornless rhinoceros that stood nearly 5 meters high had evolved. It is the largest land mammal yet known to have existed. As time approached the present, many other types evolved to a large size as well; in fact, more so than now exist. Many of these large forms were common as recently as 11,000 years ago. However, a wave of late Pleistocene extinctions rapidly eliminated these animals from the landscape. In North America, huge relatives of the elephant, the mastodon and mammoth, became extinct (Figure 9.26). In addition, sabertoothed cats, giant beavers, large ground sloths, horses, camels, giant bison, and others died out. In Europe, late Pleistocene extinctions included wooly rhinos, large cave bears, and the Irish elk. The reason for this recent wave of large

*One minor group of mammals, the monotremes, still lays eggs. The two species in this group, the duck-billed platypus and the spiny anteater, are found only in Australia. Moreover, although modern reptiles are "cold blooded," some paleontologists believe that dinosaurs may have been "warm blooded."

animal extinctions puzzles scientists. Since these animals had survived several major glacial advances and interglacial periods, it is difficult to ascribe these extinctions solely to climatic change. Some scientists believe that early man hastened the decline of these mammals by selectively hunting large forms. Although this hypothesis is advanced more than any other, it is not yet accepted by all.

FIGURE 9.26
Most Pleistocene mammals in this mural, including the mastodon, mammoth, and giant bison in the center, are now extinct. (Peabody Museum of Natural History, Yale University)

REVIEW QUESTIONS

1 Why did Aristotle have a "bad" influence on the science of geology?

2 Contrast the philosophies of catastrophism and uniformitarianism. How did the proponents of each perceive the age of the earth?

3 If a radioactive isotope of thorium (atomic number 90, mass number 232) emits 6 alpha particles and 4 beta particles during the course of radioactive decay, what is the atomic number and mass number of the stable daughter product?

4 Why is radiometric dating the most reliable method of dating the geologic past?

5 A hypothetical radioactive isotope has a half-life of 10,000 years. If the ratio of radioactive parent to stable daughter product is 1:3, how old is the rock containing the radioactive material?

6 In order to yield a reliable radiometric date, a mineral must remain a closed system from the time of its formation until the present. Why is this true?

7 Assume that the age of the earth is 5 billion years.
   (a) What fraction of geologic time is represented by recorded history (assume 5000 years for the length of recorded history)?
   (b) The first abundant fossil evidence does not appear until the beginning of the Cambrian period (600 million years ago). What percent of geologic time is represented by abundant fossil evidence?

8   Distinguish between absolute and relative dating.

9   What is the law of superposition? How are cross-cutting relationships used in relative dating?

10 When you observe an outcrop of steeply inclined sedimentary layers, what principle allows you to assume that the beds were tilted after they were deposited?

11 A mass of granite is in contact with a layer of sandstone but does not cut across it. Using a principle described in this chapter, explain how you might determine whether the sandstone was deposited on top of the granite or the granite was intruded from below after the sandstone was deposited.

12 Distinguish among angular unconformity, disconformity, and nonconformity.

13 What is meant by the term *correlation*?

14 Describe several types of fossils. What organisms have the best chance of being preserved as fossils?

15 Describe William Smith's important contribution to the science of geology.

16 Why are fossils such useful tools in correlation?

17 In addition to being important aids in dating and correlating rocks, how else are fossils helpful in geologic investigations?

18 What subdivisions make up the geologic time scale? What is the primary basis for differentiating the eras?

19 Explain the lack of a detailed time scale for the vast span known as the Precambrian.

20 Briefly describe the difficulties in assigning absolute dates to layers of sedimentary rock.

21 Provide the correct period or era for each of the following:
    (a) The era known as the "age of flowering plants."
    (b) By the _____ period large, tropical coal swamps existed.
    (c) The era known as the "age of dinosaurs."
    (d) The _____ period is sometimes called the "golden age of trilobites."
    (e) Mammals became the dominant land animals during the _____ era.
    (f) The period known as the "age of fishes."
    (g) The _____ period was a time of major extinctions, including 75 percent of amphibian families.

KEY TERMS

catastrophism	cross-cutting	principle of faunal succession
uniformitarianism	inclusions	
radioactivity	conformable layers	index fossil
radiometric dating	unconformity	era
half-life	angular unconformity	Paleozoic
absolute date	disconformity	Mesozoic
relative dating	nonconformity	Cenozoic
law of superposition	correlation	period
principle of original horizontality	paleontology	epoch
	fossil	Precambrian

PART TWO

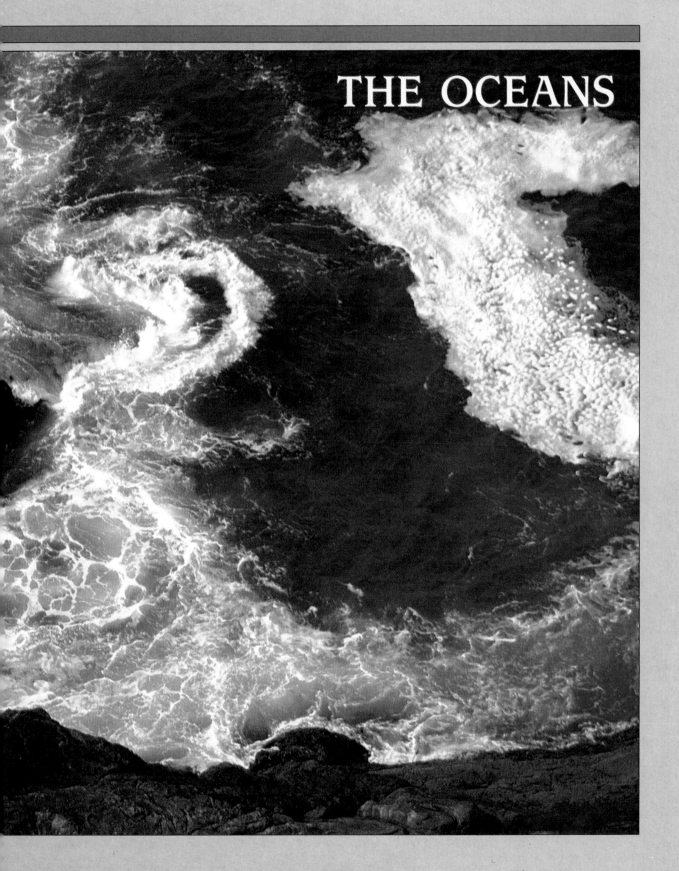

THE OCEANS

10

OCEAN WATERS
AND THE
OCEAN FLOOR

How deep is the ocean? How much of the earth is covered by the global sea? These are basic questions, but they went unanswered for thousands of years. Although people's interest in the oceans undoubtedly dates back to ancient times, it was not until rather recently that these seemingly simple questions began to be answered. The beginning of this chapter deals with these answers. Then, after examining the composition of seawater and the resources extracted from it, we shall take a brief glimpse at the history of oceanography. The remainder of the chapter provides a look at the earth beneath the sea. Suppose that all of the water were drained from the ocean. What would we see? Plains? Mountains? Canyons? Plateaus? Indeed, the ocean conceals all of these features, and more. In fact, the topography of the ocean floor is as varied as that of any continent. And what about the carpet of sediment that mantles the sea floor? Where did it come from and what can we learn by examining it?

A scenic view of the Pacific Ocean from Kauai, Hawaii. (Photo by Stephen Trimble)

273

Calling the earth the "water planet" is certainly appropriate, because nearly 71 percent of its surface is covered by the global ocean. Although the ocean comprises a much greater percentage of the earth's surface than the continents, it was only in the relatively recent past that the ocean became an important focus of study. Recently there has been a virtual explosion of data about the oceans, and with it, oceanography has grown dramatically.

Oceanography is actually not a science in itself; rather, it involves the application of all sciences in a comprehensive and interrelated study of the oceans in all of their aspects and relationships. A brief discussion of some of the major subdivisons of oceanography follows. *Physical oceanography* is concerned primarily with energy transmission through ocean water, specifically with such items as wave formation and propagation, currents, tides, energy exchange between ocean and atmosphere, and penetration of light and sound. *Chemical oceanography* is a study of the chemical properties of seawater, of the cause and effect of variation of these properties with time and from place to place, and of the means of measuring these properties. *Biological oceanography* is the study of the interrelationship of marine life with its oceanic environment. The study includes the distribution, life cycles, and population fluctuations of marine organisms. Finally, *geological oceanography* deals with the ocean floor and shore, and embraces such subjects as submarine topography, geological structure, erosion, and sedimentation. The interrelationship of all the sciences is a chief characteristic of oceanography.*

THE SEVEN SEAS?

The term *sea* is used two different ways. It is often used interchangeably with *ocean* in referring to bodies of salt water. On the other hand, a sea is also considered to be part of an ocean, or to be a body of water that is substantially smaller than an ocean.

The familiar expression *seven seas* dates back to ancient times and refers to the seas known to the Mohammedans prior to the fifteenth century. These included the Mediterranean Sea, the Red Sea, the East African Sea, the West African Sea, the China Sea, the Persian Gulf, and the Indian Ocean. In more recent times Rudyard Kipling popularized the term *seven seas* by using it as the title for a book of poetry. In fact, since Kipling's time there has been a tendency to divide the world ocean into seven parts to retain this legendary number. The popular division is the North Atlantic, South Atlan-

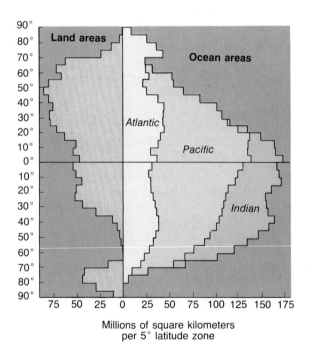

Millions of square kilometers
per 5° latitude zone

FIGURE 10.1
Distribution of land and water in each 5-degree latitude belt. (After M. Grant Gross, *Oceanography: A View of the Earth,* 2nd ed., Englewood Cliffs, N.J.: Prentice-Hall, 1977; and M. Grant Gross, *Oceanography,* 4th ed., Columbus, Ohio: Charles E. Merrill, 1980. Data from G. Wüst, W. Brogmus, and E. Noodt, "Die Zonale Verteilung von Salzgehalt, Niederschlag, Verdunstung, Temperatur und Dichte an der Oberfläche der Ozeane," *Kieler Meeresforschungen,* Band X, 1954:139)

*Adapted from H. Dubach and R. Tabor, *Questions About the Oceans,* National Oceanographic Data Center Publication G-13, p. 67.

tic, North Pacific, South Pacific, Indian, Arctic, and Antarctic. Although some speak of an Antarctic Ocean, most (including the International Hydrographic Bureau) regard these waters of the South Polar region merely as extensions of the Atlantic, Pacific, and Indian oceans. Actually, of course, all limits of oceans are arbitrary, as there is only one global ocean.

EXTENT OF THE OCEANS

The area of the earth is about 510 million square kilometers (197 million square miles). Of this total, approximately 360 million square kilometers (140 million square miles), or 71 percent, is represented by oceans and marginal seas. The continents comprise the remaining 29 percent, or 150 million square kilometers (58 million square miles).

By studying a globe or world map, it is readily apparent that the continents and oceans are not evenly divided between the Northern and South-

ern hemispheres. When we compute the percentages of land and water in the Northern Hemisphere, we find that nearly 61 percent of the surface is water, while about 39 percent is land. In the Southern Hemisphere, on the other hand, almost 81 percent of the surface is water, and only 19 percent is land. It is no wonder then that the Northern Hemisphere is called the *land hemisphere,* and the Southern Hemisphere the *water hemisphere.*

Figure 10.1 shows the distribution of land and water for each 5-degree zone in the Northern and Southern hemispheres. Between latitudes 45 degrees north and 70 degrees north there is actually more land than water, while between 40 degrees south and 65 degrees south there is almost no land to interrupt the oceanic and atmospheric circulation.

The volume of the ocean basins is many times greater than the volume of the continents above sea level. In fact, the volume of all land above sea level is only $1/18$ that of the ocean. Figure 10.2 helps to illustrate this point. Note that the mean

FIGURE 10.2
Elevations of the earth's crust. (From Foster, *Physical Geology,* 4th ed., Columbus, Ohio: Charles E. Merrill, 1983, p. 302.

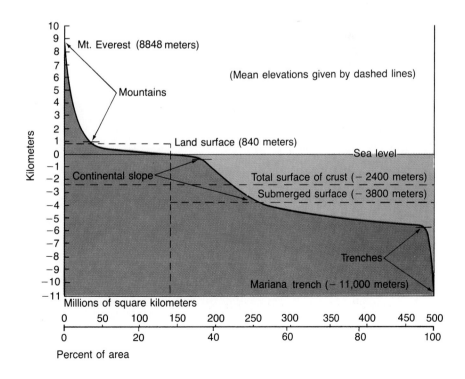

elevation of the land surface is 840 meters (2755 feet), while the average depth of the ocean is more than 4.5 times this figure—3800 meters (12,465 feet). If the solid earth was perfectly smooth (level) and round, the oceans would cover it to a depth of more than 2000 meters.

A comparison of the three major oceans reveals that the Pacific is by far the largest; it is nearly as large as the Atlantic and Indian oceans combined (Figure 10.1). The Pacific contains slightly more than half of the water in the world ocean, and because it includes few shallow seas along its margins, it has the greatest average depth—3940 meters (12,900 feet).

The Atlantic, bounded by almost parallel continental margins, is a relatively narrow ocean when compared to the Pacific. When the Arctic Ocean is included, the Atlantic has the greatest north-south extent and connects the two polar regions. Because the Atlantic has many shallow adjacent seas, including the Caribbean, Gulf of Mexico, Baltic, and Mediterranean, as well as wide continental shelves along its borders, it is the shallowest of the three oceans, with an average depth of 3310 meters (10,860 feet).

COMPOSITION OF SEAWATER

Seawater is a complex solution of salts, consisting of about 3.5 percent (by weight) dissolved mineral substances. Although the percentage of salts may seem small, the actual quantity is huge. If all of the water were evaporated from the oceans, a layer of salt approaching 60 meters (200 feet) thick would cover the entire ocean floor. The **salinity,** that is, the proportion of dissolved salts to pure water, is commonly expressed in parts per thousand rather than as a percentage (parts per hundred). The symbol used to express parts per thousand is ‰. Thus, the average salinity of the ocean is about 35 ‰.

The principal elements that contribute to the ocean's salinity are shown in Figure 10.3. If we attempted to make our own seawater, we could come reasonably close by following the recipe shown in Table 10.1. From this table it is evident that most of the salt is sodium chloride (common table salt), not a surprising revelation when we consider the proportions of elements shown in Figure 10.3. Sodium chloride together with the next four most abundant salts comprise 99 percent of

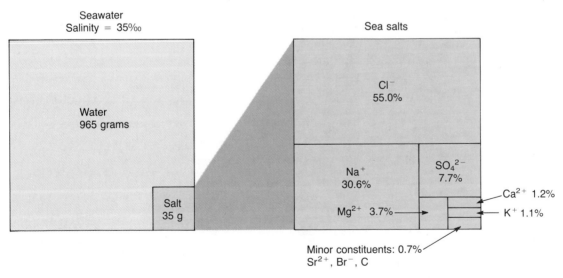

FIGURE 10.3
Relative proportions of water and dissolved salts in seawater.

TABLE 10.1
Recipe for artificial seawater.

MIX:	
Sodium chloride (NaCl)	23.48 grams
Magnesium chloride (MgCl₂)	4.98
Sodium sulfate (Na₂SO₄)	3.92
Calcium chloride (CaCl₂)	1.10
Potassium chloride (KCl)	0.66
Sodium bicarbonate (NaHCO₃)	0.192
Potassium bromide (KBr)	0.096
Hydrogen borate (H₃BO₃)	0.026
Strontium chloride (SrCl₂)	0.024
Sodium fluoride (NaF)	0.003
ADD:	
Water (H₂O) to form 1000 grams of solution.	

the salt in the sea. Although only seven elements make up these five most abundant salts, seawater contains all of the earth's other naturally occurring elements. Despite their presence in minute quantities, many of these elements are very important in maintaining the necessary chemical environment for life in the sea.

The relative abundances of the major components in sea salt are essentially constant, no matter where the ocean is sampled. Variations in salinity, therefore, are primarily a consequence of changes in the water content of the solution. As a result, high salinities are found where evaporation is high, as is the case in the dry subtropics. Conversely, where heavy precipitation dilutes ocean waters, as in the mid-latitudes and near the equator, lower salinities prevail (Figure 10.4). While salinity variations in the open ocean normally range from 33 ‰ to 37 ‰, some seas demonstrate extraordinary extremes. For example, in the restricted waters of the Persian Gulf and the Red Sea, where evaporation far exceeds precipitation, salinities may exceed 42 ‰. Conversely, very low salinities occur where large quantities of fresh water are supplied by rivers and precipitation. Such is the case for the Baltic Sea, where salinity varies from 2 ‰ to 7 ‰.

What are the sources for the vast quantities of salts in the ocean? The products of the chemical weathering of rocks on the continents constitute one source. These soluble materials are delivered to the oceans by streams at an estimated rate of more than 2.5 billion tons annually. The second major source of elements found in ocean water is the earth's interior. Through volcanic eruptions, large quantities of water and dissolved gases have been emitted during much of geologic time. This process, called **outgassing,** is thought to be the principal source of water in the oceans as well as in the atmosphere. Certain elements, notably chlorine, bromine, sulfur, and boron, are much more abundant in the ocean than in the earth's crust. Since they result from outgassing, their abundance in the sea tends to confirm the hypothesis that volcanic action is largely responsible for the present oceans.

Although rivers and volcanic activity continually contribute materials to the oceans, the salinity of seawater is not increasing. In fact, many oceanographers believe that the composition of seawater has been relatively consistent for a large span of geologic time. Why doesn't the sea get saltier? Obviously, material is being removed just as rapidly as it is added. Some elements are withdrawn from seawater by plants and animals as they build

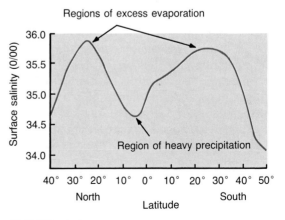

FIGURE 10.4
Variations in surface salinity with latitude.

FIGURE 10.5
Harvesting salt produced by the evaporation of seawater. (Courtesy of the Leslie Salt Company)

shells and skeletons. Others are removed when chemically precipitated as sediment. The net effect is that the overall makeup of seawater remains relatively constant.

RESOURCES FROM SEAWATER

We have seen that the ocean is a great storehouse of dissolved minerals. Some are presently being extracted from seawater while others remain as untapped potential resources. Commercial products obtained from seawater include common salt (sodium chloride), the light-weight metal magnesium, and bromine, which is used principally in gasoline additives and in the manufacture of fireproofing materials. For many years seawater provided a high percentage of the world's magnesium and bromine. In the late 1960s, 70 percent of the world's bromine and 61 percent of its magnesium came from the chemical processing of seawater. Since that time, the production of these two elements has shifted away from seawater for economic reasons. Today the primary sources of magnesium and bromine are water from Utah's Great

Salt Lake, and brine wells and evaporite deposits in the United States and other areas. In countries lacking large brine and evaporite deposits, such as Japan and the Scandinavian countries, seawater is still the primary source of these elements.

Since ancient times the ocean has been an important source of salt for human consumption, and the sea remains a significant supplier. To obtain pure sodium chloride, a series of evaporating ponds is used. The idea is that different salts precipitate from seawater at different salinities. In the first pond, evaporation proceeds until the salt concentration is such that calcium carbonate and calcium sulfate precipitate. Then the remaining brine is transferred to another pond where evaporation continues. Practically pure sodium chloride precipitates there. When most sodium chloride has precipitated, the remaining liquid (now enriched in potassium and magnesium) is drained off and the salt is harvested (Figure 10.5). In this manner about 30 percent of the world's salt is produced.

People have long been intrigued by the prospect of extracting gold from ocean water. For example, following World War I Germany seriously considered extracting gold from seawater to pay its war debt. In fact, a major goal of the famous

Meteor expedition was to investigate the feasibility of gold recovery from the ocean.* Unfortunately, what they found was that the concentration of gold in ocean water is extremely low. To extract a single ounce of gold, 200,000 tons of seawater must be processed. Further, the gold would then have to be separated from the remaining 7000 tons of solids. Thus, even though millions of tons of gold may exist in the waters of the global ocean, no one is now enjoying this wealth because the cost of recovering the gold far exceeds its value.

One of the resources derived from seawater is fresh water. The removal of salts and other chemicals from seawater is termed **desalination.** The purpose of this process is to extract low-salinity ("fresh") water suitable for drinking, industry, and agriculture from water that is so saline that it is not suitable for these purposes. Although hundreds of desalination plants are now operating and others are being planned, the cost of desalinized water is still high, and the total production remains relatively small. In 1980 desalination yielded about 4 billion liters (1 billion gallons) of water per day. Although such a volume appears to be large, it would not be sufficient to meet the daily needs of New York City.

Although fresh water produced by various desalination technologies may become more important as a source of water for human and even industrial use, it is unlikely to be an important supply for agricultural purposes. Since human water requirements are quite low, the cost of supplying areas without adequate natural supplies can often be justified. In addition, the cost of water may represent only a small fraction of the ultimate cost of many industrial goods and hence can be absorbed even in competitive markets. However, the cost of water for irrigation represents a large percentage of the total cost of the crop being produced. For this reason, the cost of desalinized water is still much too high for crop production. Consequently, making the deserts "bloom" by irrigating them with desalinized seawater is only a dream and, for economic reasons, is likely to remain so in the near future.

---

*This historic expedition is discussed on page 281.

THE OCEAN'S LAYERED STRUCTURE

By sampling ocean waters, oceanographers have found that temperature and salinity vary with depth. Generally, they recognize a three-layered structure in the open ocean: a shallow surface mixed zone, a transition zone, and a deep zone (Figure 10.6).

Since solar energy is received at the ocean surface, it is here that water temperatures are highest. The mixing of these waters by waves as well as the turbulence from currents create a rapid vertical heat transfer. Hence, this mixed surface zone is characterized by nearly uniform temperatures. The thickness and temperature of this layer vary, depending upon latitude and season. The zone may attain a thickness of 450 meters (1500 feet) or more, and temperatures ranging between 21°C and 26°C (70°F and 80°F) are common. In equatorial latitudes the temperatures may be even higher. Below the sun-warmed zone of mixing, the temperature falls abruptly with depth. This layer of rapid temperature change, known as the **thermocline,** marks the transition between the warm surface layer and the deep zone of cold water below. Below the thermocline temperatures fall only a few more degrees. At depths greater than about 1500 meters (5000 feet) ocean-water temperatures are consistently less than 4°C (39°F).

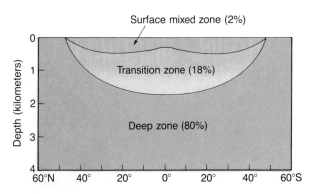

FIGURE 10.6
Layered structure of the ocean.

Since the average ocean depth is approximately 3800 meters (12,500 feet), it is evident that the temperature of most seawater is not much above freezing. In the high polar latitudes, surface waters are cold and temperature changes with depth are slight. Consequently, the three-layered structure is not present.

Salinity variations with depth correspond to the general three-layered system described for temperatures. Generally, in the low and middle latitudes a surface zone of higher salinity is created when fresh water is removed by evaporation. Below the surface zone salinity decreases rapidly. This layer of rapid change, the **halocline**, corresponds closely to the thermocline. Below the halocline salinity variations are small.

A BRIEF HISTORY OF OCEANOGRAPHY

The modern science of the oceans is often dated back to the time of American naval officer Matthew Fontaine Maury (1806–1873). In 1855 Maury published what many consider the first textbook on oceanography, entitled *The Physical Geography of the Sea.* The volume includes chapters on the Gulf Stream, the atmosphere, currents, ocean depths, winds, climates, drifts, storms, and more. The book was the end result of a great deal of data gathering and indicated that its author had a profound grasp of the concept that the sea is a single dynamic mechanism. After an injury kept Maury from sailing, he began accumulating and compiling data from ships' logs on currents and winds. In his pursuit of information, Maury enlisted the aid of mariners from all types of ships and of many nationalities. Up to this time there had been little correlated knowledge of wind, weather, tides, and currents; each sailor learned the hard way. Maury's organization of these data revealed previously unknown patterns of ocean currents, and his charts helped reduce sailing times considerably. For example, the time required to sail from the east coast of the United States to Rio de Janeiro, Brazil, was reduced by as much as 10 days, and the trip to California by way of Cape Horn was shortened by 30 days. He also produced the first bathymetric map of the North Atlantic, which showed ocean depths at 1000-fathom intervals (1 fathom equals

FIGURE 10.7
H.M.S. *Challenger.* (From C. W. Thomson and Sir John Murray, *Report on the Scientific Results of the Voyage of the H.M.S. Challenger,* Vol. 1. Great Britain: Challenger Office, 1895, Plate 1)

1.8 meters, or 6 feet, the approximate distance from fingertip to fingertip of a person with outstretched arms).

Another event of great significance to modern oceanography was the historic 3½-year voyage of the H.M.S. *Challenger* (Figure 10.7). Beginning in December 1872, and ending in May 1876, the *Challenger* expedition made the first, and perhaps most comprehensive, study of the global ocean attempted by one agency. The 110,000-kilometer (69,000-mile) trip took the ship and its crew of scientists to every ocean except the Arctic. Periodically they sampled the total depth of water, temperatures at various depths, weather conditions, and the rate and direction of surface and subsurface currents. Samples of water and life were collected at various levels, and the life and sediment of the bottom were sampled with dredges. Over 700 new genera and 4000 new species of animal and plant life were discovered. The data collected clearly indicated that the oceans are teeming with undiscovered life and showed without a question that life exists at great depths. Previously, many had believed that an *azoic* ("without life") *zone* existed within the ocean beyond a depth of 550 meters (1800 feet). The scientific results of the voyage were published over a 15-year period and filled 50 large volumes. The "*Challenger* Deep-Sea Exploring Expedition," as it was officially called, opened a great descriptive era of oceanography. It is perhaps fitting that from this voyage the term *oceanography* was born.

A remarkable person who contributed to our knowledge of the oceans was the Norwegian adventurer and scientist Fridtjof Nansen. Although best known to oceanographers as a polar explorer, Nansen was also an artist, zoologist, and winner of the Nobel Peace Prize. After studying the waters of the Arctic basin, Nansen developed an hypothesis concerning surface currents in the ice-choked water body. After convincing his government and the British Royal Geographical Society to help him test his ideas, a special ship, the *Fram,* was built. The *Fram* was designed so that expanding ice would not crush it, but would instead force it up to the surface. For three years, from September 1893, to August 1896, the *Fram* and its crew

drifted with the ice. During this pioneering voyage Nansen not only made important observations about ocean currents but also found that this ice-covered sea reached to true oceanic depths. Further, the drift of the *Fram* disproved the commonly held notion that a continent existed in this northern ocean and showed that the ice covering the polar area throughout the year was not of glacial origin but was a freely moving ice pack that formed directly upon the water surface. Following the voyage, Nansen, realizing the need for more accurate measurements of salinity and temperature, designed an instrument that came to be called the *Nansen bottle.* For many years the Nansen bottle was a standard oceanographic sampling device.

The voyage during 1925–1927 of the German ship *Meteor* has been considered a sequel to the *Challenger* expedition. This voyage was perhaps the first modern oceanographic trip because it had the advantage of more modern equipment such as the Nansen bottle and revolutionary electronic depth-sounding equipment **(echo sounder).** Prior to the echo sounder, ocean depth measurements had to be taken with a weighted line. In deep water the time required for the line to reach bottom can exceed an hour and a half, while the task of hauling in the line is even more time-consuming. Only a limited number of such soundings were practicable; consequently, our knowledge of the sea floor remained very slight. With the development of the echo sounder, depths could be determined in a matter of seconds or minutes, even while the vessel was in motion (Figure 10.8).

The echo sounder apparatus works by transmitting sound waves from the ship toward the ocean bottom. A delicate receiver catches the echo from the bottom, and a highly accurate clock measures the time interval in small fractions of a second. By knowing the velocity of the sound waves in water (about 1500 meters, or 4900 feet, per second), the depth can be calculated very precisely. Since ultrasonic waves are often used because they can be distinguished from the audible sounds made during the operation of the ship, the "sound" is not detectable by human ears.

Since World War II interest in and research

FIGURE 10.8
Ocean floor profile. **A.** An echo sounder determines water depth by measuring the time required for a sonic wave to travel from a ship to the sea floor and back. The speed of sound in water is 1500 m/sec. Therefore, water depth = ½(1500 m/sec × echo travel time). **B.** Sea-floor profile made by an echo sounder. (Courtesy of Woods Hole Oceanographic Institution)

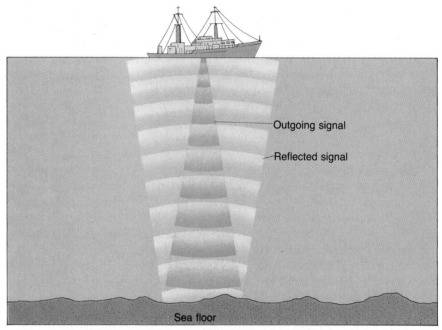

A.

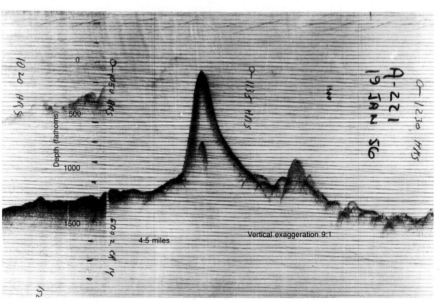

B.

concerning the ocean have grown swiftly. Today we live in a period of analytical oceanography, which has replaced the period when purely descriptive studies were the rule. Although sampling and systematic observations are still important, more emphasis is given to theories and hypotheses. Data are gathered which specifically relate to certain ideas, with the goal of proving or disproving these ideas. There are now hundreds of vessels engaged in oceanographic research, some having

A.

B.

FIGURE 10.9

A. *Glomar Challenger,* the drilling ship of the Deep Sea Drilling Project. This pioneering vessel was a significant technological breakthrough because it was capable of lowering drill pipe thousands of meters to the ocean floor and then drilling hundreds of meters into the sediments and underlying crust. During its 15 years of operation, the *Glomar Challenger* logged more than 600,000 kilometers on 96 voyages across every ocean. The drilling of 1092 holes yielded more than 96 kilometers (60 miles) of invaluable core samples. Although this historic ship has been retired, the important work of sampling the floors of the ocean basins continues. B. Amidships is the towering 42-meter- (140-foot-) high derrick. This is a view looking straight down from near the top of the derrick. (Photos courtesy of Victor S. Sotelo, Deep Sea Drilling Project)

a special purpose, others equipped to do many tasks. These ships are able to measure the global oceanic system. Although there seems to have been a great emphasis on ocean study over the past century, there is still much more to learn, for every question that is answered about the oceans leads to many others.

In addition, technological breakthroughs are advancing oceanographic study. The drilling ships *Glomar Challenger* and *JOIDES Resolution,* for example, have added a vast array of new information about the sea floor and its sediments (Figure 10.9). Moreover, deep-diving manned submersibles are expanding our knowledge of the floor of the ocean by allowing scientists to get a firsthand view of the previously unseen world below.*

---

*The accomplishments of these research vessels are detailed in Chapter 6, and deep-diving submersibles are examined more thoroughly later in this chapter.

THE EARTH BENEATH THE SEA

If all water were removed from the ocean basins, what kind of surface would be revealed? It would not be the quiet, subdued topography as was once thought, but a surface characterized by great diversity—towering mountain chains, deep canyons, and flat plains. In fact, the scenery would be just as varied as that on the continents (Figure 10.10).

Oceanographers studying the topography of the ocean basins have delineated three major units: continental margins, the ocean basin floor, and

FIGURE 10.10

The topography of the earth's solid surface is shown on the following two pages. (Copyright © by Marie Tharp) →

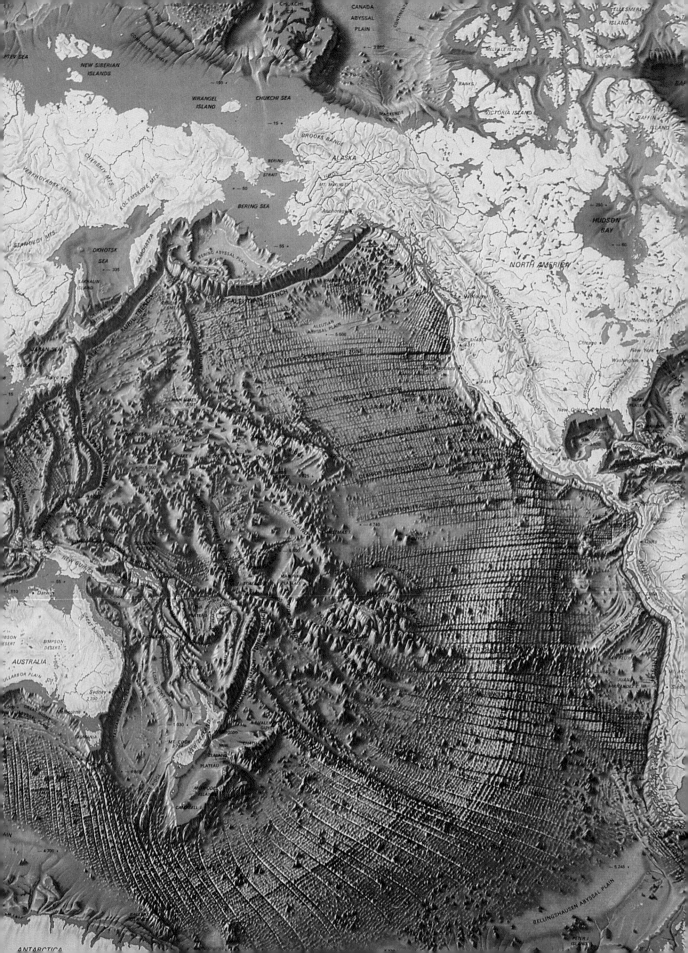

FIGURE 10.11
Major topographic divisions
of the North Atlantic and a
profile from New England to
the coast of North Africa.
(After B. C. Heezen,
M. Tharp, and M. Ewing.
"The Floors of the Oceans,"
*Geological Society of Amer-
ica Special Paper 65,* p. 16)

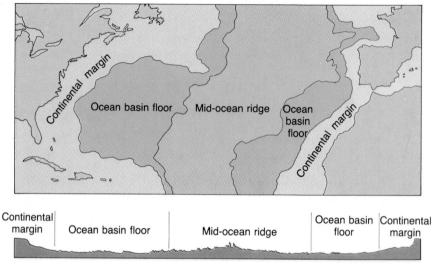

mid-ocean ridges. The map in Figure 10.11 outlines these provinces for the North Atlantic, and
the profile at the bottom of the illustration shows
the varied topography. Such profiles usually have
their vertical dimension exaggerated many times—
40 times in this case—to make topographic features more conspicuous. Because of this, the
slopes in the profile of the sea floor in Figure 10.11
appear to be much steeper than they actually are.

CONTINENTAL MARGINS

The zones that collectively make up the **continental margin** include the continental shelf, continental slope, and continental rise (Figure 10.12). The
first of these parts, the **continental shelf,** is a
gently sloping submerged surface extending from
the shoreline toward the deep-ocean basin. Since
it is underlain by continental-type crust, it is
clearly a flooded extension of the continents. The
continental shelf varies greatly in width. Almost
nonexistent along some continents, the shelf may
extend seaward as far as 1500 kilometers (900
miles) along others. On the average, the continental shelf is about 80 kilometers (50 miles) wide
and 130 meters (425 feet) deep at the seaward
edge. The average inclination of the continental
shelf is less than one-tenth of one degree, a drop

FIGURE 10.12
Schematic profile showing
the provinces of the continental margin. (Vertical exaggeration 135:1)

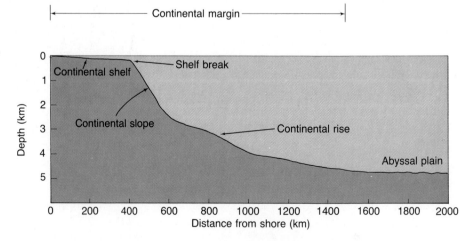

of only about 2 meters per kilometer (10 feet per mile). The slope is so slight that it would appear to an observer to be a flat surface.

The continental shelves represent 7.5 percent of the total ocean area, which is equivalent to about 18 percent of the earth's total land area. These areas have taken on increased economic and political significance since they have been found to be sites of important mineral deposits, including large reservoirs of petroleum and natural gas, as well as huge sand and gravel deposits. Of course, the waters of the continental shelf contain many important fishing grounds that are significant sources of food.

When compared with many parts of the deep-ocean floor, the surface of the continental shelf is relatively featureless. This is not to say that the shelves are completely smooth. The most profound features are long valleys running from the coastline into deeper waters. Many of these valleys are seaward extensions of river valleys on the adjacent landmass. Such valleys were excavated during the Pleistocene epoch (Ice Age). During this time great quantities of water were tied up in vast ice sheets on the continents, causing sea level to drop by 90 to 120 meters (300 to 400 feet) and exposing large portions of the continental shelves. Due to this drop in sea level, rivers extended their courses, and land-dwelling plants and animals inhabited the newly exposed portions of the continents. Today these areas are again covered by the sea and inhabited by marine organisms. Dredging along the eastern coast of North America has produced the remains of numerous land dwellers, including mammoths, mastodons, and horses. Bottom sampling has also revealed that freshwater peat bogs existed, adding to the evidence that the continental shelves were once land areas.

Marking the seaward edge of the continental shelf is the **continental slope,** which leads into deep water and has a steep gradient compared to the continental shelf. While the gradient varies from place to place, it has an average drop of about 70 meters per kilometer (370 feet per mile). The continental slope represents the true edge of the continent.

Along some mountainous coasts the continental slope descends abruptly into deep-ocean trenches, which intervene between the continent and ocean basin. In such cases, the shelf is very narrow or does not exist at all. The side of the trench and the continental slope are essentially the same feature and grade into the adjacent mountains which tower thousands of meters above sea level. This situation occurs along the west coast of South America (see Figure 10.10). Here the vertical distance from the high peaks of the Andes Mountains to the floor of the deep Peru-Chile trench bordering the continent exceeds 12,200 meters (40,000 feet).

Perhaps the most significant features associated with continental slopes are deep, steep-sided valleys known as **submarine canyons.** Originating on the outer continental shelf or on the continental slope, submarine canyons may reach water depths of 3 kilometers (2 miles). Although some appear to be seaward extensions of valleys that were carved on the continental shelf during the Ice Age, others are not oriented in this manner. Furthermore, the canyons reach depths far below the maximum lowering of sea level, which indicates that they were created by some process that operates below the ocean surface (Figure 10.13). Most

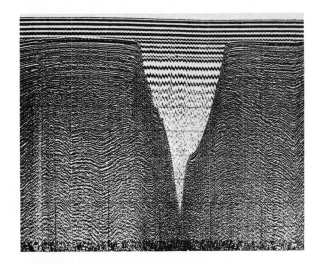

FIGURE 10.13
An echo-sounding profile of the Congo submarine canyon off the west coast of Africa. The bottom of the canyon is about 3 kilometers below the level of the sea floor of the continental shelf. Across the top, the canyon is more than 10 kilometers wide. (Courtesy of K. O. Emery, Woods Hole Oceanographic Institution)

available information seems to favor the view that these deep scars on the continental slope are excavated largely by turbidity currents, which are dense slurries of sediment and water that flow downslope.

In regions where trenches do not exist, the steep continental slope merges into a more gradual incline known as the **continental rise.** Here the gradient lessens to between 4 and 8 meters per kilometer (20 to 40 feet per mile). While the width of the continental slope averages about 20 kilometers (12 miles), the continental rise may reach for hundreds of kilometers. This feature consists of a thick accumulation of sediment that moved downslope from the continental shelf to the deep-ocean floor. More specifically, the sediments comprising the continental rise are delivered to the base of the continental slope by turbidity currents that follow submarine canyons. When these muddy currents emerge from the mouth of a canyon onto the relatively flat ocean floor, they deposit sediment that forms a **deep-sea fan.** Deep-sea fans have the same basic shape as alluvial fans, which form at the foot of steep mountain slopes on land. As fans from adjacent submarine canyons grow, they coalesce to produce the continuous apron of sediment at the base of the continental slope.

TURBIDITY CURRENTS

Turbidity currents are downslope movements of dense, sediment-laden water. They are created when sand and mud on the continental shelf and slope are dislodged, perhaps by an earthquake, and are thrown into suspension. Since the muddy water is denser than the clearer water above, it flows down the slope, eroding and accumulating more sediment as it continues to gain speed. The erosional work repeatedly carried on by these muddy torrents eventually excavates submarine canyons.

Turbidity currents usually originate along the continental slope and continue across the continental rise, still cutting channels. Eventually they lose momentum and come to rest along the ocean basin floor. As these currents slow, suspended sediments begin to settle out. First, the coarser sand is dropped, followed by successively finer accumulations of silt and then clay. Consequently, these deposits, called **turbidites,** are characterized by a decrease in sediment grain size from bottom to top, a phenomenon known as **graded bedding.**

For many years the existence of turbidity currents in the ocean was a matter of considerable debate among marine geologists. Not until the 1950s did the speculation begin to subside. Two lines of evidence helped establish turbidity currents as important mechanisms of submarine erosion and sediment transportation. The first important evidence came from records of a rather severe earthquake that took place off the coast of Newfoundland in 1929 and resulted in the breakage of 13 transatlantic telephone and telegraph cables. At the time it was presumed that the tremor had caused the multiple breaks. However, when the data were examined, it appeared that this was not the case. After plotting the locations of the breaks on a map, it was seen that all the breaks had occurred along the steep continental slope and the gentler continental rise. Since the time of each break was known from information provided by automatic recorders, a pattern of what had happened could be deduced. The breaks high up on the continental slope took place first, almost concurrently with the earthquake. The other breaks happened in succession, the last occurring 13 hours later, some 720 kilometers (450 miles) from the source of the quake (Figure 10.14). The breaks downslope had obviously taken place too long after the tremor to have been caused by the shock of the earthquake. The existence of a turbidity current, triggered by the quake, thus appeared as a plausible alternative. As the avalanche of sediment-choked water raced downslope it snapped the cables in its path. Investigators calculated that the current reached speeds approaching 80 kilometers (50 miles) per hour on the steep slopes and about 24 kilometers (15 miles) per hour on the gentler slopes below. Subsequent investigations of cable breaks in other areas revealed a similar sequence of events.

FIGURE 10.14
Profile of the sea floor showing the events of the November 18, 1929, earthquake off the shores of Newfoundland. Arrows point to cable breaks; numbers show times of breaks in hours and minutes after the earthquake. The vertical scale is greatly exaggerated. (After B. C. Heezen and M. Ewing, "Turbidity Currents and Submarine Slump and the 1929 Grand Banks Earthquake," *American Journal of Science* 250: 867)

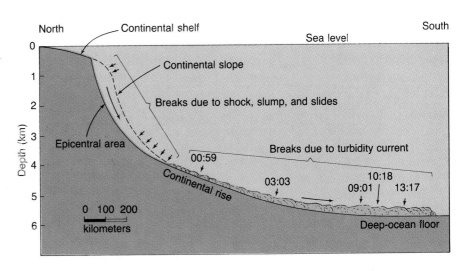

A second compelling line of evidence relating turbidity currents to submarine erosion and transportation of sediment came from the examination of deep-sea sediment samples. These cores show that extensive graded beds of sand, silt, and clay exist in the quiet waters of the deep ocean. Some samples also include fragments of plants and animals that live only in the shallower waters of the continental shelves. No mechanism other than turbidity currents could explain the existence of these deposits.

Although there is still much to be learned about the complex workings of turbidity currents, it has been well established that they are a very important mechanism of sediment transport in the ocean. By the action of turbidity currents, submarine canyons are created and sediments are carried to the deep-ocean floor.

FEATURES OF THE OCEAN BASIN FLOOR

Between the continental margin and the oceanic ridge system lies the ocean basin floor (see Figure 10.11). The size of this region—almost 30 percent of the earth's surface—is roughly comparable to

the percentage of the surface that projects above the sea as land. Here we find ocean trenches, which are dramatically deep grooves in the ocean floor; remarkably flat regions, known as abyssal plains; and steep-sided volcanic peaks, called seamounts.

DEEP-OCEAN TRENCHES

Deep-ocean trenches are long, relatively narrow features that represent the deepest parts of the ocean. Several in the western Pacific approach or exceed depths of 10,000 meters (32,800 feet), and at least a portion of one, the Challenger Deep in the Mariana trench, is more than 11,000 meters (36,000 feet) below sea level.

Although deep-ocean trenches represent only a very small portion of the ocean floor area, they are nevertheless very significant geological features. Trenches are the sites where moving crustal plates are destroyed as they plunge back into the mantle (see Figure 6.11). In addition to the earthquakes created as one plate descends beneath another, igneous activity is also associated with trench regions. Trenches in the open ocean are paralleled by volcanic island arcs, while volcanic mountains, such as the Andes, may be found paralleling trenches that are adjacent to continents. As was mentioned in Chapter 6, the melting of a descending plate produces the molten rock that leads to this igneous activity.

ABYSSAL PLAINS

Abyssal plains are incredibly flat features; in fact, these regions are likely the most level places on the earth. The abyssal plain found off the coast of Argentina, for example, has less than 3 meters (10 feet) of relief over a distance exceeding 1300 kilometers (800 miles). The monotonous topography of abyssal plains will occasionally be interrupted by the protruding summit of a buried volcanic structure.

By employing seismic profilers, instruments whose signals penetrate far below the ocean floor, researchers have shown that abyssal plains consist of thick accumulations of sediment that were deposited atop the low, rough portions of the ocean floor. The nature of the sediment indicates that these plains consist primarily of sediments transported far out to sea by turbidity currents. The turbidite deposits are interbedded with sediments composed of minute clay-sized particles that continuously settle onto the ocean floor.

Abyssal plains are found as part of the sea floor in all of the oceans. However, they are more widespread where there are no deep-ocean trenches adjacent to the continents. Since the Atlantic Ocean has fewer trenches to act as traps for the sediments carried down the continental slope, it has more extensive abyssal plains than the Pacific.

SEAMOUNTS

Dotting the ocean floors are isolated volcanic peaks called **seamounts** that may rise hundreds of meters above the surrounding topography. These steep-sided conical peaks are found on the floors of all the oceans, but the greatest number and density have been identified in the Pacific.

Many of these undersea volcanoes begin to rise near oceanic ridges, divergent plate boundaries where the plates of the lithosphere move apart (see Chapter 6). They continue to grow as they ride along on the moving plate. If the volcano rises fast enough, it emerges as an island. Examples in the Atlantic include the Azores, Ascension, Tristan da Cunha, and St. Helena. During the time they exist as islands, some of these volcanoes are eroded to near sea level by running water and wave action. Over a span of millions of years the islands gradually sink as the moving plate slowly carries them from the oceanic ridge area. These submerged, flat-topped seamounts are called **guyots.** In other instances, guyots may be remnants of eroded volcanic islands that were formed away from the ridge crest, possibly by hot spot activity. Here subsidence occurs after the volcanic activity ceases and the sea floor cools and contracts.

MID-OCEAN RIDGES

Mid-ocean ridges are found in all major oceans and represent more than 20 percent of the earth's surface. They are certainly the most prominent topographic features in the oceans, for they form an almost continuous mountain range, which extends for about 65,000 kilometers (40,000 miles) in a manner similar to the seam on a baseball. Although ocean ridges stand high above the adjacent deep-ocean basins, they are much different from the mountains found on the continents. Rather than thick sequences of folded and faulted sedimentary rocks, oceanic ridges consist of layer upon layer of basaltic rocks that have been faulted and uplifted. The term *ridge* may also be misleading since these features are not narrow, but have widths from 500 to 5000 kilometers and, in places, may occupy as much as one-half of the total ocean floor area. Ridge crests are marked by deep clefts, or **rifts,** and are flanked by ridges and lines of peaks that extend outward for hundreds of kilometers (see Figure 10.10). Axes of the ridges are marked by frequent earthquakes and characterized by a much higher heat flow through the crust. The rifts at the center of the ridges are the sites where new magma wells up from the asthenosphere below, continually creating new oceanic crust. The rifts therefore represent divergent plate boundaries where sea-floor spreading is taking place.

The primary reason for the elevated position of a ridge system is the fact that newly created oceanic crust is hot, and therefore occupies more volume than cooler rocks of the deep-ocean basin. As the young lithosphere travels away from the spreading center, it gradually cools and contracts. This thermal contraction accounts in part for the greater ocean depths that exist away from the ridge. Almost 100 million years must pass before cooling and contraction cease completely. By this time, rock that was once a part of a majestic oceanic mountain system is located in the deep-ocean basin, where it is mantled by thick accumulations of sediment.

A CLOSE-UP VIEW OF THE OCEAN FLOOR

Although much has been (and continues to be) learned about the floor of the ocean from echo sounders and other remote sensing equipment, as well as from drilling and sampling by surface ships, oceanographers in the 1970s became aware that direct manned observation was essential to bring about a true understanding of many deep-sea phenomena. What was needed was a firsthand view of the previously unseen world below.

Today, the names and accomplishments of deep-diving manned submersibles such as the *Alvin* are common knowledge among oceanographers (Figure 10.15). Manned submersibles are now extending the coverage provided by traditional oceanographic research vessels by allowing scientists to investigate the fine-scale features that previously eluded detection.

One of the pioneering research projects that used deep-diving submersibles was a cooperative venture called Project FAMOUS (French-American Mid-Ocean Undersea Study). In 1974, after three years of preliminary surveying and study by surface ships, two French submersibles and one American vessel made a total of forty-four dives to the floor of the Atlantic. The primary purpose of the project was to examine the structure of the rift valley in the Mid-Atlantic Ridge. The data collected by the three vessels proved invaluable and led to more realistic explanations of how the spreading process works in creating new ocean floor (Figure 10.16).

More recently, dives were made by the *Alvin* to a spreading center at 21°N latitude on the East

FIGURE 10.15
The deep-diving submersible *Alvin* is 7.6 meters long, weighs 16 tons, has a cruising speed of 1 knot, and can reach depths as great as 4000 meters. A pilot and two scientific observers are along during a normal 6- to 10-hour dive. (Courtesy of Woods Hole Oceanographic Institution)

FIGURE 10.16

A photograph taken from the *Alvin* during Project FAMOUS shows lava extrusions in the rift valley of the Mid-Atlantic Ridge. Large toothpastelike extrusions such as the one in this photograph were common features. A mechanical arm is sampling an adjacent blisterlike extrusion. (Courtesy of Woods Hole Oceanographic Institution)

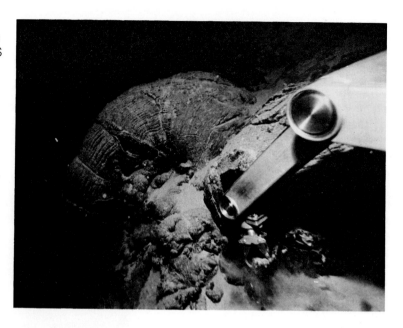

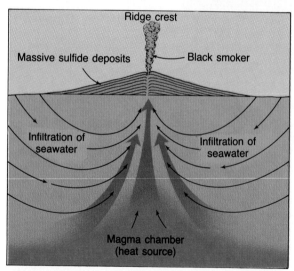

FIGURE 10.17

Massive sulfide deposits can result from the circulation of seawater through the oceanic crust along active spreading centers. As seawater infiltrates the hot basaltic crust, it leaches sulfur, iron, copper, and other metals. The hot, enriched fluid returns to the sea floor near the ridge axis along faults and fractures. Some metal sulfides may be precipitated in these channels as the rising fluid begins to cool. When the hot liquid emerges from the sea floor and mixes with cold seawater, the sulfides precipitate to form massive deposits.

Pacific Rise near the mouth of the Gulf of California. Here, in addition to gathering large quantities of basic data, the scientists aboard the *Alvin* discovered the existence of spectacular geyserlike hot springs. They witnessed two- to five-meter high chimney-like structures spewing dark, mineral-rich hot (350°C–400°C) water (Figure 10.17). As the heated solutions hit the surrounding 2°C seawater, sulfides of copper, iron, and zinc precipitated immediately, forming mounds of minerals around the steaming vents. In addition to viewing firsthand the formation of massive sulfide deposits, the scientists aboard the *Alvin* found communities of exotic, bottom-dwelling animals living near cooler (20°C) hot springs. The discovery of an animal community thriving more than three kilometers below the surface where no light can reach was totally unexpected. An analysis of the sulfur-rich vent water as well as the stomach contents of some animals revealed that the base of the food chain was sulfur-oxidizing bacteria.

Dives such as the ones briefly highlighted here have demonstrated the value of deep-diving manned submersibles in detecting and studying the fine-scale features of the ocean floor. These vessels now appear to occupy a permanent and important place as tools in oceanographic research.

CORAL REEFS AND ATOLLS

Coral reefs are among the most picturesque features found in the ocean. They are constructed primarily from the calcareous (calcite-rich) skeletal remains and secretions of corals and certain algae. The term *coral reef* is somewhat misleading in that it makes no mention of the skeletons of many small animals and plants found inside the branching framework built by the corals; nor of the fact that limy secretions of algae help bind the entire structure together.

Coral reefs are confined largely to the warm waters of the Pacific and Indian oceans, although a few occur elsewhere. Since reef-building corals grow best in waters with an average annual temperature of about 24°C (75°F), their location is in part the result of their need for warm water. They can survive neither sudden temperature changes nor prolonged exposure to temperatures below 18°C (65°F). In addition, these reef-builders require clear sunlit water. For this reason, the limiting depth of active reef growth is about 45 meters (150 feet).

FIGURE 10.18
View from space of a group of atolls in the Pacific Ocean. (Courtesy of NASA)

From 1831 to 1836 the naturalist Charles Darwin was aboard the British ship *Beagle* on a surveying expedition that circumnavigated the globe. One outcome of Darwin's studies during the five-year voyage was the development of an hypothesis on the formation of coral islands, or **atolls.** As Figure 10.18 illustrates, atolls consist of a continuous or broken ring of coral reef surrounding a central lagoon. From the time that Darwin first studied them until shortly after World War II, their manner of origin challenged people's curiosity.

Darwin's proposal explained what seemed to be a paradox; that is, how can corals, which require warm, shallow, sunlit water no deeper than 45 meters to live, create structures that reach thousands of meters to the floor of the ocean? Commenting on this in *The Voyage of the Beagle,* Darwin stated:

> . . . from the fact of the reef-building corals not living at great depths, it is absolutely certain that throughout these vast areas, wherever there is now an atoll, a foundation must have originally existed within a depth of from 20 to 30 fathoms from the surface.

The essence of Darwin's hypothesis was that coral reefs form on the flanks of sinking volcanic islands. As an island slowly sinks, the corals continue to build the reef complex upward (Figure 10.19, page 294):

> For as mountain after mountain, and island after island slowly sank beneath the water, fresh bases would be successively afforded for the growth of the corals.

Thus atolls, like guyots, are thought to owe their existence to the gradual sinking of oceanic crust. In succeeding years there were numerous challenges to Darwin's proposal. These arguments were not completely put to rest until after World War II when the United States made extensive studies of two atolls (Eniwetok and Bikini) that were going to become sites for testing atomic bombs. Drilling operations at these atolls revealed that volcanic rock did indeed underlie the thick coral reef structure. This finding was a striking confirmation of Darwin's explanation.

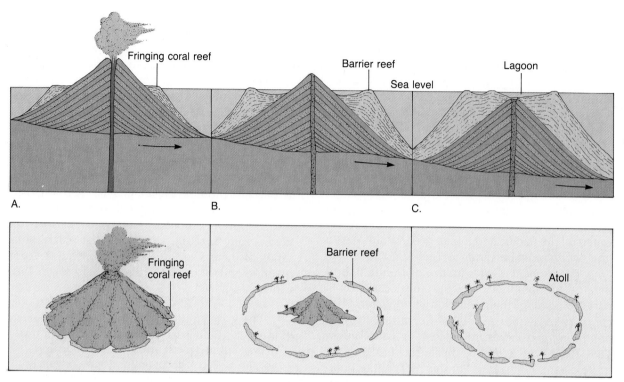

FIGURE 10.19
Cross-sectional views and map views of the formation of a coral atoll.

SEA-FLOOR SEDIMENTS

Except for a few areas, such as near the crests of mid-ocean ridges, the ocean floor is mantled with sediment. Part of this material has been deposited by turbidity currents, and the rest has slowly settled to the bottom from above. The thickness of this carpet of debris varies greatly. In some trenches, which act as traps for sediments originating on the continental margin, accumulations may approach 10 kilometers (6 miles). In general, however, sediment accumulations are considerably less. In the Pacific Ocean, uncompacted sediment measures about 600 meters or less, whereas on the floor of the Atlantic, the thickness varies from 500 to 1000 meters.

Although accumulations of sand-sized particles are found on the deep-ocean floor, mud is the most common sediment covering this region. Muds also predominate on the continental shelf and slopes, but the sediments in these areas are coarser overall because of greater quantities of sand. Sampling has shown that sands are generally deposited on the continental shelf, forming beaches along the shore. However, in some cases this coarse sediment, which is expected to be found near the shore, occurs in irregular patches at greater depths near the seaward limits of the continental shelves. While some of the sand may have been deposited by local currents that are capable of moving coarse sediment far from shore, much of it appears to result from sand deposition on ancient beaches. Such beaches formed during the Ice Age, when sea level was much lower than it is today. These patches of sand were then submerged as sea level rose again.

TYPES OF SEA-FLOOR SEDIMENTS

Sea-floor sediments can be classified according to their origin into three broad categories: (1) Lithogenous ("derived from rocks") sediment; (2)

Biogenous ("derived from organisms") sediment; and (3) Hydrogenous ("derived from water") sediment. Although each category is discussed separately, it should be remembered that all sea-floor sediments are mixtures. No body of sediment comes entirely from a single source.

Lithogenous sediment consists primarily of mineral grains which were weathered from continental rocks and transported to the ocean. The sand-sized particles settle near shore. However, since the very smallest particles take years to settle to the ocean floor, they may be carried for thousands of kilometers by ocean currents. As a consequence, virtually every area of the ocean receives some lithogenous sediment. However, the rate at which this sediment accumulates on the deep-ocean floor is indeed very slow. From 5000 to 50,000 years are necessary for a 1-centimeter layer to form. Conversely, on the continental margins near the mouths of large rivers, lithogenous sediment accumulates rapidly. In the Gulf of Mexico, for example, the sediment has reached a depth of many kilometers.

Since fine particles remain suspended in the water for a very long time, there is ample opportunity for chemical reactions to occur. Because of this, the colors of the deep-sea sediments are often red or brown. This results when iron on the particle or in the water reacts with dissolved oxygen in the water and produces a coating of iron oxide (rust).

Biogenous sediment consists of shells and skeletons of marine animals and plants (Figure 10.20). This debris is produced mostly by micro-

FIGURE 10.20
Enlarged photomicrographs of typical calcareous and siliceous materials from minute animals and plants that compose biogenous sediments. (Courtesy of Deep Sea Drilling Project, Scripps Institution of Oceanography)

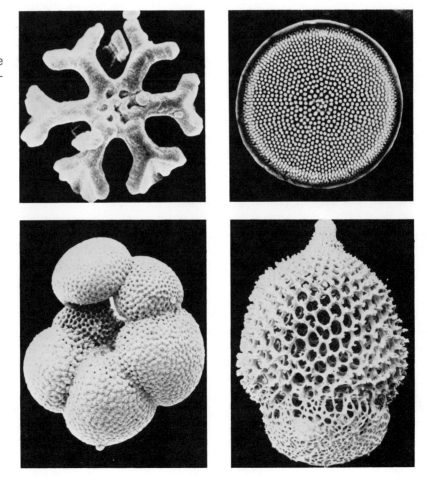

scopic organisms living in the sunlit waters near the ocean surface. The remains continually "rain" down upon the sea floor.

The most common biogenous sediments are known as *calcareous* ($CaCO_3$) *oozes,* and as the name implies, they have the consistency of thick mud. These sediments are produced by organisms that inhabit warm surface waters. When calcareous hard parts slowly sink through a cool layer of water, they begin to dissolve. This results because cold seawater contains more carbon dioxide and is thus more acidic than warm water. In seawater deeper than about 4500 meters (15,000 feet), calcareous shells will completely dissolve before they reach bottom. Consequently, calcareous ooze does not accumulate where depths are great.

Other examples of biogenous sediments are *siliceous* (SiO_2) *oozes* and phosphate-rich materials. The former is composed primarily of opaline skeletons of diatoms (single-celled algae) and radiolaria (single-celled animals), while the latter is derived from the bones, teeth, and scales of fish and other marine organisms.

Hydrogenous sediment consists of minerals that crystallize directly from seawater through various chemical reactions. For example, some limestones are formed when calcium carbonate precipitates directly from the water; however, most limestone is composed of biogenous sediment.

One of the principal examples of hydrogenous sediment, and one of the most important sediments on the ocean floor in terms of economic potential, are **manganese nodules.** These rounded blackish lumps are composed of a complex mixture of minerals that form very slowly on the floor of the ocean basins (Figure 10.21). In fact, their formation rate represents one of the slowest chemical reactions known. By analyzing the radioactive elements continually incorporated into growing nodules, researchers have determined that the growth rates vary from 0.001 to 0.2 millimeter per 1000 years. Some portions of the sea floor are littered with these deposits whereas others lack them altogether. The presence or absence of nodules has been correlated with the sedimentation rate. If sediment accumulates too rapidly

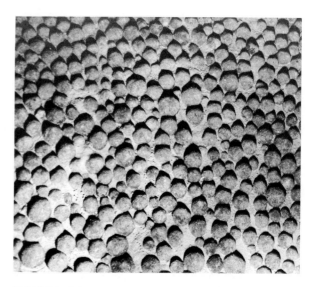

FIGURE 10.21
Manganese nodules photographed at a depth of 2909 fathoms (5323 meters) beneath the *Robert Conrad* south of Tahiti. (Courtesy of Lawrence Sullivan, Lamont-Doherty Geological Observatory)

(at a rate exceeding about 7 millimeters per 1000 years), newly forming nodules are buried and growth ceases. Since nodule growth is exceedingly slow, why are nodules not buried even where sediment accumulates at less than 7 millimeters per 1000 years? Some scientists suggest that benthic animals living in and on the sea floor are responsible for keeping the nodules at the surface. By stirring the sediment, burrowing organisms are thought to produce a slight lifting effect which, in combination with small surface animals that consume newly arrived sediment from nodule surfaces, keeps nodules from being buried.

Although manganese nodules may contain more than 20 percent manganese, the interest in them as a potential resource lies in the fact that other more valuable metals may be enriched in them. In addition to manganese, nodules may contain significant quantities of iron, copper, nickel, and cobalt. All regions containing nodules, however, are not equally good potential sites for mining. Possible mining locations must have abundant nodules (more than 5 kilograms per square meter) and contain the economically optimum mix of cobalt, copper, and nickel. Sites meeting these criteria are relatively limited. Furthermore, before such areas prove to be valuable commercial sources for these metals, the logistics of extracting nodules from the floor of the deep-ocean basins must be worked out.

SEA-FLOOR SEDIMENTS AND CLIMATIC CHANGE

Due to the fact that instrumental climate records go back only a couple of hundred years (at best), how do scientists find out about climates and climatic changes prior to that time? The obvious answer is that they must reconstruct past climates from indirect evidence; that is, they must examine and analyze phenomena that respond to and reflect changing atmospheric conditions. One of the more interesting and important techniques for analyzing the earth's climatic history is the study of sediments from the ocean floor.

Although sea-floor sediments are of many types, most contain the remains of organisms that once lived near the sea surface (the ocean-atmosphere interface). When such near-surface organisms die, their shells slowly settle to the ocean floor where they become part of the sedimentary

record. One reason that sea-floor sediments are useful recorders of worldwide climatic change is that the numbers and types of organisms living near the sea surface change as the climate changes. This principle is explained by Richard Foster Flint as follows:

> . . . we would expect that in any area of the ocean/atmosphere interface the average annual temperature of the surface water of the ocean would approximate that of the contiguous atmosphere. The temperature equilibrium established between surface seawater and the air above it should mean that . . . changes in climate should be reflected in changes in organisms living near the surface of the deep sea. . . . When we recall that the sea-floor sediments in vast areas of the ocean consist mainly of shells of pelagic foraminifers, and that these animals are sensitive to variations in water temperature, the connection between such sediments and climatic change becomes obvious.*

Thus, in seeking to understand climatic change as well as other environmental transformations, scientists have become increasingly interested in the huge reservoir of data in sea-floor sediments.

---

*Glacial and Quaternary Geology (New York: Wiley, 1971), p. 718.

REVIEW QUESTIONS

1 Define oceanography. Briefly distinguish several subdivisions of this broad discipline.

2 To what does the familiar expression "the Seven Seas" refer today? Did it always have its present-day meaning?

3 How does the area covered by the oceans compare with that of the continents? Describe the distribution of land and water on earth.

4 Answer the following questions about the Atlantic, Pacific, and Indian oceans:
   (a) Which is the largest in area? Which is smallest?
   (b) Which has the greatest north-south extent?
   (c) Which is shallowest? Explain why its average depth is least.

5 How does the average depth of the ocean compare to the average elevation of the continents (see Figure 10.2)?

6 What is meant by *salinity*? What is the average salinity of the ocean?

7 Why do variations in salinity occur? Give some specific examples to illustrate your answer.

8 What is the origin of the minerals dissolved in seawater?

9 Since the ocean contains millions of tons of gold, why is it not extracted commercially? Name some minerals that are extracted from seawater.

10 Why is desalination not likely to be a significant source of water for agriculture in the foreseeable future?

11 Briefly describe the contributions to oceanography of the following: Matthew Fontaine Maury, Fridtjof Nansen, the *Challenger* expedition, the *Meteor* expedition, deep-diving submersibles such as the *Alvin*.

12 Assuming that the average speed of sound waves in water is 1500 meters per second, determine the water depth if the signal sent out by an echo sounder requires 6 seconds to strike bottom and return to the recorder (see Figure 10.8A).

13 List the three major subdivisions of the continental margin. Which subdivision is considered a flooded extension of the continent? Which has the steepest slope?

14 How does the continental margin along the west coast of South America differ from the continental margin along the east coast of North America?

15 Defend or rebut the statement "Most submarine canyons were formed during the Ice Age when rivers extended their courses seaward."

16 What are turbidites? What is meant by the term *graded bedding*?

17 Discuss the evidence that helped confirm the existence of turbidity currents in the ocean and establish them as significant mechanisms of erosion and sediment transport.

18 Why are abyssal plains more extensive on the floor of the Atlantic than on the floor of the Pacific?

19 How are mid-ocean ridges and deep-ocean trenches related to sea-floor spreading?

20 What is an atoll? Describe Darwin's theory on the origin of atolls. Was the theory ever confirmed?

21 Distinguish among the three basic types of sea-floor sediment.

22 If you were to examine recently deposited biogenous sediment taken from a depth in excess of 4500 meters (15,000 feet), would it more likely be rich in calcareous materials or siliceous materials? Explain.

23 Why are sea-floor sediments useful in studying climates of the past?

KEY TERMS

oceanography	continental rise	mid-ocean ridge
salinity	submarine canyon	rift
outgassing	deep-sea fan	coral reef
desalination	turbidity current	atoll
thermocline	turbidite	lithogenous sediment
halocline	graded bedding	biogenous sediment
echo sounder	deep-ocean trench	hydrogenous sediment
continental margin	abyssal plain	manganese nodule
continental shelf	seamount	
continental slope	guyot	

11

THE RESTLESS OCEAN

The restless waters of the ocean are constantly in motion. Winds generate surface currents, the moon and sun produce tides, and density differences create deep-ocean circulation. Further, waves carry the energy from storms to distant shores, where their impact erodes the land. This chapter examines the movements of ocean waters and their importance.

Wave action modifying the shoreline of Hawaii.
(Photo by Stephen Trimble)

SURFACE CURRENTS

> There is a river in the ocean. In the severest droughts it never fails, and in the mightiest floods . . . it never overflows. Its banks and its bottom are of cold water, while its current is of warm. The Gulf of Mexico is its fountain, and its mouth is in the Arctic Sea. It is the Gulf Stream. There is in the world no other such majestic flow of waters.

This quote from Matthew Fontaine Maury's 1855 book, *The Physical Geography of the Sea,* describes perhaps the best-known and most-studied of the surface ocean currents. However, the Gulf Stream is just a portion of a huge, slowly moving circular whirl, or **gyre,** that begins near the equator. Similar gyres are found in the South Atlantic, and in the North and South Pacific (Figure 11.1). What creates these large surface current systems? Although our knowledge of the processes that produce and maintain ocean currents is far from complete, we do have a general understanding of the principal factors involved.

The drag exerted by winds blowing steadily across the ocean causes the surface layer of water to move. Thus, because winds are the primary driving force of surface currents, there is a relationship between the oceanic circulation and the general atmospheric circulation. A comparison of Figures 11.1 and 14.12 illustrates this. A further clue to the influence of winds on ocean circulation is provided by the currents in the northern Indian Ocean, where there are seasonal wind shifts known as the summer and winter monsoons. When the winds change direction, the surface currents also reverse direction.

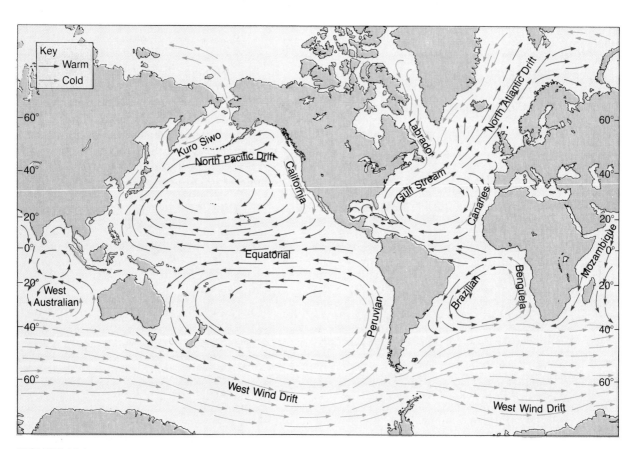

FIGURE 11.1
Average position and extent of the principal surface ocean currents.

Although winds are important in generating surface currents, other factors also influence the movement of ocean waters. The most significant of these is the **Coriolis effect.** Due to the earth's rotation, currents are deflected to the right of their path of motion in the Northern Hemisphere and to the left in the Southern Hemisphere. The Coriolis effect is greater in high latitudes and diminishes toward the equator (for a more complete discussion of the Coriolis effect, see page 387 in Chapter 14). As a consequence the direction of surface currents does not coincide with the wind direction. In general, the difference between wind direction and surface-current direction varies from about 15 degrees along shallow coastal areas to a maximum of 45 degrees in the deep oceans.

A closer look at the circulation of one of the major gyres—that of the North Atlantic—will serve to illustrate the general pattern taken by surface currents (Figure 11.1).

North and south of the equator are two westward-moving currents, the North and South Equatorial Currents, which derive their energy principally from the trade winds that blow from the northeast and southeast, respectively, toward the equator. Because of the deflection created by the Coriolis effect, the currents move almost due west. Between these westward-flowing currents there is a weaker, oppositely directed eastward flow, called the Equatorial Countercurrent. The North and South Equatorial Currents pile water up against the eastern coast of South America, and because the trade winds are weak along the equator, some of the piled-up water flows back "downhill" (eastward), creating the countercurrent.

As the North Equatorial Current continues west, the South American continent and the Coriolis effect turn the water to the north. The water then moves northwest through the Caribbean to the mouth of the Gulf of Mexico at the Straits of Yucatan. From here a portion flows parallel to but at some distance from the shore of the Gulf of Mexico, and part curves more sharply toward the east and flows directly toward the north coast of Cuba. These two parts reunite in the Straits of Florida to form the Gulf Stream. A tremendous volume of water flows northward in the Gulf Stream. It can often be distinguished by its deep indigo blue

color, which contrasts sharply with the dull green of the surrounding water. When the Gulf Stream encounters the cold waters of the Labrador Current, in the vicinity of the Grand Banks southeast of Newfoundland, there is little mixing of the waters. Rather, the junction is marked by a sharp temperature change. When the warm Gulf Stream water encounters cold air with a low water-holding capacity in this region, the result is a fog that has the appearance of smoke rising from the water.

As the Gulf Stream moves along the east coast of the United States, it is strengthened by the prevailing westerly winds and is deflected to the east (to the right) between 35 degrees north and 45 degrees north latitude. As it continues northeastward beyond the Grand Banks, it gradually widens and decreases speed until it becomes a vast, slowly moving current known as the North Atlantic Drift. As the North Atlantic Drift approaches Western Europe, it splits, with part moving northward past Great Britain and Norway. The other part is deflected southward as the cool Canaries Current. As the Canaries Current moves south, it eventually merges into the North Equatorial Current.

The clockwise circulation of the North Atlantic leaves a large central area which has no well-defined currents. This zone of calmer waters is known as the Sargasso Sea, named for the large quantities of *Sargassum,* a type of seaweed encountered there.

In the South Atlantic, surface ocean circulation is very much the same as that in the North Atlantic. The major exception is the circulation of the cold surface waters north of Antarctica. Uncomplicated by large landmasses, the currents in this region move easterly in a globe-circling pattern around the ice-covered continent. They are driven by the prevailing winds from the northwest and deflected to the left by the earth's rotation. The circulation of the Pacific generally parallels that of the Atlantic, and although the monsoons complicate the circulation of the Indian Ocean, the currents there generally coincide with the currents of the South Atlantic.

In addition to producing surface currents, winds may also cause vertical water movements. **Upwelling,** the rising of cold water from deeper layers to replace warmer surface water, is a

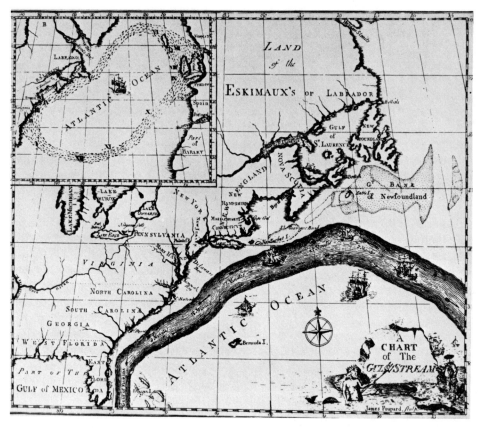

FIGURE 11.2
To aid ships crossing the Atlantic, Benjamin Franklin with the assistance of his
cousin, Timothy Folger, produced the first detailed chart of the Gulf Stream about
1769 or 1770. (Courtesy of NOAA)

common wind-induced vertical movement that is most characteristic along the eastern shores of the oceans, most notably along the coasts of California, Peru, and West Africa. Upwelling occurs in these areas when winds blow toward the equator parallel to the coast. Due to the Coriolis effect, the surface water movement is directed away from the shore. As the surface layer moves away from the coast, it is replaced by water that "upwells" from below the surface. This slow upward flow from depths of 50 to 300 meters (165 to 1000 feet) brings water which is cooler than the original surface water and creates a characteristic zone of lower temperatures near the shore. Coastal upwelling also brings to the ocean surface greater concentrations of dissolved nutrients, such as nitrates and phosphates. These nutrient-enriched waters from below promote the growth of plankton, which in turn supports extensive populations of fish.

THE IMPORTANCE OF OCEAN CURRENTS

Since ocean waters are constantly in motion, anyone who navigates the oceans needs to be aware of the horizontal movements we call currents. By understanding the position and strength of ocean currents, sailors soon realize that their voyage times can be reduced. For example, the first good chart of the Gulf Stream, which was printed by Benjamin Franklin, was developed as a result of this need to know (Figure 11.2). P. L. Richardson describes the situation as follows:

While he was in London as Deputy Postmaster General for the American colonies, Franklin was consulted on the question of why the mail packets took a fortnight longer to sail to America than the merchant ships. In October 1768 Franklin discussed this problem with his cousin Timothy Folger, a Nantucket ship captain then visiting London. Folger told him the packet captains were ignorant of the Gulf Stream and frequently sailed in this current. . . . Folger sketched the Gulf Stream on a chart and added written notes on how to avoid the Gulf Stream and Franklin had the chart printed in 1769 or 1770.*

Even today, the Franklin-Folger chart remains a good summary of the average path of the Gulf Stream.

In addition to being significant considerations in ocean navigation, currents have an important effect on climates. It is known that for the earth as a whole the gains in solar energy equal the losses to space of heat radiated from the earth. When most latitudes are considered individually, however, this is not the case. There is a net gain of energy in the low latitudes and a net loss at higher latitudes. Since the tropics are not becoming progressively warmer, nor the polar regions colder, there must be a large-scale transfer of heat from areas of excess to areas of deficit. This is indeed the case. The transfer of heat by winds and ocean currents equalizes these latitudinal energy imbalances. Ocean water movements account for about a quarter of this total heat transport, and winds the remaining three-quarters.

The moderating effect of poleward-moving warm ocean currents is well known. The North Atlantic Drift, an extension of the warm Gulf Stream, keeps Great Britain and much of northwestern Europe warmer than one would expect for their latitudes. The prevailing westerly winds carry the moderating effects far inland. For example, Berlin (52 degrees north latitude) has an average January temperature similar to that experienced at New York City, which lies 12 degrees farther south, while the January mean at London (51 degrees north latitude) is 4.5°C (8°F) higher than that at New York City.

---

* P. L. Richardson, "Benjamin Franklin and Timothy Folger's First Printed Chart of the Gulf Stream," *Science,* 207 (4431): 643.

In contrast to warm ocean currents whose effects are felt most in the middle latitudes in winter, the influence of cold currents is most pronounced in the tropics or during summer months in the middle latitudes. Cool currents, such as the Benguela Current off the west coast of southern Africa, moderate the tropical heat. For example, Walvis Bay (23 degrees south latitude), a town adjacent to the Benguela Current, is 5°C (9°F) cooler in summer than Durban, which is 6 degrees latitude farther poleward but on the eastern side of South Africa, away from the influence of the current.

In addition to influencing temperatures of adjacent land areas, cold currents have other climatic influences. For example, where tropical deserts exist along the west coasts of continents, cold ocean currents have a dramatic impact. The principal west coast deserts are the Atacama in Peru and Chile, and the Namib in southern Africa. The aridity along these coasts is intensified because the lower air is chilled by cold offshore waters. When this occurs, the air becomes very stable and resists the upward movements necessary to create precipitation-producing clouds. In addition, the presence of cold currents causes temperatures to approach and often reach the dew point, the temperature at which water vapor condenses. As a result, these areas are characterized by high relative humidities and much fog. Thus, not all tropical deserts are hot with low humidities and clear skies. Rather, the presence of cold currents transforms some tropical deserts into relatively cool, damp places that are often shrouded in fog.

DEEP-OCEAN CIRCULATION

Deep-ocean circulation cannot readily be compared with the movements of surface currents. Unlike the wind-induced movements of surface and near-surface waters, deep-ocean circulation is governed by gravity and driven by density differences. Two factors—temperature and salinity—are most significant in creating a dense mass of water. Seawater becomes denser with decreased temperature, increased salinity, or both. Consequently, deep-ocean circulation is called **thermohaline** (*thermo*—"heat", *haline*—"salt") **circulation.** After leaving the surface of the ocean, waters will not reappear at the surface for an average of 500–2000 years.

The deep circulation of the Atlantic Ocean has been the most intensively studied. A simplified cross-sectional view of its deep circulation pattern is seen in Figure 11.3. Arctic and Antarctic waters represent the two major regions where dense water masses are created. Antarctic waters are chilled during the winter. The temperatures here are low enough to form sea ice, and since sea salts are excluded from the ice, the remaining water becomes saltier. The result is the densest water in all of the oceans. This cold saline brine slowly sinks to the sea floor, where it becomes Antarctic Bottom Water (ABW in Figure 11.3). Moving northward along the ocean floor, the ABW crosses the equator and reaches as far as 20 degrees north latitude.

North Atlantic Deep Water (NADW in Figure 11.3) is thought to form when warm and highly saline Gulf Stream waters reach the Arctic region near Greenland. The high salinity of the warm surface current results from high evaporation in the low latitudes. In the Arctic these waters are chilled and sink to the bottom of the North Atlantic Basin. From here, the cold, dense water moves south, overriding the denser Antarctic Bottom Waters. North Atlantic Deep Water has been traced almost as far as the Antarctic region, where it becomes obscured by mixing.

Other important subsurface water masses have also been identified. Antarctic Intermediate Water (AIW in Figure 11.3) is formed when the very salty waters of the Brazil Current are chilled as they move south toward the Antarctic. Still another water mass (MW in Figure 11.3) has its source in the Mediterranean Sea. In the warm and dry climate of this region salinity is high because evaporation is great. As these highly saline waters are chilled during the winter they become even denser and sink. The water moves along the bottom to the west, passing into the Atlantic at the shallow Strait of Gibraltar. In order to replace the water lost in this outflow, less dense surface water enters the Mediterranean at Gibraltar. A similar water mass forms in the Red Sea. Since these waters are less dense than the bottom waters, they reach equilibrium after sinking to intermediate depths.

The circulation of deep waters in the Pacific and Indian oceans is different when compared to that of the Atlantic Ocean. Deep water from the Antarctic predominates in these oceans because the shallow barrier at the Bering Strait does not allow Arctic water to flow southward. Consequently, some deep water from the South Polar region may reach as far north as Japan and California.

TIDES

Tides are periodic changes in the elevation of the ocean surface at a specific location. Their rhyth-

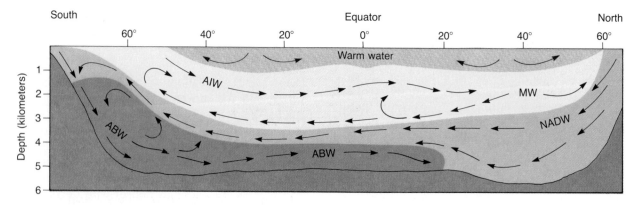

FIGURE 11.3
Cross section of the deep circulation of the Atlantic Ocean. (After Gerhard Neumann and Willard J. Pierson, Jr., *Principles of Physical Oceanography*, © 1966. Reprinted by permission of Prentice-Hall, Inc., Englewood Cliffs, N.J.)

FIGURE 11.4
High tide and low tide on Nova Scotia's Minas Basin in the Bay of Fundy.
(Courtesy of Nova Scotia Dept. of Tourism)

mic rise and fall along coastlines have been known since antiquity, and other than waves, they are the easiest ocean movements to observe (Figure 11.4). Although known for centuries, tides were not explained satisfactorily until Sir Isaac Newton applied the law of gravitation to them. Newton showed that there is a mutual attractive force between two bodies, and that since oceans are free to move, they are deformed by this force. Hence tides result from the gravitational attraction exerted upon the earth by the moon, and to a lesser extent by the sun.

To illustrate how tides are produced, we will assume that the earth is a rotating sphere covered to a uniform depth with water (Figure 11.5A). It is easy to see how the moon's gravitational force can cause the water to bulge on the side of the earth nearest the moon. In addition, however, an equally large tidal bulge is produced on the side of the earth directly opposite the moon. Both tidal bulges are caused, as Newton discovered, by the pull of gravity, a force inversely proportional to the square of the distance between two objects. In this case the two objects are the moon and the earth. Because the force of gravity decreases as distance increases, the moon's gravitational pull on the earth is slightly greater on the near side of the earth than on the far side. The result of this differential pulling is to stretch (elongate) the earth. Although the solid earth is stretched by the moon's gravitational pull, the amount of elongation is very

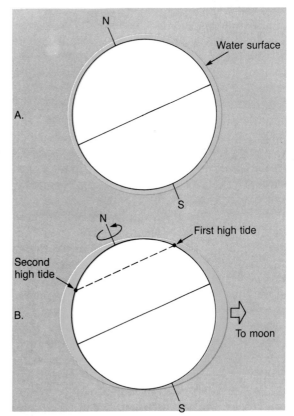

FIGURE 11.5
Tides on an earth that is covered to a uniform depth with water. Depending upon the moon's position, tidal bulges may be inclined to the equator. In this situation an observer will experience two unequal high tides.

slight. However, the world's oceans, which are mobile, are deformed quite dramatically by this effect to produce the two opposing tidal bulges.

Since the position of the moon changes slightly in a single day, it is the tidal bulges that remain in place while the earth rotates beneath them. For that reason, the earth will carry an observer at any given location alternately into areas of deeper and shallower water. When the person is carried into regions of deeper water, the tide rises, and as the person is carried away, the tide falls. Therefore during one day the observer would experience two high tides and two low tides. In addition to the earth rotating, the tidal bulges also move as the moon revolves about the earth every 28 days. As a result, the tides, like the time of moonrise, occur about 50 minutes later each day. After 28 days one cycle is complete and a new one begins.

There may be an inequality between the high tides during a given day. Depending upon the moon's position, the tidal bulges may be inclined to the equator as in Figure 11.5B. This figure illustrates that the first high tide experienced by an observer in the Northern Hemisphere is considera-

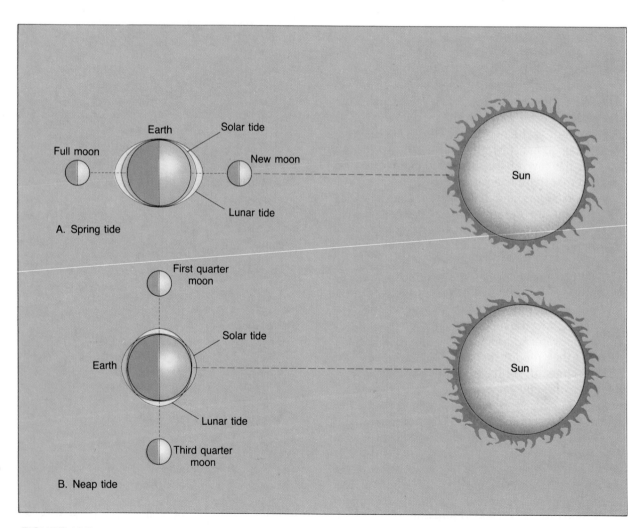

FIGURE 11.6
Relationship of the moon and sun to the earth during **A.** spring tides and **B.** neap tides.

bly lower than the high tide half a day later. On the other hand, a Southern Hemisphere observer would experience the opposite effect.

The sun also influences the tides, but because it is so far away, the effect is considerably less than that of the moon. In fact, the tide-generating potential of the sun is slightly less than half that of the moon. Near the times of new and full moons, the sun and moon are aligned and their forces are added together (Figure 11.6A). Accordingly, the two tide-producing bodies cause higher tidal bulges (high tides) and lower tidal troughs (low tides). These are called the **spring tides.** Spring tides create the largest daily tidal range, that is, the largest variation between high and low tides. Conversely, at about the time of the first and third quarters of the moon, the gravitational forces of the moon and sun on the earth are at right angles, and each partially offsets the influence of the other (Figure 11.6B). As a result, the daily tidal range is less. These are the **neap tides.**

The discussion thus far explains the basic causes and patterns of tides, but keep in mind these theoretical considerations cannot be used to predict either the height or the time of actual tides at a particular place. The shape of coastlines and the configuration of ocean basins greatly influence the tide. Consequently, tides at various locations respond differently to the tide-producing forces. This being the case, the nature of the tide at any place can be determined most accurately by actual observation. The predictions in tidal tables and the tidal data on nautical charts are based upon such observations.

Tides are classified as one of three types according to the tidal pattern occurring at a particular locale. The *semidiurnal* type fits the twice-daily pattern described earlier. There are two high and two low tides each tidal day, with a relatively small difference in the high and low water heights. Tides along the Atlantic coast of the United States are representative of the semidiurnal type, which is illustrated by the tidal curve for Boston in Figure 11.7A. *Diurnal* tides are characterized by a single high and low water height each tidal day (Figure 11.7B). Tides of this type occur along the northern shore of the Gulf of Mexico, among other locations. More common than the diurnal tide, the *mixed* type of tide is characterized by a large in-

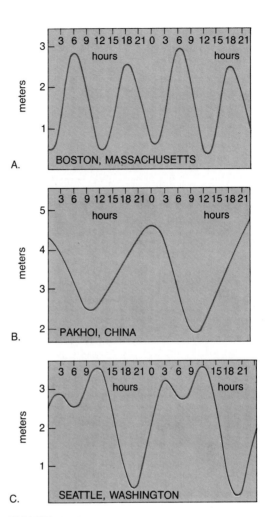

FIGURE 11.7
Types of tides. **A.** Semidiurnal. **B.** Diurnal. **C.** Mixed.

equality in high water heights, low water heights, or both (Figure 11.7C). In this case, there are usually two high and two low waters each day. Such tides are prevalent along the Pacific coast of the United States and in many other parts of the world.

Tidal current is the term used to denote the horizontal flow of water accompanying the rise and fall of the tide. As the tide rises water flows in toward the shore as a **flood tide,** submerging the low-lying coastal zone. When the tide falls, a reverse flow, the **ebb tide,** again exposes the drowned portion of the shore. The areas affected by these alternating tidal currents are called **tidal flats.** Depending upon the nature of the coastal zone, tidal flats vary in size from narrow strips lying seaward of the beach to zones that may extend for several kilometers.

Although tidal currents are not important in the open sea, tides may create rapid currents in bays, river estuaries, straits, and other narrow places. Off the coast of Brittany, for example, tidal currents which accompany a high tide of 12 meters (40 feet) may attain a speed of 20 kilometers (12 miles) per hour. While it is generally believed that tidal currents are not major agents of erosion and sediment transport, notable exceptions occur where tides move through narrow inlets. Here they scour the small entrances to many good harbors that would otherwise be blocked.

TIDAL POWER

With increased public interest in the rising costs and eventual depletion of petroleum, greater attention is being focused upon alternate energy sources. Although several methods of generating

electrical energy from the oceans have been proposed, the ocean's energy potential remains largely untapped. The development of tidal power is the principal example of energy production from the ocean.

Tides have been used as a source of power for centuries. Beginning in the twelfth century, water wheels driven by the tides were used to power gristmills and sawmills. During the seventeenth and eighteenth centuries, much of Boston's flour was produced at a tidal mill. Today, far greater energy demands must be met and more sophisticated ways of using the force created by the perpetual rise and fall of the ocean must be employed.

Tidal power is harnessed by constructing a dam across the mouth of a bay or an estuary in a coastal area having a large tidal range (Figure 11.8). The narrow opening between the bay and the open ocean magnifies the variations in water level that occur as the tides rise and fall. The

FIGURE 11.8
Simplified diagram showing the principle of the tidal dam. (After John J. Fagan, Jr., *Earth Environment,* © 1974, Prentice-Hall, Inc., Englewood Cliffs, N.J.)

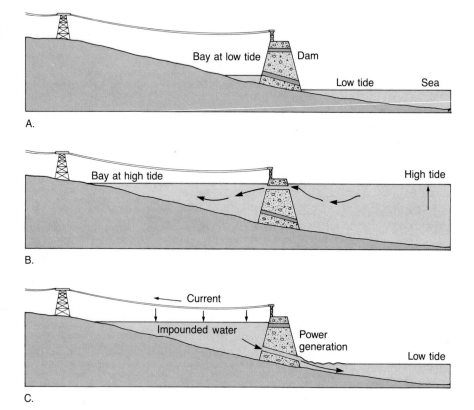

FIGURE 11.9
The world's first tidal power station to produce electricity was built across the Rance River estuary in 1966. The tide which rushes up this estuary on the northern coast of Britanny is one of the highest in the world, reaching 13.5 meters (44 feet). (Courtesy of Phototeque/Electricite de France)

strong in-and-out flow that results at such a site is then used to drive turbines and electrical generators.

Tidal energy utilization is exemplified by the tidal power plant at the mouth of the Rance River in France (Figure 11.9). By far the largest yet constructed, this plant went into operation in 1966 and produces enough power to satisfy the needs of Brittany and also contribute to the demands of other regions. Much smaller experimental facilities near Murmansk in the Soviet Union and near Taliang in China are also being used to generate electricity. The United States has not yet tapped its tidal power potential, although a site at Passamaquoddy Bay in Maine, where the tidal range approaches 15 meters (50 feet), has been under review for more than 50 years.

Along most of the world's coasts it is not possible to harness tidal energy. If the tidal range is less than 8 meters (25 feet) or narrow, enclosed bays are absent, tidal power development is uneconomical. For this reason, the tides will never provide a very high proportion of our ever-increasing electrical energy requirements. Nevertheless, the development of tidal power may be worth pursuing as Paul R. Ryan points out:

> Although total tidal power potential represents only a relatively small proportion of world energy requirements, its realization would nevertheless save a significant amount of fossil fuels. Tidal projects worldwide have been estimated to have a potential energy output of 635,000 gigawatts, the equivalent of more than a billion barrels of oil, a year.*

In addition to the fact that electricity produced by the tides consumes no exhaustible fuels (and hence creates no noxious wastes), such facilities disturb the landscape much less than the large reservoirs that are created when rivers are dammed.

WAVES

The waters of the ocean are constantly in motion. The restless nature of the water is most noticeable along the shore—the dynamic interface between land and sea. Here we can observe the rhythmic rise and fall of tides and see waves constantly rolling in and breaking. Sometimes the waves are low and gentle. At other times, they pound the shore with an awesome fury.

Although it may not be readily apparent to the occasional visitor, the shoreline is constantly being shaped and modified by the moving ocean waters. However, the nature of present-day shorelines is not just the result of the relentless attack of the land by the sea. Indeed, the shore is a complex zone whose unique character results from many geologic processes. For example, practically all coastal areas were affected by the worldwide rise

---

*"Harnessing Power from the Tides: State of the Art," *Oceanus* 22 (1980): 64.

in sea level that accompanied the melting of glaciers at the close of the Pleistocene epoch. As the sea edged landward, the shoreline became superimposed upon landscapes that resulted from such processes as stream erosion, glaciation, volcanic activity, and the forces of mountain building.

Wind-generated waves provide most of the energy that shapes and modifies shorelines. Where the land and sea meet, waves that may have traveled unimpeded for hundreds of kilometers suddenly encounter a barrier that will not allow them to advance farther. Stated another way, the shore is where a practically irresistible force confronts an almost immovable object. The conflict that results is never-ending and sometimes dramatic.

The undulations of the water surface, called waves, derive their energy and motion from the wind. If a breeze of less than 3 kilometers (2 miles) per hour starts to blow across still water, small wavelets appear almost instantly. When the breeze dies, the ripples disappear as suddenly as they formed. However, if the wind exceeds 3 kilometers per hour, more stable waves gradually form and progress with the wind.

All waves are described in terms of the characteristics illustrated in Figure 11.10. The tops of the

waves are the *crests,* which are separated by *troughs.* The vertical distance between trough and crest is called the **wave height,** and the horizontal distance separating successive crests is the **wave length.** The **wave period** is the time interval between the passage of successive crests at a stationary point. The height, length, and period that are eventually achieved by a wave depend upon three factors: (1) Wind speed; (2) Length of time the wind has blown; and (3) **Fetch,** the distance that the wind has traveled across the open water. As the quantity of energy transferred from the wind to the water increases, the heights of the waves transmitting it increase. In the open ocean, wave heights of 1 to 4 meters (3 to 13 feet) are common, although storms may produce much higher waves. Winds are often gusty and turbulent; for that reason, waves covering the ocean surface are quite irregular in height and length.

When the wind stops or changes direction, or the waves leave the stormy area where they were created, they continue on without relation to local winds. The waves also undergo a gradual change to swells that are lower in height and longer in length and may carry the storm's energy to distant shores. Because many independent wave systems exist at the same time, the sea surface acquires a complex and irregular pattern. Hence the sea waves we watch from the shore are usually a mixture of swells from faraway storms and waves created by local winds.

It is important to realize that in the open sea the motion of the wave is different from the motion of the water particles within it. It is the wave form that moves forward, not the water itself. Each water particle moves in a circular path during the passage of a wave (Figure 11.11). As a wave

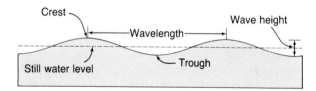

FIGURE 11.10
Characteristics of a wave.

FIGURE 11.11
Movement of water particles with the passage of a wave.

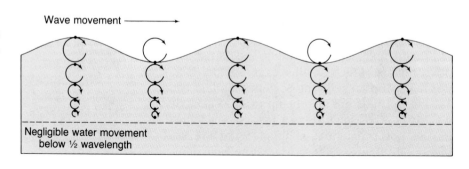

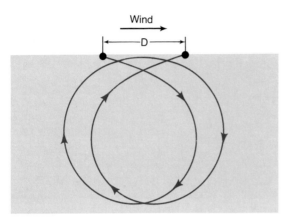

FIGURE 11.12
Circular motion and displacement (D) of a particle on the surface of deep water during two wave periods.

passes, a water particle returns almost to its original position. This is demonstrated by observing the behavior of a floating cork as a wave passes. The cork merely seems to bob up and down and sway slightly to and fro without advancing appreciably from its original position. For this reason, waves in the open sea are called **waves of oscillation.**

Yet, as Figure 11.12 illustrates, the wind does drag the water slightly forward, causing the surface circulation of the oceans. The energy contributed by the wind to the water is transmitted not only along the surface of the sea but downward as well. Due to friction, there is a progressive loss of energy with an increase in depth until at a depth equal to about one-half the wave length the movement of water particles becomes negligible. This is shown by the rapidly diminishing diameters of water-particle orbits in Figure 11.11.

As long as a wave is in deep water it is unaffected by water depth. However when a wave approaches the shore the water becomes shallower and influences wave behavior. The wave begins to "feel bottom" at a water depth equal to about one-half its wave length. Since some energy is used in moving small particles of sediment back and forth, the wave slows. As the wave continues to advance toward the shore, the slightly faster seaward waves catch up, decreasing the wave length. As the speed and length of the wave diminish, the wave steadily grows higher. Finally a critical point is reached when the steep wave front is unable to support the wave, and it collapses, or *breaks* (Figure 11.13). What had been a wave of oscillation now becomes a **wave of translation** in which the water advances up the shore. The turbulent water created by breaking waves is called **surf.** Following the uprush of water onto the beach, a seaward backwash occurs. The water from expended waves most commonly moves seaward in a broad sheet that produces the undertow so often felt by swimmers. Sometimes the backwash occurs in narrow, localized channels and if strong enough, it may pull swimmers into deep water.

WAVE EROSION

During periods of calm weather, wave action is at a minimum. However just as streams do most of their work during floods, so waves do most of their work during storms. The impact of high, storm-induced waves against the shore can at times be awesome in its violence. Each breaking wave may hurl thousands of tons of water against the land, sometimes causing the earth to literally tremble. The pressures exerted by Atlantic waves, for

FIGURE 11.13
Changes that occur when a wave moves onto shore.

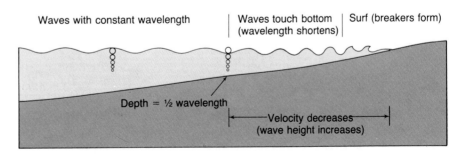

Waves with constant wavelength | Waves touch bottom (wavelength shortens) | Surf (breakers form)

Depth = ½ wavelength

Velocity decreases (wave height increases)

example, average nearly 10,000 kilograms per square meter (more than 2000 pounds per square foot) in winter. The force during storms is even greater. During one such storm, for instance, a 1350-ton portion of a steel and concrete breakwater was ripped from the rest of the structure and moved to a useless position toward the shore at Wick Bay, Scotland. Five years later the 2600-ton unit that replaced the first met a similar fate. There are many such stories that demonstrate the great force of breaking waves. It is no wonder then that cracks and crevices are quickly opened in cliffs, seawalls, breakwaters, and anything else that is subjected to these enormous shocks. Water is forced into every opening, causing air in the cracks to become highly compressed by the thrust of crashing waves. When the wave subsides, the air expands rapidly, dislodging rock fragments and enlarging and extending pre-existing fractures.

In addition to the erosion caused by wave impact and pressure, **abrasion,** the sawing and grinding action of the water armed with rock fragments, is also important. In fact, abrasion is probably more intense in the surf zone than in any other environment. Smooth and rounded stones and pebbles along the shore are obvious reminders of the grinding action of rock against rock in the surf zone. Further, such fragments are used as "tools" by the waves as they cut horizontally into the land.

Along shorelines composed of unconsolidated material rather than hard rock, the rate of erosion by breaking waves can be extraordinary. In parts of Britain, where waves have the easy task of eroding glacial deposits of sand, gravel, and clay, the coast has been worn back 3–5 kilometers (2–3 miles) since Roman times, sweeping away many villages and ancient landmarks. A similar retreat may be seen along the cliffs of Cape Cod, which in places are retreating at a rate of up to 1 meter (3 feet) per year.

WAVE REFRACTION

Most waves approach a shoreline at an angle. When they reach the shallow water of a smoothly sloping bottom, however, they are bent and tend

FIGURE 11.14
Wave bending around the end of a beach at Stinson Beach, California. (Photo by James E. Patterson)

to become parallel to the shore. Such bending of waves is called **refraction** (Figure 11.14). The part of the wave nearest the shore touches bottom and slows first, while the end that is still in deep water continues forward at its regular speed. The net result is a wave front that may approach nearly parallel to the shore regardless of the original direction of the wave.

Due to refraction, wave impact is concentrated against the sides and ends of headlands projecting into the water, while wave attack is weakened in bays. This differential wave attack along irregular coastlines is illustrated in Figure 11.15. Since the waves reach the shallow water in front of the headland sooner than they do in adjacent bays, they are bent more nearly parallel to the protruding land and strike it from all three sides. Over a period of time the effect of this process is to straighten irregular coastlines.

BEACH DRIFT AND LONGSHORE CURRENTS

Although waves are refracted, most still reach the shore at an angle, however slight. Consequently, the uprush of water from each breaking wave is oblique. Nevertheless, the backflow is straight down the slope of the beach. The effect of this pattern of water movement is to transport particles of

FIGURE 11.15
Wave refraction along an irregular shoreline.

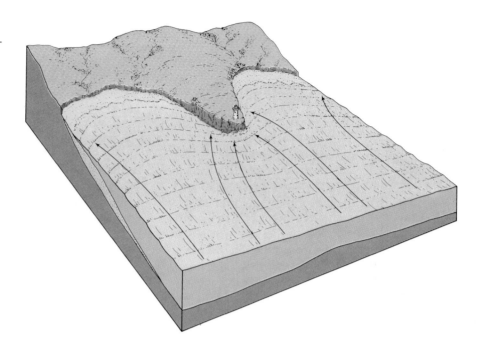

FIGURE 11.16
Beach drift, caused by the
uprush of water from oblique
waves.

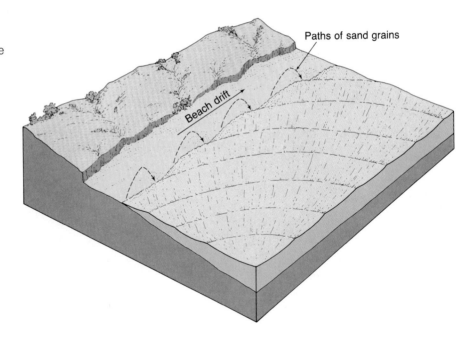

sediment in a zigzag pattern along the beach (Figure 11.16). This movement, called **beach drift,**
can transport sand and pebbles hundreds or even
thousands of meters each day.

Oblique waves also produce currents within the
surf zone that flow parallel to the shore. Since the
water here is turbulent, these **longshore currents**
easily move the fine suspended sand as well as
roll larger sand and gravel along the bottom.
When the sediment transported by longshore

currents is added to the quantity moved by beach drift, the total amount can be very large. At Sandy Hook, New Jersey, for example, the quantity of sand transported along the shore over a 48-year period averaged almost 750,000 tons per year. For a 10-year period at Oxnard, California, more than 1.5 million tons of sediment moved along the shore each year.

There should be little wonder that beaches have been characterized as "rivers of sand." At any point along a beach there is likely to be more sediment that was derived elsewhere than material eroded from the cliff immediately behind it. It is also worth noting that much of the sediment composing beaches is not wave-eroded debris. Rather, in many areas sediment-laden rivers that discharge into the ocean are the major source of material. For that reason, if it were not for beach drift and longshore currents, many beaches would be nearly sandless.

SHORELINE FEATURES

Whether along the rugged and irregular New England coast or along the steep shorelines of the West Coast, the effects of wave erosion are often easily seen. **Wave-cut cliffs,** as their name implies, originate by the cutting action of the surf against the base of coastal land. As erosion progresses, rocks overhanging the notch at the base of the cliff crumble into the surf and the cliff retreats. A relatively flat, benchlike surface, the **wave-cut platform,** is left behind by the receding cliff (Figure 11.17). The platform broadens as wave attack continues. Some debris produced by the breaking waves remains along the water's edge as part of the beach, while the remainder is transported farther seaward.

Headlands that extend into the sea are vigorously attacked by the waves because of refraction. The surf erodes the rock selectively, wearing away

FIGURE 11.17
Elevated wave-cut platform along the California coast north of San Francisco. A new platform is being created at the base of the cliff. (Photo by John S. Shelton)

FIGURE 11.18
Sea arch and sea stack along the coast of Iceland. (Photo by Bruce F. Molnia, courtesy of Terraphotographics/BPS)

the softer or more highly fractured rock at the fastest rate. At first, sea caves may form. When two caves on opposite sides of a headland unite, a **sea arch** results (Figure 11.18). Finally the arch falls in, leaving an isolated remnant, or **sea stack,** on the wave-cut platform (Figure 11.18). Eventually it too will be consumed by the action of the waves.

Where beach drift and longshore currents are active, several features related to the movement of sediment along the shore may develop (see Figure 11.20). **Spits** are elongated ridges of sand that project from the land into the mouth of an adjacent bay. Often the end in the water hooks landward in response to wave-generated currents. The term **baymouth bar** is applied to a sand bar that completely crosses a bay, sealing it off from the open ocean. Such a feature tends to form across bays where currents are weak, allowing a spit to extend to the other side. A **tombolo,** a ridge of sand that connects an island to the mainland or to

another island, forms in much the same manner as does a spit.

The gently sloping coastline found along the Gulf Coast and much of the eastern shore of the United States south of New York City is frequently characterized by **barrier islands,** which are low offshore ridges of sand that parallel the coast (Figure 11.19). The lagoons that separate these narrow islands from the shore represent zones of relatively quiet water that allow small craft traveling between New York and northern Florida to avoid the rough waters of the North Atlantic.

How barrier islands originate is still not certain. Quite possibly they form in three or more different ways. Some are thought to have originated as spits that were subsequently severed from the mainland by wave erosion or by the general rise in sea level following the last episode of glaciation. It is also possible that some barrier islands are created when turbulent waters in the line of breakers heap up sand that has been scoured from the bottom. Since these sand barriers rise above normal sea level, the piling up of sand likely resulted from the work of storm waves at high tide. Finally, some studies suggest that barrier islands may be former sand dune ridges that originated along the shore

FIGURE 11.19
This satellite image of New Jersey shows the well-developed barrier islands that parallel the coast. Barrier islands are common features along the Atlantic and Gulf coasts of the United States. (LANDSAT image courtesy of Phillips Petroleum Company, Exploration Projects Section)

FIGURE 11.20
Development of an initially
irregular coastline.

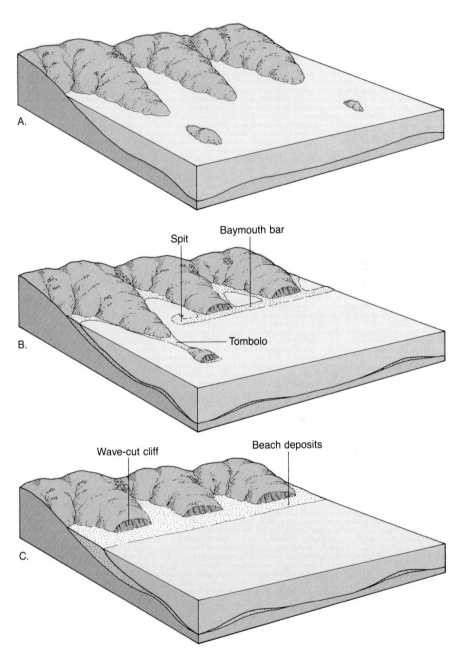

during the last glacial period, when sea level was lower. As the ice sheets melted, sea level rose and flooded the area behind the beach-dune complex.

There is little question that a shoreline soon undergoes modification regardless of its initial configuration. At first most coastlines are irregular, although the degree of and reason for the irregularity may vary considerably from place to place. Along a coastline that is characterized by varied geology the pounding surf may at first increase its irregularity because the waves will erode the weaker rocks more easily than the stronger ones. Be that as it may, it is commonly agreed that if a shoreline remains stable, marine erosion and deposition will eventually produce a more regular coast. Figure 11.20 illustrates the evolution of an initially irregular coast. As waves erode the headlands, creating cliffs and a wave-cut platform, sediment is carried along the shore. Some material is deposited in the bays, while other debris is formed into spits and baymouth bars. At the same time rivers fill the bays with sediment; ultimately a smooth coast results.

SHORELINE EROSION PROBLEMS

Compared with other natural hazards such as earthquakes, volcanic eruptions, and hurricanes, shoreline erosion is generally a more continuous and predictable process that appears to cause relatively modest damage to limited areas. Nonetheless, erosion along our coasts has caused and will continue to cause significant property damage. Large sums are spent annually not only to repair damage, but also to prevent or control erosion. Already a problem at many sites, shoreline erosion is certain to become an increasingly serious problem as extensive development in coastal areas continues.

The shorelines along the Pacific Coast of the United States are strikingly different from those characterizing the Atlantic and Gulf coast regions. The differences are related to plate tectonics. The West Coast represents the leading edge of the North American plate and because of this, experiences active uplift and deformation. By contrast, the East Coast is a tectonically quiet region that is far from any active plate margin. Because of this basic geological difference, the nature of shoreline erosion problems along America's opposite coasts is different.

The discussion that follows will focus first on the problems faced along the East Coast of the United States, especially its vulnerable barrier islands. This will be contrasted with a look at the equally serious erosion problems occurring at many West Coast locations.

GULF AND ATLANTIC COASTS

During this century growing affluence and increasing demands for recreation have brought unprecedented development to many coastal areas. Much of this development has occurred on barrier islands. Typically, barrier islands consist of a wide beach that is backed by dunes and sepa-rated from the mainland by marshy lagoons. The broad expanses of sand and exposure to the ocean have made barrier islands exceedingly attractive sites for development.

Because barrier islands face the open ocean, they receive the full force of major storms that strike the coast. When a storm occurs, the barriers absorb the energy of the waves primarily through the movement of sand. Frank Lowenstein describes this process and the dilemma that results:

> Waves may move sand from the beach to offshore areas or, conversely, into the dunes; they may erode the dunes, depositing sand onto the beach or carrying it out to sea; or they may carry sand from the beach and the dunes into the marshes behind the barrier, a process known as overwash. The common factor is movement. Just as a flexible reed may survive a wind that destroys an oak tree, so the barriers survive hurricanes and nor'easters not through unyielding strength but by giving before the storm.
>
> This picture changes when a barrier is developed for homes or a resort. Storm waves that previously rushed harmlessly through gaps between the dunes now encounter buildings and roadways. Moreover, since the dynamic nature of the barriers is readily perceived only during storms, homeowners tend to attribute damage to a particular storm, rather than to the basic mobility of coastal barriers. With their homes or investments at stake, local residents are more likely to seek to hold the sand in place and the waves at bay than to admit that development was improperly placed to begin with.*

Protecting property from storm-induced waves as well as from the ongoing movement of sand by longshore currents has been a significant concern in developed coastal areas for many years. Attempts at controlling the dynamic beach environment include building such artificial structures as jetties, breakwaters, groins, and seawalls. Such interference can create many new problems and result in unwanted changes that are difficult and expensive to correct.

From relatively early in America's history a principal goal in coastal areas was the development

---

*"Beaches or Bedrooms—The Choice as Sea Level Rises," *Oceanus*, Vol. 28, No. 3 (Fall 1985): p. 22.

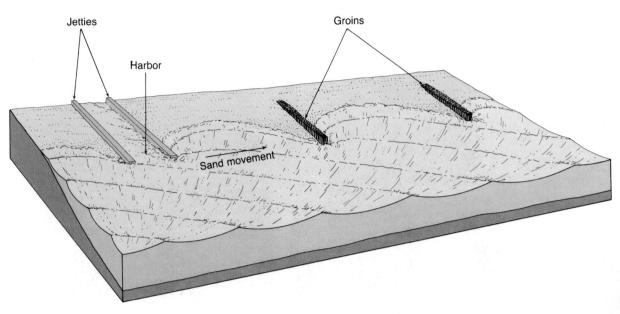

FIGURE 11.21
Jetties and groins trap sand that would otherwise be moved down the beach by
wave action.

and maintenance of harbors. In many cases, this involved the construction of jetty systems. **Jetties** are usually built in pairs and extend into the ocean at the entrances to rivers and harbors. By confining the flow of water to a narrow zone, the ebb and flow caused by the rise and fall of the tides keep the sand in motion and prevent deposition in the channel. However, as illustrated in Figure 11.21, the jetty may act as a dam against which the longshore current and beach drift deposit sand. At the same time, wave activity removes sand on the other side. Since the other side is not receiving any new sand, there is soon no beach at all.

To maintain or widen beaches that are losing sand, **groins** are sometimes constructed. A groin is a barrier built at a right angle to the beach for the purpose of trapping sand that is moving parallel to the shore (Figure 11.21). The result is an irregular but wider beach. These structures often do their job so effectively that the longshore current beyond the groin is sand deficient. As a result, the current removes sand from the beach on the leeward side of the groin. To offset this effect,

property owners downcurrent from the structure may erect a groin on their property. In this manner, the number of groins multiplies. An example of such proliferation is the shoreline of New Jersey, where more than three hundred such structures have been built. Since it has been shown that groins often do not provide a satisfactory solution, they are no longer the preferred method of keeping beach erosion in check.

In some coastal areas a **breakwater** may be constructed parallel to the shoreline. The purpose of such a structure is to protect boats from the force of large breaking waves by creating a quiet water zone near the shore. However, when this is done, the reduced wave activity along the shore behind the structure may allow sand to accumulate. If this happens, the marina will eventually fill with sand while the downstream beach erodes and retreats. At Santa Monica, California, where the building of a breakwater created such a problem, the city had to install a dredge to remove sand from the protected quiet water zone and deposit it down the beach where longshore currents and

A.

B.

FIGURE 11.22
A. The shoreline at Santa Monica pier as it appeared in 1931. **B.** The same area
in 1949. Construction of the breakwater disrupted longshore transport and
caused the seaward growth of the beach. (Photos courtesy of Fairchild Aerial
Photography Collection, Whittier College)

beach drift could recirculate the sand (Figure
11.22).

As development has moved ever closer to the
beach, seawalls represent yet another structure
that is built to defend property from the force of
breaking waves. **Seawalls** are simply massive barriers intended to prevent waves from reaching the
areas behind the wall. Waves expend much of
their energy as they move across an open beach.
Seawalls cut this process short by reflecting the
force of unspent waves seaward. As a consequence, the beach experiences significant erosion
and may, in some instances, be eliminated entirely. Once the width of the beach is reduced, the
seawall is subjected to even greater pounding by
the waves. Eventually this battering will cause the
wall to fail and a larger, more expensive wall must
be built to take its place.

Recently the wisdom of building temporary
protective structures has been questioned with
greater frequency. The feelings of many coastal

scientists is expressed in the following excerpt
from a position paper that grew out of a confer-
ence on America's Eroding Shoreline:

> It is now clear that halting the receding shoreline
> with protective structures benefits only a few and
> seriously degrades or destroys the natural beach
> and the value it holds for the majority. Protective
> structures divert the ocean's energy temporarily
> from private properties, but usually refocus that
> energy on the adjacent natural beaches. Many in-
> terrupt the natural sand flow in coastal currents,
> robbing many beaches of vital sand replacement.*

Of the various approaches to stabilizing the
sands of barrier islands, **beach nourishment** is
now considered the most acceptable method. As
the term implies, this practice simply involves the
addition of large quantities of sand to the beach
system. By building the beaches seaward, beach

---
* "Strategy for Beach Preservation Proposed," *Geotimes,* Vol. 30,
No. 12, (December 1985): p. 15.

quality and storm protection are both improved. The source of the sand may be the bottom of a nearby lagoon or inland dunes. In some cases, sand is trucked in and added to the beach. In other instances, the sand is added at an upstream location to be distributed down the coast by wave activity. Beach nourishment, however, is not a permanent solution to the problem of shrinking beaches. A case in point is Virginia Beach, Virginia, one of the largest seaside resort communities on the Atlantic Coast. Each year since 1952, this community has added the equivalent of 30,000 dump trucks' worth of sand along the shore in order to maintain just a thin strip of beach. The price is about $1.5 million per year. When beach nourishment was used to renew 24 kilometers of Miami Beach, the cost was $64 million. Furthermore, in some instances, beach nourishment can lead to unwanted environmental effects. For example, beach replenishment at Waikiki Beach, Hawaii, involved replacing coarse calcareous sand with softer, muddier calcareous sand. Destruction of the soft beach sand by breaking waves increased the water's turbidity and killed offshore coral reefs. At Miami Beach, where quartz sand was replaced by calcareous sand, the increased turbidity damaged local coral communities.

Beach nourishment appears to be an economically viable long-range solution to the beach preservation problem only in areas where there are dense development, large supplies of sand, relatively low wave energy, and reconcilable environmental issues. Unfortunately, few areas possess all of these attributes.

SEA LEVEL IS RISING

The shifting, dynamic nature of barrier islands and the ineffectiveness of most shoreline protection measures are now relatively well-established facts. Unfortunately, recent research, which indicates that sea level is rising, has compounded this already distressing situation. One study shows that there has been a 10-cm rise over the past century. Furthermore, some investigators predict an accelerated sea-level rise in the years to come—as much as 30 cm or more by the middle of the next century. Although such a vertical change may seem modest, many coastal geologists believe that

any given rise in sea level along the gently sloping Atlantic and Gulf coasts will cause from 10 to 1000 times as much horizontal shoreline retreat.

The belief that sea level will continue to rise in the coming decades is linked to the results of climatic studies that predict a global warming trend. Such predictions are based upon the now well-established fact that the carbon dioxide (CO_2) content of the atmosphere has been rising at an accelerating rate for more than half a century. The CO_2 is added primarily as a by-product of the combustion of ever-increasing quantities of fossil fuels. If we assume that the use of fossil fuels will continue to rise at projected rates, current estimates indicate that the atmosphere's CO_2 content will grow by an additional 40 percent or more by sometime in the second half of the next century. The importance of CO_2 lies in the fact that it traps a portion of the radiation emitted by the earth and thereby keeps the air near the earth's surface warmer than it would be without CO_2. In other words, CO_2 is an important heat-absorbing gas; it follows logically that an increase in the air's CO_2 content should lead to higher atmospheric temperatures.*

How is a warmer atmosphere related to a global rise in sea level? The most significant connection appears to be that higher air temperatures raise the temperature of the upper layers of the ocean. This, in turn, causes the water to expand and sea level to rise. If the atmospheric warming is sufficiently great, melting glaciers may contribute to a higher sea level as well. It is also believed that a warmer ocean may spur storm development. Of course, an increase in storm activity would compound an already serious problem in many coastal areas.

Since rising sea level is a gradual phenomenon, it may be overlooked by coastal residents as a significant contributor to shoreline erosion problems. Rather, the blame is assigned to other forces, especially storm activity. Although a given storm may be the immediate cause, the magnitude of its destruction may result from the relatively small sea level rise that allowed the storm's power to cross a much greater land area.

---

*The impact of rising CO_2 levels on global temperatures is treated in considerable detail in Chapter 16.

PACIFIC COAST

In contrast to the broad, gently sloping coastal plains of the East, much of the Pacific Coast is characterized by relatively narrow beaches that are backed by steep cliffs and mountain ranges. Recall that America's western margin is a more rugged and tectonically active region than the East. Since uplift continues, the apparent rise in sea level in the West is negligible or even nonexistent. Nevertheless, like the shoreline erosion problems facing the East's barrier islands, West Coast difficulties also stem largely from the alteration of a natural system by people.

A major problem facing the Pacific shoreline, and especially portions of southern California, is a significant narrowing of many beaches. The bulk of the sand on many of these beaches is supplied by rivers that transport it from the mountainous regions to the east. Over the years this natural flow of material to the coast has been interrupted by dams built for irrigation and flood control. The reservoirs effectively trap the sand that would otherwise nourish the beach environment. When the beaches were wider, they served to protect the cliffs behind them from the force of storm waves. Now, however, the waves move across the narrowed beaches without losing much energy and cause more rapid erosion of the sea cliffs.

Although the retreat of the cliffs provides materials to replace some of the sand impounded behind dams, it also endangers homes and roads built on the bluffs. In addition, development atop the cliffs aggravates the problem. Urbanization increases runoff which, if not carefully controlled, can result in serious bluff erosion. Watering lawns and gardens adds significant quantities of water to the slope. This water percolates downward toward the base of the cliff, where it may emerge in small seeps. This action reduces the slope's stability and facilitates mass wasting.

Shoreline erosion along the Pacific Coast varies considerably from one year to the next, largely because of the sporadic occurrence of storms. As a consequence, when the infrequent but serious episodes of erosion occur, the damage is often blamed on the unusual storms and not on coastal development or the sediment-trapping dams that may be great distances away. If, as predicted, sea level rises at an increasing rate in the years to come, increased shoreline erosion and sea cliff retreat should be expected along many parts of the Pacific Coast.

EMERGENT AND SUBMERGENT COASTS

The great variety of present-day shorelines suggests that they are complex areas. Indeed, to understand the nature of any particular coastal area, many factors must be considered, including rock types, size and direction of waves, number of storms, tidal range, and submarine profile. Moreover, recent tectonic events and changes in sea level must also be taken into account. These many variables make shoreline classification difficult.

One way that many geologists classify coasts is based upon changes that have occurred with respect to sea level. This commonly used, although incomplete, classification divides coasts into two categories: emergent and submergent. **Emergent coasts** develop either because an area has been uplifted or as a result of a drop in sea level. Conversely, **submergent coasts** are created when sea level rises or the land adjacent to the sea subsides.

In some areas the coast is clearly emergent because rising land or a falling water level expose wave-cut cliffs and platforms above sea level. Excellent examples include portions of coastal California where uplift has occurred in the recent geological past. The elevated wave-cut platform shown in Figure 11.17 illustrates this. In the case of the Palos Verdes Hills, south of Los Angeles, seven different terrace levels exist, indicating seven episodes of uplift. The ever-persistent sea is now cutting a new platform at the base of the cliff. If uplift follows, it too will become an elevated marine terrace.

Other examples of emergent coasts include regions that were once buried beneath great ice sheets. When glaciers were present, their weight depressed the crust and when the ice melted, the crust began to gradually spring back. As a result, prehistoric shoreline features today are found high above sea level. The Hudson Bay region of Canada is one such area, portions of which are still rising at a rate of more than one centimeter per year.

FIGURE 11.23
Satellite image of a portion of the East Coast showing Chesapeake Bay, an estuary created when the lower portion of a river valley was submerged by the rise in sea level that followed the end of the Ice Age. (Courtesy of Earth Satellite Corporation)

In contrast to the preceding examples, other coastal areas show definite signs of submergence. The shoreline of a coast that has been submerged in the relatively recent past is often highly irregular because the sea typically floods the lower reaches of river valleys flowing into the ocean. The ridges separating the valleys, however, remain above sea level and project into the sea as headlands. These drowned river mouths, which are often called **estuaries,** characterize many coasts today. Along the Atlantic coast, the Chesapeake and Delaware bays are examples of estuaries created by submergence (Figure 11.23, page 325). The picturesque coast of Maine, particularly in the vicinity of Acadia National Park, is another excellent example of an area that was flooded by the post-glacial rise in sea level and transformed into a highly irregular submerged coastline.

It should be kept in mind that most coasts have a rather complicated geologic history. With respect to sea level, many have at various times emerged and then submerged again. Each time they retain some of the features created during the previous situation.

REVIEW QUESTIONS

1 Although winds are the primary cause of surface currents, ocean currents do not coincide precisely with the wind direction. Explain this phenomenon.

2 Briefly describe the North Atlantic gyre. How does the surface circulation of the South Atlantic and Indian oceans differ from that of the North Atlantic?

3 Describe the process of coastal upwelling.

4 How do ocean currents influence climate?

5 What is meant by *thermohaline circulation*?

6 Discuss the origin of ocean tides.

7 Explain why an observer can experience two unequal high tides during a day (see Figure 11.5).

8 How does the sun influence tides?

9 How do semidiurnal, diurnal, and mixed tides differ?

10 What is meant by flood tide? Ebb tide?

11 Why will tidal power never contribute substantially to filling the world's growing energy needs? What advantages does tidal power offer?

12 List three factors that determine the height, length, and period of a wave.

13 Describe the motion of a water particle as a wave passes (see Figure 11.11).

14 Explain what happens when a wave breaks.

15 How do waves cause erosion?

16 What is wave refraction? What is the effect of this process along irregular coastlines?

17 Why are beaches often called "rivers of sand"?

18 Describe the formation of the following features: wave-cut cliff, wave-cut platform, sea stack, spit, baymouth bar, tombolo.

19 List three possible ways that barrier islands form.

20 For what purpose is a groin built? Why might the building of one groin lead to the building of others?

21 How was the beach at Santa Monica, California, affected when a breakwater was constructed (see Figure 11.22)?

22 How can a seawall increase beach erosion?

23 It is believed that global temperatures will be increasing in the decades to come. How can a warmer atmosphere lead to a rise in sea level?

24 Relate the damming of rivers to the shrinking of beaches at many locations along the West Coast of the United States.

25 What observable features would lead you to classify a coastal area as emergent?

26 Are estuaries associated with submergent or emergent coasts? Why?

KEY TERMS

gyre	wave length	sea stack
Coriolis effect	wave period	spit
upwelling	fetch	baymouth bar
thermohaline circulation	wave of oscillation	tombolo
	wave of translation	barrier island
tide	surf	jetty
spring tide	abrasion	groin
neap tide	refraction	breakwater
tidal current	beach drift	seawall
flood tide	longshore current	beach nourishment
ebb tide	wave-cut cliff	emergent coast
tidal flat	wave-cut platform	submergent coast
wave height	sea arch	estuary

PART THREE

THE ATMOSPHERE

12

COMPOSITION, STRUCTURE, AND TEMPERATURE

The study of the atmosphere—the earth's gaseous envelope—goes back thousands of years. Present knowledge suggests that no other planet has the exact mixture of gases or the heat and moisture conditions necessary to sustain life as we know it. The gases that make up the earth's atmosphere and the controls to which they are subject are vital to our very existence. In this chapter we begin our examination of the ocean of air in which we all must live. We shall attempt to answer a number of basic questions. What is the composition of the atmosphere? At what point do we leave the atmosphere and enter outer space? What causes the seasons? How is air heated? What factors control temperature variations over the globe?

The scattering of certain wavelengths of light by the atmosphere can produce colorful sunsets. (Photo by Stephen Trimble)

331

eople's interest in the atmosphere is probably as old as the history of humankind. **Meteorology,** the science of the atmosphere
and its weather, goes back many centuries. The
term is derived from the title of Aristotle's four-
volume treatise, *Meteorologica,* literally "discourse
on things above." The study of meteorology has
progressed through many phases over the centuries, evolving from the days when weather lore
and superstition guided explanations to what it is
today—a scientific study using the most modern
instruments and equipment.

WEATHER AND CLIMATE

If you were to search through the library for books
about the atmosphere, many of the titles you
would find would contain the term *weather,* while
others would have the word *climate.* What is the
difference between these two terms? **Weather** is a
word used to denote the state of the atmosphere
at a particular place for a short period of time.
Weather is constantly changing—hourly, daily,
and seasonally. **Climate,** on the other hand, might
best be described as an aggregate or composite of
weather. Stated another way, the climate of a
place or region is a generalization of the weather
conditions over a long period of time. Therefore, a
climatic description is possible only after weather
records have been kept for many years.

Although weather and climate are not identical,
the nature of both is expressed in terms of the
same *elements*—temperature, moisture, pressure, and wind. We shall study these elements
separately at first. Even so, keep in mind that they
are very much interrelated. A change in any one of
the elements will often bring about changes in the
others.

COMPOSITION OF THE ATMOSPHERE

In the days of Aristotle all things were thought to
be a combination of four fundamental substances—fire, air, earth (soil), and water. Today
we know that matter is much more complex. For
example, the earth is composed of minerals,

TABLE 12.1
Composition of clean, dry air.

Component	Chemical Symbol	Percent of Air by Volume
Nitrogen	N_2	78.08
Oxygen	O_2	20.95
Argon	Ar	0.93
Carbon dioxide	CO_2	0.03
All others		Trace

which, in turn, are made of numerous elements
and compounds. Furthermore, air, which appears
to be a unique substance, is really a mixture of
many different gases and tiny solid particles.

The composition of air is not constant; it varies
from time to time and from place to place. This
variability is readily observable when we drive
from the rural countryside into the city and see
increases in dust, smoke, and sometimes cloud
cover. Be that as it may, if the water vapor, dust,
and other variable components were removed
from the atmosphere, we would find that its composition is quite uniform. Clean, dry air sampled
anywhere on earth is composed almost entirely of
the two gases nitrogen and oxygen. As can be seen
in Table 12.1, nitrogen makes up about 78 percent
of the atmosphere, and oxygen represents about
21 percent. Although these gases are the most
plentiful components of air and are of great significance to life on earth, they are of minor importance in affecting weather phenomena. The remaining percent of dry air is mostly the inert gas
argon plus tiny quantities of a number of other
gases. Carbon dioxide, although present in only
minute amounts, is nevertheless an important
constituent of air, because it has the ability to absorb heat energy radiated by the earth and thus
helps keep the atmosphere warm.

If meteorologists had to choose the most significant component of the atmosphere, they would
undoubtedly select water vapor. The amount of
water vapor in the air varies considerably, from
practically none at all up to about 4 percent by
volume. Why is such a small fraction of the atmo-

sphere so significant? Certainly the fact that water vapor is the source of all clouds and precipitation would be enough to justify the choice. However, water vapor has other roles. Like carbon dioxide, it has the ability to absorb heat energy given off by the earth as well as some solar energy. It is therefore an important factor to consider when we examine the heating of the atmosphere. When water changes from one state to another (see Figure 13.1), it absorbs or releases heat energy (termed *latent heat*). As we shall see in later chapters, latent heat plays an important role in the transport of heat from one region to another, and it is the energy source that helps drive many storms. Most of us probably think of dust as small, barely visible bits of dirt. Nevertheless, from a meteorological standpoint, dust is much more than that, for it includes many microscopic particles that are invisible to the naked eye, among them organic materials like pollen, spores, and seeds. Dust particles are most numerous in the lower part of the atmosphere near their primary source, the earth's surface. Still, the upper atmosphere is not free of them, because some dust is carried to great heights by rising currents of air, while other particles are given to the upper atmosphere by meteors which disintegrate as they pass through the earth's envelope of air. Some particles act as surfaces upon which water vapor may condense. This function is very basic to the formation of clouds and fog. In addition, dust may intercept and reflect incoming solar radiation. Thus when dust loading of the atmosphere is great, as it may be following an explosive volcanic eruption, the amount of sunlight reaching the earth's surface can be measurably reduced. Finally, dust in the air contributes to a phenomenon we all have observed—the red and orange colors of sunrise and sunset.

Another important component of the atmosphere is *ozone* (O_3), the triatomic form of oxygen. Ozone is not the same as the oxygen we breathe, which has two atoms per molecule (O_2). There is very little of this gas in the atmosphere. If all of the ozone in the atmosphere were brought down to the earth's surface, it would form a layer only about 0.4 centimeter thick. Furthermore, its distribution is not uniform. In the lowest portion of the

atmosphere ozone represents less than one part in 100 million. It is concentrated well above the surface between 10 and 50 kilometers (6 and 31 miles) with a peak near the altitude of 25 kilometers. In this altitude range, oxygen molecules (O_2) are split into single atoms of oxygen after absorbing ultraviolet radiation emitted by the sun. Ozone is then created when an atom of oxygen (O) and a molecule of oxygen (O_2) collide in the presence of a third, neutral molecule that acts as a catalyst by allowing the reaction to take place without itself being consumed in the process. Ozone is concentrated in the 10- to 50-kilometer height range because it is there that a crucial balance exists: The availability of ultraviolet radiation from the sun is in sufficient amounts to produce atomic oxygen, and there is an atmospheric density that is great enough to bring about the required collisions between atomic and molecular oxygen.

The presence of the ozone layer in our atmosphere is of vital importance to those of us on earth. The reason lies in the capability of ozone to absorb the burning ultraviolet radiation from the sun. If ozone did not act to filter a great deal of the ultraviolet radiation and if the rays were allowed to reach the surface of the earth, our planet would likely be uninhabitable for most life as we know it. Thus, anything that would act to reduce the amount of ozone in the atmosphere could affect the well-being of life on earth. Just such a concern has been raised.

THE OZONE PROBLEM

Since the early 1970s there has been continuing public concern over the possibility that the amount of ozone in the atmosphere is being reduced as a result of human activities. If true, the consequences may be serious.

The greatest human impact on the ozone layer is believed to be from a group of chemicals known as chlorofluorocarbons (abbreviated CFC).* CFCs are used as propellents for aerosol sprays, in the production of certain plastics, and in air conditioning

---

* CFCs are also called chlorofluoromethanes (CFMs) or halocarbons, and go by the trade name Freons.

FIGURE 12.1
Refrigerant leakage from the
air conditioning units in
these junkyard vehicles is
one way that chlorofluorocarbons are released into the
atmosphere. (Photo by E. J.
Tarbuck)

and refrigeration equipment (Figure 12.1). Due to the fact that CFCs are practically inert (that is, not chemically active) in the lower atmosphere, a portion of these gases gradually makes its way upward to the ozone layer, where sunlight separates the chemicals into their constituent atoms. The chlorine atoms released in this way would, by a complex series of reactions, have the net effect of converting some ozone into oxygen.

Since ozone filters out most of the damaging ultraviolet radiation in sunlight, a decrease in ozone concentration would permit more of these harmful wavelengths to reach the earth's surface. Two important questions are thus raised: (1) What will be the magnitude of the ozone depletion? and (2) What will be the effects of the increased ultraviolet radiation?

A report from the National Academy of Sciences indicates that the release of CFCs could lead to an eventual ozone depletion of between 2 and 4 percent by late in the next century.* Such figures, however, should be viewed very cautiously because as refinements occur in atmospheric mod-

*Causes and Effects of Changes in Stratospheric Ozone: Update 1983, (Washington D.C.: National Academy Press, 1984).

els and measurements, the estimates could change. For instance, over the five-year period 1979–1984, the projected risk to atmospheric ozone was reduced in successive reports from the National Academy of Sciences. A 1979 report indicated that the reduction in stratospheric ozone might approach or exceed 16.5 percent, whereas a 1982 report suggested an ozone depletion of between 5 and 9 percent.

Each 1-percent decrease in the concentration of stratospheric ozone increases the amount of ultraviolet radiation that reaches the earth's surface by about 2 percent. Therefore, because ultraviolet radiation is known to induce certain types of skin cancer, ozone depletion could seriously affect human health. Since up to one-half million cases of these cancers occur in the United States each year, ozone depletion could ultimately lead to many thousands of additional cases each year. The effects of increased ultraviolet radiation on animal and plant life may be as important as the direct effects on human health. Too little data exist for scientists to make specific predictions about effects on particular crops or ecosystems. However, analyses by the U.S. Environmental Protection Agency indicate that more aquatic organisms would die and that crop yields and quality could be adversely affected in many cases.

What can be done to protect the atmospheric ozone layer? The obvious solution is to curtail the production of CFCs. This is being done: In December 1978, the United States government banned the manufacture of all nonessential aerosol products using CFC propellents. As a result, world production of CFCs dropped to pre-1973 levels. Although this action was a positive step, the problem persists because domestic production for other uses as well as production outside the United States continue. In the absence of further regulatory actions, the potential threat will undoubtedly continue. Recognizing the dangers of not curbing CFC emissions, representatives from twenty-eight countries signed the Vienna Convention for the Protection of the Ozone Layer in 1985. The United Nations Environmental Program held the convention to frame an international strategy for protecting the ozone layer.

EXTENT AND STRUCTURE OF THE ATMOSPHERE

To state that the atmosphere begins at the earth's land-sea surface and extends upward is obvious. However, where does the atmosphere end and outer space begin? To help understand the vertical extent of the atmosphere, let us examine the changes in atmospheric pressure with height. Atmospheric pressure is simply the weight of the air above. At sea level, the average weight of the atmosphere, that is, the average pressure, is slightly more than 1 kilogram per square centimeter (14.7 pounds per square inch). Obviously the pressure at higher altitudes is less. This fact was demonstrated in a dramatic way by some nineteenth-century balloonists who were attempting to study the upper atmosphere. Several passed out upon reaching heights in excess of 6 kilometers; some even perished in the rarified upper air.

By examining Table 12.2 you can see that one-half of the atmosphere lies below an altitude of 5.6 kilometers (3.5 miles). At about 16 kilometers (10 miles), 90 percent of the atmosphere has been traversed, and above 100 kilometers (62 miles), only

TABLE 12.2
Percent of sea level pressure encountered at selected altitudes.

Altitude (kilometers)	Altitude (miles)	Percent of Sea Level Pressure
0	0	100
5.6	3.5	50
16.2	10	10
31.2	19.3	1
48.1	29.8	0.1
65.1	40.4	0.01
79.2	49.1	0.001
100	62	0.00003

0.00003 percent of all the gases composing the atmosphere remains. At this latter altitude, the atmosphere is so tenuous that the density of air is less than can be found in the best artificial vacuum at the earth's surface. Even so, traces of our atmosphere extend far beyond this altitude. Thus, to say where the atmosphere ends and outer space begins is quite arbitrary and to a large extent depends upon what phenomena we are studying. There is no definitive, sharp boundary. Certainly the data on vertical pressure changes reveal that the vast bulk of the gases composing the atmosphere are very near the earth's surface and that they gradually merge with the emptiness of space.

The atmosphere is divided vertically into four layers on the basis of temperature (Figure 12.2). The bottom layer, where temperature decreases with an increase in altitude, is known as the **troposphere.** The term literally means the region where air "turns over," a reference to the appreciable vertical mixing of the air in this lowermost zone. Practically all clouds and precipitation, as well as all violent storms, form in and are restricted to this layer of the atmosphere. There should be little doubt why the troposphere is often called the "weather sphere." The average temperature decrease in the troposphere is 6.5°C for each kilometer of altitude increase (3.5°F per 1000 feet), a figure known as the **normal lapse rate.** The rate is not constant, however, and may vary considerably from the average (normal) value. Thus, to

FIGURE 12.2
Thermal structure of the atmosphere to a height of about 110 kilometers (68 miles).

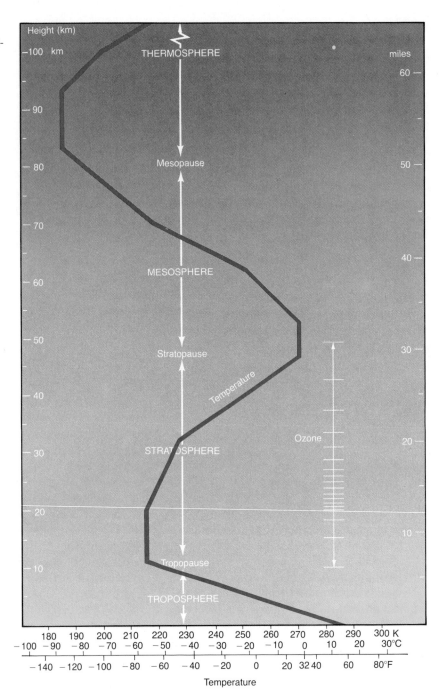

determine the lapse rate for any particular time and place, as well as gather information about vertical changes in pressure, wind, and humidity, radiosondes are used. The radiosonde is attached to a balloon and transmits data by radio as it ascends through the atmosphere (Figure 12.3). Although the thickness of the troposphere is not the same everywhere, the lapse rate continues to an average height of approximately 12 kilometers (7.4 miles).

Beyond the troposphere lies the **stratosphere,** and the boundary between them is called the **tropopause.** In the stratosphere, the temperature remains constant to a height of about 20 kilometers and then begins a gradual increase which continues until the **stratopause,** at a height of about 50 kilometers above the earth's surface. Below the tropopause atmospheric properties are readily transferred by large-scale turbulence and mixing; above it, in the stratosphere, they are not. The reason for the increased temperatures in the stratosphere is that the atmosphere's ozone is concentrated in this layer. Recall that ozone absorbs ultraviolet radiation from the sun. As a consequence, the stratosphere is heated.

In the **mesosphere** temperatures again decrease with height until, at the **mesopause,** approximately 80 kilometers (50 miles) above the surface, the temperature approaches −90°C. Extending upward from the mesopause and having no well-defined upper limit is the **thermosphere,** a layer which contains only a minute fraction of the atmosphere's mass. In the extremely rarified air of this outermost layer, temperatures again increase due to the absorption of very short wave solar energy by atoms of oxygen and nitrogen. While temperatures rise to extremely high values of more than 1000°C, such temperatures are not strictly comparable to those experienced near the earth's surface. Temperature is defined in terms of the average speed at which molecules move. Since the gases of the thermosphere are moving at very high speeds, the temperature is obviously very high. Even so, the gases are so sparse that only a very few of these fast-moving air molecules collide with a foreign body and hence only an insignificant quantity of energy is transferred. For this reason, the temperature of a satellite orbiting the earth in the thermosphere is determined chiefly by the amount of solar radiation it absorbs and not by the temperature of the surrounding air. If an astronaut inside were to expose his or her hand, it would not feel hot.

FIGURE 12.3
Atmospheric soundings using radiosondes supply data on vertical changes in temperature, pressure, and humidity. (Courtesy of Qualimetrics, Inc.)

EARTH-SUN RELATIONSHIPS

The earth intercepts only a minute percentage of the energy given off by the sun—less than one two-billionth. This may seem to be an insignificant amount until we realize that it is several hundred thousand times the electrical-generating capacity of the United States. Energy from the sun is undoubtedly the most important control of our weather and climate. Solar radiation represents more than 99.9 percent of the energy that heats the earth, drives the oceans and creates winds. Therefore, in order to have a basic understanding of atmospheric processes, we must understand what causes the time and space variations in the amount of solar energy reaching the earth.

MOTIONS OF THE EARTH

The earth has two principal motions—rotation and revolution. **Rotation** is the spinning of the earth about its axis, which is an imaginary line running through the poles. Our planet rotates once every 24 hours, producing the daily cycle of daylight and darkness. Half the earth is always experiencing daylight, and the other half darkness. The line

337

separating the dark half of the earth from the lighted half is called the **circle of illumination. Revolution** refers to the movement of the earth in its orbit around the sun. Hundreds of years ago, most people believed that the earth was stationary in space. We now know that the earth is traveling at more than 107,000 kilometers per hour in an elliptical orbit about the sun.

The distance between the earth and sun averages about 150 million kilometers (93 million miles). Because the earth's orbit is not perfectly circular, the distance varies during the course of a year. Each year, on about January 3, our planet is 147 million kilometers from the sun, closer than at any other time. This position is called **perihelion.** About six months later, on July 4, the earth is 152 million kilometers from the sun, farther away than at any other time. This position is called **aphelion.** Variations in the amount of solar radiation received by the earth as the result of its slightly elliptical orbit, are nevertheless slight and of little consequence when explaining major seasonal temperature variations. By way of illustration, consider that the earth is closest to the sun during the Northern Hemisphere winter.

THE SEASONS

We know that it is colder in winter than in summer, but if the variations in solar distance do not cause this, what does? All of us have made adjustments for the continuous change in the length of daylight that occurs throughout the year by planning our outdoor activities accordingly. The grad-

ual but significant change in length of daylight certainly accounts for some of the difference we notice between summer and winter. Further, a gradual change in the **altitude** (angle above the horizon) of the noon sun during the course of a year's time is evident to most people. At midsummer the sun is seen high above the horizon as it makes its daily journey across the sky. But as summer gives way to autumn, the noon sun appears lower and lower in the sky and sunset occurs earlier each evening.

The seasonal variation in the altitude of the sun affects the amount of energy received at the earth's surface in two ways. First, when the sun is directly overhead (90-degree angle), the solar rays are most concentrated. The lower the angle, the more spread out and less intense is the solar radiation that reaches the surface. This idea is illustrated in Figure 12.4. Second, and of lesser importance, the angle of the sun determines the amount of atmosphere the rays must traverse (Figure 12.5). When the sun is directly overhead, the rays pass through a thickness of only 1 atmosphere, while rays entering at a 30-degree angle travel through twice this amount, and 5-degree rays travel through a thickness roughly equal to 11 atmospheres. The longer the path, the greater are the chances for absorption, reflection, and scattering by the atmosphere, which reduce the intensity at the surface. These same effects account for the fact that the midday sun can be literally blinding, while the setting sun can be a sight to behold.

It is important to remember that the earth is spherical. Hence, on any given day only places lo-

FIGURE 12.4
Changes in the sun angle cause variations in the amount of solar energy reaching the earth's surface. The higher the angle, the more intense the solar radiation.

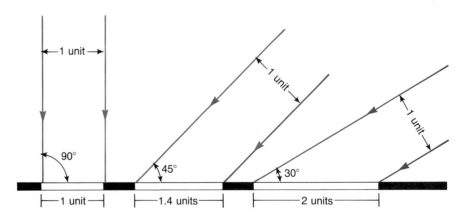

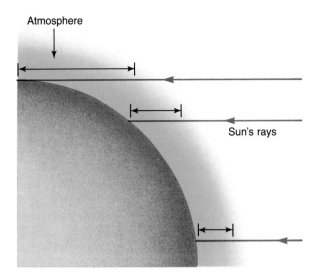

FIGURE 12.5
Rays striking at a low angle must traverse more of the atmosphere than rays striking at a higher angle and thus are subject to greater depletion by reflection and absorption.

cated at a particular latitude receive vertical (90-degree) rays from the sun. As we move either north or south of this location, the sun's rays strike at an ever-decreasing angle. Thus, the nearer a place is to the latitude receiving the vertical rays of the sun, the higher will be its noon sun.

What causes the yearly fluctuations in the sun angle and length of daylight? They occur because the earth's orientation to the sun continually changes. The earth's axis is not perpendicular to the plane of its orbit around the sun, but instead is tilted 23½ degrees from the perpendicular. This is termed the **inclination of the axis,** and as we shall see, if the axis were not inclined, we would have no seasonal changes. In addition, because the axis remains pointed in the same direction (toward the North Star) as the earth journeys around the sun, the orientation of the earth's axis to the sun's rays is constantly changing (Figure 12.6). On one day each year the axis is such that the Northern Hemisphere is "leaning" 23½ degrees toward

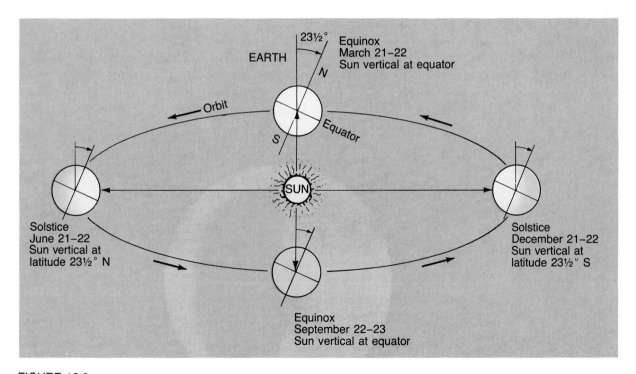

FIGURE 12.6
Earth-sun relationships.

the sun. Six months later, when the earth has moved to the opposite side of its orbit, the Northern Hemisphere leans 23½ degrees away from the sun. On days between these extremes, the earth's axis leans at amounts less than 23½ degrees to the rays of the sun. This change in orientation causes the vertical rays of the sun to make a yearly migration from 23½ degrees north of the equator to 23½ degrees south of the equator. This migration in turn causes the altitude of the noon sun to vary by as much as 47 degrees (23½ + 23½) during the course of a year at places located poleward of latitude 23½ degrees. For example, a mid-latitude city like New York (about 40 degrees north latitude) has a maximum noon sun angle of 73½ degrees when the sun's vertical rays reach their farthest northward location and a minimum noon sun angle of 26½ degrees six months later.

Historically, four days each year have been given special significance based on the annual migration of the direct rays of the sun and its importance to the yearly weather cycle. On June 21 or 22 the earth is in a position such that the axis in the Northern Hemisphere is tilted 23½ degrees toward the sun (Figure 12.7). At this time the vertical rays of the sun strike 23½ degrees north latitude (23½ degrees north of the equator), a line of latitude known as the **Tropic of Cancer.** For people in the Northern Hemisphere, June 21 or 22 is known as the **summer solstice.** Six months later,

on about December 21 or 22, the earth is in the opposite position, with the sun's vertical rays striking at 23½ degrees south latitude. This line is known as the **Tropic of Capricorn.** For those in the Northern Hemisphere, December 21 or 22 is the **winter solstice.** However, at the same time in the Southern Hemisphere, people are experiencing just the opposite—the summer solstice.

The equinoxes occur midway between the solstices. September 22 or 23 is the date of the **autumnal equinox** in the Northern Hemisphere, and March 21 or 22 is the date of the **vernal,** or **spring, equinox.** On these dates, the vertical rays of the sun strike at the equator (0 degrees latitude) because the earth is in such a position in its orbit that the axis is tilted neither toward nor away from the sun.

Further, the length of daylight versus darkness is also determined by the position of the earth in its orbit. The length of daylight on June 21, the summer solstice in the Northern Hemisphere, is greater than the length of night. This fact can be established from Figure 12.7 by comparing the fraction of a given latitude that is on the "day" side of the circle of illumination with the fraction on the "night" side. The opposite is true for the winter solstice, when the nights are longer than the days (Table 12.3). Again for comparison let us consider New York City, which has 15 hours of daylight on June 21 and only 9 hours on December 21. Also

FIGURE 12.7
Characteristics of the summer solstice (Northern Hemisphere).

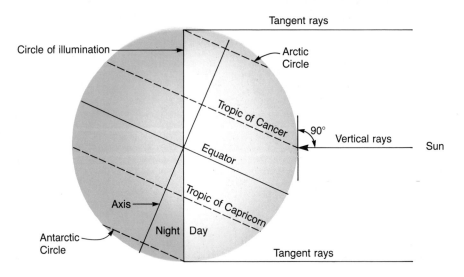

TABLE 12.3
Length of daylight.

Latitude (degrees)	Summer Solstice	Winter Solstice	Equinoxes
0	12 h	12 h	12 h
10	12 h 35 min	11 h 25 min	12
20	13 12	10 48	12
30	13 56	10 04	12
40	14 52	9 08	12
50	16 18	7 42	12
60	18 27	5 33	12
70	2 mo	0 00	12
80	4 mo	0 00	12
90	6 mo	0 00	12

note from Table 12.3 that on June 21 the farther you are north of the equator the longer is the period of daylight, until the Arctic Circle is reached, where the length of daylight is 24 hours. During an equinox (meaning "equal night") the length of daylight is 12 hours everywhere on earth, for the circle of illumination passes directly through the poles, dividing the latitudes in half.

As a review of the characteristics of the summer solstice for the Northern Hemisphere, examine Figure 12.7 and Table 12.3 and consider the following facts:

1 The date of occurrence is June 21 or 22.
2 The vertical rays of the sun are striking the Tropic of Cancer (23½ degrees north latitude).
3 Locations in the Northern Hemisphere are experiencing their greatest length of daylight. (Opposite for the Southern Hemisphere.)
4 Locations north of the Tropic of Cancer are experiencing their highest noon sun angles. (Opposite for places south of the Tropic of Capricorn.)
5 The farther you are north of the equator, the longer the period of daylight is, until the Arctic Circle is reached, where the day is 24 hours long. (Opposite for the Southern Hemisphere.)

The facts about the winter solstice will be just the opposite. It should now be apparent why a mid-latitude location is warmest in the summer, for it is then that days are longest and the sun's altitude is highest.

All locations at the same latitude have identical sun angles and lengths of daylight. If the earth-sun relationships just described were the only controls of temperature, we would expect these places to have identical temperatures as well. Obviously this is not the case. Although the altitude of the sun is the main control of temperature, it is not the only control, as we shall see.

In summary, seasonal fluctuations in the amount of solar energy reaching places on the earth's surface are caused by the migrating vertical rays of the sun and the resulting variations in sun angle and length of daylight.

RADIATION

From our everyday experience we know that the sun emits light and heat as well as the rays which give us a suntan. Although these forms of energy comprise a major portion of the total energy that radiates from the sun, they are only a part of a large array of energy called **radiation,** or **electromagnetic radiation.** This array or spectrum of electromagnetic energy is shown in Figure 12.8. All radiation, whether x rays, radio waves, or heat waves, is capable of transmitting energy through the vacuum of space at 300,000 kilometers (186,000 miles) per second and only slightly

FIGURE 12.8
The electromagnetic spectrum.

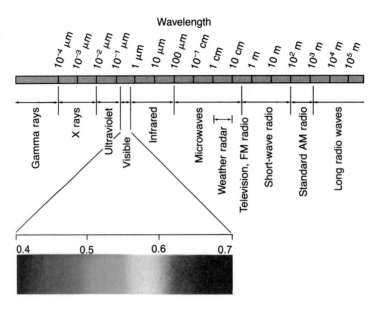

slower through our atmosphere. Nineteenth century physicists were so puzzled by the seemingly impossible task of energy traveling without a medium to transmit it that they assumed that a material, which they named ether, existed between the sun and the earth. This medium was thought to transmit radiant energy in much the same way that air transmits sound waves produced by the vibration of one person's vocal cords to another person's eardrums. Today we know that, like gravity, radiation requires no material to transmit it. Yet, physicists still do not know how this is possible.

In some respects, the transmission of radiant energy parallels the motion of the gentle swells in the open ocean. Not unlike ocean swells, these electromagnetic waves, as they are called, come in various sizes. For our purpose, the most important difference between electromagnetic waves is their wavelength, or distance from one crest to the next. Radio waves have the longest wavelengths, ranging to tens of kilometers, whereas gamma waves are the very shortest, being less than one billionth of a centimeter long. **Visible light,** as the name implies, is the only portion of the spectrum we can see. We often refer to visible light as white light since it appears "white" in color. However, it is easy to show that white light is really an array of colors, each color corresponding to a particular

wavelength. Using a prism, we can divide white light into the colors of the rainbow, violet having the shortest wavelength—0.4 micrometer (1 micrometer is 0.0001 centimeter)—and red having the longest—0.7 micrometer (see Figure 18.10, page 483). Located adjacent to red, and having a longer wavelength, is **infrared** radiation, which we cannot see but which we can detect as heat. The closest invisible waves to violet are called **ultraviolet** rays and are responsible for sunburn after an intense exposure to the sun. Although we divide radiant energy into groups based on our ability to perceive them, all forms of radiation are basically the same. When any form of radiant energy is absorbed by an object, the result is an increase in molecular motion, which causes a corresponding increase in temperature.

To better appreciate how the sun's radiant energy interacts with the earth's atmosphere and surface, we must have a general understanding of the basic laws governing radiation. Although the mathematical implications of these laws are beyond the scope of this book, the concepts themselves are well within the student's grasp.

1 All objects, at whatever temperature, emit radiant energy. Hence, not only hot objects like the sun but also the earth, including its polar ice caps, continually emit energy.

2 Hotter objects radiate more total energy per unit area than colder objects. The sun, with a surface temperature of 6000 K,* emits hundreds of thousands of times more energy than the earth, which has an average surface temperature of 288 K.

3 The hotter the radiating body, the shorter the wavelength of maximum radiation. For example, a very hot metal rod will emit visible radiation, producing a white glow. Upon cooling, it will emit more of its energy in longer wavelengths and glow a reddish color. Eventually no light will be given off, but if you place your hand near the rod, the still-longer infrared radiation will be detectable as heat. The sun, with a surface temperature of 6000 K, radiates maximum energy at 0.5 micrometer, which is in the visible range. The maximum radiation for the earth occurs at a wavelength of 10 micrometers, well within the infrared (heat) range. Because the maximum earth radiation is roughly 20 times longer than the maximum solar radiation, terrestrial radiation is often called long-wave radiation, and solar radiation is called short-wave radiation.

4 Objects that are good absorbers of radiation are good emitters as well. Technically, a perfect emitter is any object that radiates, for every wavelength, the maximum intensity of radiation possible for that temperature. The earth's surface and the sun approach being perfect radiators since they absorb and radiate with nearly 100 percent efficiency for their respective temperatures. On the other hand, gases are selective absorbers and radiators. Thus the atmosphere, which is nearly transparent to (does not absorb) certain wavelengths of radiation, is nearly opaque (a good absorber) to others. Our experience tells us that the atmosphere is transparent to visible light; hence, it readily reaches the earth's surface. This is not the case for long-wave terrestrial radiation, as we shall see.

*To convert kelvins to degrees Celsius, add the degree symbol and subtract 273.

MECHANISMS OF HEAT TRANSFER

Three mechanisms of heat transfer are recognized: radiation, conduction, and convection. Since radiation is the only one of these that can travel through the relative emptiness of space, the vast majority of energy coming to and leaving the earth must be in this form. Radiation also plays an important role in transferring heat from the earth's land-sea surface to the atmosphere and vice versa.

Conduction is familiar to most of us. Anyone who has attempted to pick up a metal spoon that was left in a hot pan has discovered that heat was conducted through the spoon. Conduction is the transfer of heat through matter by molecular activity. The energy of molecules is transferred through collisions from one molecule to another, with the heat flowing from the higher temperature to the lower temperature. The ability of substances to conduct heat varies considerably. Metals are good conductors, as those of us who have touched a hot spoon have quickly learned. Air, on the other hand, is a very poor conductor of heat. Consequently, conduction is only important between the earth's surface and the air directly in contact with the surface. As a means of heat transfer for the atmosphere as a whole, conduction is the least significant and can be disregarded when considering most meteorological phenomena.

Heat gained by the lowest layer of the atmosphere from radiation or conduction is most often transferred by convection. **Convection** is the transfer of heat by the movement of a mass or substance from one place to another. It can only take place in liquids and gases. Convective motions in the atmosphere are responsible for the redistribution of heat from equatorial regions to the poles and from the surface upward. The term **advection** is usually reserved for horizontal convective motions such as winds, while *convection* is used to describe vertical motions in the atmosphere.

Figure 12.9 summarizes the various mechanisms of heat transfer. The heat produced by the campfire passes through the pan by conduction, warming the water at the bottom of the pan. Convection currents carry the warmed water through-

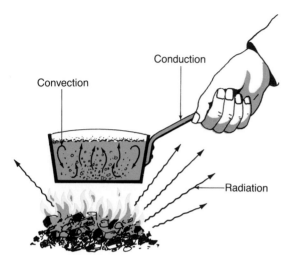

FIGURE 12.9
Illustration of conduction, convection, and radiation.

out the container, heating the remaining water. Meanwhile, the camper is warmed by radiation from the fire and the pan. Furthermore, since metals are good conductors, the camper's hand is

likely to be burned if a pot holder is not used. In most situations involving heat transfer, conduction, convection, and radiation occur simultaneously.

HEATING THE ATMOSPHERE

Although the atmosphere is quite transparent to incoming solar radiation, less than 25 percent of this radiation penetrates directly to the earth's surface without some sort of interference by the atmosphere. The remainder is either *absorbed* by the atmosphere, *scattered* about until it reaches the earth's surface or returns to space, or is *reflected* back to space (Figure 12.10). What determines whether radiation will be absorbed, scattered, or reflected outward? As we shall see, it depends greatly on the wavelength of the energy being transmitted, as well as on the size and nature of the intervening material.

When light is scattered by very small particles, primarily gas molecules, it is distributed in all directions, forward as well as backward. Some of the

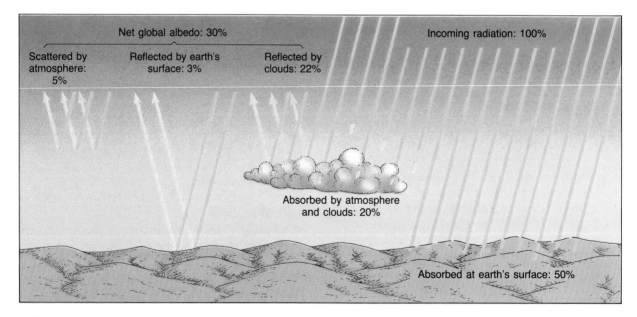

FIGURE 12.10
Solar radiation budget of the earth and atmosphere. More solar energy is absorbed by the earth's surface than by the atmosphere. Consequently, the air is not heated directly by the sun, but indirectly from the earth's surface.

light that is back-scattered is lost to space, while the remainder continues downward, where it interacts with other molecules which scatter it further by changing the direction of the light beam, but not its wavelength.

The light that reaches the earth's surface after having its direction changed is called **diffused light.** It is diffused light that keeps a shaded area or room lighted when direct sunlight is absent. Scattering also produces the blue color of the sky. Air molecules are more effective scatterers of the shorter wavelength (blue and violet) portion of "white" sunlight than the longer wavelength (red and orange) portion. Thus, when we look in a region of the sky away from the direct solar rays, we see predominantly blue light, which was more readily scattered. On the other hand, the sun appears to have a yellowish to reddish tint when viewed near the horizon. When the sun is in this position, the solar beam must travel through a great deal of atmosphere before it reaches your eye. For this reason, most of the blues and violets will be scattered out, leaving a beam of light composed mostly of reds and yellows. This phenomenon is particularly pronounced on a day when fine dust or smoke particles are present.

About 30 percent of the solar energy reaching the outer atmosphere is reflected back to space. Included in this figure is the amount sent skyward by back-scattering. This energy is lost to the earth and does not play a role in heating the atmosphere.

The fraction of the total radiation encountered that is reflected by a surface is called its **albedo.** Thus the albedo for the earth as a whole (planetary albedo) is 30 percent. However, the albedo from place to place as well as from time to time in the same locale varies considerably, depending upon the amount of cloud cover and particulate matter in the air, as well as upon the angle of the sun's rays and the nature of the surface. A lower angle means that more atmosphere must be penetrated, thus making the "obstacle course" longer, and therefore the loss of solar radiation greater (see Figure 12.5). Table 12.4 gives the albedo for various surfaces and clouds. Note that the angle at which the sun's rays strike a water surface greatly affects its albedo.

TABLE 12.4
Albedo of various surfaces.

Surface	Percent Reflected
Clouds, stratus	
<150 meters thick	25–63
150–300 meters thick	45–75
300–600 meters thick	59–84
Average of all types and thicknesses	50–55
Concrete	17–27
Crops, green	5–25
Forest, green	5–10
Meadows, green	5–25
Ploughed field, moist	14–17
Road, blacktop	5–10
Sand, white	30–60
Snow, fresh-fallen	80–90
Snow, old	45–70
Soil, dark	5–15
Soil, light (or desert)	25–30
Water	8*

*Typical albedo value for a water surface. The albedo of a water suface varies greatly depending upon the sun angle. If the sun angle is greater than 30 degrees, the albedo is less than 5 percent. When the sun is near the horizon (sun angle less than 3 degrees), the albedo is more than 60 percent.

As stated earlier, gases are selective absorbers, meaning that they absorb strongly in some wavelengths, moderately in others, and only slightly in still others. When a gas molecule absorbs light waves, this energy is transformed into internal molecular motion, which is detectable as a rise in temperature. Nitrogen, the most abundant constituent in the atmosphere, is a rather poor absorber of all types of incoming radiation. Oxygen (O_2) and ozone (O_3) are efficient absorbers of ultraviolet radiation. Oxygen removes most of the shorter ultraviolet radiation high in the atmosphere, and ozone absorbs longer wavelength ultraviolet rays in the stratosphere between 10 and 50 kilometers. The absorption of ultraviolet radiation in the stratosphere accounts for the high temperatures experienced there. The only other significant absorber of incoming solar radiation is water vapor, which along with oxygen and ozone accounts for most of the solar radiation absorbed within the atmosphere.

For the atmosphere as a whole, none of the gases are effective absorbers of radiation with wavelengths between 0.3 and 0.7 micrometer. This region of the spectrum corresponds to the visible range, to which a large fraction of solar radiation belongs. This explains why most visible radiation reaches the ground and why we say that the atmosphere is transparent to incoming solar radiation. Thus, direct solar energy is a rather ineffective "heater" of the earth's atmosphere. The fact that the atmosphere does not acquire the bulk of its energy directly from the sun but rather from reradiation from the earth's surface is of the utmost importance to the dynamics of the weather machine.

Approximately 50 percent of the solar energy that strikes the top of the atmosphere reaches the earth's surface directly or indirectly (diffused) and is absorbed. Most of this energy is then reradiated skyward. Since the earth has a much lower surface temperature than the sun, terrestrial radiation is emitted in longer wavelengths than solar radiation. The bulk of terrestrial radiation has wavelengths between 1 and 30 micrometers, placing it well within the infrared range. The atmosphere as a whole is a rather efficient absorber of radiation between 1 and 30 micrometers (terrestrial radiation). Water vapor and carbon dioxide are the principal absorbing gases in that range. Water vapor absorbs roughly five times more terrestrial radia-

tion than do all the other gases combined and accounts for the warm temperatures found in the lower troposphere, where it is most highly concentrated. Because the atmosphere is quite transparent to shorter wavelength solar radiation and more readily absorbs longer wavelength terrestrial radiation, the atmosphere is heated from the ground up rather than vice versa. This explains the general drop in temperature with increasing altitude experienced in the troposphere. The farther from the "radiator," the colder it becomes.

When the gases in the atmosphere absorb terrestrial radiation, they warm, but they eventually radiate this energy away. Some travels upward, where it may be reabsorbed by other gas molecules, a possibility less likely with increasing height because the concentration of water vapor decreases with altitude. The remainder travels downward and is again absorbed by the earth. For this reason, the earth's surface is continually being supplied with heat from the atmosphere as well as from the sun. Without these absorptive gases in our atmosphere, the earth would not be a suitable habitat for humans and numerous other life forms.

This very important phenomenon has been termed the *greenhouse effect* because it was once thought that greenhouses were heated in a similar manner (Figure 12.11). The gases of our atmosphere, especially water vapor and carbon dioxide,

FIGURE 12.11
The heating of a greenhouse is analogous to the heating of the atmosphere. Like the atmosphere, the glass in the greenhouse allows much of the shorter wavelength solar radiation to pass through, where it is absorbed by objects within. These objects, in turn, emit energy in the form of longer wavelength terrestrial radiation. The glass, like the atmosphere, acts to slow the loss of heat from the greenhouse, thus keeping it warmer than it would otherwise be.

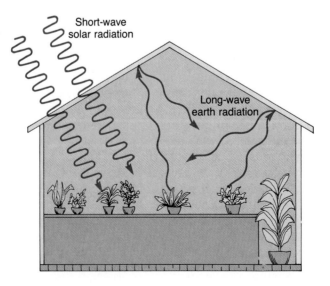

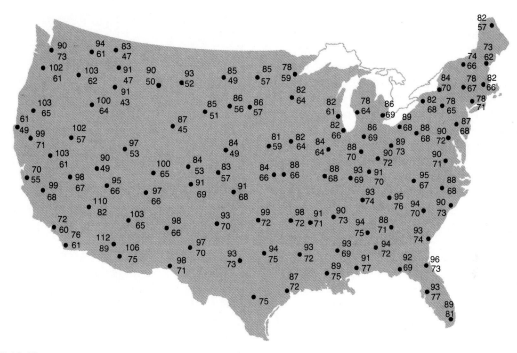

FIGURE 12.12
Daily maximum and minimum temperatures (°F) for a typical July day.
°C = (°F − 32)/1.8.

act very much like the glass in the greenhouse. They allow shorter wavelength solar radiation to enter, where it is absorbed by the objects inside. These objects in turn reradiate the heat, but at longer wavelengths, to which glass is nearly opaque. The heat therefore is trapped in the greenhouse. However, a more important factor in keeping a greenhouse warm is the fact that the greenhouse itself prevents mixing of air inside with cooler air outside. Nevertheless, the term greenhouse effect is still used.

The role that these absorbing gases play in keeping the atmosphere warm is well known to those residing in mountainous regions. More radiant energy is received on mountaintops than in the valleys below because there is less atmosphere to hinder its arrival. Yet the decrease in water vapor content with altitude allows much of the heat to escape these lofty peaks. This loss more than compensates for the extra radiation received. As a result, valleys remain warmer than the adjacent mountains even though they receive less solar radiation.

The importance of water vapor in keeping our atmosphere warm is easily demonstrated. Figure 12.12 shows the maximum and minimum temperatures for various cities in the United States on a typical July day. By comparing the temperatures for cities in the arid Southwest (Nevada and southern Arizona, for example) with temperatures in the more humid East, the following can be seen:

1 Maximum temperatures are, for the most part, higher in the arid regions.
2 Minimum temperatures often are lower in the dry Southwest as compared with the humid East.
3 The difference between the maximum and minimum temperatures (a calculation known as the **daily temperature range**) is in almost every case greater in the arid areas.

The higher maximum temperatures in arid regions are easily explained, for with little or no cloud cover, a maximum amount of sunshine is received. Further, the records of the stations in the dry regions indicate that these places cool rapidly at night, resulting in lower minimum temperatures. A hot, muggy day in a humid location often means a warm night as well, because the water vapor in the air traps some of the heat being radiated from the earth's surface. In the desert, however, there is very little water vapor in the air. Therefore, the terrestrial radiation escapes, and the air near the surface cools rapidly. From this example, it is quite evident that water vapor plays an important role in absorbing heat, and when it is lacking, the air cools rapidly, resulting in a high daily temperature range.

TEMPERATURE MEASUREMENT AND DATA

Changes in air temperature are probably noticed by people more often than changes in any other element of weather. At a weather station, the temperature is read on a regular basis from instruments mounted in an instrument shelter (Figure 12.13). The shelter protects the instruments from direct sunlight and allows a free flow of air. In addition to a standard mercury thermometer, the shelter is likely to contain a thermograph, which is an instrument that makes a continuous record of temperature, and a set of maximum-minimum thermometers. As their name implies, these thermometers record the highest and lowest temperatures during a given day.

The daily maximum and minimum temperatures are the bases for much of the temperature data compiled by meteorologists:

1 By adding the maximum and minimum temperatures and then dividing by two, the **daily mean** is calculated.
2 The daily temperature range is computed by finding the difference between the maximum and minimum temperatures for a given day.
3 The **monthly mean** is calculated by adding together the daily means for each day of the month and dividing by the number of days in the month.

FIGURE 12.13
Standard instrument shelter. A shelter protects instruments from direct sunlight and allows for the free flow of air. (Courtesy of Qualimetrics, Inc.)

4 The **annual mean** is an average of the twelve monthly means.
5 The **annual temperature range** is computed by finding the difference between the highest and lowest monthly means.

Mean temperatures are particularly useful for making comparisons, whether on a daily, monthly, or annual basis. It is quite common to hear a weather reporter state that, "Last month was the hottest July on record," or "Today Chicago was ten degrees warmer than Miami." Temperature ranges are also useful statistics, because they give an indication of extremes.

CONTROLS OF TEMPERATURE

The controls of temperature are those factors that cause variations in temperature from place to place. Earlier in this chapter we examined the single greatest cause for temperature variations—differences in the receipt of solar radiation. Since variations in sun angle and length of daylight are a function of latitude, they are responsible for warm temperatures in the tropics and colder temperatures at more poleward locations. However, latitude is not the only control of temperature; if it were, we would expect that all places along the same parallel would have identical temperatures. This is clearly not the case. For example, Eureka, California, and New York City are both coastal cities at about the same latitude and both have an average annual mean temperature of 11°C (52°F). However, New York City is 9°C (16°F) warmer than Eureka in July and 10°C cooler in January. Quito and Guayaquil, two cities in Ecuador, are within a relatively few kilometers of each other, yet the mean annual temperatures at these two cities differ by 12°C (21°F). To explain these situations and countless others, we must realize that factors other than variations in solar radiation also exert a strong influence upon temperatures. Among the most important of these other factors are the differential heating of land and water, ocean currents,* altitude, and geographic position.

LAND AND WATER

The heating of the earth's surface directly influences the heating of the air above. Therefore, in order to understand variations in air temperature, we must examine the nature of the surface. Different land surfaces absorb varying amounts of incoming solar energy, which in turn cause variations in the temperature of the air above. The largest contrast, however, is not between different land surfaces, but between land and water. Land heats more rapidly and to higher temperatures than water and cools more rapidly and to lower temperatures than water. Temperature variations,

*For a discussion of the effects of ocean currents on temperature, see page 305 in Chapter 11.

therefore, are considerably greater over land than over water.

Among the reasons for the differential heating of land and water are the following:

1. The specific heat (amount of energy needed to raise 1 gram of a substance 1°C) is far greater for water than for land. Thus, water requires a great deal more heat to raise its temperature the same amount as an equal quantity of land.
2. Land surfaces are opaque, so heat is absorbed only at the surface. Water, being more transparent, allows heat to penetrate to a depth of many meters.
3. The water that is heated often mixes with water below, thus distributing the heat through an even larger mass.
4. Evaporation (a cooling process) from water bodies is greater than that from land surfaces.

Monthly temperature data for two cities will demonstrate the moderating influence of a large water body and the extremes associated with land (Table 12.5). Vancouver, British Columbia, is located along a windward coast, whereas Winnipeg, Manitoba, is in a continental position far from the

TABLE 12.5
Monthly mean temperatures (°C) for Vancouver, British Columbia, and Winnipeg, Manitoba.

Month	Vancouver	Winnipeg
Jan	2.3	−17.7
Feb	4.2	−15.5
Mar	5.8	−7.9
Apr	9.1	3.3
May	12.6	11.3
June	15.2	16.5
July	17.6	20.2
Aug	17.0	18.9
Sept	14.3	12.8
Oct	10.1	6.2
Nov	6.0	−4.8
Dec	3.9	−12.9
Annual	9.8	2.5

influence of water. Both cities are at about the same latitude and thus experience similar sun angles and lengths of daylight. Winnipeg, however, has a mean January temperature that is 20°C lower than Vancouver's. Conversely, Winnipeg's July mean is 2.6°C higher than Vancouver's. Although their latitudes are nearly the same, Winnipeg, which has no water influence, experiences much greater temperature extremes than Vancouver, which does.

On a different scale, the moderating influence of water may also be demonstrated when temperature variations in the Northern and Southern hemispheres are compared. Sixty-one percent of the Northern Hemisphere is covered by water, and land accounts for the remaining 39 percent. However, the figures for the Southern Hemisphere (81 percent water, 19 percent land) reveal why it is correctly called the water hemisphere. Table 12.6 portrays the considerably smaller annual temperature variations in the water-dominated Southern Hemisphere as compared with the Northern Hemisphere.

ALTITUDE

The two cities in Ecuador mentioned earlier—Quito and Guayaquil—demonstrate the influence of altitude upon mean temperatures. Although both cities are near the equator and relatively close to each other, the annual mean at Guayaquil is 25°C (77°F) as compared to Quito's mean of 13°C (55°F). The difference is explained largely by the difference in the cities' elevations. Guayaquil is only 12 meters (40 feet) above sea level, whereas

TABLE 12.6
Variation in mean annual temperature range (°C) with latitude.

Latitude	Northern Hemisphere	Southern Hemisphere
0	0	0
15	3	4
30	13	7
45	23	6
60	30	11
75	32	26
90	40	31

Quito is high in the Andes Mountains at 2800 meters (9200 feet). Recall that temperatures drop an average of 6.5°C per kilometer in the troposphere; thus cooler temperatures are to be expected at greater heights. Even so, the magnitude of the difference is not explained completely by the normal lapse rate. If the normal lapse rate is used, we would expect Quito to be about 18°C cooler than Guayaquil; the difference, however, is only 12°C. Places at high elevations, such as Quito, are warmer than the value calculated using the normal lapse rate to the same altitude in the free atmosphere because of the absorption and reradiation of solar energy by the ground surface.

GEOGRAPHIC POSITION

The geographic setting may greatly influence the temperatures experienced at a particular locale. For example, a windward coastal location, that is, a place subject to prevailing ocean winds, experiences considerably different temperatures from a coastal location where the prevailing winds are directed from the land toward the ocean. In the first situation, the place will experience the full moderating influence of the ocean—cool summers and mild winters, compared with an inland station at the same latitude. However, a leeward coastal site will have a more continental temperature pattern because the winds do not carry the ocean's influence onshore. Eureka, California, and New York City, the two cities mentioned in the section on temperature controls, illustrate this aspect of geographic position. The annual temperature range at New York City is 19°C (34°F) higher than that at Eureka.

Seattle and Spokane, both in the state of Washington, illustrate a second aspect of geographic position—mountains that act as barriers. Although Spokane is only about 360 kilometers (220 miles) east of Seattle, the towering Cascade Range separates the cities. Consequently, while Seattle's temperatures show a marked marine influence, Spokane's are more typically continental. Spokane is 7°C (13°F) cooler than Seattle in January and 4°C (7°F) warmer than Seattle in July. The annual range at Spokane is 11°C (20°F) greater than at Seattle. The Cascade Range effectively cuts Spokane off from the moderating influence of the Pacific Ocean.

WORLD DISTRIBUTION OF TEMPERATURE

Temperature distribution is shown on a map by using **isotherms,** which are lines that connect places of equal temperature. On world maps that depict global patterns, temperatures are often corrected to sea level to eliminate the complications caused by altitude variations. January (Figure 12.14) and July (Figure 12.15, page 352) are selected most often for analysis because, for most locations, they represent temperature extremes.

On Figures 12.14 and 12.15 the isotherms generally trend east and west and show a decrease in temperatures poleward from the tropics, illustrating one of the most fundamental and best-known aspects of the world distribution of temperature—that the effectiveness of incoming solar radiation

in heating the earth's surface and the atmosphere above is largely a function of latitude. Further, there is a latitudinal shifting of temperatures caused by the seasonal migration of the sun's vertical rays.

The added effect of the differential heating of land and water is also reflected on the January and July temperature maps. The warmest and coldest temperatures are found over land. Hence, since temperatures do not fluctuate as much over water as over land, the north-south migration of isotherms is greater over the continents than over the oceans. In addition, it is clear that the isotherms in the Southern Hemisphere, where there is little land and the oceans predominate, are much more regular than in the Northern Hemisphere, where they bend sharply northward in July and southward in January over the continents.

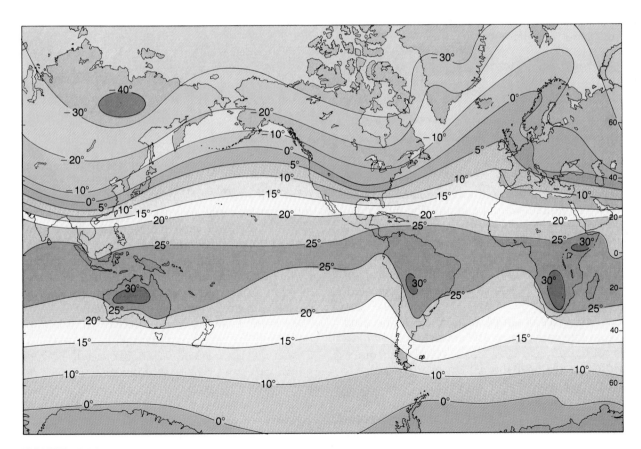

FIGURE 12.14
World distribution of mean temperatures (°C) for January.

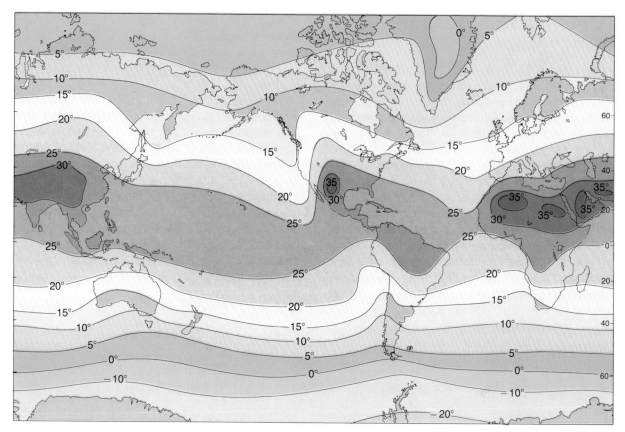

FIGURE 12.15
World distribution of mean temperatures (°C) for July.

Isotherms also reveal the presence of ocean currents. Warm currents cause isotherms to be deflected toward the pole, whereas cold currents cause an equatorward bending. The horizontal transport of water poleward warms the overlying air and results in air temperatures that are higher than otherwise would be expected for the latitude. Conversely, currents moving toward the equator produce air temperatures cooler than expected.

Since Figures 12.14 and 12.15 show the seasonal extremes of temperature, they can be used to evaluate variations in the annual range of temperature from place to place. A comparison of the two maps shows that a station near the equator will record a very small annual range because it experiences little variation in the length of daylight and always has a relatively high sun angle. By contrast, a station in the middle latitudes experiences much wider variations in sun angle and length of daylight, and thus larger variations in temperature. Therefore, we can state that the annual temperature range increases with an increase in latitude. Moreover, land and water also affect seasonal temperature variations, especially outside the tropics. A continental location must endure hotter summers and colder winters than a coastal location. Consequently, the annual range will increase with an increase in continentality.

A classic example of the effect of latitude and continentality on annual temperature range is Yakutsk, a city in Siberia, approximately 60 degrees north latitude and far from the influence of water. As a result, Yakutsk has an average annual temperature range of 62.2°C (112°F), one of the highest in the world.

REVIEW
QUESTIONS

1 How are the studies of weather and climate different? In what ways are they similar?

2 What are the major components of clean, dry air (Table 12.1)? Are they important meteorologically?

3 Why are water vapor and dust important constituents of our atmosphere?

4 **(a)** Why is ozone important to life on earth?

   (b) What are CFCs and what is their connection to the ozone problem?

   (c) If the ozone layer is reduced, what will be the probable impact on human health?

5 What percentage of the atmosphere is found below the following altitudes: 5.6 kilometers, 16.2 kilometers, 31.2 kilometers (see Table 12.2)?

6 Why is the troposphere called the "weather sphere"?

7 Why do temperatures rise in the stratosphere?

8 Are the variations in the amount of solar energy received by the earth because of its elliptical orbit important to understanding seasonal temperature variations? Explain your answer.

9 After examining Table 12.3, write a generalization that relates the season, the latitude, and the length of daylight.

10 Describe the relationship between the temperature of a radiating body and the wavelengths it emits.

11 Distinguish among the three basic mechanisms of heat transfer.

12 Figure 12.10 illustrates what happens to incoming solar radiation. The percentages shown, however, are only global averages. In particular, the amount of solar radiation reflected (albedo) may vary considerably. What factors might cause variations in albedo?

13 How does the earth's atmosphere act as a "greenhouse"?

14 On a warm summer day, one city had a daily temperature range of 25°C (45°F), while another experienced a range of only 8.3°C (15°F). One of these cities is located in Nevada, and the other in Indiana. Which location likely had the highest daily temperature range? Explain.

15 How are the following temperature data computed: Daily mean, daily range, monthly mean, annual mean, annual range?

16 Quito, Ecuador is located on the equator and is not a coastal city. It has an average annual temperature of only 13°C (55°F). What is the likely cause for this low average temperature?

17 In what ways can geographic position be considered a control of temperature?

18 Yakutsk, U.S.S.R. is located in Siberia at about 60 degrees north latitude. This city has one of the highest average annual temperature ranges in the world: 62.2°C (112°F). Explain the reasons for the very high annual temperature range.

KEY TERMS

meteorology	perihelion	infrared
weather	aphelion	ultraviolet
climate	altitude (of the sun)	conduction
troposphere	inclination of the axis	convection
normal lapse rate	Tropic of Cancer	advection
stratosphere	summer solstice	diffused light
tropopause	Tropic of Capricorn	albedo
stratopause	winter solstice	daily temperature range
mesosphere	autumnal equinox	daily mean
mesopause	vernal (spring) equinox	monthly mean
thermosphere	radiation (electromagnetic radiation)	annual mean
rotation		annual temperature range
circle of illumination	visible light	isotherms
revolution		

13

MOISTURE

As you observe day-to-day weather changes, many questions may come to mind concerning the role of water in the air. What is humidity, and how is it measured? Why do clouds form on some occasions but not on others? What processes produce clouds and precipitation? Why are some clouds thin, white, and harmless, while others are towering, gray, and ominous? What is the difference between sleet and hail? What is fog? Are all fogs alike? This chapter investigates these and other questions as the topic of moisture in the air is examined.

Turbulent clouds over the Grand Tetons. (Photo by Steven Trimble)

An understanding of the role that water plays in the atmosphere is very important. In addition to exerting a strong influence on the heating of air (Chapter 12), water vapor is the source of all clouds and precipitation. Areas within the Atacama Desert of South America may go years without a drop of rain; in fact, legend has it that some areas have not received rain since the arrival of the Spaniards, some 400 years ago! By contrast, Cherrapunji, India, has received 930 centimeters (366 inches) in a single month, and 2644 centimeters (over 86 feet) in a single year. Indeed, the large variations in the amount of precipitation from place to place as well as local differences from time to time have a significant impact not only on the nature of the physical landscape but also on people's life-styles.

CHANGES OF STATE

Water vapor is an odorless, colorless gas that mixes freely with the other gases of the atmosphere. Unlike oxygen and nitrogen, the two most abundant components of the atmosphere, water vapor changes from one state of matter to another at the temperatures and pressures experienced near the surface of the earth. It is because of this occurrence, which allows water to leave the oceans as a gas and return again as a liquid, that the vital hydrologic cycle exists. The processes that involve a change of state require that heat be absorbed or released (Figure 13.1). This heat energy is measured in calories. One **calorie** is the amount of heat required to raise the temperature of 1 gram of water 1°C.

The process of converting a liquid to a vapor is termed **evaporation**. Approximately 600 calories of energy are needed to convert just 1 gram of water to water vapor. The energy absorbed by the water molecules during evaporation is used solely to give them the motion needed to escape the surface of the liquid and become a gas. Because this energy is subsequently released as heat when the vapor changes back to a liquid, it is generally called **latent heat** (meaning hidden, or stored, heat).

Since the evaporating molecules require energy to escape, the remaining liquid must be cooled by an equivalent amount; hence the common expression "evaporation is a cooling process." You have undoubtedly experienced this cooling effect upon stepping dripping wet out of a swimming pool. On a larger scale is the cooling effect of evaporating rainwater during a summer shower.

Condensation denotes the process of water vapor changing to the liquid state. During condensation, the water molecules release energy (*latent heat of condensation*) equivalent to that which was absorbed during evaporation. This energy plays an important role in producing violent weather and can act to transfer great quantities of heat energy from tropical oceans to more poleward locations.

FIGURE 13.1
Changes of state.

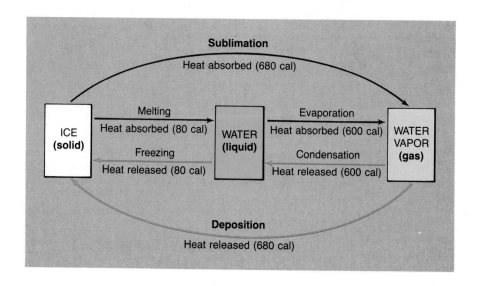

When condensation occurs in the atmosphere, it results in such phenomena as fog and clouds.

Melting is the process by which a solid is changed to a liquid. Approximately 80 calories of energy are required to melt one gram of ice. **Freezing,** the reverse process, releases these 80 calories per gram as *latent heat of fusion.*

The final processes illustrated in Figure 13.1 are sublimation and deposition. The term **sublimation** is used to describe the conversion of a solid directly to a gas, without passing through the liquid state. You may have observed this change as you watched the sublimation of dry ice (solid carbon dioxide). The term **deposition** is used to denote the reverse process, by which a vapor is changed to a solid.* This change occurs, for example, during the formation of frost. As shown in Figure 13.1, sublimation and deposition involve an amount of energy equal to the total of the other two processes.

HUMIDITY

Humidity is the general term used to describe the amount of water vapor in air. Several methods are used to quantitatively express humidity. Among these are specific humidity and relative humidity.

Before we consider these humidity measures individually, it is important to understand the concept of **saturation.** Imagine a closed jar half full of water and overlain with dry air at the same temperature. As the water begins to evaporate from the water surface, a small increase in pressure can be detected in the air above. This increase is the result of the motion of the water vapor molecules that were added to the air through evaporation. In the open atmosphere this pressure is termed **vapor pressure** and is defined as that part of the total atmospheric pressure that can be attributed to the water vapor content. In the closed container, as more and more molecules escape from the water surface, the steadily increasing vapor pressure in the air above forces more and more of these molecules to return to the liquid. Eventually the number of vapor molecules returning to the surface will balance the number leaving. At that

---

* The term *sublimation* may also be used for the vapor-to-solid phase change.

TABLE 13.1
Water vapor capacity (at average sea level pressure).

Temperature (°C)	Grams/kg
−40	0.1
−30	0.3
−20	0.75
−10	2
0	3.5
5	5
10	7
15	10
20	14
25	20
30	26.5
35	35
40	47

point, the air is said to be saturated, or filled to capacity. However, if we increase the temperature of the water and air in the jar, more water will evaporate before a balance is reached. Consequently, at higher temperatures more moisture is required for saturation. Stated another way, the water vapor capacity of air is temperature dependent, with warm air having a much greater capacity than cold air. The amount of water vapor required for saturation at various temperatures is shown in Table 13.1.

One method used to express humidity, specific humidity, specifies the amount of water vapor contained in a unit of air. **Specific humidity** is expressed as the weight of water vapor per weight of a chosen mass of air, including the water vapor. Since it is measured in units of weight (usually in grams per kilogram), specific humidity is not affected by changes in pressure or temperature.

The most familiar, and perhaps the most misunderstood, term used to describe the moisture content of air is relative humidity. Stated in an admittedly oversimplified manner, **relative humidity** is the ratio of the air's water vapor content to its water vapor capacity at a given temperature. From Table 13.1, we see that at 25°C, the capacity of the air is 20 grams per kilogram. If on a 25°C day the air contains 10 grams per kilogram, the relative humidity is expressed as 10/20, or 50

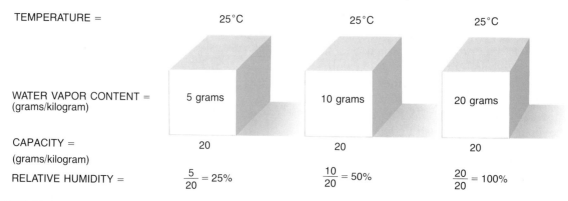

FIGURE 13.2
At a constant temperature, relative humidity will increase as water vapor is added to the air. Here the capacity remains constant at 20 grams per kilogram, while the relative humidity rises from 25 percent to 100 percent as the water vapor content increases.

percent. When air is saturated, the relative humidity is 100 percent.

Since relative humidity is based on the air's water vapor content as well as on its capacity, relative humidity can be changed in either of two ways. First, if moisture is added by evaporation, the relative humidity will increase (Figure 13.2). The addition of moisture by this means occurs mainly over the oceans, but plants, soil, and smaller bodies of water do make contributions. The second method, illustrated by Figure 13.3, involves a change in temperature. We can general-

ize this concept as follows: With the specific humidity at a constant level, a decrease in air temperature will result in an increase in relative humidity, and an increase in temperature will cause a decrease in the relative humidity. In Figure 13.4 the variations in temperature and relative humidity during a typical day demonstrate rather well the relationship just described.

Another important idea related to relative humidity is the dew-point temperature. **Dew point** is the temperature to which air would have to be cooled in order to reach saturation. Note that in

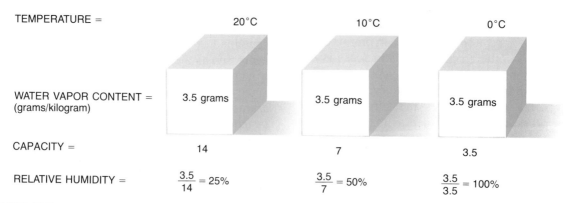

FIGURE 13.3
When the water vapor content (specific humidity) remains constant, the relative humidity may be changed by increasing or decreasing the air temperature. In this example the specific humidity remains at 3.5 grams per kilogram. The reduction in temperature from 20°C to 0°C causes a decrease in capacity and thus an increase in the relative humidity from 25 percent to 100 percent.

FIGURE 13.4
Typical daily variations in temperature and relative humidity during a spring day at Washington, D.C.

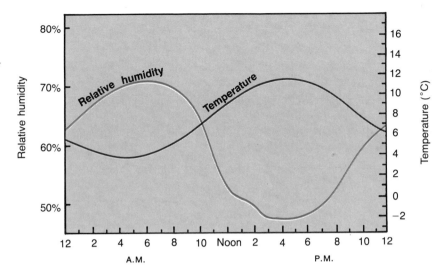

Figure 13.3 unsaturated air at 20°C is cooled to 0°C before saturation occurs. Therefore, 0°C would be the dew-point temperature for this air. If this same parcel of air were cooled further, the air's capacity would be exceeded, and the excess vapor would condense.

Although relative humidity is used exclusively in our daily weather reports, it does have limitations if improperly interpreted. Recall from our earlier discussion that the capacity of air is temperature dependent. Table 13.1 illustrates that the capacity of air at 35°C is 35 grams per kilogram, whereas the capacity of air at 10°C is only 7 grams per kilogram. It should be apparent that when relative humidities are equal, air at 35°C contains five times as much water vapor as air at 10°C. This explains why cold winter air with a high relative humidity may be described as dry when compared to warm air that has an equally high relative humidity. In summary, relative humidity indicates how near the air is to being saturated, and specific humidity denotes the quantity of water vapor contained in that air.

HUMIDITY MEASUREMENT

Specific humidity is difficult to measure directly. Nevertheless it may be readily computed by consulting an appropriate table or graph if the air temperature and relative humidity are known.

Relative humidity is most commonly measured using a psychrometer or a hygrometer. The **psychrometer** consists of two identical thermometers mounted side by side (Figure 13.5). One

FIGURE 13.5
Sling psychrometer. This instrument is used to determine relative humidity and dew point. The dry-bulb thermometer gives the current air temperature. The thermometers are spun until the temperature of the wet-bulb thermometer stops declining. Then the thermometers are read and the data used in conjunction with Tables 13.2 and 13.3. (Photo by E. J. Tarbuck)

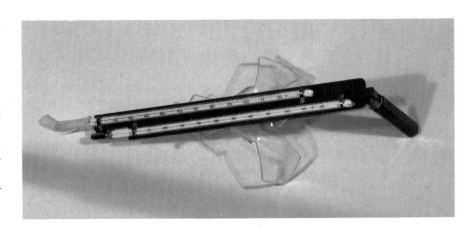

thermometer, the *dry bulb,* gives the present air temperature. The other, called the *wet-bulb* thermometer, has a thin muslin wick tied around the end. To use the psychrometer, the cloth sleeve is saturated with water and a continuous current of air is passed over the wick, either by swinging the instrument freely in the air or by fanning air past it. As a consequence, water evaporates from the wick and the temperature of the wet-bulb drops. The loss of heat that was required to evaporate water from the wet bulb lowers the thermometer reading.

The amount of cooling that takes place is directly proportional to the dryness of the air. The drier the air, the more the cooling. Therefore, the larger the difference between the thermometers, the lower the relative humidity; the smaller the difference, the higher the relative humidity. If the

TABLE 13.2
Relative humidity (percent).*

Air (dry-bulb) Temperature (°C)	Dry-bulb Temperature − Wet-bulb Temperature = Depression of the Wet Bulb																					
	1	2	3	4	5	6	7	8	9	10	11	12	13	14	15	16	17	18	19	20	21	22
−20	28																					
−18	40																					
−16	48	0																				
−14	55	11																				
−12	61	23																				
−10	66	33	0																			
−8	71	41	13																			
−6	73	48	20	0																		
−4	77	54	32	11																		
−2	79	58	37	20	1																	
0	81	63	45	28	11																	
2	83	67	51	36	20	6																
4	85	70	56	42	27	14																
6	86	72	59	46	35	22	10	0														
8	87	74	62	51	39	28	17	6														
10	88	76	65	54	43	33	24	13	4													
12	88	78	67	57	48	38	28	19	10	2												
14	89	79	69	60	50	41	33	25	16	8	1											
16	90	80	71	62	54	45	37	29	21	14	7	1										
18	91	81	72	64	56	48	40	33	26	19	12	6	0									
20	91	82	74	66	58	51	44	36	30	23	17	11	5	0								
22	92	83	75	68	60	53	46	40	33	27	21	15	10	4	0							
24	92	84	76	69	62	55	49	42	36	30	25	20	14	9	4	0						
26	92	85	77	70	64	57	51	45	39	34	28	23	18	13	9	5						
28	93	86	78	71	65	59	53	47	42	36	31	26	21	17	12	8	4					
30	93	86	79	72	66	61	55	49	44	39	34	29	25	20	16	12	8	4				
32	93	86	80	73	68	62	56	55	46	41	36	32	27	22	19	14	11	8	4			
34	93	86	81	74	69	63	58	52	48	43	38	34	30	26	22	18	14	11	8	5		
36	94	87	81	75	69	64	59	54	50	44	40	36	32	28	24	21	17	13	10	7	4	
38	94	87	82	76	70	66	60	55	51	46	42	38	34	30	26	23	20	16	13	10	7	5
40	94	89	82	76	71	67	61	57	52	48	44	40	36	33	29	25	22	19	16	13	10	7

*To determine the relative humidity and dew point, find the air (dry-bulb) temperature on the vertical axis (far left) and the depression of the wet bulb on the horizontal axis (top). Where the two meet, the relative humidity or dew point is found. For example, use a dry-bulb temperature of 20°C and a wet-bulb temperature of 14°C. From Table 13.2, the relative humidity is 51 percent, and from Table 13.3, the dew point is 10°C.

air is saturated, no evaporation will occur, and the two thermometers will have identical readings.

To determine the precise relative humidity from the thermometer readings, a standard table is used (Table 13.2). With the same information, but using a different table (Table 13.3), the dew-point temperature may also be calculated.

The second commonly used instrument for measuring relative humidity, the **hygrometer,** can be read directly, without the use of tables. The hair hygrometer operates on the principle that hair or certain synthetic fibers change their length in proportion to changes in the relative humidity, lengthening as relative humidity increases and shrinking as the relative humidity drops. The tension of a bundle of hairs is linked mechanically to an indicator that is calibrated between 0 and 100 percent. Thus we need only glance at the dial to

TABLE 13.3
Dew-point temperature (°C).*

Air (dry-bulb) Temperature (°C)	Dry-bulb Temperature − Wet-bulb Temperature = Depression of the Wet Bulb																					
	1	2	3	4	5	6	7	8	9	10	11	12	13	14	15	16	17	18	19	20	21	22
−20	−33																					
−18	−28																					
−16	−24																					
−14	−21	−36																				
−12	−18	−28																				
−10	−14	−22																				
−8	−12	−18	−29																			
−6	−10	−14	−22																			
−4	−7	−12	−17	−29																		
−2	−5	−8	−13	−20																		
0	−3	−6	−9	−15	−24																	
2	−1	−3	−6	−11	−17																	
4	1	−1	−4	−7	−11	−19																
6	4	1	−1	−4	−7	−13	−21															
8	6	3	1	−2	−5	−9	−14															
10	8	6	4	1	−2	−5	−9	−14	−28													
12	10	8	6	4	1	−2	−5	−9	−16													
14	12	11	9	6	4	1	−2	−5	−10	−17												
16	14	13	11	9	7	4	1	−1	−6	−10	−17											
18	16	15	13	11	9	7	4	2	−2	−5	−10	−19										
20	19	17	15	14	12	10	7	4	2	−2	−5	−10	−19									
22	21	19	17	16	14	12	10	8	5	3	−1	−5	−10	−19								
24	23	21	20	18	16	14	12	10	8	6	2	−1	−5	−10	−18							
26	25	23	22	20	18	17	15	13	11	9	6	3	0	−4	−9	−18						
28	27	25	24	22	21	19	17	16	14	11	9	7	4	1	−3	−9	−16					
30	29	27	26	24	23	21	19	18	16	14	12	10	8	5	1	−2	−8	−15				
32	31	29	28	27	25	24	22	21	19	17	15	13	11	8	5	2	−2	−7	−14			
34	33	31	30	29	27	26	24	23	21	20	18	16	14	12	9	6	3	−1	−5	−12	−29	
36	35	33	32	31	29	28	27	25	24	22	20	19	17	15	13	10	7	4	0	−4	−10	
38	37	35	34	33	32	30	29	28	26	25	23	21	19	17	15	13	11	8	5	1	−3	−9
40	39	37	36	35	34	32	31	30	28	27	25	24	22	20	18	16	14	12	9	6	2	−2

*See footnote to Table 13.2.

determine the relative humidity. Unfortunately, the hair hygrometer is less accurate than the psychrometer. Furthermore, it requires frequent calibration and is slow in responding to changes in humidity, especially at low temperatures.

A different type of hygrometer is used in remote-sensing instrument packages such as radiosondes which transmit upper-air observations back to ground stations. The electric hygrometer contains an electrical conductor coated with a moisture-absorbing chemical. It works on the principle that the passage of current varies as the relative humidity varies.

NECESSARY CONDITIONS FOR CONDENSATION

As we learned earlier in this chapter, condensation occurs when water vapor in the air changes to a liquid. The result of this process may be dew, fog, or clouds. Although these forms of condensation are quite different, they have two things in common. First, for any form of condensation to occur, the air must be saturated. Saturation occurs either when air is cooled to its dew point, which most commonly happens, or when water vapor is added to the air. Second, there generally must be a surface on which the water vapor may condense. When dew occurs, objects at or near the ground serve this purpose. When condensation occurs in the air above the ground, tiny bits of particulate matter known as **condensation nuclei** serve as surfaces for water vapor condensation. The importance of these nuclei should be noted, since in their absence a relative humidity well in excess of 100 percent is needed to produce clouds. However, condensation nuclei such as microscopic dust, smoke, and salt particles are profuse in the lower atmosphere. Due to this abundance of particles, relative humidity rarely exceeds 101 percent. Some particles, such as salt from the ocean, are particularly good nuclei because they absorb water. These particles are termed **hygroscopic** ("water-seeking") **nuclei.**

When condensation takes place, the initial growth rate of cloud droplets is rapid but diminishes quickly because the excess water vapor is readily consumed by the numerous competing particles. This results in the formation of a cloud consisting of millions upon millions of tiny water droplets, all so fine that they remain suspended in air. The slow growth of these cloud droplets by additional condensation and the immense size difference between cloud droplets and raindrops suggest that condensation alone is not responsible for the formation of drops large enough to fall as rain. We will first examine condensation and then return to the question of how precipitation forms.

CONDENSATION ALOFT: ADIABATIC TEMPERATURE CHANGES

During cloud formation, and often in the formation of fog, air is cooled to its dew point. Near the earth's surface, heat is readily exchanged between the ground and the air above. This accounts for the cooling involved in the formation of some types of fog. However, because air is a poor conductor of heat, this exchange is virtually nonexistent above a few thousand meters. Thus some other mechanism must operate during cloud formation. This mechanism is easily understood if you have ever pumped up a bicycle tire and noticed that the pump barrel became quite warm. The heat you felt was the consequence of the work you did on the air to compress it. When energy is used to compress air, an equivalent amount of energy is released as heat. Conversely, air that is allowed to escape from a bicycle tire cools as it expands. This results because the expanding air pushes (does work on) the surrounding air and must cool by an amount equivalent to the energy expended. You probably have experienced the cooling effect of an expanding gas while applying a spray deodorant. As the compressed gas propellent in the aerosol can is released, it quickly expands and cools. This drop in temperature occurs even though heat is neither added or subtracted. Such variations are known as **adiabatic temperature changes** and result when air is compressed or allowed to expand. In summary, when air is allowed to expand, it cools, and when it is compressed, it warms.

Any time air moves upward, it passes through regions of successively lower pressure. As a result, the ascending air expands and cools adiabatically.

Unsaturated air cools at the rather constant rate of 10°C for every 1000 meters of ascent (1°C per 100 meters). Conversely, descending air comes under increasingly higher pressures, compresses, and is heated 10°C for every 1000 meters of descent. This rate of cooling or heating applies only to unsaturated air and is known as the **dry adiabatic rate.** If air rises high enough, it will cool sufficiently to cause condensation. From this point on along its ascent, latent heat stored in the water vapor will be liberated. Although the air will continue to cool after condensation begins, the released latent heat works against the adiabatic process, thereby reducing the rate at which the air cools. This slower rate of cooling caused by the addition of latent heat is called the **wet adiabatic rate** of cooling. Since the amount of latent heat released depends upon the quantity of moisture present in the air, the wet adiabatic rate varies from 5°C per 1000 meters for air with a high moisture content to 9°C per 1000 meters for dry air. Figure 13.6 illustrates the role of adiabatic cooling in the formation of clouds. Note that from the surface up to the condensation level the air cools at the dry adiabatic rate. The wet adiabatic rate commences at the condensation level.

STABILITY

As we have learned, if air rises, it will cool and eventually produce clouds. Why does air rise on some occasions, but not on others? Why do the size of clouds and the amount of precipitation vary so much when air does rise? The answers to these questions are closely related to the stability of the air. Imagine, if you will, a large bubble of air with a thin flexible cover which allows it to expand but prevents it from mixing with the surrounding air. If the imaginary bubble were forced to rise, its temperature would decrease because of expansion. By comparing the bubble's temperature to that of the surrounding air, we can determine the stability of the bubble. If the bubble's temperature is lower than that of its environment, it will be heavier, and if allowed to move freely, it would sink to its original position. Air of this type, termed *stable air,* resists vertical displacement. On the other hand, if our imaginary bubble were warmer, and therefore lighter, than the surrounding air, it would continue to rise until it reached an altitude having the same temperature, much as a hot-air balloon rises as long as it is lighter than the surrounding air. This type of air is called *unstable air.*

FIGURE 13.6
Rising air cools at the dry adiabatic rate of 10°C per 1000 meters, until the air reaches the dew point and condensation (cloud formation) begins. As air continues to rise, the latent heat released by condensation reduces the rate of cooling. The wet adiabatic rate is therefore always less than the dry adiabatic rate.

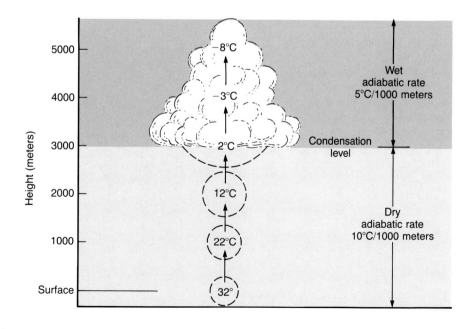

DETERMINATION OF STABILITY

In an actual situation, the stability of the air is determined by examining the temperature of the atmosphere at various heights. As you recall, this measure is termed the *lapse rate.* It is important not to confuse the lapse rate, which is the temperature of the atmosphere as determined from observations made by balloons and airplanes, with adiabatic temperature changes. The latter measure is the change in temperature due to expansion or compression as a parcel of air rises or falls.

For illustration, we will examine a situation where the prevailing lapse rate is 5°C per 1000 meters (Figure 13.7). Under this condition, when air at the surface has a temperature of 25°C, the air at 1000 meters will be 5 degrees cooler, or 20°C, the air at 2000 meters will have a temperature of 15°C, and so forth. At first glance it appears that the air at the surface is lighter than the air at 1000 meters since it is 5 degrees warmer. However, if the air near the surface is unsaturated and were to rise to 1000 meters, it would expand and

cool at the dry adiabatic rate of 10°C per 1000 meters. Therefore, upon reaching 1000 meters, its temperature would have dropped 10°C. Being 5 degrees cooler than its environment, it would be heavier and tend to sink to its original position. Hence, we say that the air near the surface is potentially cooler than the air aloft and therefore will not rise. Using similar reasoning, if the air at 1000 meters subsided, adiabatic heating would increase its temperature 10 degrees by the time it reached the surface, making it warmer than the surrounding air, so its buoyancy would cause it to return. The air just described is stable and resists vertical movement.

Stated quantitatively, **absolute stability** prevails when the lapse rate is less than the wet adiabatic rate. Figure 13.8 depicts this situation using a lapse rate of 5°C per 1000 meters and a wet adiabatic rate of 6°C per 1000 meters. Note that at 1000 meters the temperature of the surrounding air is 15°C, while the rising parcel of air has cooled to 10°C and is therefore the heavier air. Even if this stable air were to be forced above the condensa-

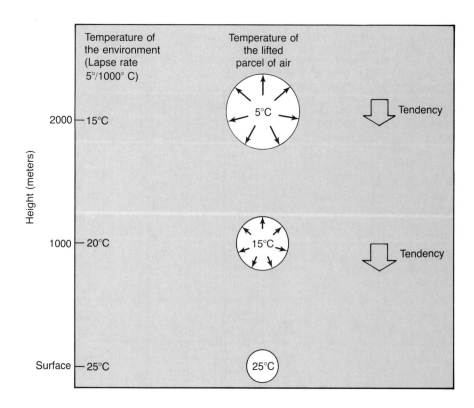

FIGURE 13.7
In a stable atmosphere, as an unsaturated parcel of air is lifted, it expands and cools at the dry adiabatic rate of 10°C per 1000 meters. Because the temperature of the rising parcel of air is lower than that of the surrounding environment, it will be heavier and, if allowed to do so, will sink to its original position.

FIGURE 13.8
Absolute stability prevails
when the lapse rate is less
than the wet adiabatic rate.
The rising parcel of air is
therefore always cooler and
heavier than the surrounding
air. When stable air is forced
to rise, it spreads out, producing flat layered clouds.

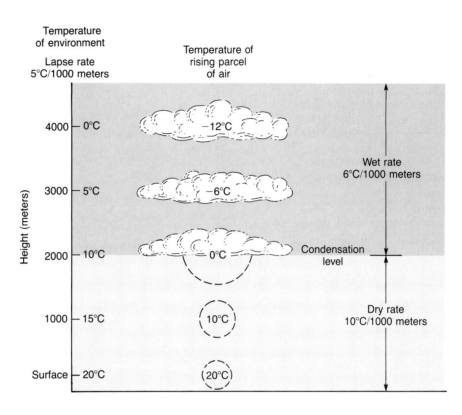

tion level, it would remain cooler and heavier than
its environment and would have a tendency to return to the surface.

At the other extreme, air is said to exhibit **absolute instability** when the lapse rate is greater than
the dry adiabatic rate. As shown in Figure 13.9, the
ascending parcel of air is always warmer than its
environment and will continue to rise because of
its own buoyancy. However, absolute instability is
generally limited to a narrow zone near the earth's

FIGURE 13.9
Absolute instability illustrated
using a lapse rate of 12°C
per 1000 meters. The rising
air is always warmer and
therefore lighter than the surrounding air.

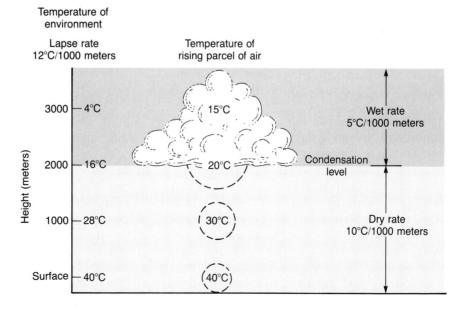

FIGURE 13.10
Conditional instability illustrated using a lapse rate of 8°C per 1000 meters, which lies between the dry adiabatic rate and the wet adiabatic rate. The rising parcel of air is cooler than the surrounding air below 4000 meters and warmer above 4000 meters.

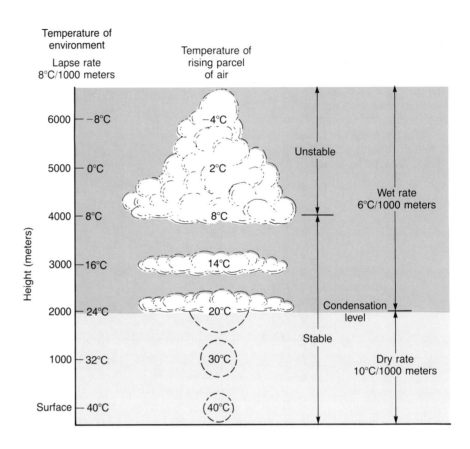

surface. On hot, sunny days the air above some surfaces is heated more than the air over adjacent surfaces. These invisible pockets of more intensely heated air, being less dense than the air aloft, will rise like a hot air balloon. This phenomenon produces the small fluffy clouds we associate with fair weather. Occasionally, when the surface air is considerably warmer than the air aloft, clouds.with great vertical development can form. As we shall see, clouds of this type are associated with heavy precipitation.

Another situation in the atmosphere is called **conditional instability.** This occurs when moist air has a lapse rate between the dry and wet adiabatic rates (between 5°C and 10°C per 1000 meters). Referring to Figure 13.10, notice that for the first 4000 meters the rising parcel of air is cooler than the surrounding air and is therefore considered stable. However, with the addition of latent heat above the condensation level, the parcel eventually becomes warmer than the surrounding air. From this point on along its ascent, the parcel

will continue to rise without an outside force and is considered unstable. Conditionally unstable air can be described as air that begins its ascent as stable air but at some point above the condensation level becomes unstable. The word *conditional* is used because the air can only become unstable if it is forced upward initially. Conditional instability is perhaps the most common type of instability.

STABILITY AND DAILY WEATHER

From the previous discussion we can conclude that stable air resists vertical movement, whereas unstable air ascends freely because of its own buoyancy. But how do these facts manifest themselves in our daily weather? Since stable air resists upward movement, we might conclude that clouds will not form when stable conditions prevail in the atmosphere. Although this seems reasonable, processes do exist which force air aloft. These will be discussed in the following section. On occasions when stable air is forced aloft, the clouds that form are widespread and have little vertical thickness when compared to their horizontal dimension, and precipitation, if any, is light-to-moderate. By contrast, clouds associated with unstable air are towering and are usually accompanied by heavy precipitation. For this reason we can conclude that on a dreary, overcast day with light drizzle, stable air is forced aloft. On the other hand, during a day when cauliflower-shaped clouds appear to be growing as if bubbles of hot air are surging upward, we can be fairly certain that the ascending air is unstable.

In summary, the role of stability in determining our daily weather is very important. To a large degree stability determines the type of clouds that develop and whether precipitation will come as a gentle shower or a heavy downpour.

FORCEFUL LIFTING

Earlier we demonstrated that stable and conditionally unstable air will not rise on their own; they require some mechanism to trigger the vertical

movement. Three such mechanisms are convergence, orographic lifting, and frontal wedging.

Whenever air converges (flows together), it results in general upward movement. This occurs because as air converges it occupies a smaller and smaller area, which requires that the height of the air column increase. Consequently, the air within the column must move upward, enhancing instability. The Florida peninsula provides an excellent example of the role convergence plays in initiating instability. On warm days the air flow is off the ocean along both coasts of Florida, causing general convergence over the peninsula. This convergence and associated uplift, aided by the intense solar heating, causes more mid-afternoon thunderstorms in this part of the United States than in any other area.

Orographic lifting occurs when sloping terrain, such as mountains, acts as a barrier to the flow of air, forcing the air to ascend. Many of the rainiest places in the world are located on windward mountain slopes. A station at Mt. Waialeale, Hawaii, for example, records the highest average

FIGURE 13.11
A. When stable air is lifted, layered clouds usually result. **B.** When warm, unstable air is forced to rise over cooler air, "towering" clouds develop.

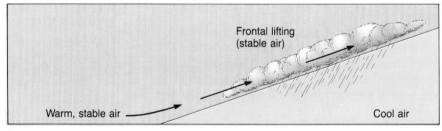

A.

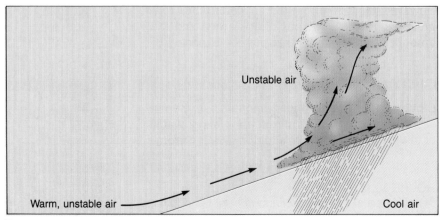

B.

annual rainfall in the world, some 1168 centimeters (38 feet). The station is located on the windward (northeast) coast of the island of Kauai, at an elevation of 1523 meters (4955 feet).

Besides providing the lift to render air unstable, mountains further remove more than their share of moisture in other ways. By slowing the horizontal air flow, they cause convergence as well as retard the passage of storm systems. Also, the irregular topography of mountains enhances differential heating and surface instability. These combined effects account for the generally higher precipitation we associate with mountainous regions as compared to the surrounding lowlands.

By the time air reaches the leeward side of a mountain, much of the moisture has been lost, and if the air descends, it warms, making condensation and precipitation even less likely. The result often is a **rainshadow desert.** The Great Basin Desert of the western United States lies only a few hundred kilometers from the Pacific Ocean but is effectively cut off by the imposing Sierra Nevada. The Gobi Desert of Mongolia, the Takla Makan of China, and the Patagonia Desert of Argentina are other examples of deserts found on the leeward sides of mountains.

Frontal wedging occurs when cool air acts as a barrier over which warmer, lighter air rises. This phenomenon is quite common throughout the continental United States and is responsible for the bulk of precipitation in many areas, as we shall see later. Figure 13.11, page 369, illustrates frontal wedging of stable and unstable air. As shown, forceful lifting is important in producing clouds. However, the stability of the air determines to a great extent the types of clouds formed and the amount of precipitation that may be expected.

CLOUDS

Clouds are a form of condensation best described as visible aggregates of minute droplets of water or of tiny crystals of ice. In addition to being prominent and sometimes spectacular features in the sky, clouds are of continual interest to meteorologists, because they provide a visible indication of what is going on in the atmosphere. Anyone who observes clouds with the hope of recognizing different types often finds that there is a bewildering variety of these familiar white and gray masses streaming across the sky. Still, once one comes to know the basic classification scheme for clouds, most of the confusion vanishes.

Clouds are classified on the basis of their appearance and height (Figure 13.12). Three basic forms are recognized: cirrus, cumulus, and stratus. **Cirrus** clouds are high, white, and thin. They are separated, or detached, and form delicate veil-like patches or extended wispy fibers that often have a feathery appearance. The **cumulus** form consists of globular individual cloud masses. Normally they exhibit a flat base and have the appearance of rising domes or towers. Such clouds are frequently described as having a cauliflowerlike

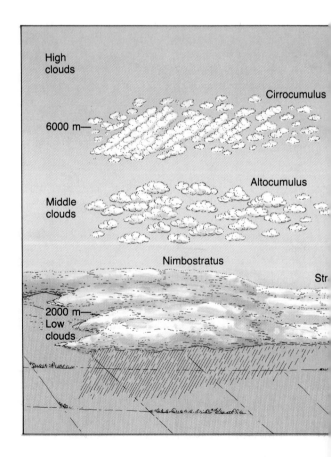

FIGURE 13.12
Classification of clouds according to height and form. (After Ward's Natural Science Establishment, Inc., Rochester, N.Y.)

structure. **Stratus** clouds are best described as sheets or layers that cover much or all of the sky. While there may be minor breaks, there are no distinct individual cloud units. All other clouds reflect one of these three basic forms or are combinations or modifications of them.

Three levels of cloud heights are recognized: high, middle, and low. **High clouds** normally have bases above 6000 meters; **middle clouds** generally occupy heights from 2000 to 6000 meters; and **low clouds** form below 2000 meters. The altitudes listed for each height category are not hard and fast. There is some seasonal as well as latitudinal variation. For example, at high latitudes or during cold winter months in the mid-latitudes, high clouds are often found at lower altitudes.

As a result of the low temperatures and small quantities of water vapor found at high altitudes, all of the high clouds are thin, white, and composed of ice crystals. Since much more water vapor is available at lower altitudes, middle and low clouds are thicker and denser.

Layered clouds in any of these height ranges generally indicate that the air is stable. Normally wè might not expect clouds to grow or persist in stable air. However, cloud growth of this type is common when air is forced to rise, as along a front or near the center of a cyclone, where converging winds cause air to ascend. Such forced ascent of stable air leads to the formation of a stratified cloud layer that is large horizontally when compared to its depth.

Some clouds do not fit into any one of the three height categories mentioned. Such clouds have their bases in the low height range but often extend upward into the middle or high altitudes. Consequently, clouds in this category are called **clouds of vertical development.** These clouds are all related to one another and are associated with unstable air. While cumulus clouds are often connected with "fair" weather, they may grow dramatically under the proper circumstances. Once

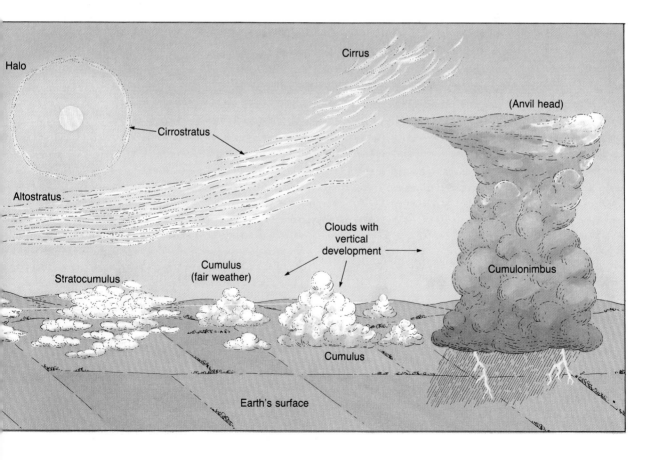

TABLE 13.4
Cloud types and characteristics.

Cloud Family and Height	Cloud Type	Characteristics
High clouds— above 6000 meters (20,000 feet)	Cirrus	Thin, delicate, fibrous ice-crystal clouds. Sometimes appear as hooked filaments called "mares' tails." (Figure 13.13A)
	Cirrocumulus	Thin, white ice-crystal clouds in the form of ripples, waves, or globular masses all in a row. May produce a "mackerel sky." Least common of the high clouds. (Figure 13.13B)
	Cirrostratus	Thin sheet of white ice-crystal clouds that may give the sky a milky look. Sometimes produce halos around the sun or moon. (Figure 13.13C)
Middle clouds— 2000–6000 meters (6500–20,000 feet)	Altocumulus	White to gray clouds often composed of separate globules; "sheep-back" clouds. (Figure 13.13D)
	Altostratus	Stratified veil of clouds that are generally thin and may produce very light precipitation. When thin, the sun or moon may be visible as a "bright spot," but no halos are produced.
Low clouds— below 2000 meters (6500 feet)	Stratocumulus	Soft, gray clouds in globular patches or rolls. Rolls may join together to make a continuous cloud.
	Stratus	Low uniform layer resembling fog but not resting on the ground. May produce drizzle.
	Nimbostratus	Amorphous layer of dark gray clouds. One of the chief precipitation-producing clouds. (Figure 13.13E)
Clouds of vertical development— 500–18,000 meters (1600–60,000 feet)	Cumulus	Dense, billowy clouds often characterized by flat bases. May occur as isolated clouds or closely packed. (Figure 13.13F)
	Cumulonimbus	Towering cloud sometimes spreading out on top to form an "anvil head." Associated with heavy rainfall, thunder, lightning, hail, and tornadoes. (Figure 13.13G)

upward movement is triggered, acceleration is powerful and clouds with great vertical extent form. The end result is often a towering cloud that may produce rainshowers or a thunderstorm.

Since definite weather patterns can often be associated with particular clouds or certain combinations of cloud types, it is important to become familiar with cloud descriptions and characteristics. Table 13.4 lists the ten basic cloud types that are recognized internationally and gives some characteristics of each. The series of photos in Figure 13.13 depicts common forms of several cloud types.

FORMATION OF PRECIPITATION

If all clouds contain water, why do some produce precipitation while others drift placidly overhead? This seemingly simple question perplexed meteorologists for many years. First, cloud droplets are very small, averaging less than 10 micrometers in diameter (for comparison, a human hair is about 75 micrometers in diameter). Because of their small size, cloud droplets fall incredibly slowly. Theoretically, an average cloud droplet falling from a cloud base at 1000 meters would require about 48 hours to reach the ground. Of course, it

A.

B.

C.

D.

F.

G.

E.

FIGURE 13.13
A. Cirrus. B. Cirrocumulus. C. Cirrostratus with halo.
D. Altocumulus at sunrise. E. Nimbostratus. F. Cumulus. G. Cumulonimbus. (Photos A, B, E, and F by
E. J. Tarbuck; photos C and D courtesy of Ward's
Natural Science Establishment, Inc., Rochester, N.Y.;
photo G courtesy of Ron Holle)

would never complete its journey. Even falling through humid air, a cloud droplet would evaporate before it fell a few meters below the cloud base. In addition, clouds are made up of many billions of these droplets, all competing for the available water vapor; thus, their continued growth via condensation is very slow.

A raindrop large enough to reach the ground without evaporating contains roughly a million times more water than a cloud droplet. Therefore, for precipitation to form, millions of cloud droplets must somehow coalesce (join together) into drops large enough to sustain themselves during their descent. Two mechanisms have been proposed to explain this phenomenon: the Bergeron process and the collision-coalescence process.

The **Bergeron process,** named after its discoverer, relies on two interesting properties of water. First, cloud droplets do not freeze at 0°C as expected. In fact, pure water suspended in air does not freeze until it reaches a temperature of nearly −40°C. Water in the liquid state below 0°C is generally referred to as **supercooled.** Supercooled water will readily freeze if sufficiently agitated. This explains why airplanes collect ice when they pass through a liquid cloud composed of supercooled droplets. In addition, supercooled droplets will freeze upon contact with solid particles that have a crystal form closely resembling that of ice. These materials are termed **freezing nuclei.** The need for freezing nuclei to initiate the freezing process is similar to the requirement for condensation nuclei in the process of condensation. However, in contrast to condensation nuclei, freezing nuclei are very sparse in the atmosphere and do not generally become active until the temperature reaches −10°C or less. Only at temperatures well below freezing will ice crystals begin to form in clouds, and even at that, they will be few and far between. Once ice crystals form, they are in direct competition with the supercooled droplets for the available water vapor.

This brings us to the second interesting property of water. When air is saturated (100 percent relative humidity) with respect to water, it is supersaturated (relative humidity greater than 100 percent) with respect to ice. Table 13.5 shows that at −10°C, when the relative humidity is 100 per-

TABLE 13.5
Relative humidity with respect to ice when relative humidity with respect to water is 100 percent.

Temperature (°C)	Relative Humidity with Respect to:	
	Water (%)	Ice (%)
0	100	100
−5	100	105
−10	100	110
−15	100	116
−20	100	121

cent with respect to water, the relative humidity with respect to ice is nearly 110 percent. Thus, ice crystals cannot coexist with water droplets, because the air always "appears" supersaturated to the ice crystals. So the ice crystals begin to consume the "excess" water vapor, which lowers the relative humidity near the surrounding droplets. In turn, the water droplets evaporate to replenish the diminishing water vapor, thereby providing a continual source of vapor for the growth of the ice crystals (Figure 13.14).

Because the level of supersaturation with respect to ice can be quite great, the growth of ice crystals is generally rapid enough to generate crystals large enough to fall. During their descent, these ice crystals enlarge as they intercept cloud drops, which freeze upon them. Air movement will sometimes break up these delicate crystals and the fragments will serve as freezing nuclei. A chain reaction develops, producing many ice crystals, which by accretion form into large crystals called snowflakes. When the surface temperature is above 4°C (39°F), snowflakes usually melt before they reach the ground and continue their descent as rain. Even a summer rain may have begun as a snowstorm in the clouds overhead.

Cloud seeding to produce precipitation utilizes the Bergeron process just described. By adding freezing nuclei (commonly silver iodide) to supercooled clouds, the growth of these clouds can be markedly changed. This process is discussed in greater detail in Chapter 16.

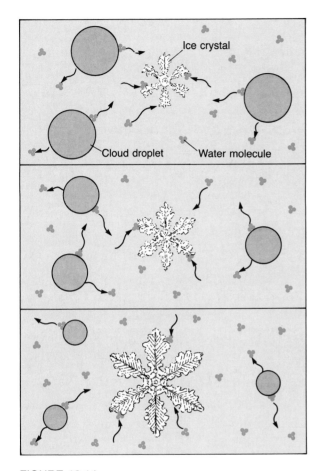

FIGURE 13.14
The Bergeron process. Ice crystals grow at the expense of cloud droplets until they are large enough to fall. The size of these particles has been greatly exaggerated.

Thirty years ago, meteorologists believed that the Bergeron process was responsible for the formation of most precipitation, with the exception of light drizzle. However, it was discovered that copious rainfall is often associated with clouds located well below the freezing level (warm clouds), particularly in the tropics. This led to the proposal of a second mechanism thought to produce precipitation—the **collision-coalescence process.**

Clouds composed entirely of liquid droplets must contain droplets larger than 20 micrometers if precipitation is to form. These large droplets form when "giant" condensation nuclei are present and on occasion when hygroscopic particles

such as sea salt exist. Hygroscopic particles begin to remove water vapor from the air at relative humidities under 100 percent and can grow quite large. Since the rate at which drops fall is size dependent, these "giant" droplets fall most rapidly. As such, they collide with the smaller, slower droplets and coalesce. Becoming larger in the process, they fall even more rapidly (or in an updraft, rise more slowly), increasing their chances of collision and rate of growth (Figure 13.15). After

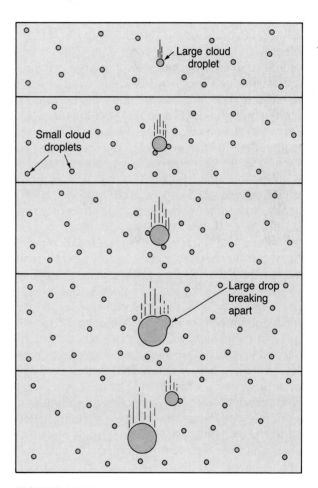

FIGURE 13.15
The collision-coalescence process. Because large cloud droplets fall more rapidly than smaller droplets, they are able to sweep up the smaller ones in their path and grow. Most cloud droplets are so small that the motion of the air keeps them suspended. Even if these small cloud droplets were to fall, they would evaporate long before reaching the surface.

a great many such collisions they are large enough to fall to the surface without completely evaporating. Due to the number of collisions required for growth to raindrop size, droplets in clouds with great vertical thickness and abundant moisture have a better chance of reaching the required size. Updrafts also aid in this process since they allow the droplets to traverse the cloud repeatedly. Raindrops can grow to a maximum size of 5 millimeters when they fall at the rate of 30 kilometers per hour. At this size and speed the water's surface tension, which holds the drop together, is surpassed by the drag imposed by the air, which in turn succeeds in pulling the drops apart. The resulting breakup of a large raindrop produces numerous smaller drops which begin anew the task of sweeping up cloud droplets. Drops that are less than 0.5 millimeter upon reaching the ground are termed drizzle and require about ten minutes to fall from a cloud 1000 meters (3280 feet) overhead.

Keep in mind that the collision-coalescence process is not quite as simple as described. First, as the larger droplets descend, they produce an air stream around them similar to that produced by an automobile when it is driven rapidly down the highway. If an automobile is driven at night and we use the bugs that are often out to be analogous to the cloud droplets, it is easy to visualize how most cloud droplets are swept aside. The larger the cloud droplet (or bug), the better chance it will have of colliding with the giant droplet (or car). Second, collision does not guarantee coalescence. Experimentation has indicated that the presence of atmospheric electricity may be the key to what holds these droplets together once they collide. If a droplet with a negative charge should collide with a positively charged droplet, their electrical attraction may bind them together.

SLEET, GLAZE, AND HAIL

While rain and snow are the most common and familiar forms of precipitation, other forms do exist and are worthy of mention. Sleet, glaze, and hail fall into this category. Although limited in occurrence and sporadic in both time and space,

FIGURE 13.16
Glaze forms when supercooled raindrops freeze on contact with objects. (Photo by E. J. Tarbuck)

these forms, especially the latter two, may on occasion cause considerable damage.

Sleet is a wintertime phenomenon and refers to the fall of small, clear-to-translucent particles of ice. For sleet to be produced, a layer of air with temperatures above freezing must overlie a subfreezing layer near the ground. When the raindrops leave the warmer air and encounter the colder air below, they solidify, reaching the ground as small pellets of ice no larger than the raindrops from which they formed.

On some occasions, when the vertical distribution of temperatures is similar to that associated with the formation of sleet, freezing rain or **glaze** results instead. In such situations, the subfreezing air near the ground is not thick enough to allow the raindrops to freeze. The raindrops, however, do become supercooled as they fall through the cold air and consequently turn to ice upon colliding with solid objects. The result can be a thick coating of ice that is sufficiently heavy to break tree limbs and down power lines as well as make walking or motoring extremely hazardous (Figure 13.16).

Hail is precipitation in the form of hard, rounded pellets or irregular lumps of ice. Furthermore, large hailstones often consist of a series of nearly concentric shells of differing densities and degrees of opaqueness (Figure 13.17). Most frequently, hailstones have a diameter of about 1 centimeter, but they may vary in size from 5 millimeters to more than 10 centimeters in diameter. The largest hailstone on record fell on Coffeyville, Kansas, September 3, 1970. With a 14-centimeter diameter and a circumference of 44 centimeters, this "giant" weighed 766 grams (over 1½ pounds)! The destructive effects of heavy hail are well known, especially to farmers whose crops can be devastated in a few short minutes and to persons whose windows are shattered.

Hail is produced only in cumulonimbus clouds where updrafts are strong and where there is an abundant supply of supercooled water. First, rain is lifted above the freezing level by the rapidly ascending air. Once frozen, these small ice granules grow by collecting supercooled cloud droplets as they fall through the cloud. If they encounter another strong updraft, they may be carried upward again and begin the downward journey anew. Each trip above the freezing level may be represented by an additional layer of ice.

Hailstones, however, may also form from a single descent through an updraft. In this situation, the layered structure is attributed to variations in the rate at which supercooled droplets accumulate and freeze, which, in turn, is related to differences in the amount of supercooled water in different parts of the cumulonimbus tower.

In either case, hailstones grow by the addition of supercooled water upon growing ice pellets. The ultimate size of the hailstone depends primarily upon three factors: the strength of the updrafts, the concentration of supercooled water, and the length of the path through the cloud.

FOG

Fog is generally considered to be an atmospheric hazard. When it is light, visibility is reduced to 2 or 3 kilometers. However, when it is dense, visibility may be cut to a few tens of meters or less, making travel by any mode not only difficult but often dangerous as well. Officially, visibility must be reduced to 1 kilometer or less before fog is reported. While this figure is arbitrary, it does permit a more objective criterion for comparing fog frequencies at different locations.

Fog is defined as a cloud with its base at or very near the ground. Physically, there is no basic difference between a fog and a cloud; the appearance and structure of both are the same. The essential difference is the method and place of formation. While clouds result when air rises and cools adiabatically, fogs (with the exception of upslope fogs) are the consequence of radiation

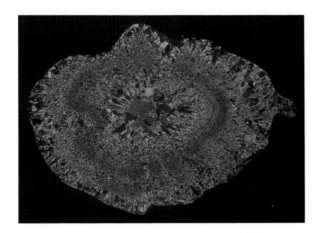

FIGURE 13.17
Cross section of a large hailstone photographed in polarized light. Note the concentric layers of ice. (Courtesy of the National Center for Atmospheric Research/National Science Foundation)

FIGURE 13.18
Radiation fog in a valley.
(Courtesy of Richard Anthes)

FIGURE 13.19
Patches of upslope fog forming on the sides of a
mountain valley. (Courtesy of Ward's Natural Science
Establishment, Inc., Rochester, N.Y.)

cooling or the movement of air over a cold surface.
In other circumstances, fogs form when enough
water vapor is added to the air to bring about saturation (evaporation fogs).

FOGS CAUSED BY COOLING

When warm, moist air is blown over a cool surface, the result may be a blanket of fog called **advection fog.** Examples of such fogs are very common. The foggiest location in the United States,
and perhaps in the world, is Cape Disappointment,
Washington. The name is indeed appropriate,
since this station averages 2552 hours of fog each
year. The fog experienced at Cape Disappointment, as well as that at other West Coast locations,
is produced when warm, moist air from the Pacific
Ocean moves over the cold California Current and
is then carried onto shore by the prevailing winds.
Advection fogs are also quite common in the winter season when warm air from the Gulf of Mexico
is blown over cold, often snow-covered surfaces of
the Midwest and East.

    Radiation fog forms on cool, clear, calm nights,
when the earth's surface cools rapidly by radiation. As the night progresses, a thin layer of air in
contact with the ground is cooled below its dew
point. As the air cools and becomes heavier, it

drains into low areas, resulting in "pockets" of fog. The largest "pockets" are often river valleys, where rather thick accumulations may occur (Figure 13.18).

As its name implies, **upslope fog** is created when relatively humid air moves up a gradually sloping plain or, in some cases, up the steep slopes of a mountain (Figure 13.19). Due to the upward movement, air expands and cools adiabatically. If the dew point is reached, an extensive layer of fog may form. In the United States, the Great Plains offers an excellent example. When humid easterly or southeasterly winds move westward from the Mississippi River toward the Rocky Mountains, the air gradually rises, resulting in an adiabatic decrease of about 13°C. When the difference between the air temperature and dew point of westward-moving air is less than 13°C, an extensive fog often results in the western plains.

EVAPORATION FOGS

When cool air moves over warm water, enough moisture may evaporate from the water surface to produce saturation. As the rising water vapor meets the cold air, it immediately recondenses and rises with the air that is being warmed from below. Since the water has a steaming appearance, the phenomenon is called **steam fog** (Figure 13.20). Steam fog is fairly common over lakes and rivers in the fall and early winter, when the water may still be relatively warm and the air is rather crisp. Steam fog is often quite shallow because as the steam rises it re-evaporates in the unsaturated air above.

FIGURE 13.20
Early-morning steam fog on a small lake. (Courtesy of Ward's Natural Science Establishment, Inc., Rochester, N.Y.)

When frontal wedging occurs, warm air is lifted over colder air. If the resulting clouds yield rain, and the cold air below is near the dew point, enough rain will evaporate to produce fog. A fog formed in this manner is called **frontal fog,** or **precipitation fog.** The result is a more or less continuous zone of condensed water droplets reaching from the ground up through the clouds.

In summary, both steam fog and frontal fog result from the addition of moisture to a layer of air. As we learned, the air is usually cool or cold and already near saturation. Since air's capacity to hold water vapor at low temperatures is small, only a relatively modest amount of evaporation is necessary to produce saturated conditions and fog.

REVIEW QUESTIONS

1 Summarize the processes by which water changes from one state to another. Indicate whether heat energy is absorbed or liberated.

2 After studying Table 13.1, write a generalization relating temperature and the capacity of air to hold water vapor.

3 How do relative and specific humidity differ?

4 Referring to Figure 13.4, answer the following questions:
 (a) During a typical day, when is the relative humidity highest? Lowest?
 (b) At what time of day would dew most likely form?
 Write a generalization relating air temperature and relative humidity.

5 If the temperature remains unchanged and the specific humidity decreases, how will relative humidity change?

6 On a cold winter day when the temperature is −10°C and the relative humidity is 50 percent, what is the specific humidity (refer to Table 13.1)? What is the specific humidity for a day when the temperature is 20°C and the relative humidity is 50 percent?

7 Explain the principle of the sling psychrometer. The hair hygrometer.

8 Using the standard tables (Tables 13.2 and 13.3), determine the relative humidity and dew-point temperature if the dry-bulb thermometer reads 16°C and the wet-bulb thermometer reads 12°C. How would the relative humidity and dew point change if the wet-bulb thermometer read 8°C?

9 On a warm summer day when the relative humidity is high, it may seem even warmer than the thermometer indicates. Why do we feel so uncomfortable on a "muggy" day?

10 What is the function of condensation nuclei in cloud formation? The function of the dew point?

11 As you drink an ice-cold beverage on a warm day, the outside of the glass or bottle becomes wet. Explain.

12 Why does air cool when it rises through the atmosphere?

13 Explain the difference between lapse rate and adiabatic cooling.

14 If unsaturated air at 23°C were to rise, what would its temperature be at 500 meters? If the dew-point temperature at the condensation level were 13°C, at what altitude would clouds begin to form?

15 Why does the adiabatic rate of cooling change when condensation begins? Why is the wet adiabatic rate not a constant figure?

16 The contents of an aerosol can are under very high pressure. When you push the nozzle on such a can, the spray feels cold. Explain.

17 How do orographic lifting and frontal wedging act to force air to rise?

18 Explain why the Great Basin area of the western United States is so dry. What term is applied to such a situation?

19 How does stable air differ from unstable air? Describe the general nature of the clouds and precipitation expected with each.

20 What is the basis for the classification of clouds?

21 Why are high clouds always thin?

22 Which cloud types are associated with the following characteristics: thunder, halos, precipitation, hail, mackerel sky, lightning, mares' tails?

23 What is the difference between precipitation and condensation?

24 List the forms of precipitation and the circumstances of their formation.

25 List five types of fog and discuss the details of their formation.

KEY TERMS

calorie

evaporation

latent heat

condensation

melting

freezing

sublimation

deposition

humidity

saturation

vapor pressure

specific humidity

relative humidity

dew point

psychrometer

hygrometer

condensation nuclei

hygroscopic nuclei

adiabatic temperature change

dry adiabatic rate

wet adiabatic rate

absolute stability

absolute instability

conditional instability

orographic lifting

rainshadow desert

frontal wedging

clouds

cirrus

cumulus

stratus

high clouds

middle clouds

low clouds

clouds of vertical development

Bergeron process

supercooled

freezing nuclei

collision-coalescence process

sleet

glaze

hail

fog

advection fog

radiation fog

upslope fog

steam fog

frontal (precipitation) fog

14

PRESSURE AND WIND

We have already dealt with the two elements of weather and climate that generally are of the greatest interest to people—temperature and moisture. In this chapter we shall investigate the remaining two elements—pressure and wind—which may not seem as important, but indeed are. It is the wind that often brings changes in temperature and moisture conditions, and it is pressure differences that drive the wind. Among the questions we will try to answer in this chapter are: How can the weight of the air be measured? What are the factors that control the winds? What is a prevailing wind? Why are "highs" and "lows" always shown on the weather map? How can the weather be predicted by checking the barometer?

The spiraling cloud patterns in this satellite image show the characteristic inward and counterclockwise circulation around a low in the Northern Hemisphere. (Courtesy of Robert Sheets, National Hurricane Research Laboratory)

Of the various elements of weather and climate, changes in air pressure are the least noticeable. When listening to a weather report, we are generally interested in moisture conditions (humidity and precipitation), temperature, and perhaps wind. It is the rare individual, however, who wonders about air pressure. Although the hour-to-hour and day-to-day variations in air pressure are not perceptible to human beings, they are very important in producing changes in our weather. Variations in air pressure from place to place are responsible for the movement of air (wind), as well as being a significant factor in weather forecasting. As we shall see, air pressure is tied very closely to the other elements of weather in a cause-and-effect relationship.

PRESSURE MEASUREMENT

In Chapter 12 we saw that air has weight; at sea level, it exerts a pressure of 1 kilogram per square centimeter (14.7 pounds per square inch). The air pressure at a particular place is simply the force exerted by the weight of the air above. With an increase in altitude the weight of the air above, and thus the pressure, decreases rapidly at first, then much more slowly (see Table 14.1).

When meteorologists measure atmospheric pressure, they employ a unit called the *millibar.* Standard sea level pressure is expressed as 1013.2

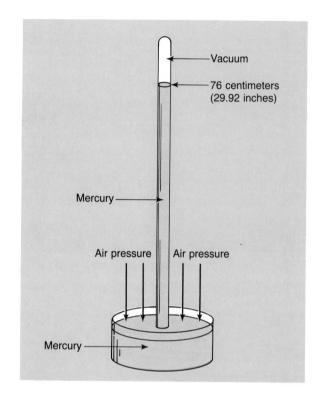

FIGURE 14.1
Simple mercurial barometer. The weight of the column of mercury is balanced by the pressure exerted on the dish of mercury by the air above. If the pressure decreases, the column of mercury falls; if the pressure increases, the column rises.

TABLE 14.1
Pressure changes with altitude.

Altitude (kilometers)	Altitude (miles)	Pressure (millibars)
0	0	1013
1.0	0.6	899
2.0	1.2	795
3.0	1.9	701
4.0	2.5	617
5.0	3.1	540
10.0	6.2	265
20.0	12.4	55
30.0	18.6	12
40.0	24.8	3

millibars. Although the millibar has been the unit of measure on all United States weather maps since January 1940, you might be better acquainted with the expression "inches of mercury," which is used by the media to describe atmospheric pressure. In the United States the National Weather Service converts millibar values to inches of mercury for public and aviation use.

The use of mercury for measuring air pressure dates from 1643, when Torricelli, a student of the famous Italian scientist Galileo, invented the **mercurial barometer.** Torricelli correctly described the atmosphere as a vast ocean of air that exerts pressure on us and all objects about us. To measure this force he filled a glass tube that was closed at one end with mercury. The tube was then inverted into a dish of mercury (Figure 14.1). Tor-

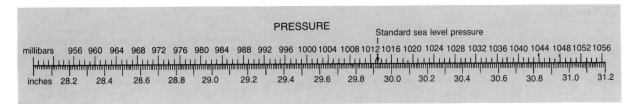

FIGURE 14.2
Millibars and inches. This scale compares two of the most common units of pressure measurement. Standard sea level pressure is equal to 29.92 inches or 1013.2 millibars. (After NOAA)

ricelli found that the mercury flowed out of the tube until the weight of the column was balanced by the pressure that the atmosphere exerted on the surface of the mercury. In other words, the weight of mercury in the column equalled the weight of a similar diameter column of air that extended from the ground to the top of the atmosphere. Torricelli noted that when air pressure increased, the mercury in the tube rose; conversely, when air pressure decreased so did the height of the column of mercury. The length of the column of mercury, therefore, became the measure of the air pressure. With some refinements the mercurial barometer invented by Torricelli is still the standard pressure-measuring instrument used today. Standard atmospheric pressure at sea level equals 29.92 inches of mercury. A scale comparing millibars and inches of mercury is shown in Figure 14.2.

The need for a smaller and more portable instrument for measuring air pressure led to the development of the **aneroid** ("without liquid") **barometer.** Based on a different principle from the mercurial barometer, this instrument consists of partially evacuated metal chambers that have a spring inside, keeping them from collapsing. The metal chambers, being very sensitive to air pressure variations, change shape, compressing as the pressure increases and expanding as the pressure decreases. Aneroids are often used in making **barographs,** instruments that continuously record pressure changes.

An important adaptation of the aneroid is its use as an **altimeter** in aircraft. Recall that air pressure decreases with altitude and that the pressure

distribution with height is well established. An altimeter is an aneroid barometer calibrated in meters rather than millibars. For example, at an altitude of 5 kilometers (3 miles) a barometer will indicate a pressure of 540 millibars (Table 14.1). This occurs because roughly one-half of the earth's atmosphere is found above this altitude and, therefore, exerts one-half of the pressure experienced at sea level. Thus, whenever an aircraft is at an altitude corresponding to 540 millibars of pressure, its altimeter will indicate an altitude of 5 kilometers.

FACTORS AFFECTING WIND

We have discussed the upward movement of air and its importance in cloud formation. As important as vertical motion is, far more air is involved in horizontal movement, the phenomenon we call **wind.** Although we know that air will move vertically if it is warmer, and consequently more buoyant, than the surrounding air, what causes air to move horizontally? Simply stated, wind is the result of horizontal differences in air pressure. Air flows from areas of higher pressure to areas of lower pressure. You may have experienced this when opening a vacuum-packed can of coffee. The noise you hear is caused by air rushing from the higher pressure outside the can to the lower pressure inside. Wind is nature's attempt to balance similar inequalities in air pressure. Because unequal heating of the earth's surface generates these pressure differences, solar radiation is the ultimate driving force of wind.

FIGURE 14.3
Isobars are lines used to connect places of equal barometric pressure. They show the distribution of pressure on weather maps. The lines usually curve and often join where cells of high and low pressure exist. The "arrows" indicate the expected airflow surrounding cells of high and low pressure and are plotted as "flying" with the wind. Wind speed is indicated by flags and feathers as shown along the right-hand side of this drawing.

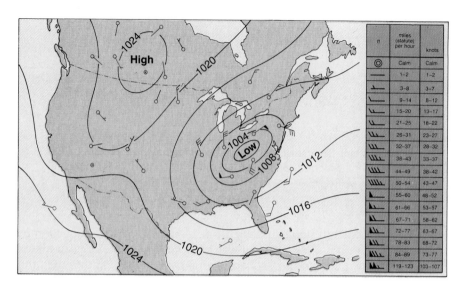

ff	miles (statute) per hour	knots
◎	Calm	Calm
—	1–2	1–2
⌐	3–8	3–7
⌐	9–14	8–12
⌐	15–20	13–17
⌐	21–25	18–22
⌐	26–31	23–27
⌐	32–37	28–32
⌐	38–43	33–37
⌐	44–49	38–42
⌐	50–54	43–47
⌐	55–60	48–52
⌐	61–66	53–57
⌐	67–71	58–62
⌐	72–77	63–67
⌐	78–83	68–72
⌐	84–89	73–77
⌐	119–123	103–107

If the earth did not rotate, and if there were no friction, air would flow directly from areas of higher pressure to areas of lower pressure. But because both of these factors exist, wind is controlled by the following combination of forces: (1) the pressure gradient force; (2) Coriolis effect; (3) friction, and (4) the tendency of a moving object to continue moving in a straight line. The last factor is often referred to as centrifugal force. The magnitude of centrifugal force is small compared to the other forces and thus of minor importance, except in rapidly rotating storms such as tornadoes and hurricanes. Discussions of the other factors follow.

PRESSURE GRADIENT FORCE

Pressure differences create wind, and the greater these differences, the greater the wind speed. Over the earth's surface, variations in air pressure are determined from barometric readings taken at hundreds of weather stations. These pressure data are shown on a weather map using **isobars,** lines that connect places of equal air pressure (Figure 14.3). The spacing of isobars indicates the amount of pressure change occurring over a given distance and is expressed as the **pressure gradient.**

You might find it easier to visualize a pressure gradient if you think of it as being analogous to the

FIGURE 14.4
Pressure gradient force. Closely spaced isobars indicate a steep pressure gradient and high wind speeds, whereas widely spaced isobars indicate a weak pressure gradient and low wind speeds.

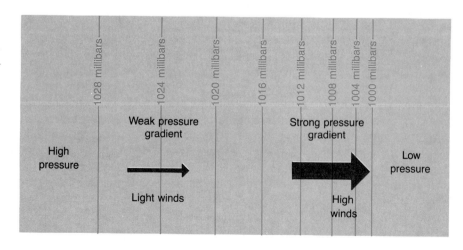

slope of a hill. A steep pressure gradient, like a steep hill, causes greater acceleration of an air parcel than does a weak pressure gradient. Thus, the relationship between wind speed and the pressure gradient is rather simple: Closely spaced isobars indicate a steep pressure gradient and high winds, whereas widely spaced isobars indicate a weak pressure gradient and light winds. Figure 14.4 illustrates the relationship between the spacing of isobars and wind speed.

The pressure gradient is the driving force of wind, and it has both magnitude and direction. While its magnitude is determined from the spacing of isobars, the direction of force is always from areas of higher pressure to areas of lower pressure and at right angles to the isobars. Once the air starts to move, the Coriolis effect and friction come into play, but then only to modify the movement, not to produce it.

CORIOLIS EFFECT

Figure 14.3 shows the typical air movements associated with high- and low-pressure systems. As expected, the air moves out of the regions of higher pressure and into the regions of lower pressure. However, the wind does not cross the isobars at right angles as the pressure gradient force directs. This deviation is the result of the earth's rotation and has been named the **Coriolis effect** after its discoverer. All free-moving objects, including the wind, are deflected to the right of their path of motion in the Northern Hemisphere and to the left in the Southern Hemisphere. The reason for this deflection can be illustrated by imagining the path of a rocket launched from the North Pole toward a target located on the equator (Figure 14.5). If the rocket took an hour to reach its target, during its flight the earth would have rotated 15 degrees to the east. To someone standing on the earth, it would look as if the rocket veered off its path and hit the earth 15 degrees west of its target. The true path of the rocket was straight and would appear so to someone out in space looking down at the earth. It was the earth turning under the rocket that gave it its *apparent* deflection. The

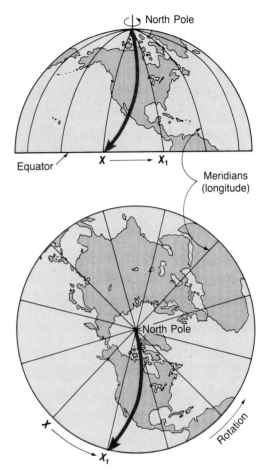

FIGURE 14.5
The Coriolis effect. During the rocket's flight from the North Pole to point X, the earth rotates eastward, moving point X to point X_1. The rotation gives the rocket's trajectory a curved path when plotted on the earth's surface. This deflection is termed the Coriolis effect.

same deflection is experienced by wind regardless of the direction it is moving.

For convenience, we attribute this apparent shift in wind direction to the Coriolis effect. This deflection: (1) is always directed at right angles to the direction of airflow; (2) affects only wind direction, not wind speed; and (3) is affected by wind speed. The stronger the wind, the greater the deflection.

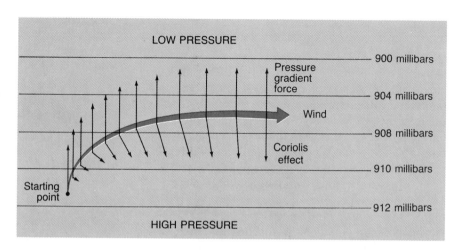

FIGURE 14.6
The geostrophic wind. Upper-level winds are deflected by the Coriolis effect until the Coriolis effect just balances the pressure gradient force. Above 600 meters (1970 feet), where friction is negligible, these winds will flow nearly parallel to the isobars and are called geostrophic winds.

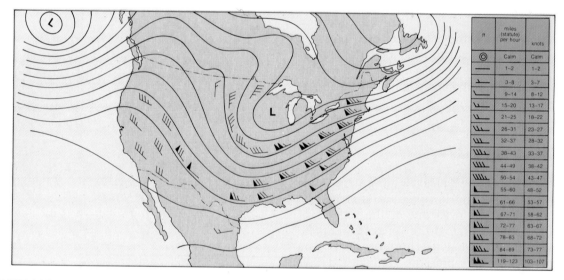

FIGURE 14.7
Upper-air winds. This map shows the direction and speed of the upper-air wind for a particular day. Note that the airflow is nearly parallel to the contours. These isolines are height contours for the 500-millibar level.

FIGURE 14.8
Comparison between upper-level winds and surface winds showing the effects of friction on airflow. Friction slows surface wind speed, which weakens the Coriolis effect, causing the winds to cross the isobars.

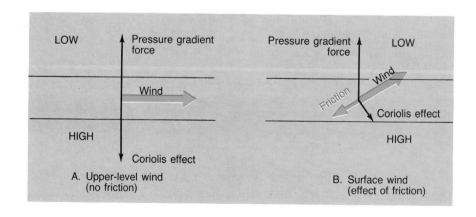

FRICTION

Friction as a factor affecting wind is only important within the first few kilometers of the earth's surface. It acts to slow the movement of air and, as a consequence, alters wind direction. To illustrate friction's effect on wind direction, let us look first at a situation in which it has no role. Above the friction layer, the pressure gradient force and the Coriolis effect are directing the flow of air. Under these conditions, the pressure gradient force will cause the air to start moving across the isobars. As soon as the air starts to move, the Coriolis effect will act at right angles to this motion. The faster the wind speed, the greater the deflection. Eventually, the Coriolis effect will balance the pressure gradient force and the wind will blow parallel to the isobars (Figure 14.6). Upper-air winds generally take this path and are called **geostrophic winds.** Due to the lack of friction, geostrophic winds travel at higher speeds than surface winds (Figure 14.7).

The most prominent features of upper-level flow are the **jet streams.** First encountered by high-flying bombers during World War II, these fast-moving "rivers" of air travel between 120 and 240 kilometers (75 and 150 miles) per hour in a west-to-east direction. One such stream is situated over the polar front, which is the zone separating cool polar air from warm subtropical air.

Below 600 meters (2000 feet), friction complicates the airflow just described. Recall that the Coriolis effect is proportional to wind speed. By lowering the wind speed, friction reduces the Coriolis effect. Since the pressure gradient force is not affected by wind speed, it wins the tug-of-war shown in Figure 14.8. The result is a movement of air at an angle across the isobars toward the area of low pressure. The roughness of the terrain determines the angle of airflow across the isobars. Over the smooth ocean surface, friction is low and the angle is small. Over rugged terrain, where friction is higher, the angle that air makes as it flows across the isobars can be as great as 45 degrees. In summary, upper airflow is nearly parallel to the isobars, while the effect of friction causes the surface winds to move more slowly and cross the isobars at an angle.

CYCLONES AND ANTICYCLONES

Among the most common features on any weather map are areas designated as pressure centers. **Cyclones,** or **lows,** are centers of low pressure, and **anticyclones,** or **highs,** are high-pressure centers. As Figure 14.3 illustrates, the pressure decreases from the outside toward the center in a cyclone, while in an anticyclone, just the opposite is the case—the pressure increases from the outside toward the center. By knowing just a few basic facts about centers of high and low pressure, you can greatly increase your understanding of current and forthcoming weather.

CYCLONIC AND ANTICYCLONIC WINDS

From the preceding section, we learned that the two most significant factors that affect wind are pressure differences and the Coriolis effect. Winds move from higher pressure to lower pressure and are deflected to the right or left by the earth's rotation. When these controls of airflow are applied to pressure centers in the Northern Hemisphere, the result is that winds blow inward and counterclockwise around a low and outward and clockwise around a high (Figure 14.3). Of course, in the Southern Hemisphere the Coriolis effect deflects the winds to the left, and therefore winds about a low are blowing clockwise, and winds around a high are moving counterclockwise. However, in whatever hemisphere, friction causes a net inflow **(convergence)** around a cyclone and a net outflow **(divergence)** around an anticyclone.

WEATHER GENERALIZATIONS ABOUT HIGHS AND LOWS

Rising air is associated with cloudy conditions and precipitation, whereas subsidence produces adiabatic heating and clearing conditions. In this section we will learn how the movement of air can itself create pressure change and hence generate winds. Upon doing so, we will examine the interrelationship between horizontal and vertical flow, and their effects on the weather.

Let us first consider the situation around a surface low-pressure system where the air is spiraling inward. Here the net inward transport of air causes a shrinking of the area occupied by the air mass, a process which is termed *horizontal convergence.* Whenever air converges horizontally, it must pile up, that is, increase in height to allow for the decreased area it now occupies. This generates a "taller" and therefore heavier air column. Yet a surface low can exist only as long as the column of air exerts less pressure than that occurring in surrounding regions. We seem to have encountered a paradox—low-pressure centers cause a net accumulation of air, which increases their pressure. Consequently, a surface cyclone should quickly eradicate itself, in a manner not unlike what happens when a vacuum-packed can is opened.

In light of the preceding discussion, it should be apparent that for a surface low to exist for any appreciable time, compensation must occur at some layer aloft. For example, surface convergence could be maintained if divergence (spreading out) aloft occurred at a rate equal to the inflow below. Figure 14.9 shows diagrammatically the relationship between surface convergence (inflow) and divergence (outflow) aloft that is needed to maintain a low-pressure center. Note that surface convergence about a cyclone causes a net upward

movement. The rate of this vertical movement is quite slow, generally less than 1 kilometer per day. Nevertheless, since rising air cools adiabatically, cloudy conditions and precipitation are often associated with the passage of a low-pressure system. On occasion, divergence aloft may even exceed surface convergence, resulting in intensified surface inflow and increased vertical motion. For that reason, divergence aloft can intensify these storm centers as well as maintain them.

Just as with cyclones, anticyclones, which are associated with surface divergence, must also be maintained from above. The mass outflow near the surface is accompanied by convergence aloft and general subsidence of the air column (Figure 14.9). Since descending air is compressed and warmed, cloud formation and precipitation are unlikely in an anticyclone, and "fair" weather can usually be expected with the approach of a high.

For reasons which should now be obvious, it has been common practice to print on barometers intended for household use the words "stormy" at the low end and "fair" on the high end. By noting whether the pressure is rising, falling, or steady, we have a good indication of what the forthcoming

FIGURE 14.9
A. Cross-sectional view of a high. Highs, or anticyclones, are associated with descending air and diverging winds. As a result, clear skies and "fair" weather may be expected with the approach of such a system.
B. Cross-sectional view of a low. Converging winds and rising air are associated with a low, or cyclone. Consequently, clouds and rain are often associated with the passage of such a system.

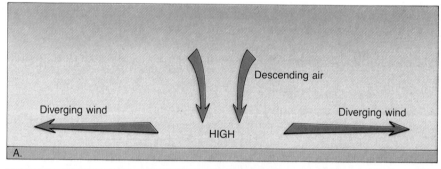

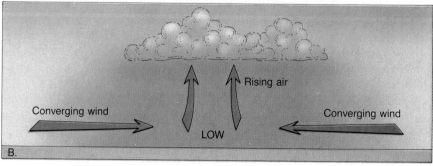

weather will be. Such a determination, called the **pressure,** or **barometric, tendency,** is a very useful aid in short-range weather prediction. The generalizations relating cyclones and anticyclones to the weather conditions just considered are stated rather poetically in the following proverb. Note that *glass* refers to the barometer.

When the glass falls low,
Prepare for a blow;
When it rises high,
Let all your kites fly.

In conclusion, you should now be better able to understand why local television weather broadcasters emphasize the positions and projected paths of cyclones and anticyclones. The "villain" on these weather programs is always the cyclone, which produces "bad" weather in any season. Lows move in roughly a west-to-east direction across the United States and require a few days to more than a week for the journey. Their paths can be somewhat erratic; thus accurate prediction of their migration is difficult, although essential, for short-range forecasting. Meteorologists must also determine if the flow aloft will intensify an embryo storm or act to suppress its development. Due to the close tie between conditions at the surface and those aloft, a great deal of emphasis has been placed on the importance and understanding of the total atmospheric circulation, particularly in the mid-latitudes. Once we have examined the workings of the general circulation, we will again consider the structure of the cyclone in light of these findings.

GENERAL CIRCULATION OF THE ATMOSPHERE

The underlying cause of wind is the unequal heating of the earth's surface. In tropical regions more solar radiation is received than is radiated back to space. In polar regions the opposite is true—less solar energy is received than is lost. Attempting to balance these differences, the atmosphere acts as a giant heat transfer system, moving warm air poleward and cool air equatorward. On a smaller scale, but for the same reason, ocean currents also

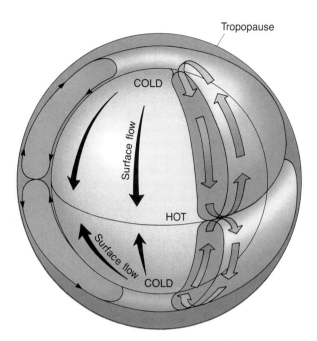

FIGURE 14.10
Global circulation on a nonrotating earth. A simple convection system is produced by unequal heating of the surface on a nonrotating planet.

contribute. The general circulation is very complex, and there is a great deal which has yet to be explained. We can, however, approximate its major components by first considering the circulation that would occur on a nonrotating earth having a uniform surface. We will then modify this system to fit observed patterns.

On a hypothetical nonrotating planet with a smooth surface of either all land or all water, two large thermally produced cells would form (Figure 14.10). The heated equatorial air would rise until it reached the tropopause, which, acting like a lid, would deflect the air poleward. Eventually, this upper-level airflow would reach the poles, sink, and spread out in all directions at the surface and move back toward the equator. Once there, it would be reheated and start its journey over again. This hypothetical circulation system has upper-level air flowing poleward and surface air flowing equatorward.

If we add the effect of rotation, this simple convection system will break down into smaller cells.

FIGURE 14.11
Idealized global circulation.

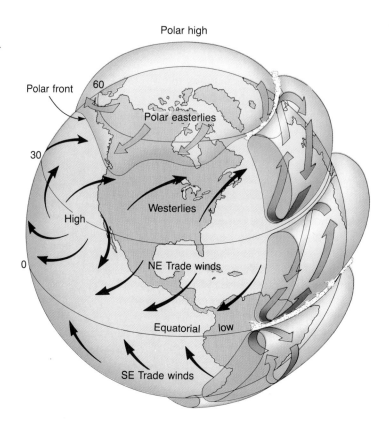

Figure 14.11, illustrates the three pairs of cells proposed to carry on the task of heat redistribution on a rotating planet. The polar and tropical cells retain the characteristics of the thermally generated convection described earlier. The nature of the mid-latitude circulation is complex and will be discussed in more detail in the next section.

Near the equator, the rising air is associated with the pressure zone known as the **equatorial low**—a region marked by abundant precipitation. As the upper-level flow from the equatorial low reaches 20–30 degrees latitude, north or south, it will have cooled enough to sink toward the surface. This subsidence and associated adiabatic heating produce the hot, arid regions in this latitude range. The center of this zone of subsiding dry air is the **subtropical high,** which encircles the globe near 30 degrees latitude (Figure 14.11). Located here are extensive arid and semiarid regions. The great deserts of Australia, Arabia, and North Africa, for example, are dry primarily because of the stable conditions associated with the subtropical highs. At the surface, airflow is outward from the center of this high-pressure system. Some of the air travels equatorward and is deflected by the Coriolis effect, producing the rather constant **trade winds.** The remainder travels poleward and is also deflected, generating the prevailing **westerlies** of the mid-latitudes. As the westerlies move poleward, they encounter the cool **polar easterlies** in the region of the **subpolar low.** The interaction of these warm and cool winds produces the stormy belt known as the **polar front.** The source region for the variable polar easterlies is the **polar high.** Here, polar air is subsiding and spreading equatorward.

In summary, this simplified global circulation is dominated by four pressure zones. The subtropical and polar highs are areas of dry subsiding air

FIGURE 14.12   ⟶
Average surface barometric pressure in millibars for
A. July and **B.** January with associated winds.

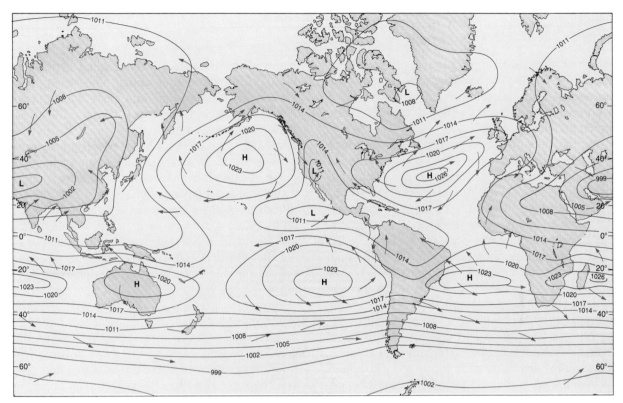

A. July

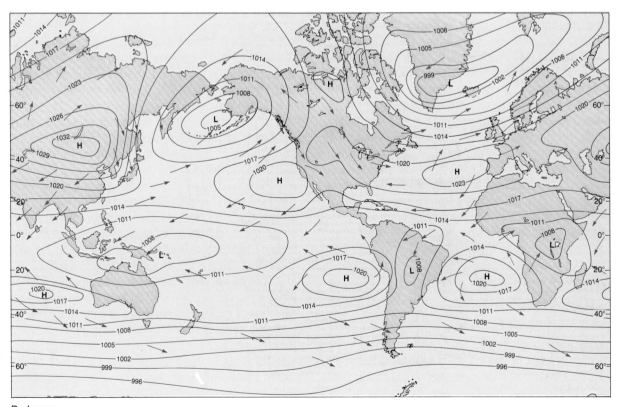

B. January

which flows outward at the surface, producing the prevailing winds. The low-pressure zones of the equatorial and subpolar regions are associated with inward and upward airflow accompanied by precipitation.

Up to this point, we have described the surface pressure and associated winds as continuous belts around the earth. The only true zonal distribution of pressure exists in the subpolar low of the Southern Hemisphere, where the ocean is continuous. At other locations, particularly in the Northern Hemisphere, where the bulk of the land exists, large seasonal temperature differences disrupt this zonal pattern. Figure 14.12, page 393, shows the resulting pressure and wind patterns for July and January. The circulation over the oceans is dominated by semipermanent cells of high pressure in the subtropics and cells of low pressure over the subpolar regions. The subtropical highs are responsible for the trade winds and westerlies, as mentioned earlier. The large landmasses, on the other hand, particularly Asia, become cold in the winter and develop a seasonal high-pressure system from which surface flow is directed off the land (Figure 14.12). In the summer, the opposite occurs; the landmasses are heated and develop a low-pressure cell, which permits air to flow onto the land. These seasonal changes in wind direction are known as the **monsoons.** During warm months, areas such as India experience a flow of warm, water-laden air from the Indian Ocean, which produces the rainy summer monsoon. The winter monsoon is dominated by dry continental air. A similar situation exists, but to a lesser extent, over North America. In summary, the general circulation is produced by semipermanent cells of high and low pressure over the oceans and is complicated by seasonal pressure changes over land.

CIRCULATION IN THE MID-LATITUDES

The circulation in the mid-latitudes, the zone of the westerlies, is complex and does not fit the convection system proposed for the tropics. Between 30 and 60 degrees latitude, the general west-to-east flow is interrupted by the migration of cyclones, low-pressure systems often associated with precipitation, and anticyclones, high-pressure systems associated with clear skies. Do not confuse these with the much larger semipermanent pressure systems that comprise the general circulation. In the Northern Hemisphere these cells move in an eastward direction around the globe, creating an anticyclonic (clockwise) flow or a cyclonic (counterclockwise) flow in their area of influence. A close correlation exists between the paths taken by these surface pressure systems and the position of the upper-level airflow, indicating that the upper air is responsible for directing the movement of cyclonic and anticyclonic systems.

Although a great deal is still unknown about the wavy upper-level flow of the westerlies, some of its basic features are understood with some degree of certainty. Among the most obvious features of the flow aloft are the seasonal changes. The change in wind speed is reflected on upper-air charts by more closely spaced contour lines in the cool season. The seasonal fluctuation of wind speeds is a consequence of the seasonal variation of the temperature gradient. The steep temperature gradient across the middle latitudes in the winter months corresponds to stronger flow aloft. In addition, the polar jet stream fluctuates seasonally such that its mean position migrates southward with the approach of winter and northward as summer nears. By midwinter, the jet core may penetrate as far south as central Florida. Since the paths of cyclonic systems are guided by the flow aloft, we can expect the southern tier of states to experience most of their severe storms in the winter season. During the hot summer months, the storm track is across the northern states, and some cyclones never leave Canada. The northerly storm track associated with summer also applies to Pacific storms, which move toward Alaska during the warm months, thus producing a rather long dry season for much of our west coast. The number of cyclones generated is seasonal as well, with the largest number occurring in the cooler months when the temperature gradients are greatest. This fact is in agreement with the role of cyclonic storms in the distribution of heat across the mid-latitudes.

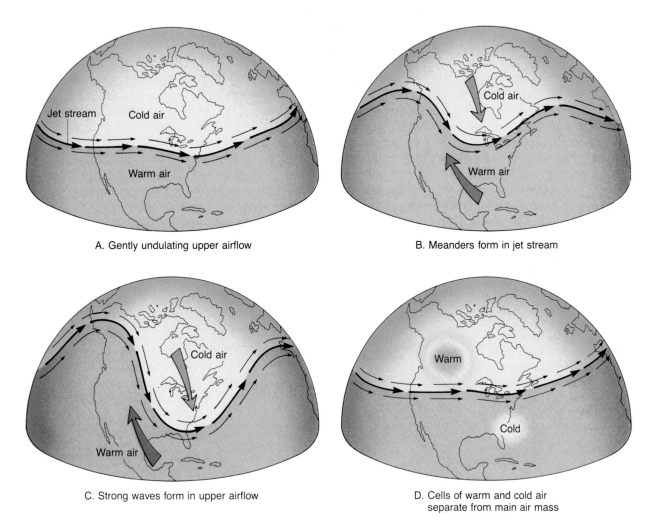

A. Gently undulating upper airflow

B. Meanders form in jet stream

C. Strong waves form in upper airflow

D. Cells of warm and cold air
separate from main air mass

FIGURE 14.13
Cyclic changes that occur in the upper-level airflow of the westerlies. The flow,
which has the jet stream as its axis, starts out nearly straight, then develops
meanders, which are eventually cut off. (After J. Namias, NOAA)

However, even in the cool season, the westerly
flow goes through an irregular cyclic change.
There may be periods of a week or more when the
flow is nearly west to east. Under these conditions,
relatively mild temperatures occur, and few disturbances are experienced in the region south of
the jet stream (Figure 14.13A). Then, without
warning, the upper flow begins to meander, producing large-amplitude waves and a general
north-to-south flow (Figure 14.13B, C, and D). This

change allows for an influx of cold air southward,
which intensifies the temperature gradient and
the flow aloft. During these periods, cyclonic activity dominates the weather picture. For a week or
more, the cyclonic storms redistribute large quantities of heat across the mid-latitudes by moving
cold air southward and warm air northward. This
redistribution eventually results in a weakened
temperature gradient and a return to a flatter flow
aloft and less intense weather at the surface.

These cycles, consisting of alternating periods of calm and stormy weather, can last from one to six weeks.

Because of the rather irregular behavior of the circulation patterns of the flow aloft, long-range weather prediction still remains essentially beyond the forecaster's reach. Nonetheless, numerous attempts are being made to more accurately predict changes in the upper-level flow on a long-term basis. It is hoped that this research will answer such questions as, "Will next winter be colder than normal?" and, "Will California experience a drought next year?"

LOCAL WINDS

SEA AND LAND BREEZES

In coastal areas during the warm summer months, the land heats more intensely than the adjacent body of water (see Chapter 12). As a result, the air above the land surface heats and expands, creating an area of low pressure. A **sea breeze** then develops, blowing from the water (higher pressure) toward the land (lower pressure) (Figure 14.14). The sea breeze begins to develop shortly before noon and generally reaches its greatest intensity during the mid- to late afternoon. These

FIGURE 14.14
Sea and land breezes. During the daytime, land heats more intensely than water. A low is created over the land, and the air moves from the water (higher pressure) to the land (lower pressure). At night, the land cools more rapidly, resulting in higher pressure over land and a reversal of the wind.

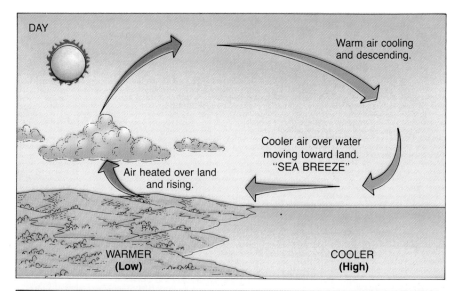

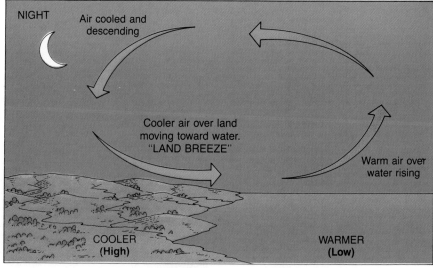

cool winds are a significant moderating influence in coastal areas. At night, the land cools rapidly, eventually resulting in higher pressure. As a consequence, a **land breeze** may develop (Figure 14.14). Small-scale sea breezes can also develop along the shores of large lakes. People who live in a city near the Great Lakes, such as Chicago, recognize the lake effect, especially in the summer. These people are reminded daily by weather reports of the cool temperatures near the lake as compared to warmer outlying areas.

VALLEY AND MOUNTAIN BREEZES

Valley and mountain breezes are created for much the same reason as land and sea breezes, namely differential heating. On warm, sunny days, the floor and slopes of a valley are intensely heated. The heated air rises up the sides of the valley, creating a **valley breeze.** At night, on the other hand, the air in close proximity to the sides of the valley cools. The cooler air drains downslope into the valley, creating a **mountain breeze.**

CHINOOK AND SANTA ANA WINDS

Warm, dry winds are common on the eastern slopes of the Rockies, where they are called **chinooks.** Such winds are created when air descends the leeward side of a mountain and warms by compression. Since condensation may have occurred as the air ascended the windward side, releasing latent heat, the air descending the leeward slope will be warmer and drier than it was at a similar elevation on the windward side. Although the temperature of these winds is generally less than 10°C (50°F), which is not particularly warm, they generally occur in the winter and spring when the affected area may be experiencing below-freezing temperatures. Thus, by comparison, these dry, warm winds often bring a drastic change. When the ground has a snow cover, these winds are known to melt it in short order. The word *chinook* literally means "snow-eater."

Another chinooklike wind that occurs in the United States is the **Santa Ana.** Found in southern California, these hot, desiccating winds greatly increase the threat of fire in this already dry area.

WIND MEASUREMENT

Two basic wind measurements, direction and speed, are particularly significant to the weather observer. Winds are always labeled by the direction *from* which they blow. A north wind blows from the north toward the south, an east wind from the east toward the west. The instrument most commonly used to determine wind direction is the **wind vane** (Figure 14.15). This instrument,

FIGURE 14.15
Wind vane and cup anemometer. (Courtesy of Qualimetrics, Inc.)

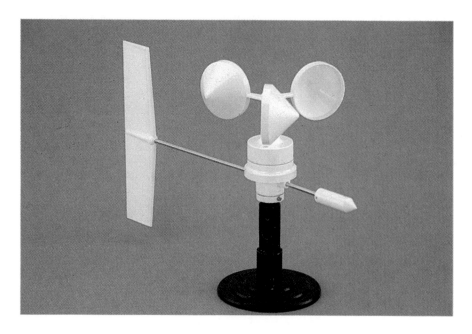

which is a common sight on many buildings, always points into the wind. Often the wind direction is shown on a dial that is connected to the wind vane (Figure 14.16). The dial will indicate the direction of the wind either by points of the compass, that is N, NE, E, SE, etc., or by a 0–360-degree scale. On the latter scale, 0 degrees or 360 degrees are both north, 90 degrees is east, 180 degrees is south, and 270 degrees is west. When the wind consistently blows more often from one direction than from any other, it is termed a **prevailing wind.** Wind speed is commonly measured using a **cup anemometer** (Figure 14.15). The wind speed is read from a dial much like the speedometer of an automobile.

By knowing the location of cyclones and anticyclones in relation to where you are, you can predict the changes in wind direction that will be experienced as the pressure center moves past. Since changes in wind direction often bring changes in temperature and moisture conditions, the ability to predict the winds can be very useful. In the Midwest, for example, a north wind may bring cool, dry air from Canada, while a south wind may bring warm, humid air from the Gulf of Mexico. Sir Francis Bacon summed it up nicely when he wrote, "Every wind has its weather."

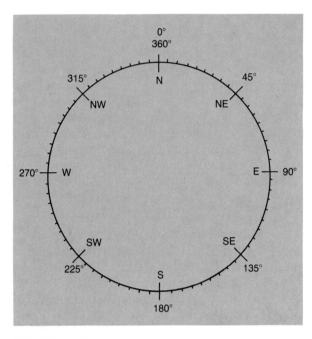

FIGURE 14.16
Wind direction. Wind direction may be expressed using the points of the compass or a scale of 0–360 degrees. Winds are always labeled according to the direction *from* which they are blowing.

REVIEW QUESTIONS

1   What is standard sea level pressure in millibars? In inches of mercury? In pounds per square inch?

2   Mercury is 13 times heavier than water. If you built a barometer using water rather than mercury, how tall would it have to be in order to record standard sea level pressure (in centimeters of water)?

3   If a very flexible balloon rose to 31.2 kilometers, how many times larger would it be because of expansion than it was at sea level? (See Table 12.2, p. 335.)

4   Why do your ears "pop" when you quickly descend a steep hill?

5   What force is responsible for generating wind?

6   How does the Coriolis effect modify air movement?

7   Contrast surface winds and upper-air winds in terms of speed and direction.

8   Describe the weather that usually accompanies a drop in barometric pressure. A rise.

9   Draw a diagram showing the winds associated with cyclones and anticyclones in the Southern Hemisphere.

10 If you live in the Northern Hemisphere and are directly west of a cyclone, what most probably will be the wind direction? West of an anticyclone?

11 Which global pressure system produces the tropical deserts of the world?

12 Using Figures 14.9A and 14.11, explain why some describe the polar regions as deserts.

13 Which global pressure systems are associated with precipitation (Figure 14.11)?

14 Describe the jet stream.

15 What influence does upper-level airflow seem to have on surface pressure systems?

16 If unequal heating produces pressure differences, which in turn drive the wind, why does the wind blow at night?

17 Describe the monsoon circulation of India (Figure 14.12).

18 Why are sea breezes in the mid-latitudes most pronounced during the summer months?

19 A northeast wind is blowing *from* the _____ (direction) *toward* the _____ (direction).

20 The wind direction is 225 degrees. From what compass direction is the wind blowing (Figure 14.16)?

KEY TERMS

mercurial barometer	convergence	monsoon
aneroid barometer	divergence	sea breeze
barograph	pressure (barometric) tendency	land breeze
altimeter		valley breeze
wind	equatorial low	mountain breeze
isobar	subtropical high	chinook
pressure gradient	trade winds	Santa Ana
Coriolis effect	westerlies	wind vane
geostrophic winds	polar easterlies	prevailing wind
jet stream	subpolar low	cup anemometer
cyclone (low)	polar front	
anticyclone (high)	polar high	

15

WEATHER PATTERNS AND SEVERE STORMS

Tornadoes and hurricanes rank high among nature's most destructive forces. Each spring, newspapers report the death and destruction left in the wake of a "band" of tornadoes. During the summer, we begin hearing the names of hurricanes in the news—Allen, Betty, Camille, Doria, Frederic—the list goes on and on. Thunderstorms, although less intense and more common than tornadoes and hurricanes, will also be part of our discussion in this chapter on the nature of severe weather disturbances. Before looking at violent weather, however, we shall study those atmospheric phenomena that most often affect our day-to-day weather: air masses, fronts, and traveling middle-latitude cyclones. Here we shall see the interplay of the elements of weather discussed earlier.

Storm clouds near sunset. (Photo by E. J. Tarbuck)

AIR MASSES

When a portion of the lower troposphere moves slowly or stagnates over a relatively uniform surface, the air will assume the distinguishing features of that area, particularly with regard to temperature and moisture conditions. A body of air that forms in this fashion is termed an air mass. An **air mass** is a very large body of air, usually 1600 kilometers (1000 miles) or more across, and perhaps several kilometers thick, which is characterized by a homogeneity of temperature and moisture at any given altitude. When this air moves out of its region of origin, it will carry these temperatures and moisture conditions elsewhere, eventually affecting a large portion of a continent.

An air mass may require several days to traverse an area. As a result, the region under its influence will probably experience fairly constant weather, a situation called **air-mass weather.** Certainly there may be some day-to-day variations, but the events will be quite unlike those in an adjacent air mass. For that reason the boundary between two adjoining air masses having contrasting characteristics, called a *front,* marks a change in weather.

SOURCE REGIONS

The area where an air mass acquires its characteristic properties of temperature and moisture is called its **source region.** The source regions that produce air masses which influence North America most are shown in Figure 15.1.

Air masses are classified according to their source region. **Polar (P)** air masses originate in high latitudes, whereas those that form in low latitudes are called **tropical (T).** The designation polar or tropical gives an indication of the temperature characteristics of the air mass. *Polar* indicates cold, and *tropical* indicates warm. In addition, air masses are classified according to the nature of the surface in the source region. **Continental (c)** designates land, and **maritime (m)** indicates water. The designation *continental* or *maritime* thus suggests the moisture characteristics of the air mass. Continental air is likely to be dry and maritime air humid. The four basic types of air

masses according to this scheme of classification are continental polar (cP), continental tropical (cT), maritime polar (mP), and maritime tropical (mT).

WEATHER ASSOCIATED WITH AIR MASSES

Continental polar and maritime tropical air masses influence the weather of North America most, especially east of the Rocky Mountains. Continental polar air masses originate in northern Canada, interior Alaska, and the Arctic—areas that are uniformly cold and dry in winter and cool and dry in summer. These air masses are usually associated with high pressure and typically have few clouds. In winter, an invasion of continental polar air brings the clear skies and cold temperatures we associate with a cold wave as it moves southward through Canada into the United States. In summer, this air mass may bring a few days of cooling relief.

Although cP air masses are not normally associated with heavy precipitation, those that cross the Great Lakes in winter sometimes bring snow to the leeward shores. As continental polar air crosses a lake in winter, the air is heated and acquires moisture from the relatively warm lake surface. By the time it reaches the southern and eastern shore, the cP air is uncharacteristically humid and unstable. The result can be heavy "lake effect" snow showers.

Maritime tropical air masses affecting North America most often originate in the Gulf of Mexico, the Caribbean Sea, or the adjacent Atlantic Ocean. As we might expect, these air masses are warm, moisture-laden, and usually unstable. Maritime tropical air is the source of much, if not most, of the precipitation received in the eastern two-thirds of the United States. In summer, when this air mass invades the central and eastern United States, and occasionally southern Canada, it brings the heat and oppressive humidity typically associated with its source region. The tropical Pacific is also a source region for maritime tropical air. However, this maritime tropical air seldom enters the continent. When it does, it may

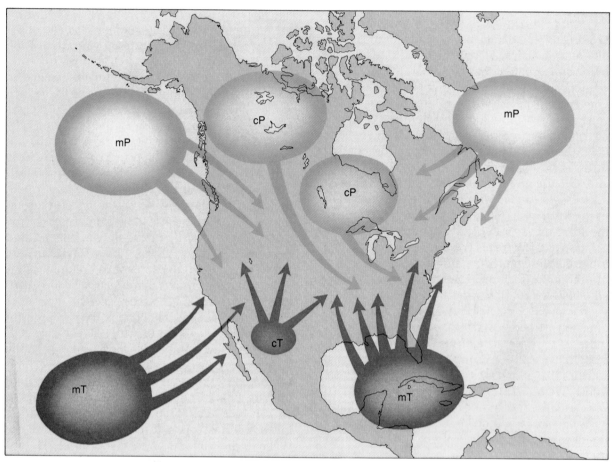

FIGURE 15.1
Air masses are classified on the basis of their source region. The designation
continental (c) or maritime (m) gives an indication of moisture content, whereas
polar (P) and tropical (T) indicate temperature conditions.

bring orographic rainfall to the southwestern
United States and northern Mexico.

Of the two remaining air masses, maritime
polar and continental tropical, the latter has the
least influence upon the weather of North America. Hot, dry continental tropical air, originating in
the Southwest and Mexico during the summer,
seldom affects the weather outside its source
region.

During the winter, maritime polar air masses
coming from the North Pacific often originate as
continental polar air masses in Siberia. As they
move across the Pacific, they warm and gradually
accumulate moisture. Upon entering North America, these air masses drop much of their moisture
because of orographic lifting in the western mountains. Maritime polar air also originates off the
coast of eastern Canada and occasionally influences the weather of the northeastern United
States. When New England is on the northern or
northwestern edge of a passing low, the cyclonic
winds draw in maritime polar air. The result is a
storm characterized by snow and cold temperatures, known locally as a *northeaster*.

FRONTS

Fronts are boundaries that separate air masses of different densities, one warmer and often higher in moisture content than the other. Ideally, fronts can form between any two contrasting air masses. Considering the vast size of the air masses involved, these 15- to 200-kilometer-wide bands of discontinuity are relatively narrow. On the scale of a weather map, they are generally narrow enough to be represented satisfactorily by a broad line.

Above the ground, the frontal surface slopes at a low angle so that warmer air overlies cooler air. In the ideal case, the air masses on both sides of the front move in the same direction and at the same speed. Under this condition, the front acts as a barrier with which the air masses must move, but through which they cannot penetrate. Generally, however, the pressure field across a front is such that air on one side is moving faster in the direction perpendicular to the front than the air mass on the other side of the front. Thus, one air mass actively advances into another and "clashes" with it. The boundaries were thus tagged *fronts* during World War I by Norwegian meteorologists, who visualized them as analogous to battle lines.

As one air mass moves into another, some mixing does occur along the frontal surface, but for the most part the air masses retain their identity as one air mass is displaced upward over the other. No matter which air mass is advancing, it is always the warmer, less dense air that is forced aloft, while the cooler, denser air acts as the wedge upon which lifting takes place.

WARM FRONTS

When the surface position of a front moves so that warm air occupies territory formerly covered by cooler air, it is called a **warm front.** On a weather map, the surface position of a warm front is denoted by a line with semicircles extending into the cooler air. East of the Rockies, warm tropical air often enters the United States from the Gulf of Mexico and overruns receding cool air. As the cold wedge retreats, friction slows the advance of the surface position of the front more so than its position aloft; for this reason, the boundary separating these air masses acquires a small slope. The aver-age slope of a warm front is about 1:200, which means that if you are 200 kilometers ahead of the surface location of a warm front, you will find the frontal surface at a height of 1 kilometer.

As warm air ascends the retreating wedge of cold air, it cools by adiabatic expansion to produce clouds and frequently, precipitation. The sequence of clouds shown in Figure 15.2A typically precedes a warm front. The first sign of the approach of a warm front is the appearance of cirrus clouds overhead. These high clouds form 1000 kilometers or more ahead of the surface front where the overrunning warm air has ascended high up the wedge of cold air. As the front nears, cirrus clouds grade into cirrostratus, which blend into denser sheets of altostratus. About 300 kilometers ahead of the front, thicker stratus and nimbostratus clouds appear and rain or snow begins. Usually warm fronts produce several hours of moderate-to-gentle precipitation over a large region. This is in agreement with the relatively gentle slope of a warm front, which does not generally encourage convectional activity. However, on some occasions, warm fronts do produce cumulonimbus clouds and thunderstorm activity (Figure 15.2B). This occurs when the overrunning air is inherently unstable and the front is rather sharp. When these conditions exist, cirrus clouds are generally followed by cirrocumulus clouds, giving us the familiar "mackerel sky," which sailors saw as a warning of an impending storm, as indicated by the following proverb:

> Mackerel scales and mares' tails
> Make lofty ships carry low sails.

At the other extreme, a warm front associated with a dry air mass could pass practically unnoticed by those of us at the surface.

A gradual increase in temperature occurs with the passage of a warm front. As we would expect, the increase is most apparent when there is a large temperature difference between the adjacent air masses. Furthermore, a wind shift from the east to the southwest is often detectable. The reason for this shift will be evident later. The moisture content and stability of the encroaching warm air mass largely determine when clear skies will return. During the summer, cumulus, and occasion-

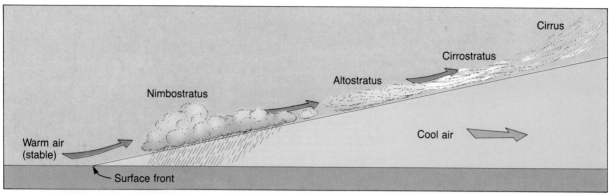

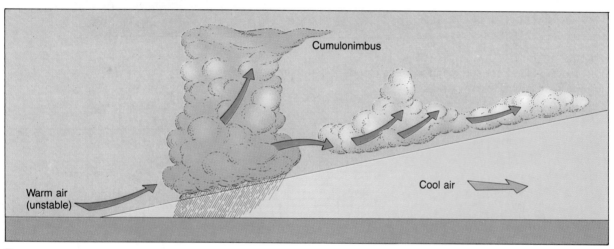

FIGURE 15.2
A. Warm front with stable air and associated stratiform clouds. Precipitation is moderate and occurs within a few hundred kilometers of the surface front.
B. Warm front with unstable air and cumuliform clouds. Precipitation is heavy near the surface front.

ally cumulonimbus, are embedded in the warm unstable air mass that follows the front. Precipitation from these clouds is usually sporadic and not extensive.

COLD FRONTS

When cold air is actively advancing into a region occupied by warmer air, the zone of discontinuity is called a **cold front** (Figure 15.3). As with warm fronts, friction tends to slow the surface position of

a cold front more so than its position aloft. However, because of the relative positions of the adjacent air masses, the cold front steepens as it moves. On the average, cold fronts are about twice as steep as warm fronts, having a slope of perhaps 1:100. In addition, cold fronts advance more rapidly than warm fronts. These two differences—rate of movement and steepness of slope—largely account for the more violent nature of cold-front weather. The displacement of air along a cold front is often rapid enough that the released latent heat

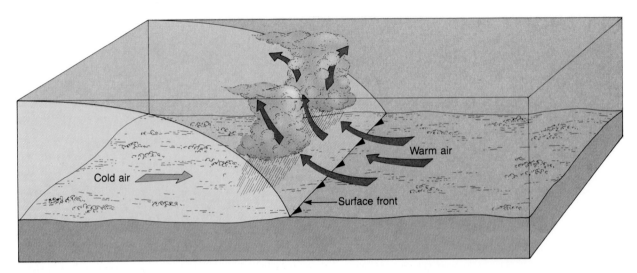

FIGURE 15.3
Fast-moving cold front and cumulonimbus clouds. Often thunderstorms occur if
the warm air is unstable.

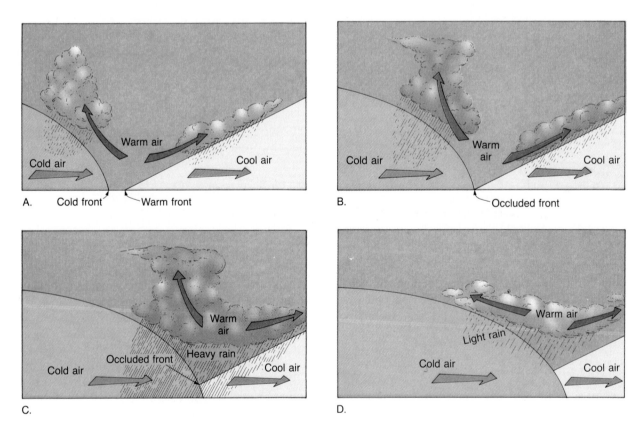

FIGURE 15.4
Stages in the formation and eventual dissipation of an occluded front.

appreciably increases the air's buoyancy. This frequently results in the sudden downpours and vigorous gusts of wind associated with mature cumulonimbus clouds. Since a cold front produces roughly the same amount of lifting as a warm front, but over a shorter distance, the intensity of precipitation is greater, but the duration is shorter.

A cold front is sometimes preceded by altocumulus clouds. As the front approaches, generally from the west or northwest, towering clouds are often seen in the distance. Near the front, a dark band of ominous clouds foretells the ensuing weather. Usually a marked temperature drop and a wind shift from the south to west or northwest accompany the passage of the front. On a weather map, the sometimes violent weather and sharp temperature contrast are indicated by a line with triangle-shaped points that extend into the warmer air mass.

The weather behind a cold front is dominated by a subsiding and relatively cold air mass. Hence, clearing conditions prevail after the front passes. Although general subsidence causes some adiabatic heating, this has a minor effect on surface temperatures. In the winter, the clear skies associated with these cold outbreaks further reduce surface temperatures because of more rapid radiation cooling at night. If the continental polar air mass, which most frequently accompanies a cold front, moves into a relatively warm and humid area, surface heating can produce shallow convection, which in turn may generate low cumulus or stratocumulus clouds behind the front.

OCCLUDED FRONTS

Another commonly occurring front is the **occluded front.** Here an active cold front overtakes a warm front (Figure 15.4). As the advancing cold air wedges the warm front upward, a new front emerges between the advancing cold air and the air over which the warm front is gliding. The weather of an occluded front generally is quite complex. Most of the precipitation is associated with the warm air being forced aloft. However, when conditions are suitable, the newly formed front can produce precipitation of its own.

THE MIDDLE-LATITUDE CYCLONE

The cold fronts and warm fronts just described continually influence the weather in the mid-latitudes. These fronts are usually associated with a low-pressure system called a **middle-latitude,** or **wave, cyclone.**

LIFE CYCLE OF A WAVE CYCLONE

Wave cyclones form along fronts where they change in a somewhat predictable way. This life cycle can last for a few hours or for several days. Figure 15.5, page 408, is a schematic representation of the stages in the development of a "typical" wave cyclone. As the figure shows, cyclones originate along a front where air masses of different densities (temperatures) are moving parallel to the front in opposite directions. In the classic model, continental polar air associated with the polar easterlies would be north of the front, and maritime tropical air of the westerlies south of the front. The result of this opposing airflow is counterclockwise (cyclonic) rotation. (To better visualize this effect, place a pencil between the palms of your hands. Now move your right hand ahead of your left hand and notice that your pencil rotates in a counterclockwise fashion.) Under the correct conditions, the frontal surface will take on a wave shape. These waves are analogous to the waves produced on the surface of a water body by moving air, except the scale is different. The waves generated between two contrasting air masses are usually several hundred kilometers long. Some waves tend to dampen out while others become unstable and grow in amplitude. The latter ones change in shape with time much as a gentle ocean swell does as it moves into shallow water and becomes a tall breaking wave.

Once a small wave forms, warm air invades this weak spot along the front and extends itself poleward, while the surrounding cold air moves equatorward. This change causes a readjustment in the pressure field, which results in nearly circular isobars with the low pressure centered at the apex of the wave. The creation of the low encourages the inflow (convergence) of air and general vertical lifting, particularly where warm air is overrunning

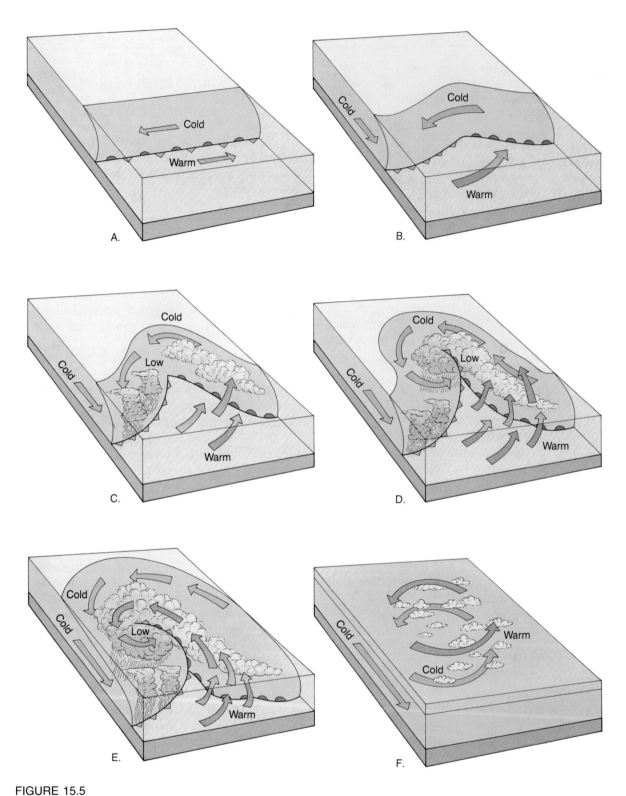

FIGURE 15.5
Schematic representation (map view) of the life cycle of a hypothetical middle-
latitude cyclone. **A.** Front develops. **B.** Wave appears along front. **C.** Cyclonic
circulation is well developed. **D.** Occlusion begins. **E.** Occluded front is fully developed. **F.** Cyclone dissipates. (As proposed by J. Bjerknes)

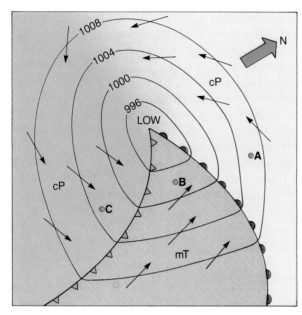

FIGURE 15.6
Idealized circulation of a mature cyclone. Study this figure to determine the wind shifts that would be expected for a city as the storm moved so that the city's relative position shifted from location A to B, and then from B to C.

colder air. We can see in Figure 15.6 that the air in the warm sector is flowing from the southwest toward colder air flowing from the southeast. Since the warm air is moving faster than the cold air in a direction perpendicular to this front, we can conclude that warm air is invading a region formerly occupied by cold air; hence, this must be a warm front. Similar reasoning indicates that in the rear of the cyclonic disturbance, cold air is underrunning the air of the warm sector, generating a cold front there. Generally, the position of the cold front advances faster than the warm front and begins to close the warm sector as shown in Figure 15.5. This process, called **occlusion,** results in an occluded front with the displaced warm sector located aloft. The cyclone enters maturity (maximum intensity) when it reaches this stage in its development. A steep pressure gradient and strong winds develop as lifting continues. Eventually all of the warm sector is forced aloft and cold air surrounds the cyclone at low levels. Once the sloping discontinuity (front) between the air

masses no longer exists, the pressure gradient weakens. At this point, the cyclone has exhausted its source of energy, and the storm comes to an end.

IDEALIZED WEATHER OF A WAVE CYCLONE

The wave cyclone model provides a useful tool for examining the weather patterns of the middle latitudes. Figure 15.7, page 410, illustrates the distribution of clouds and thus the regions of possible precipitation associated with a mature wave cyclone. Guided by the westerlies aloft, cyclones generally move eastward across the United States, so we can expect the first signs of their arrival in the west. However, often in the region of the Mississippi valley, cyclones begin a more northeasterly trajectory and occasionally move directly northward. A mid-latitude cyclone typically requires two to four days to pass over a given region. During that short time, rather abrupt changes in atmospheric conditions may be experienced. This is particularly true in the spring, when the largest temperature contrasts occur across the mid-latitudes.

Using Figure 15.7 as a guide, we will now consider these weather producers and what we should expect from them as they pass an area during the spring. To facilitate our discussion, profiles are provided along lines A–E and F–G. First, imagine the change in weather as you move along profile A–E. At point A the sighting of high cirrus clouds would be the first sign of the approaching cyclone. These high clouds can precede the surface front by 1000 kilometers or more and they generally will be accompanied by falling pressure. As the warm front advances, a lowering and thickening of the cloud deck is noticed. Usually within 12 to 24 hours after the first sighting of cirrus clouds, light precipitation begins (point B). As the front nears, the rate of precipitation increases, a rise in temperature is noticed, and winds begin to change from an easterly to a southerly flow. With the passage of the warm front, the area is under the influence of the maritime tropical air mass of the warm sector (point C). Generally the region affected by this sector of the cyclone experiences warm temperatures, southerly winds, and generally clear skies, although fair-weather cumulus or altocumulus are not uncommon here. The rather pleasant weather of the warm sector passes quickly and is replaced by gusty winds and precipitation generated along the cold front. The approach of a rapidly advancing cold front is marked by a wall of dark clouds (point D). Severe weather accompanied by heavy precipitation, hail, and an occasional tornado is a definite possibility at this time of year. The passage of the cold front is easily detected by a wind shift; the southerly flow is replaced by winds from the west to northwest and by a pronounced drop in temperature. Also, rising

pressure hints of the subsiding cool, dry air behind the front. Once the front passes, the skies clear quickly as the cooler air invades the region (point E). Usually a day or two of almost cloudless deep-blue skies can be expected unless another cyclone is edging into the region.

A very different set of weather conditions will prevail in those regions that encounter the portion of the cyclone containing the occluded front as shown along profile F–G. Here the temperatures remain cool during the passage of the storm; however, a continual drop in pressure and increas-

FIGURE 15.7
Distribution of clouds associated with an idealized mature wave cyclone.

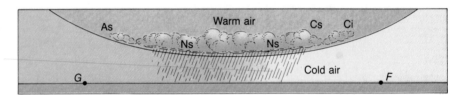
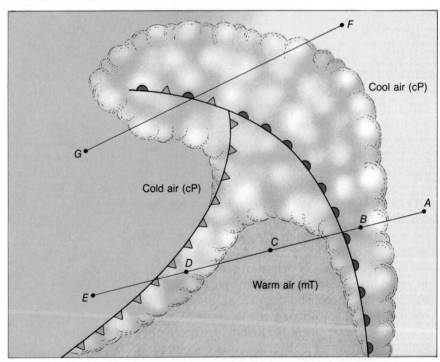
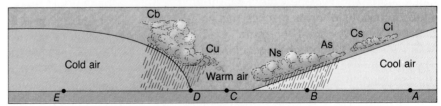

ingly overcast conditions strongly hint at the approach of the low-pressure center. This sector of the cyclone most often generates snow or icing storms during the cool months. Further, the occluded front often moves more slowly than the other fronts; hence, the wishbone-shaped frontal structure shown in Figure 15.7 rotates in a counterclockwise manner, such that the occluded front appears to "bend over backwards." This effect adds to the misery of the region influenced by the occluded front since it remains over the area longer than the other fronts. The storm also reaches its greatest intensity during occlusion; consequently, the area affected by the developing occluded front receives the brunt of the storm's fury.

CYCLONE FORMATION

When the earliest studies of cyclones were made, little was known about the nature of the airflow in the middle and upper troposphere. Since then a close relationship between surface disturbances and the flow aloft has been established. Whenever the flow aloft is relatively straight, that is, from west to east, very little cyclonic activity occurs at the surface. However, when the upper air begins to meander widely in a north-to-south direction, high-amplitude waves consisting of alternating troughs and ridges are produced, and surface cyclonic activity intensifies.

Before we discuss how cyclone formation is believed to be aided by the flow aloft, let us first review the nature of cyclonic and anticyclonic winds. Recall that the airflow about a cyclone (low) is inward. This leads to mass convergence (coming together). The resulting accumulation of air must be accompanied by a corresponding increase in surface pressure. Consequently, we might expect a low-pressure system to "fill" rapidly and be eliminated, just as the vacuum in a coffee can is quickly dissipated when we open it. However, this does not occur. On the contrary, cyclones often exist for extended periods. In order for this to happen, surface convergence must be offset by a mass outflow at some higher level (see Figure 14.9). As long as divergence (spreading out) aloft is equal to, or greater than, the surface

inflow, the low pressure and its accompanying convergence can be sustained.

Because cyclones are bearers of stormy weather, they have received far more attention than anticyclones. Nevertheless, a close relationship exists, which makes it difficult to separate any discussion of these two pressure systems. The surface air that feeds a cyclone, for example, generally originates as air flowing out of an anticyclone. As a rule, cyclones and anticyclones are found adjacent to one another. Like the cyclone, an anticyclone depends on the flow far above to maintain its circulation. In this instance, divergence at the surface is balanced by convergence aloft and general subsidence of the air column (see Figure 14.9).

We have seen that the flow aloft is necessary to maintain cyclonic and anticyclonic circulation, but perhaps more often than not, these surface wind systems are actually generated by upper-level flow (Figure 15.8, page 412). In cyclone development, the role of upper-level divergence is most significant. Frequently associated with the jet stream, upper-air divergence creates an environment analogous to a partial vacuum, which initiates upward flow. The fall in surface pressure which accompanies the outflow aloft will induce inward flow at the surface. The Coriolis effect will then come into play to produce the curved flow pattern indicative of a cyclone.

Divergence aloft does not involve the outward, clockwise movement of air that occurs about a surface anticyclone. Instead, the flow aloft is nearly geostrophic. Its path is from west to east and generally has sweeping curves. One of the mechanisms responsible for the mass transport of air high above the earth's surface is a phenomenon known as **speed divergence.** It has been known for some time that the wind speeds along the axis of the jet stream are not constant. Some regions experience much higher wind speeds than others. Upon entering a zone of maximum wind velocity, air accelerates and therefore diverges. Conversely, when air leaves a zone of maximum wind velocity, a pile up (convergence) results. An analogous situation occurs on a tollway in the region between toll stations. Upon exiting one toll booth and entering the zone of maximum speed, the automobiles diverge. As the automobiles slow

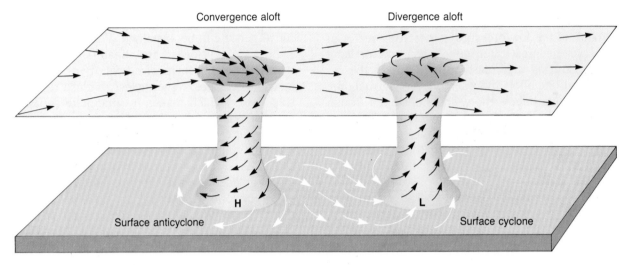

FIGURE 15.8
Illustration depicting the formation of cyclonic and anticyclonic flow at the surface when induced by divergence and convergence aloft.

to pay the next toll, they converge. Similarly, in the zone upstream from a region of maximum speed, the air experiences speed divergence, while downstream, mass convergence occurs. Other factors, for example, the spreading out of the air stream, can also contribute to divergence aloft, while confluence of an air stream under correct conditions can cause mass convergence.

In conclusion, we find that divergence aloft initiates upward air movement, reduced surface pressure, and cyclonic flow. On the other hand, convergence along the jet stream results in general subsidence of the air column, increased surface pressure, and anticyclonic surface winds.

THUNDERSTORMS

Most everyone has observed a small-scale phenomenon that is caused by the vertical motion of warm, unstable air. Perhaps you have seen a small dust devil which formed over an open field on a hot day when it whirled its dusty load to great heights. Or perhaps you have noticed a bird glide skyward effortlessly upon an invisible thermal of hot air. These examples serve to illustrate the much more dynamic thermal instability that occurs during the development of a **thunderstorm.** At any given time over the earth's surface, nearly

2000 thunderstorms are in progress, mainly in tropical regions. Thunderstorm activity is associated with cumulonimbus clouds that generate heavy rainfall, thunder, lightning, and occasionally hail.

The greatest number of thunderstorms occur in association with relatively short-lived cumulonimbus clouds that produce local precipitation. In the United States these cells typically form within warm, humid (maritime tropical) air masses that originate over the Gulf of Mexico and migrate northward. Occasionally, thunderstorms grow very large and remain active for hours. These *severe thunderstorms* produce frequent lightning and are accompanied by locally damaging winds or hail. Most severe thunderstorms in the middle latitudes form along or ahead of cold fronts. Here, forceful lifting of unstable mT air masses triggers thunderstorm development.

No matter what circumstances trigger thunderstorm development, all evolve in a similar manner. Each requires warm, moist air that when lifted releases latent heat, becomes unstable, and rises. As might be expected, the greatest thunderstorm activity in the United States occurs in the Southeast, in close proximity to the source region for maritime tropical air masses (Figure 15.9). Because instability and buoyancy are enhanced

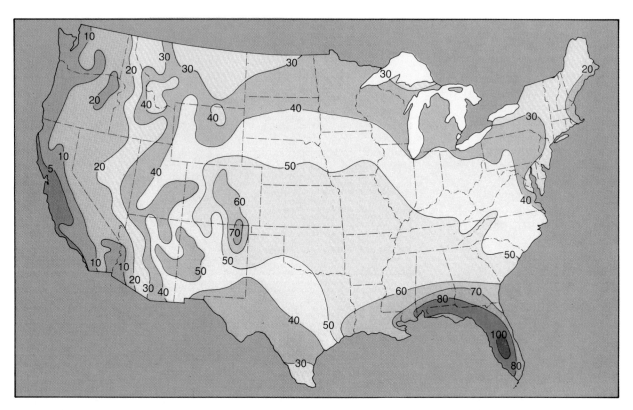

FIGURE 15.9
Average number of days per year with thunderstorms. Due to its close proximity to the source region for warm, humid, and unstable air masses, the Gulf Coast receives much of its precipitation from thunderstorms. (Source: Environmental Data Service, NOAA)

by high surface temperatures, thunderstorms are most common in the afternoon and early evening. However, surface heating is not sufficient for the growth of towering cumulonimbus clouds. A solitary cell of rising hot air produced by surface heating could, at best, produce a small cumulus cloud, which would evaporate within 10–15 minutes. The development of 12,000-meter (or on rare occasions 18,000-meter) cumulonimbus towers requires a continual supply of moist air. Each new surge of warm air rises higher than the last, adding to the height of the cloud (Figure 15.10, page 414). These updrafts must occasionally reach speeds over 100 kilometers (62 miles) per hour to accommodate the size of hailstones they are capable of carrying upward. Usually within an hour the amount and size of precipitation that has accumu-

lated is too much for the updrafts to support, and in one part of the cloud downdrafts develop, releasing heavy precipitation. This represents the most active stage of the thunderstorm. Gusty winds, lightning, heavy precipitation, and sometimes hail, are experienced. Eventually downdrafts dominate throughout the cloud. The cooling effect of falling precipitation coupled with the influx of colder air aloft mark the end of the thunderstorm activity. The life span of a cumulonimbus cell within a thunderstorm complex is only about an hour, but as the storm moves, fresh supplies of warm, water-laden air generate new cells to replace those that are dissipating.

One obvious feature of a thunderstorm is thunder. Due to the fact that thunder is produced by lightning, lightning must also be present (Figure

FIGURE 15.10
Stages in the development of a thunderstorm. During the cumulus stage, strong updrafts act to build the storm. The mature stage is marked by heavy precipitation and cool downdrafts in part of the storm. When the warm updrafts disappear completely, precipitation becomes light, and the cloud begins to evaporate.

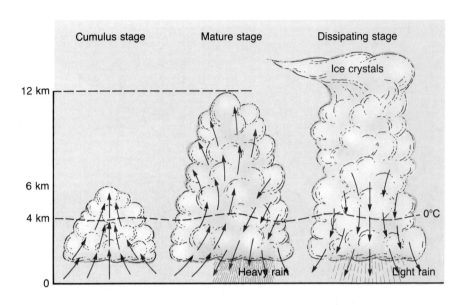

15.11 and the part-opening photo on pages 328–29). **Lightning** is similar to the electrical shock you may have experienced on a very dry day upon touching a metal object. Only the intensities differ. During the development of a cumulonimbus cloud, a similar buildup of charge is generated. The reason for this, although not fully understood, hinges on the movement of precipitation within the cloud. The upper portion of the cloud acquires a positive charge, while the lower portion of the cloud maintains an overall negative charge while containing small positively charged pockets.

These charges will build to millions, and even hundreds of millions, of volts before a lightning stroke acts to discharge the cloud by striking the earth, or possibly another cloud. The lightning we see as a single stroke is really several very rapid strokes between the cloud and the earth, which together last about only one-tenth of a second.

The electrical discharge of lightning rapidly heats the air, which, in turn, causes it to expand explosively. We hear this expansion as **thunder.** Since lightning and thunder occur simultane-

FIGURE 15.11
Lightning. (Courtesy of National Center for Atmospheric Research/National Science Foundation)

ously, it is possible to estimate the distance to the stroke. Lightning is seen instantaneously, whereas the rather slow sound waves, which travel approximately 330 meters per second, reach us some time later. Therefore, if thunder is heard 5 seconds after the lightning is seen, the lightning occurred about 1650 meters away (approximately 1 mile).

The thunder we hear as a rumble is produced along a rather long lightning path located at some distance from the observer. The sound that originates along the path nearest the observer arrives before the sound that originates farthest away. This lengthens the duration of the thunder. Reflection of the sound waves further delays their arrival and adds to this effect. When lightning occurs more than 20 kilometers away, thunder is rarely heard. This type of lightning, popularly called *heat lightning,* is no different from that which we associate with thunder.

Since lightning moves easily through good electrical conductors, it is wise during thunderstorms to avoid touching or being close to any metallic or tall structure such as an antenna or a tree. Activities in water such as swimming and boating should also be avoided. Indeed, statistics gathered by the federal government show that lightning and the fires it may cause kill more people each year than tornadoes and hurricanes combined.

TORNADOES

Tornadoes are local storms of short duration that must be ranked high among nature's most destructive forces. Their sporadic occurrence and violent winds cause many deaths each year. Tornadoes are violent windstorms that take the form of a rotating column of air that extends downward from a cumulonimbus cloud (Figure 15.12). Pressures within some tornadoes have been estimated to be as much as 10 percent lower than immediately outside the storm. Drawn by the much lower pressure in the center of the storm, air near the

FIGURE 15.12
Two views of a tornado taken a few minutes apart near Peoria, Illinois. **A.** The tornado is well developed. **B.** The funnel is beginning to dissipate. (Photos by Raymond Pierce)

A.

B.

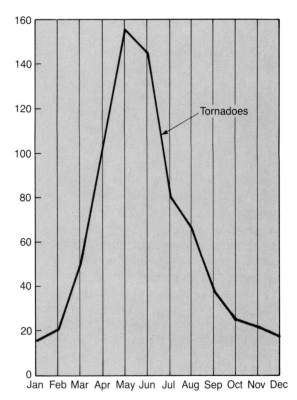

FIGURE 15.13
Average number of tornadoes each month in the
United States for a 25-year period. (After NOAA)

ground rushes into the tornado from all directions.
As the air streams inward, it is spiraled upward
around the core until it eventually merges with the
airflow of the parent thunderstorm deep in the
cumulonimbus tower. Because of the tremendous
pressure gradient assoicated with a strong tornado, maximum winds are believed to approach
480 kilometers (300 miles) per hour.

About 750 tornadoes are reported each year in
the United States, and they have been known to
occur during every month of the year. April
through June is the period of greatest tornado frequency in the United States, while the number is
lowest during December and January (Figure
15.13). During a recent 25-year period, an average
of almost five per day occurred during May. At the
other extreme, a tornado was reported only every
other day in January.

Tornadoes form in association with severe
thunderstorms that produce high winds, heavy

rainfall, and often damaging hail. Fortunately, less
than 1 percent of all thunderstorms produce tornadoes. Although weather scientists are still not
sure what triggers tornado formation, apparently
they are the product of the interaction between
strong updrafts in the thunderstorm and winds in
the troposphere. In spite of recent advances in
modeling the many variables that eventually produce a strong tornado, our knowledge is still limited. Nevertheless, the general atmospheric conditions that are most likely to develop into tornado
activity are known.

Severe thunderstorms—and hence tornadoes—
are most often spawned along the cold front of a
middle-latitude cyclone. Throughout spring, air
masses associated with middle-latitude cyclones
are most likely to have greatly contrasting conditions. Continental polar air from the Canadian Arctic may still be very cold and dry, whereas maritime tropical air from the Gulf of Mexico is warm,
humid, and unstable. The greater the contrast, the
more intense is the storm. Because these two contrasting air masses are more likely to meet in the
central United States, it is not surprising that this
region generates more tornadoes than any other
area of the country or, in fact, the world. Figure
15.14, which depicts tornado incidence in the
United States based on data for a 25-year period,
readily substantiates this fact.

The average tornado has a diameter of between
150 and 600 meters (500 and 2000 feet), travels
across the landscape at 45 kilometers (28 miles)
per hour, and cuts a path about 26 kilometers (16
miles) long. Since many tornadoes occur slightly
ahead of a cold front, in the zone of southwest
winds, most move toward the northeast. The Illinois example demonstrates this movement (Figure 15.15, page 418). What Figure 15.15 also
shows is that many tornadoes do not fit the description of the "average" tornado. Many have had
paths a great deal longer than 26 kilometers and
have traveled not at 45 kilometers per hour, but at
speeds in excess of 100 kilometers (62 miles) per
hour. Furthermore, there have been tornadoes
that have had diameters of up to 1.6 kilometers
(1 mile)—more than four times the "average" size.

The potential for tornado destruction depends
largely upon the strength of the winds generated
by the storm. One commonly used guide to tor-

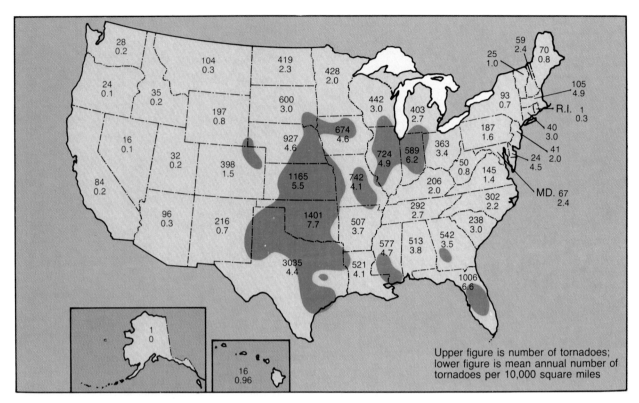

FIGURE 15.14
Tornado incidence by state and area for a 25-year period. (Data from NOAA)

nado intensity was developed by T. Theodore Fujita at the University of Chicago and is appropriately called the *Fujita intensity scale,* or simply the *F-scale* (Table 15.1). Since tornado winds cannot be measured directly, a rating on the F-scale is determined by assessing the worst damage produced by a storm.

Tornadoes take many lives each year, sometimes hundreds in a single day. When tornadoes struck an area stretching from Canada to Georgia on April 3, 1974, the death toll exceeded 300, the worst in half a century. Most tornadoes, however, do not result in a loss of life. During the 29-year period 1950–1978 there were 689 tornadoes that resulted in deaths. This figure represents slightly less than 4 percent of the total 19,312 reported storms. Although the percentage of tornadoes that result in death is small, each tornado is potentially

TABLE 15.1
Fujita intensity scale for tornadoes.

Scale	Wind Velocity (km/hr)	Wind Velocity (mi/hr)	Expected Damage
F0	<116	<72	Light damage
F1	116–180	72–112	Moderate damage
F2	181–253	113–157	Considerable damage
F3	254–332	158–206	Severe damage
F4	333–419	207–260	Devastating damage
F5	>419	>260	Incredible damage

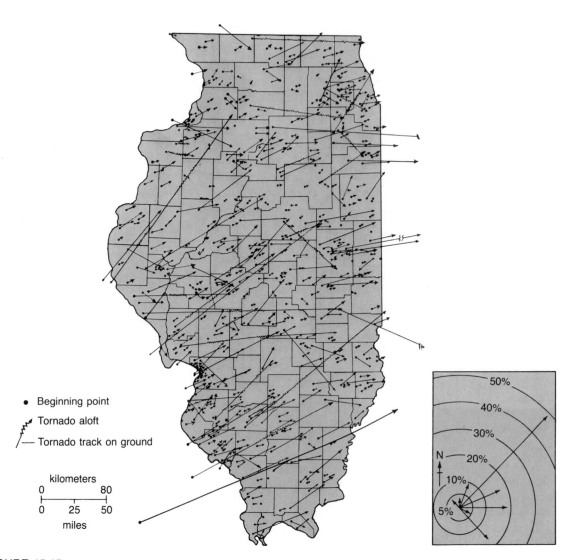

- • Beginning point
- ⚡ Tornado aloft
- /— Tornado track on ground

kilometers
0   80

0   25   50
        miles

FIGURE 15.15
Paths of Illinois tornadoes (1916–1969). Since most tornadoes occur slightly
ahead of a cold front, in the zone of southwest winds, they tend to move toward
the northeast. Tornadoes in Illinois verify this. Over 80 percent exhibited directions of movement toward the northeast through east. (After J. W. Wilson and
S. A. Changnon, Jr., *Illinois Tornadoes,* Illinois State Water Survey Circular 103,
1971, pp. 10, 24)

lethal. When tornado fatalities and storm intensities are compared, the results are quite interesting: The majority (63 percent) of tornadoes are
weak (F0 and F1) and the number of storms decreases as tornado intensity increases. The distribution of tornado fatalities, however, is just the
opposite. Although only 2 percent of tornadoes are

classified as violent (F4 and F5), they account for
nearly 70 percent of the deaths. If there is some
question as to the causes of tornadoes, there certainly is no question about the destructive effects
of these violent storms.

Because tornadoes are small and relatively
short-lived phenomena, they are among the most

difficult weather features to forecast precisely. When conditions appear favorable for tornado formation, a **tornado watch** is issued for areas covering about 65,000 square kilometers. Between 35 and 40 percent of the predictions are correct; that is, one or more tornadoes occur somewhere in the specified region. The incorrect forecasts are about evenly divided between cases when no tornadoes are sighted and cases when tornadoes occur outside, but near, the watch area.

Whereas a tornado watch is designed to alert people to the possibility of tornadoes, a **tornado warning** is issued when a funnel cloud has actually been sighted or is indicated by radar. When severe weather threatens, the radar screens are monitored for very intense echoes, which in turn are associated with heavy precipitation and the greater likelihood of hail, strong winds, and tornadoes. In addition, the echo from a tornadic storm sometimes displays a hook-shaped appendage. If the direction and approximate speed of the storm are known, the storm's most probable path can be estimated. Because tornadoes often move erratically, the warning area is fan-shaped downwind from the point where the tornado has been spotted. Since the late 1960s warnings have been given for most major tornadoes. It is believed that such warnings substantially reduce the number of deaths and serious injuries that might otherwise occur.

The tornado warning system that has been used throughout the United States for many years relies heavily on visual sightings by a few trained observers as well as the general public. Unfortunately, such a system is prone to incomplete coverage and mistakes. The errors are most likely to occur at night when tornadoes may go unnoticed or harmless clouds may be mistaken for funnel clouds. Hence, there may be a lack of adequate warning on the one hand or unnecessary warnings on the other.

Many of the difficulties that have limited the accuracy of tornado warnings will be reduced or eliminated in the future when a new advancement in radar technology, called **Doppler radar,** is put into general use. Doppler radar can not only perform the same tasks as conventional radar, but also has the ability to detect motion directly.

Within a radius of about 230 kilometers (140 miles), a Doppler radar unit can detect the initial formation and subsequent development of a **mesocyclone,** an intense rotating wind system in the lower part of a thunderstorm that precedes tornado development. Almost all mesocyclones produce damaging hail, severe winds, or tornadoes. Those that produce tornadoes (about 50 percent) can be distinguished by their stronger wind speeds and their sharper gradients of wind speeds.

In carefully planned tests conducted over several years, Doppler radar provided an average warning time of 21 minutes before tornado touchdown. By comparison, the average warning time provided by visual observation is less than two minutes. An additional advantage was a decreased false alarm rate. Predictions based on Doppler radar were correct 75 percent of the time—a considerable improvement over present standards. As a practical tool for tornado detection, it has significant advantages over a system that uses observers and conventional radar. Recognizing these advantages, the National Weather Service plans to replace its aging system of weather radars with more advanced systems that will use Doppler principles of direct wind speed and direction measurements. The new radars are part of the 10-year *Next Generation Weather Radar* (NEXRAD) program designed to improve the forecasting of severe storms, tornadoes, and flash floods.

HURRICANES

The whirling tropical cyclones that on occasion have wind speeds reaching 300 kilometers (186 miles) per hour are known in the United States as **hurricanes** and are the greatest storms on earth. Out at sea, they can generate 15-meter (50-foot) waves capable of inflicting destruction hundreds of kilometers from their source. Should a hurricane smash into land, strong winds coupled with extensive flooding can impose millions of dollars in damages and great loss of life. These storms form in all tropical waters (except those of the South Atlantic and eastern South Pacific) between the latitudes of 5 degrees and 20 degrees and are known by different names in various parts of the

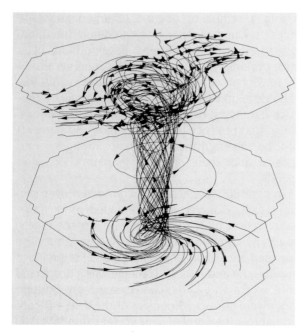

FIGURE 15.16
Circulation in a well-developed hurricane. Cyclonic winds are experienced at the surface. At the storm's core the air is whirled in great spirals upward. Outward flow dominates near the top of the storm and seems to be aided by upper-level airflow.

world. In the western Pacific, they are called *typhoons* and in the Indian Ocean, *cyclones.* The North Pacific has the greatest number of storms, averaging 20 per year. Fortunately for those living in the coastal regions of the southern and eastern United States, fewer than 5 hurricanes, on the average, develop each year in the warm sector of the North Atlantic.

Although many tropical disturbances develop each year, only a few reach hurricane status, which by international agreement requires wind speeds in excess of 119 kilometers (74 miles) per hour and a rotary circulation. Hurricanes average 600 kilometers (375 miles) in diameter and often extend 12,000 meters (40,000 feet) above the ocean surface. From the outer edge to the center, the barometric pressure has on occasion dropped 60 millibars, from 1010 millibars to 950 millibars. The lowest pressure ever recorded in the United States was 892.31 millibars, measured during a hurricane in September 1935. A steep pressure

gradient generates the rapid, inward-spiraling winds of a hurricane. As the inward rush nears the core of the storm, it is whirled upward (Figure 15.16). Upon ascending, the air condenses, creating the cumulonimbus clouds that constitute the doughnut-shaped inner structure of the hurricane called the **eye wall.** It is here that the greatest wind speeds and heaviest rainfall occur. Surrounding the eye wall are curved bands of clouds that trail away in a spiral fashion. Near the top of the hurricane the airflow is outward, carrying the rising air away from the storm center, thereby providing room for more inward flow at the surface.

At the center of the storm is the spectacular **eye** (Figure 15.17). Averaging 20 kilometers (12½ miles) in diameter, this zone of calm and scattered cloud cover is unique to the hurricane. Separating periods of torrential rains, the eye, with its brighter sky and intermittent sunlight against a background of circling clouds, which are 12 kilometers (7½ miles) high, is an awesome sight. The air within the eye slowly descends and heats by compression, making it the warmest part of the storm.

A hurricane can be described as a heat engine that is fueled by the energy (latent heat) liberated during the condensation of water vapor. The energy from an average hurricane is roughly equivalent to the total amount of electricity consumed in the United States over a six-month period. The re-

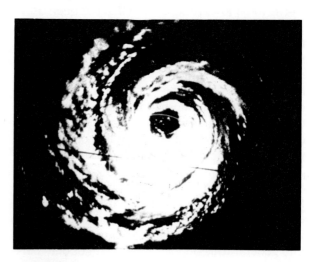

FIGURE 15.17
Radar view of Hurricane Betsy showing the eye. (Courtesy of NOAA)

lease of latent heat warms the air and provides buoyancy for its upward flight. The result is to reduce the pressure near the surface, which encourages a more rapid inward flow of air. To get this engine started, a large quantity of warm, moisture-laden air is required, and a continual supply is needed to keep it going.

Hurricanes develop most often in the late summer when water temperatures have reached 27°C (80°F) or higher and are thus able to provide the necessary heat and moisture to the air. This ocean-water temperature requirement is thought to account for the fact that hurricanes do not form over the relatively cool waters of the South Atlantic and the eastern South Pacific. For the same reason, few hurricanes form poleward of 20 degrees of latitude. Although water temperatures are sufficiently high, hurricanes do not form within 5 degrees of the equator, presumably because the Coriolis effect is too weak to initiate the necessary rotary motion.

Although the exact mechanism of formation is not completely understood, we know that smaller tropical cyclones initiate the process. These initial disturbances are regions of low-level convergence and lifting. Many tropical disturbances of this type occur each year, but only a few develop into full-fledged hurricanes. By international agreement, lesser tropical cyclones are given different names based on the strength of their winds. When a cyclone's strongest winds do not exceed 61 kilometers (38 miles) per hour, it is called a **tropical depression;** when winds are between 61 and 119 kilometers (38 and 74 miles) per hour, the cyclone is termed a **tropical storm.** Each year between 80 and 100 tropical storms develop, and of this total, half or more become hurricanes.

Once on land, a hurricane loses its punch rapidly because its source of warm, water-laden air is cut off. Also, the added frictional effect of land causes the winds to slow and move more directly into the center of the low, filling the storm and eliminating the large pressure differences.

North Atlantic hurricanes develop in the trade winds, which generally move these storms from east to west at about 25 kilometers (15 miles) per hour. Then, almost without exception, hurricanes curve poleward and are deflected into the westerlies, which increases their forward motion up to a maximum of 100 kilometers per hour. Some move toward the mainland, but their irregular paths make prediction of their movement difficult.

A location only a few hundred kilometers from a hurricane—just one day's striking distance away—may experience clear skies and virtually no wind. Prior to the age of weather satellites, such a situation made the job of warning people of impending storms very difficult.

The worst natural disaster in U.S. history resulted from a hurricane that struck an unprepared Galveston, Texas, on September 8, 1900. The strength of the storm coupled with the lack of adequate warning caught the population by surprise and cost 6000 people in the city their lives. At least 2000 more were killed elsewhere. Fortunately, hurricanes are no longer the unheralded killers they once were. Once a storm develops cyclonic flow and the spiraling bands of clouds characteristic of a hurricane, it receives continuous monitoring. For example, when Hurricane Allen, one of the strongest storms of the century, formed in August, 1980, satellites were able to identify and track the storm long before it struck land (Figure 15.18, p. 422). Consequently, by the time Allen hit the coast of Texas, people had been safely evacuated and loss of life was minimal.

Damage caused by hurricanes can be divided into three categories: (1) Wind damage; (2) Storm surge; and (3) Inland freshwater flooding. Although wind damage is perhaps the most obvious of the categories, it is not directly responsible for the greatest amount of destruction. This is not to say, however, that wind damage cannot be significant. For some structures, the force of the wind is sufficient to cause total destruction. Mobile homes are particularly vulnerable. In addition, the strong winds can create a dangerous barrage of flying debris.

Without question, the most devastating damage is caused by the storm surge. It not only accounts for a large share of coastal property losses, but is also responsible for 90 percent of all hurricane-caused deaths. A **storm surge** is a dome of water 65 to 80 kilometers (40 to 50 miles) long that sweeps across the coast near the point where the eye makes landfall. If all of the wave activity were smoothed out, the storm surge would be the height of the water above normal

FIGURE 15.18
Hurricane Allen was one of
the strongest storms of the
century. This is a satellite
view of Allen on August 9,
1980, shortly before it struck
the south Texas coast.
(Courtesy of National Environmental Satellite Service)

tide level. In addition, superimposed upon the surge is tremendous wave activity. We can easily imagine the damage this surge of water could inflict on low-lying coastal areas. In the delta region of Bangladesh, for example, the land is mostly less than 2 meters above sea level. When a storm surge superimposed upon normal high tide inundated that area on November 13, 1970, the official death toll was 200,000; unofficial estimates ran to 500,000. This was one of the worst disasters of modern times.

The torrential rains that accompany most hurricanes represent a third significant threat—flooding. Although hurricanes weaken rapidly as they move inland, the remnants of the storm can still yield 15 to 30 centimeters (6 to 12 inches) or more of rain as they move inland. A good example of such destruction is Hurricane Agnes. Although this was just a modest storm in terms of wind speeds and pressure, it was one of the costliest hurricanes of the century, responsible for more than $2 billion in damages and 122 deaths. Most destruction was attributed to flooding caused by an inordinate amount of rainfall. Agnes dropped an estimated 105 cubic kilometers or more of rainwater over the eastern United States.

In the United States, early warning systems have greatly reduced the number of deaths caused by hurricanes. At the same time, however, there has been an astronomical increase in the amount of property damage. The primary reason, of course, has been the rapid population growth in coastal areas. During the past three decades, the number of people living in beachfront subdivisions has increased far more rapidly than the overall population of the nation.

REVIEW
QUESTIONS

1 What are the characteristics of a maritime tropical air mass?

2 Describe the weather associated with a continental polar air mass in the winter. In the summer. When would this air mass be most welcome in the United States?

3 Where are the source regions for the maritime tropical air masses which affect North America? The maritime polar air masses?

4 Describe the weather along a cold front where very warm, moist air is being displaced.

5 Explain the basis for the following weather proverb:
    Rain long foretold, long last;
    Short notice, soon past.

6   The formation of an occluded front marks the beginning of the end of a wave cyclone. Why is this true?

7   Refer to Figure 15.6. Describe the weather conditions that city B will be experiencing several hours hence. Do the same for city A.

8   For each of the weather elements that follow, describe the changes that an observer experiences when a wave cyclone passes with its center north of the observer: wind direction, pressure tendency, cloud type, cloud cover, precipitation, temperature.

9   Describe the weather conditions an observer would experience if the center of a wave cyclone passed to the south.

10 What is the primary requirement for the formation of thunderstorms?

11 Based on your answer to Question 10, where would you expect thunderstorms to be most common on the earth? Where specifically in the United States?

12 How is thunder produced? How far away is a lightning stroke if thunder is heard 15 seconds after the lightning is seen?

13 Why do tornadoes have such high wind speeds?

14 Why are reliable measurements of pressure and winds in a tornado not available?

15 What general atmospheric conditions are most conducive to the formation of tornadoes?

16 When is the "tornado season"? That is, during what months is tornado activity most pronounced?

17 Distinguish between a tornado watch and a tornado warning.

18 Which has stronger winds, a tropical storm or a tropical depression?

19 Why does the intensity of a hurricane diminish rapidly when it moves onto land?

20 Hurricane damage can be divided into three broad categories. Name them. Which category is responsible for the highest percentage of hurricane-related deaths?

21 A hurricane has slower wind speeds than a tornado, yet it inflicts more total damage. How might this be explained?

KEY TERMS

air mass	occluded front	Doppler radar
air-mass weather	middle-latitude (wave) cyclone	mesocyclone
source region	occlusion	hurricane
polar (P) air mass	speed divergence	eye wall
tropical (T) air mass	thunderstorm	eye
continental (c) air mass	lightning	tropical depression
maritime (m) air mass	thunder	tropical storm
front	tornado	storm surge
warm front	tornado watch	
cold front	tornado warning	

16

HUMAN IMPACT ON THE ATMOSPHERE

AIR POLLUTION

Sources and Types of Air Pollution | Meteorological Factors Affecting Air Pollution | Acid Precipitation

THE CLIMATE OF CITIES

The Urban Heat Island | Urban-Induced Precipitation | Other Urban Effects

HUMAN IMPACT ON GLOBAL CLIMATE

Greenhouse Warming | Possible Consequences of a Greenhouse Warming

INTENTIONAL WEATHER MODIFICATION

Cloud Seeding | Fog and Cloud Dispersal | Hail Suppression | Hurricane Modification | Frost Prevention

Chapter 16 brings together meteorological principles and people. How people modify the atmospheric environment represents the major theme of this chapter. In addition to a close look at air pollution and the factors which affect the quality of the air we breathe, we will examine some of the ways that humans may inadvertently be changing the climate, both locally and globally. We will also explore some of the methods devised to intentionally alter weather processes. Can we make it rain? Can the fury of a hurricane be diminished? If so, how? These are some of the ideas that will be examined in the following pages.

Industrial complex near Peoria, Illinois, adding particulate matter and gaseous emissions to the atmosphere. (Photo by E. J. Tarbuck)

The goal of this chapter is to survey the many ways by which humans modify their atmospheric environment and to understand the consequences of that modification. We will begin by examining inadvertent changes; that is, human-induced modifications of our atmosphere that are unintentional or accidental. First among these is the growing problem of air pollution. As we shall see, air pollution not only affects human health, it also threatens other constituents of our physical and biological environment. Following this, we will examine the most apparent human impact on climate—the building of cities. The construction of each factory, road, office building, and house destroys existing microclimates and creates new ones of great complexity. After looking at the climatic changes created by cities, we will shift our attention to a far greater scale and study the possible effects of human activities on the global climate. The chapter will conclude by examining the wide variety of ways by which people deliberately attempt to modify the weather. Intentional weather modification involves human intervention to influence and improve the atmospheric processes and events that comprise the weather; in other words, to "aim" weather at human purposes. Attempts to increase precipitation, disperse fog, reduce storm damage, and prevent frost damage to crops are all examples of intentional weather modification.

AIR POLLUTION

Air pollution has become a growing threat to our health and welfare because of the *ever-increasing* emissions of air contaminants into our *never-increasing* atmosphere. An average adult requires about 13.5 kilograms (30 pounds) of air each day, as compared to about 1.2 kilograms (2.65 pounds)

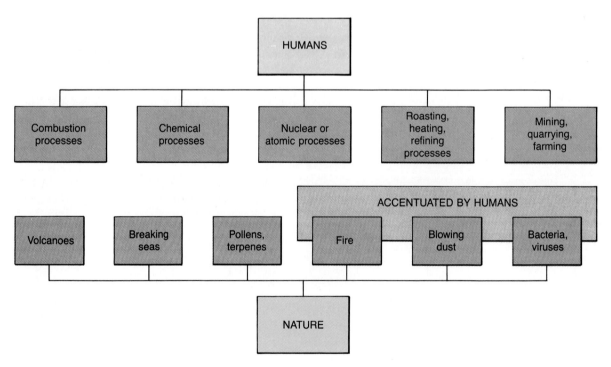

FIGURE 16.1
Sources of primary pollutants. (After Reid A. Bryson and John I. Kutzbach, *Air Pollution*. Washington, D.C.: Association of American Geographers, 1968, fig. 3, p. 9)

of food and 2 kilograms (4.5 pounds) of water. The cleanliness of the air, therefore, should certainly be as important to us as the cleanliness of our food and water.

Air is never perfectly clean. There have always been many natural sources of air pollution. Ash from volcanic eruptions, salt particles from breaking waves, pollen and spores released by plants, smoke from forest and brush fires, and windblown dust are all examples of "natural air pollution." From the time humans have been on the earth, however, they have added to the frequency and intensity of some of these natural pollutants, particularly smoke and dust (Figure 16.1). With the discovery of fire came an increased number of accidental as well as intentional burnings. Even today, in many parts of the world fire is used to clear land for agricultural purposes (the so-called slash-and-burn method), filling the air with smoke and reducing visibility. When people clear the land of its natural vegetative cover for whatever purpose, soil is exposed and blown into the air. Be that as it may, when considering the air in a modern-day industrial city, these human-accentuated forms of pollution, although significant, may seem minor by comparison.

With the Industrial Revolution came "big-time" air pollution. Rather than simply accelerating nat-ural sources, which of course continued to occur, people found many new ways to pollute the air (Figure 16.1), and many new agents to pollute it with.

This rapid increase in air pollution was not always viewed with great alarm. Instead, chimneys belching forth smoke and soot were considered a sign of growth and prosperity. For example, the following quotation from an 1880 speech by Robert Ingersoll, a well-known lawyer and orator, is reported to have elicited great cheering from the audience:

> "I want the sky to be filled with the smoke of American industry and upon that cloud of smoke will rest forever the bow of perpetual promise. That is what I am for."

With the rapid growth of the world's population and accelerated industrialization, the quantities of atmospheric pollutants increased drastically.

SOURCES AND TYPES OF AIR POLLUTION

Pollutants may be grouped into two categories—primary and secondary. **Primary pollutants** are emitted directly from identifiable sources. Table 16.1 lists major sources as well as the types and amounts of primary pollutants that characterize

TABLE 16.1
Estimated nationwide emissions (millions of metric tons/year).

Source	Carbon Monoxide	Particulates	Sulfur Oxides	Volatile Organics	Nitrogen Oxides	Total
Transportation	53.3	1.3	0.9	6.1	9.7	71.3
Fuel combustion in stationary sources	6.6	2.4	17.4	2.0	9.6	38.0
Industrial processes	4.8	2.4	3.1	7.1	0.6	18.0
Solid waste disposal	2.1	0.4	0.0	0.6	0.1	3.2
Miscellaneous	6.8	1.0	0.0	2.4	0.2	10.4
Total	73.6	7.5	21.4	18.2	20.2	140.9

SOURCE: U.S. Environmental Protection Agency Publication No. EPA-450/4-83-024, February, 1984.

each. The transportation category is most significant in that it accounts for more than one-half of our air pollution (by weight). In addition to motor vehicles, this category includes trains, ships, and airplanes. However, the tens of millions of cars and trucks on U.S. roads are, without a doubt, the greatest sources of air pollution.

Secondary pollutants are produced in the atmosphere when certain chemical reactions take place among the primary pollutants. Sulfuric acid (H_2SO_4) is one example of a secondary pollutant. It is produced when sulfur dioxide combines with oxygen, yielding sulfur trioxide, which then combines with water to form this irritating and corrosive acid. A later section on acid precipitation will discuss the effects of this in some detail.

Many reactions that produce secondary pollutants are triggered by strong sunlight and are therefore termed **photochemical reactions.** One common example occurs when nitrogen oxides absorb solar radiation and initiate a chain of complex reactions. In the presence of certain organic compounds the result is the formation of a number of undesirable secondary products that are very reactive, irritating, and toxic. Among these products is ozone. Recall from Chapter 12 that ozone is also formed by natural processes in the stratosphere where it plays a vital role because of its ability to absorb damaging ultraviolet radiation. Nevertheless, because ozone is poisonous, it is considered a pollutant when it is produced near the earth's surface, where we breathe.

Air pollution in urban and industrial areas is often termed **smog.** The word was coined in 1905 by Dr. Harold A. Des Veaux, a London physician, and was created by combining the words "smoke" and "fog." Dr. Des Veaux's term was indeed an apt description of London's principal air pollution threat, which was associated with the products of coal burning coupled with periods of high humidity. Today the term "smog" is used as a synonym for general air pollution and does not necessarily imply the smoke-fog combination. Therefore, when greater clarity is desired, we often find the term "smog" preceded by modifiers such as "London-type," "classical," "Los Angeles-type," or "photochemical." The first two refer to the original

meaning of the term, whereas the last two refer to air quality problems created by secondary pollutants.

METEOROLOGICAL FACTORS AFFECTING AIR POLLUTION

There are two ways in which air pollution and meteorology are linked. One connection concerns the influence of weather conditions on the dilution of air pollutants. The second connection is the reverse of the first, and has to do with the effect that air pollution has on weather and climate. The first of these associations will be examined in this section. The second and equally important relationship will be discussed in more detail later in this chapter.

Certainly the most obvious factor influencing air pollution is the quantity of contaminants emitted into the atmosphere. However, experience tells us that even when emissions remain relatively steady for extended periods, we often find wide variations in air quality from one day to the next. Indeed, when air pollution episodes occur, they are not generally the result of a drastic increase in the output of pollutants; instead they occur because of changes in certain atmospheric conditions.

Perhaps you have heard the following well-known phrase: "The solution to pollution is dilution." To a significant degree this is true. If the air into which the pollution is released is not dispersed, the air will become more toxic. Two of the most important atmospheric conditions affecting the dispersion of pollutants are the strength of the wind and the stability of the air. These factors are critical because they determine how rapidly pollutants are diluted by mixing with the surrounding air after leaving the source.

The manner in which wind speed influences the concentration of pollutants is shown in Figure 16.2. When winds are weak, the concentration of pollutants is much higher than when winds are strong. For example, assume that a burst of pollution leaves a smoke stack every second. If the wind speed were 10 meters per second, the distance

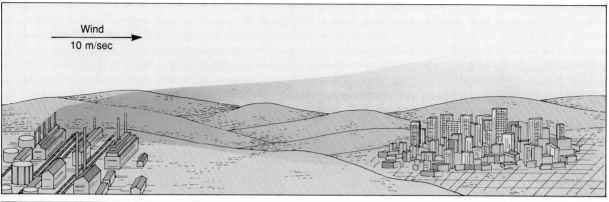

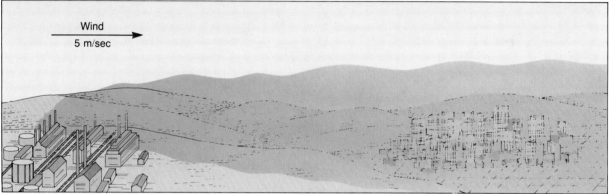

FIGURE 16.2
Effect of wind speed on the dilution of pollutants. The concentration of pollutants
increases as wind speed decreases.

between each pollution "cloud" would be 10 meters. If the wind were reduced to 5 meters per second, the distance between "clouds" would be 5
meters. Consequently, because of the direct effect
of wind speed, the concentration of pollutants is
twice as great with the 5 meters per second wind
as with the 10 meters per second wind. Therefore,
it is easy to understand why air pollution problems
are associated with periods when winds are weak
or calm and seldom occur when winds are strong.

A second aspect of wind speed also influences
air quality—the stronger the wind, the more turbulent the air. Thus, strong winds mix polluted air
more rapidly with the surrounding air to cause the
pollution to be more dilute. Conversely, when

winds are light, there is little turbulence and the
concentration of pollutants remains high.

Although wind speed governs the amount of air
into which pollutants are initially mixed, atmospheric stability determines the extent to which
vertical motions will mix the pollution with
cleaner air above. When air is unstable, smoke
and exhaust fumes are carried upward by air currents, mixed with cleaner air, and dispersed by
winds aloft. However, if the air is very stable, vertical mixing, and hence dilution, is suppressed. This
is the case when a **temperature inversion** exists.
Warm air overlying cooler air acts as a lid and prevents upward movement, leaving the pollutants
trapped below. This effect is illustrated very

A.

B.

FIGURE 16.3
A. Los Angeles City Hall on a clear day. B. A temperature inversion at 100 meters (328 feet) is visible as smog shrouds the lower portion of City Hall while the upper portion of the building is visible in the clear air above the base of the inversion. (Courtesy of the Los Angeles County Air Pollution Control District)

dramatically by the photographs of downtown Los Angeles shown in Figure 16.3.

Inversions are generally classified into one of two categories—those that form near the ground and those that form aloft. The first type, sometimes called surface inversions, develops close to the ground on clear and relatively calm nights. They form because the ground is a more effective radiator than the air above. This being the case, radiation from the ground to the clear night sky causes more rapid cooling at the surface than higher in the atmosphere. The result is that the air close to the ground is cooled more than the air above, yielding a temperature profile similar to the one shown in Figure 16.4. After sunrise the ground is heated and the inversion disappears. Although surface inversions are usually rather shallow, they may be quite deep in regions where the land surface is uneven. Due to the fact that cold air is denser (heavier) than warm air, the chilled air near the surface gradually drains from the uplands and slopes into adjacent lowlands and valleys. As might be expected, these deeper surface inversions will not dissipate as quickly after

sunrise. Thus, although valleys are often preferred sites for manufacturing because they afford easy access to water transportation, they are also more likely to experience relatively thick surface inversions which, in turn, will have a negative effect upon air quality.

Many extensive and long-lived air pollution episodes are linked to the second type of temperature inversion, which develops aloft because of the sinking air associated with a high-pressure center. As the air sinks to lower altitudes, it is compressed and because of this its temperature rises. Since turbulence is almost always present near the ground, this lowermost portion of the atmosphere is generally prevented from participating in the general subsidence. Thus an inversion develops aloft between the lower turbulent zone and the subsiding warmed layers above.

ACID PRECIPITATION

As a consequence of burning tremendous quantities of fossil fuels such as coal and oil, the United States discharges more than 40 million metric tons

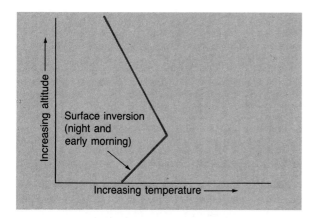

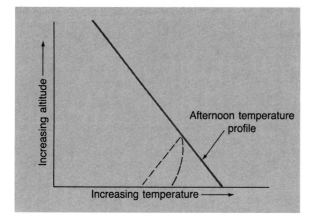

FIGURE 16.4
Surface temperature inversion. The solid line shows
the vertical temperature profile as it might appear at
night and early in the morning. The dashed lines
show changes that will occur as the ground is heated
during the day.

of sulfur dioxide and nitrogen oxides into the atmosphere each year. Through a series of complex
chemical reactions some of these pollutants are
converted into acids, which then fall to the earth's
surface as rain or snow. Another portion is deposited directly on surfaces and converted into acid
after coming in contact with rain, dew, or fog.

Acid precipitation was first identified more than
a century ago by an English chemist, and Swedish
scientists have been studying the problem since
the 1950s. However, it has only been since the
early 1970s that acid precipitation has become the
subject of intensive research in the United States
as well as a source of concern in many other parts
of the world. The concern is that acid precipitation
may have a serious ecological impact over widespread areas.

Before we can define acid precipitation in a precise way, we must become acquainted with the
pH scale. The pH is a common measure of the
degree of acidity or alkalinity of a solution. As Figure 16.5 illustrates, the scale ranges from 0 to 14,
with a value of 7 denoting a solution that is neutral. Values below 7 indicate greater acidity and
numbers above 7 indicate greater alkalinity. It is
important to note that the pH scale is logarithmic,
that is, each whole number change indicates a
ten-fold difference. Thus, pH 4 is 10 times more
acidic than pH 5 and 100 times more acidic than
pH 6. Rain is somewhat naturally acidic because
carbon dioxide dissolves in water in the atmosphere to produce carbonic acid. However, normal, unpolluted precipitation is assumed to have
a pH not lower than 5.6. **Acid precipitation** there-

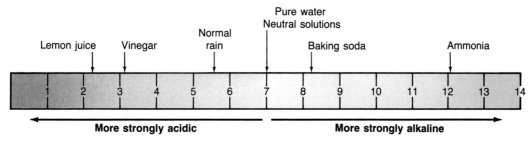

FIGURE 16.5
The pH scale.

fore is defined as rain or snow with pH values of less than 5.6.

Many pollutants remain in the atmosphere for periods as long as five days during which they may be transported great distances. Consequently, some of the acidity found in the northeastern United States may originate hundreds of kilometers away in the industrialized regions of the Midwest and Ohio Valley. A similar situation exists in Europe where Swedish scientists suggest that as much as 75 percent of the human-generated sulfur in the atmosphere over southern Sweden originates from sources outside of their country.

Unlike most atmospheric pollutants, acid precipitation has not yet been directly linked to any adverse effects on human health. Even so, its damaging effects on the environment are believed to be already considerable in some areas and imminent in others. The best-known effect of acid precipitation is the lowering of pH in thousands of lakes in Scandinavia, the northeastern United States, and Canada. Accompanying this have been substantial increases in dissolved aluminum, which is leached from the soil by the acidic waters, and which in turn is toxic to fish and wildlife. As a consequence, some lakes are virtually devoid of fish while others are approaching this condition. Further, since ecosystems are characterized by many interactions at many levels of organization, evaluating the effects of acid precipitation on these complex systems is difficult and expensive, and therefore far from complete. It should also be pointed out that, even within small areas, the effects of acid precipitation can vary significantly from one lake to another. Much of this variation is related to the nature of the soil and rock materials in the area surrounding the lake. Because minerals such as calcite in some rocks and soils can neutralize acid solutions, lakes surrounded by such materials are less likely to become acidic. On the other hand, lakes that lack this buffering material can be severely affected. Even so, over a period of time, the pH of lakes that have not yet been acidified may drop as the buffering material in the surrounding soil becomes depleted.

In addition to the thousands of lakes that can no longer support fish, preliminary research indicates that acid precipitation may also reduce agricultural crop yields and impair the productivity of forests. Acid rain not only harms the foliage but also leaches nutrient minerals from the soil. Finally, acid precipitation is known to promote the corrosion of metals and contributes to the destruction of stone structures.

Although the acid precipitation issue is very complex, Philip Abelson summarized the current state of the problem nicely when he wrote:

> Acid precipitation is increasingly an important national and international issue with strong financial and emotional overtones. Abundant examples of deleterious effects have been cited, but there is wide disagreement among sincere people as to mechanisms of damage, who is responsible, and how the problem should be ameliorated.
>
> If long-term damage from acid rain is to be reduced, it will not suffice to use a single scapegoat. Rather, there must be more conservation, better analysis of how to manage, and the development of technologies that effectively reduce emissions while not creating additional environmental problems.*

THE CLIMATE OF CITIES

As far back as the early nineteenth century, Luke Howard, the Englishman who is best remembered for his cloud classification scheme, recognized that the weather in London differed from that in the surrounding rural countryside, at least in terms of reduced visibility and increased temperature. Indeed, with the coming of the industrial revolution in the 1800s, the trend toward urbanization accelerated, leading to significant changes in the climate in and near most cities. At the beginning of the last century, only about 2 percent of the world's population lived in cities of more than 100,000 people. Today, not only is the total world population dramatically larger, but a far greater percentage resides in cities. On a worldwide basis, perhaps a quarter of the population is urban, and in many regions (including the United States, Western Europe, and Japan), the percentage is more than two or three times that figure.

---

*"Acid Rain," *Science,* 221(4606): 115.

TABLE 16.2
Average climatic changes produced by cities.

Element	Comparison with Rural Environment
Particulate matter	10 times more
Temperature	
Annual mean	0.5–1.5°C higher
Winter	1–2°C higher
Solar radiation	15–30% less
Ultraviolet, winter	30% less
Ultraviolet, summer	5% less
Precipitation	5–15% more
Thunderstorm frequency	16% more
Winter	5% more
Summer	29% more
Relative humidity	6% lower
Winter	2% lower
Summer	8% lower
Cloudiness (frequency)	5–10% more
Fog (frequency)	60% more
Winter	100% more
Summer	30% more
Wind speed	25% lower
Calms	5–20% more

SOURCE: After Landsberg, Changnon, and others.

As Table 16.2 illustrates, the climatic changes produced by urbanization involve all major surface conditions. Some of these changes are quite obvious and relatively easy to measure. Others are more subtle and sometimes difficult to measure. The amount of change in any of these elements, at any time, depends on several variables, including the extent of the urban complex, the nature of industry, site factors such as topography and proximity to water bodies, time of day, season, and existing weather conditions.

THE URBAN HEAT ISLAND

The most studied and well-documented urban climatic effect is the **urban heat island.** The term simply refers to the fact that temperatures within cities are generally higher than in rural areas. The heat island is evident when temperature data such as appear in Table 16.3 are examined. As is typical, the data for Philadelphia show the heat island is most pronounced when minimum temperatures are examined. The magnitude of the temperature differences shown by this table is probably even greater than the figures indicate, because temperatures observed at suburban airports are usually higher than those in truly rural environments. Figure 16.6, page 434, which shows the distribution of average minimum temperatures in the Washington, D.C. metropolitan area for the three-month winter period over a five-year span, also illustrates a well-developed heat island. The warmest winter temperatures occurred in the heart of the city, while the suburbs and surrounding countryside experienced average minimum temperatures that were as much as 3.3°C lower. Remember that these temperatures are averages, because on many clear, calm nights the temperature difference between the city center and the countryside was considerably greater, often 11°C (20°F) or more. Conversely, on many overcast or windy nights the temperature differential approached zero degrees.

The radical change in the surface that results when rural areas are transformed into cities is a significant cause of the urban heat island. First, the tall buildings and the concrete and asphalt of the city absorb and store greater quantities of

TABLE 16.3
Average temperatures (°C) for suburban Philadelphia Airport and downtown Philadelphia (ten-year averages).

	Airport	Downtown
Annual mean	12.8	13.6
Mean June maximum	27.8	28.2
Mean December maximum	6.4	6.7
Mean June minimum	16.5	17.7
Mean December minimum	−2.1	−0.4

SOURCE: After H. Neuberger and J. Cahir, *Principles of Climatology*. New York: Holt, Rinehart and Winston, 1969, p. 128.

FIGURE 16.6
The heat island of Washington, D.C., as shown by average minimum temperatures (°C) during the winter season (December–February). The city center had an average minimum that was more than 3°C higher than in some outlying areas. (After Clarence A. Woollum, "Notes From a Study of the Microclimatology of the Washington, D.C. Area for the Winter and Spring Seasons," *Weatherwise* 17(1964): 264)

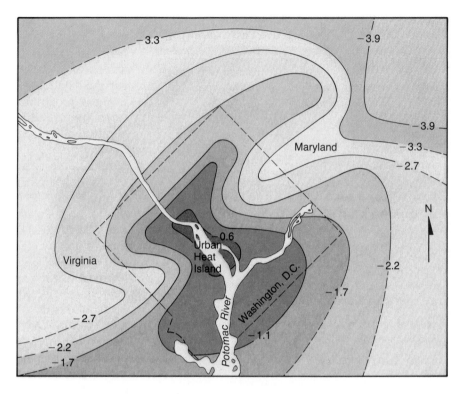

solar radiation than do the vegetation and soil typical of rural areas. In addition, because the city surface is impermeable, the runoff of water following a rain is rapid, resulting in a severe reduction in the evaporation rate. Hence, heat that once would have been used to convert liquid water to a gas now goes to further increase the surface temperature. At night, while both the city and countryside cool by radiative losses, the stonelike surface of the city gradually releases the additional heat accumulated during the day, keeping the urban air warmer than that of the outlying areas.

A portion of the urban temperature rise must also be attributed to waste heat from sources such as home heating and air conditioning, power generation, industry, and transportation. Many studies have shown that the magnitude of human-generated energy in metropolitan areas is great when compared to the amount of energy received from the sun at the surface. For example, investigations in Sheffield, England, and Berlin showed that the annual heat production in those cities was equal to approximately one-third of that received

from solar radiation. Another study of densely built-up Manhattan in New York City revealed that during the winter, the quantity of heat produced by combustion alone was 2½ times greater than the amount of solar energy reaching the ground. In summer, the figure dropped to ⅙.

There are other, somewhat less influential, causes of the heat island. For example, the "blanket" of pollutants over a city contributes to the heat island by absorbing a portion of the upward-directed long-wave radiation emitted at the surface and re-emitting some of it back to the ground. A somewhat similar effect results from the complex three-dimensional structure of the city. The vertical walls of office buildings, stores, and apartments do not allow radiation to escape as readily as in outlying rural areas where surfaces are relatively flat. As the sides of these structures emit their stored heat, a portion is reradiated between buildings rather than upward, and is therefore slowly dissipated.

In addition to retarding the heat loss from the city, tall buildings also alter the flow of air. Because of the greater surface roughness, wind speeds within an urban area are reduced. Estimates from available records suggest a decrease on the order of about 25 percent from rural values. The lower wind speeds decrease the city's ventilation by inhibiting the movement of cooler outside air which, if allowed to penetrate, would reduce the higher temperatures of the city center.

URBAN-INDUCED PRECIPITATION

Most climatologists agree that cities influence the frequency and amount of precipitation in their vicinities. Several reasons have been proposed to explain why an urban complex might be expected to increase precipitation.

1 The urban heat island creates thermally induced upward motions which act to diminish the atmosphere's stability.
2 Clouds may be modified by the addition of condensation nuclei and freezing nuclei from industrial discharges.
3 The three-dimensional nature of urban areas creates an obstacle effect that impedes the passage of weather systems. Rain-producing processes may therefore linger over a city and increase precipitation.

Several studies comparing urban and rural precipitation have concluded that the amount of precipitation over a city is about 10 percent greater than over the nearby countryside. However, more recent investigations have shown that although cities may indeed increase their own rainfall totals, the greatest effects may occur downwind of the city center.

A striking and controversial example of such a downwind effect was examined by Stanley Changnon in the late 1960s.* Records indicated that since 1925, LaPorte, Indiana, located 48 kilometers (30 miles) downwind of the large complex of industries at Chicago, experienced a notable increase in total precipitation, number of rainy days, number of thunderstorm days, and number

*S. A. Changnon, Jr., "The LaPorte Weather Anomaly: Fact or Fiction?" *Bulletin of the American Meteorological Society,* 49(1): 4–11. (NOTE: The term *anomaly* refers to a deviation in excess of normal variation.)

TABLE 16.4

Difference between average LaPorte weather values for 1951–1955 and the means at surrounding stations expressed as a percentage of the means.

Annual precipitation	+31%
Warm season precipitation	+28%
Annual number of days with precipitation ≥ 0.25 inch	+34%
Annual number of thunderstorm days	+38%
Annual number of hail days	+246%

SOURCE: After S. A. Changnon, Jr., "The LaPorte Weather Anomaly: Fact or Fiction?" *Bulletin of the American Meteorological Society,* 49(1).

of days with hail (Table 16.4). The magnitude of the changes and the absence of such change in the surrounding area led to widespread public attention. Many questioned whether the anomaly was real or simply the result of such factors as observer errors and changes in the exposure of instruments. After completing his study, however, Changnon concluded that the observed differences were real and were likely produced by the large industrial complex west of LaPorte. Among the reasons for this conclusion was that the number of days with smoke and haze (a measure of atmospheric pollution) in Chicago after 1930 corresponded quite well with the LaPorte precipitation curve. An examination of Figure 16.7 on page 436, for example, shows a marked increase in smoke-haze days after 1940, when the LaPorte graph begins its sharp rise. Further, the decrease in smoke-haze days after a peak in 1947 also generally matches a drop in the LaPorte curve. A second urban-related factor, steel production, was also found to correlate with the LaPorte precipitation curve. Records showed that seven peaks in steel production between 1923 and 1962 were associated with highs in the LaPorte curve.

The conclusions were accepted very cautiously by many and criticized by others who believed that because of the high natural rainfall variability of the area the data were neither sufficient nor accurate enough to make a sound determination. Nevertheless, this study was a pioneering effort, which illustrated the possible effect of human activity on local climates.

FIGURE 16.7
Precipitation values at LaPorte, Indiana, and smoke-haze days at Chicago, both plotted as five-year moving totals. [Data from Stanley A. Changnon, Jr., "The LaPorte Weather Anomaly: Fact or Fiction?" *Bulletin of the American Meteorological Society* 49 (1968)]

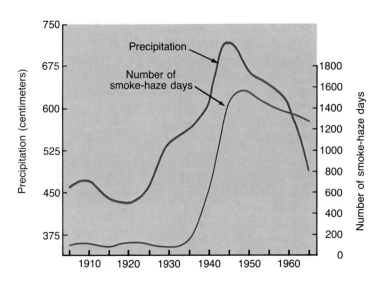

Subsequent studies of the complex problem of urban effects on precipitation have generally confirmed the view that cities increase rainfall in downwind areas. One study in the St. Louis area showed that the magnitude of the downwind precipitation increase climbed as industrial develop-

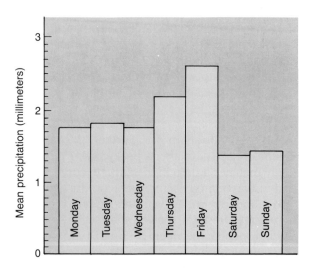

FIGURE 16.8
Average precipitation by day of the week in Paris for an eight-year period (1960–1967). Notice the gradual increase from Monday to Friday and the sharp drop in precipitation totals on Saturday and Sunday. The average for weekdays was 1.93 millimeters compared to only 1.47 millimeters on weekends, a difference of 24 percent.

ment expanded. Another line of evidence from the St. Louis study and other studies also supports the hypothesis that cities are responsible for these increases. If cities are really modifying precipitation, the effects should be greater on weekdays, when urban activities are most intense, than on weekends, when much of the activity ceases. This is indeed the case. In the St. Louis study, analysis of precipitation on weekdays versus weekends revealed a significantly greater frequency of rain per weekday than per weekend day in the affected downwind area. Figure 16.8 shows similar findings that have been reported for Paris. Here a rise in precipitation can be noted from Monday through Friday, whereas the amounts for Saturday and Sunday are considerably smaller.

OTHER URBAN EFFECTS

In earlier sections, we saw that the greater air pollution in cities (1) Contributes to the heat island by inhibiting the loss of long-wave radiation at night and (2) Has a cloud-seeding effect that is believed to increase precipitation in and downwind of cities. These influences, however, are not the only ways in which pollutants influence urban climates. For example, the blanket of particulates over most large cities significantly reduces the amount of solar radiation reaching the surface. As Table 16.2 indicates, the overall reduction in the receipt of solar energy is 15 percent, whereas short-wavelength ultraviolet is decreased by up to 30 percent. This weakening of solar radiation is, of course, variable. The decrease will be much greater during air pollution episodes than for periods when ventilation is good. Furthermore, particles are most effective in reducing solar radiation near the surface when the sun angle is low, because the length of the path through the dust increases as the sun angle drops. Hence, for a given quantity of particulate matter, solar energy will be reduced by the largest percentage at higher-latitude cities during the winter.

Relative humidities are generally lower in cities, because temperatures are higher and evaporation is reduced (see Table 16.2). Nevertheless, the frequency of fogs and the amount of cloudiness are greater. It is likely that the large quantities of condensation nuclei produced by human activities in urban areas lead to this increase in cloudiness and fog. When hygroscopic (water-seeking) nuclei are plentiful, water vapor readily condenses on them, even when the air is not yet saturated.

Finally, mention should be made of another urban-induced phenomenon, the **country breeze.** As the name implies, this circulation pattern is characterized by a light wind blowing into the city from the surrounding countryside. This breeze is best developed on relatively clear, calm nights, that is, on nights when the heat island is most pronounced. Higher city temperatures create upward air motion, which in turn initiates the country-to-city flow. One investigation in Toronto showed that the heat island created a rural-city pressure difference that was sufficient to cause an inward and counterclockwise circulation centered on the downtown area. When this circulation pattern exists, pollutants emitted near the urban perimeter tend to drift in and concentrate near the city's center.

HUMAN IMPACT ON GLOBAL CLIMATE

The proposals to explain global climatic change are many and varied. In Chapter 4 we examined some possible causes for ice-age climates. These hypotheses, which include the movement of litho-spheric plates and variations in the earth's orbit, involved natural forcing mechanisms. Another natural forcing mechanism, discussed in Chapter 7, is the possible role of explosive volcanic eruptions in modifying the atmosphere. It is important to remember that these mechanisms, as well as others, may not only have contributed to climatic changes in the geologic past, but they may also be responsible for future shifts in climate. However, when relatively recent and future changes in our climate are considered, we must also examine the possible impact of human beings. In this section we shall examine the major way in which humans are believed to contribute to global climatic change. This impact results from the addition of carbon dioxide and other gases to the atmosphere.

It should be mentioned at this point, however, that human influence on regional and global climate probably did not just begin with the onset of the modern industrial period. There is good evidence that humans have been modifying the environment over extensive areas for thousands of years. The use of fire as well as the overgrazing of marginal lands by domesticated animals has negatively affected the abundance and distribution of vegetation. By altering ground cover, such important climatological factors as surface albedo, evaporation rates, and surface winds have been, and continue to be, modified. Commenting on this aspect of human-induced climatic modification, the authors of one study state,

> In contrast to the prevailing view that only modern humans are able to alter climate, we believe it is more likely that the human species has made a substantial and continuing impact on climate since the invention of fire.*

Finally, it should be pointed out that when any hypothesis of climatic change is examined, whether it depends upon natural or human-induced causes, a degree of caution must be exercised. It is safe to say that most, if not all, hypotheses are to some degree controversial and speculative. This is to be expected if we consider the fact that at present all of our models of the earth's climate are far from complete. When the time comes

---

*Carl Sagan et al., "Anthropogenic Albedo Changes and the Earth's Climate," *Science,* 206(4425): 1367.

that our atmosphere and its changes through time are fully understood, we will likely see that many factors, both natural and human, influence climatic change.

GREENHOUSE WARMING

In Chapter 12 we learned that although carbon dioxide (CO_2) represents only 0.03 percent of the gases composing clean, dry air, it is nevertheless a meteorologically significant component. The importance of CO_2 lies in the fact that it is transparent to incoming short-wavelength solar radiation but is not transparent to some of the longer-wavelength outgoing terrestrial radiation. A portion of the energy leaving the ground is absorbed by CO_2 and subsequently re-emitted, part of it toward the surface, keeping the air near the ground warmer than it would be without CO_2. Thus, along with water vapor, CO_2 is largely responsible for the so-called greenhouse effect of the atmosphere. Since CO_2 is an important heat absorber it logically follows that any change in the atmosphere's CO_2 content should alter temperatures in the lower atmosphere.

Paralleling the rapid growth of industrialization, which began in the nineteenth century, has been the consumption of fossil fuels. The combustion of these fuels has added great quantities of CO_2 to the atmosphere. Although some of the CO_2 is dissolved in the ocean and some is used by plants, between 40 and 50 percent remains in the atmosphere. Consequently, from 1860 to the late 1980s there has been an increase of between 15 and 20 percent in the atmosphere's CO_2 content. Since 1958 scientists at Mauna Loa Observatory in Hawaii have continuously measured carbon dioxide concentrations (Figure 16.9). These measurements clearly show an upward trend from about 315 parts per million (ppm) to almost 350 ppm. This increase closely matches the growth in CO_2 emissions. Seasonal fluctuations occur because CO_2 is removed from the air by plants during the growing season and returned later when the plants decay. Naturally, we might expect that global temperatures have already increased as a result of growing carbon dioxide levels. Yet a steady temperature rise that can be attributed to carbon dioxide has not yet been detected. Most climatologists believe that the effects of the CO_2 increase are not yet large enough to show up clearly in the climatic record. However, the time is not far off when the effects will be evident.

If we assume that the use of fossil fuels will continue to increase at projected rates, current estimates indicate that the atmosphere's present carbon dioxide content of almost 350 ppm will approach 400 parts per million by the year 2000

FIGURE 16.9
Monitoring at Mauna Loa Observatory in Hawaii has revealed a rise in the concentration of carbon dioxide during the period shown here. The yearly oscillation is caused by the seasonal growth and decay of vegetation. (After NOAA.)

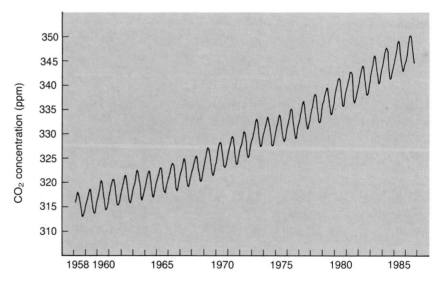

and will reach 600 ppm by some time in the second half of the next century. With such an increase in carbon dioxide, the enhancement of the greenhouse effect would be much more dramatic and measurable than in the past. When it is assumed that the atmosphere's carbon dioxide content will reach projected levels, the most realistic models predict a global surface temperature increase of between 1.5°C and 4.5°C (2.7°F and 8.1°F). Carbon dioxide is not the only gas contributing to a future global temperature increase. In recent years atmospheric scientists have come to realize that the industrial and agricultural activities of people are causing a buildup of certain trace gases that may also play a significant role. The substances are called *trace gases* because their concentrations are so much smaller than that of carbon dioxide. The trace gases that appear to be most important are methane (NH_4), nitrous oxide (N_2O), and certain types of chlorofluorocarbons. These gases absorb wavelengths of outgoing earth radiation that would otherwise escape into space. Although individually their impact is small, taken together the effects of these trace gases may be as great as CO_2 in warming the earth.

Sophisticated computer models of the atmosphere show that the warming of the lower atmosphere triggered by CO_2 and trace gases will not be the same everywhere. Rather, the temperature response in polar regions could be as much as two to three times greater than the global average. Part of the reason for such response is that the polar troposphere is very stable. This stability suppresses vertical mixing and thus limits the amount of surface heat that is transferred upward. In addition, an expected reduction in sea ice would also contribute to the greater temperature increase.

Because the atmosphere is a very complex, interactive physical system, scientists must consider many possible outcomes when one of the system's elements is altered. These various possibilities, termed **climatic feedback mechanisms,** not only complicate climatic modeling efforts, but they also add greater uncertainty to climatic predictions.

The most important and obvious of the feedback effects related to increases in carbon dioxide arises from the fact that higher surface temperatures produce greater evaporation rates. This, in turn, increases the amount of water vapor in the atmosphere. Remember that water vapor is also a powerful absorber of terrestrial radiation. Therefore, with more water vapor in the air, the temperature increase caused by carbon dioxide alone is reinforced.

Recall that the temperature increase at high latitudes is expected to be as much as two or three times greater than the global average. This assumption is based in part on the likelihood that the area covered by floating pack ice will decrease as surface temperatures rise. Since ice reflects a much larger percentage of incoming solar radiation than does open water, the melting of the pack ice would replace a highly reflective surface with a relatively dark surface. The result would be a substantial increase in the amount of solar energy absorbed at the surface. This, in turn, would feed back to the atmosphere and magnify the initial temperature increase created by higher carbon dioxide levels.

Up to this point the climate feedback mechanisms we have discussed have magnified the temperature rise caused by the buildup of carbon dioxide. The effects that reinforce an initial change are termed **positive feedback mechanisms.** As we shall see, however, other effects must be classified as **negative feedback mechanisms** because they produce results that are just the opposite of the initial change and tend to offset it.

One possible and perhaps even probable result of a global rise in temperatures would be an accompanying increase in cloud cover due to the higher moisture content of the atmosphere. Most clouds are good reflectors of solar radiation. At the same time, however, they are also good absorbers and emitters of terrestrial radiation. For this reason, clouds produce two opposite effects. They act as a negative feedback mechanism because they increase albedo and thus diminish the amount of solar energy available to heat the atmosphere. On the other hand, clouds act as a positive feedback mechanism by absorbing and emitting terrestrial radiation that would otherwise be lost from the troposphere.

Which effect, if either, is stronger? Atmospheric modeling shows that the negative effect of a higher albedo is the more dominant. Therefore, the net result of an increase in cloudiness should be a decrease in air temperature. The magnitude of this negative feedback, however, is not believed to be as strong as the positive feedback caused by added moisture and decreased pack ice. Thus, although increases in cloud cover may partly offset a global temperature increase, climatic models show that the ultimate effect of a doubling of the atmosphere's carbon dioxide will still be a temperature increase.

The problem of global warming induced by increases in the atmosphere's carbon dioxide content has been and continues to be one of the most studied and discussed aspects of climatic change. Although there are no models that can yet incorporate the full range of potential influences and feedbacks, the consensus of opinion in the scientific community is that increasing atmospheric carbon dioxide will eventually lead to a warmer planet with a different distribution of climatic regimes.

POSSIBLE CONSEQUENCES OF A GREENHOUSE WARMING

What consequences can be expected if the carbon dioxide content of the atmosphere reaches a level twice that of the mid-nineteenth century? Although climatic models contain many uncertainties, they are providing us with some possible answers to this very basic question.

As has already been noted, the magnitude of the temperature increase will not be the same everywhere. The temperature rise will probably be smallest in the tropics and increase toward the poles. Further, the models indicate that some regions will experience a significant increase in precipitation and runoff, whereas other regions will experience a decrease in runoff either because of reduced precipitation or because of increased evaporation rates brought about by higher temperatures. Such changes could have a profound impact upon the distribution of the world's water resources and hence affect the productivity of agricultural regions that depend upon rivers for irrigation water. For example, a proposed 2°C-temperature increase coupled with a 10-percent precipitation decrease in the region drained by the Colorado River could diminish the river's flow by 50 percent or more. Since the present flow of the river is barely enough to meet current demands for irrigated agriculture, the negative effect would be very serious (Figure 16.10). Many other rivers form the basis for extensive systems of irrigated agriculture and the projected reduction of their flow could have equally grave consequences. By contrast, large precipitation increases in other areas would increase the flow of some rivers and bring more frequent destructive floods to many productive agricultural regions.

The effects of increased atmospheric carbon dioxide on crops that depend upon rain for moisture is complex and difficult to estimate. Some areas will no doubt experience productivity losses due to either decreases in rainfall or increases in evaporation rates. Even so, these losses may be offset by gains elsewhere. For example, increased temperatures in the high latitudes could lengthen the growing season. This in turn could allow the expansion of agriculture into areas that are presently not suited to crop production. Additional gains may result because CO_2 is a basic plant nutrient. Experiments indicate that a higher concentration of CO_2 promotes photosynthesis and contributes to faster growth. This could lead to increased production of some crops. Further, when CO_2 levels are high, many plants tend to partly close the leaf pores that release water vapor to the atmosphere. In areas with marginal rainfall, this reduced water loss by plants would mean that the plants would be less affected by water stress as the atmosphere's CO_2 level increased.

It should be emphasized that the impact upon global climate of an increase in the atmosphere's CO_2 content is obscured by many unknowns and uncertainties. A recent report by the National Research Council suggests that no drastic effects of a CO_2-induced global warming will occur in the next few decades.* The changes that do occur will take

---

*Carbon Dioxide Assessment Committee. *Changing Climate.* (Washington, D.C.: National Academy Press, 1983).

FIGURE 16.10
Decreased rainfall and increased evaporation rates could diminish the flow of the Colorado and other rivers, forcing the abandonment of much presently productive irrigated farm land. (Photo by James E. Patterson)

the form of gradual environmental shifts that will be imperceptible to most people from year to year. What many scientists are urging is summarized in the following quotation from an article by Roger Revelle:

> It would be prudent to begin thinking now about what the changes might be and how humankind might best avoid or ameliorate the unfavorable effects and gain the most benefit from the favorable ones.*

INTENTIONAL WEATHER MODIFICATION

Certainly a desire to change the weather is far from modern. From earliest recorded times people have tried to use prayer, wizardry, dancing, and even magic to alter the weather. Until modern times, however, attempts at weather modification remained largely in the realm of the mystic. By the

*"Carbon Dioxide and World Climate," *Scientific American,* Vol. 247, No. 2, (August, 1982), p. 43.

nineteenth century such devices as smudge pots, sprinklers, and wind machines to fight frost were in use. During the Civil War, observations that rainfall apparently increased following some battles led to experiments where cannons were fired into clouds to bring more rain. Unfortunately, these experiments, as well as many others, proved unsuccessful.

Weather modification strategies fall into three broad divisions. The first relies upon the injection of energy by "brute force." The use of powerful heat sources or the intense mechanical mixing of the air (such as by helicopters) are both examples of techniques undertaken at some airports to disperse fog. The second division involves the alteration of land and water surfaces in order to change their natural interactions with the lower atmosphere. One unproven technique is to blanket a land area with a dark substance. If this were done, the amount of heat absorbed by the surface would increase and lead to stronger upward air currents that in turn could aid cloud formation. Finally, the third division involves triggering, intensifying, and/or redirecting the atmosphere's natural

441

energies. The seeding of clouds for many purposes, including precipitation enhancement, is the primary example of this category. Because it offers a relatively inexpensive and easily used technique, **cloud seeding** has been and will likely continue to be the main focus of modern weather modification technology.

CLOUD SEEDING

Recall from the discussion of precipitation formation in Chapter 13 that at temperatures between 0°C and −40°C small particles known as freezing nuclei act as catalysts to trigger the formation of ice crystals. Once ice crystals form, they grow larger at the expense of the remaining liquid droplets and upon reaching sufficient size, they fall as precipitation. Freezing nuclei, however, are not abundant in the atmosphere. For example, nuclei capable of triggering ice formation at −20°C have a typical concentration of only 0.1 to 1.0 per liter of air. By contrast, condensation nuclei, the particles on which cloud droplets form, often have concentrations of up to one million parts per liter of air.

Due to the natural rarity of freezing nuclei, the basis of cloud seeding efforts is that the addition of nuclei will spur the formation of additional precipitation. One of two agents is commonly used in cloud seeding. Dry ice may be introduced into a cloud to chill the air and cause ice crystals to form. Silver iodide crystals, however, are more com-

monly used. The similarity in the crystalline structures of silver iodide and ice accounts for silver iodide's ability to initiate the growth of ice crystals. Thus, unlike dry ice, silver iodide crystals act as freezing nuclei rather than as a cooling agent. For either substance to be successful, certain atmospheric conditions must exist. Clouds must be present, for seeding cannot generate them. In addition, at least the top portion of the cloud must be supercooled, that is, it must be composed of liquid droplets having a temperature below freezing.

The most common method of seeding clouds is to fly through or below a cloud with burning mixtures whose smoke contains large numbers of freezing nuclei (Figure 16.11). This method relies on the naturally present turbulence to diffuse the material through large volumes of air. In some experiments, freezing nuclei have been introduced into clouds by firing rockets or artillery shells. However, conflicts with aviation and concerns for public safety have made this a seldom used method. Burning silver iodide mixtures from burners on the ground is a third method of cloud seeding. Ground-based burners seem to be most effective in mountainous terrain where the upward-forced air helps the seeding agent into the clouds.

The seeding of winter clouds rising over some mountain barriers has been one of the most actively pursued examples of cloud seeding. Since only about 20 to 50 percent of the water that con-

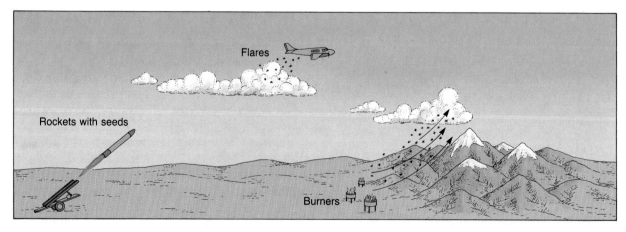

FIGURE 16.11
Methods of cloud seeding.

denses from orographic clouds actually falls as precipitation, these clouds are good candidates for seeding to increase the percentage of water falling on mountain slopes. The purpose is to increase winter snowpack, which melts and runs off during warmer months. The runoff can be collected in reservoirs for use in irrigation and hydroelectric power generation. Seeding of nonorographic cumulus clouds is more difficult because of their great variability in time and space. Nevertheless, there have been some apparently successful attempts as well as a number of unsuccessful tries. Perhaps the most successful application of this method to date has occurred in Israel. Results from two separate phases of testing indicated a 15 percent increase in rainfall. When the results became known, the Israeli government stopped experimenting and began a program of operational seeding at every opportunity.

FOG AND CLOUD DISPERSAL

Perhaps the simplest of all cloud-seeding experiments involves spreading dry-ice pellets or silver iodide particles into layers of supercooled fog and low stratus clouds. These applications transform water droplets in clouds into ice crystals. The results are easily observed and may be dramatic. Initially the ice crystals are too small to fall to the ground, but as they grow at the expense of the remaining water droplets and as turbulent air motions disperse them through nearby air, they can grow large enough to produce a snow shower. The amount of snowfall produced in this manner is minor, but visibility almost always improves and often a large hole is opened in the stratus clouds or supercooled fog. The U.S. Air Force has practiced this technology for many years at various airbases, as have commercial airlines at selected airports in the northwestern United States. The possibility of using this technology to open large holes in winter clouds to increase the amount of solar radiation reaching the ground has been discussed, but thus far has received little research attention.

Most fog is of the warm type, which means that it does not consist of supercooled water droplets. Unfortunately, these fogs are more difficult and expensive to combat because seeding will not diminish them. Successful attempts at dispersing warm fogs have involved mechanical mixing of the fog with drier warmer air from above or by heating the air. When the layer of fog is very shallow, helicopters have sometimes been used to disperse it. By flying just above the fog, the helicopter creates a strong downdraft that pulls drier air toward the surface and mixes it with the saturated air near the ground. If the air aloft is dry enough, sufficient evaporation will occur to improve visibility significantly. At some airports where warm fogs are a common hazard it has become more usual to heat and thus evaporate the fog. A sophisicated thermal fog dissipation system, called Turboclair, was installed at Orly Airport in Paris. This system consists of eight jet engines located in underground chambers alongside the upwind edge of the runway. Although expensive to install, the system is capable of improving visibility for a distance of about 900 meters (3000 feet) along the approach and touchdown zones.

HAIL SUPPRESSION

Hail is believed to form by the collecting and freezing of supercooled droplets around a nucleus. Updrafts of moist air in a cumulonimbus cloud are often so strong that small hailstones may be recirculated in the upper part of the cloud and collect additional coatings of water that freeze to enlarge the stone. Normally only a small number of hail embryos exist because freezing nuclei are relatively scarce and so the embryos grow freely until they fall from the cloud. Modern attempts at hail suppression have been based largely on the idea that introducing silver iodide crystals into appropriate clouds will interrupt the formation and growth of hailstones. It is assumed that each of these crystals, acting as a freezing nucleus, attracts a portion of the cloud's water supply and thereby increases the competition for the available supercooled water in the upper portion of the cloud. Without ample supplies of supercooled water, hailstones cannot grow large enough to be destructive when they fall.

Due to the dramatic crop and property losses, hail suppression technology has been the focus of

FIGURE 16.12
This cornfield was devastated by hail and strong winds. (Photo by James E. Patterson)

much interest and many field and laboratory studies (Figure 16.12). The results have, at best, been mixed. In France, a cloud-seeding project that used silver iodide burners to combat hail damage had encouraging results. In the Soviet Union, scientists using rockets and artillery shells to carry freezing nuclei to the clouds have claimed extraordinary success. On the other hand, the major effort in the United States, known as the National Hail Research Experiment, had disappointing results. This carefully planned and executed multi-year experiment found that no statistically significant reduction in hail was produced by cloud seeding. Commenting on current possibilities for successful hail suppression, one group of experts states, " . . . clearly the current status of hail suppression may be described as one of uncertainty."*

HURRICANE MODIFICATION

One goal of many scientists who study hurricanes is to someday find a method of reducing the intensity of these great storms and thus decrease the destruction they bring.

As the inward rush of warm, moist air approaches the center of a hurricane, wind speeds increase. If wind speeds could be lessened, de-

*S. A. Changnon, Jr. et al., "Hail Suppression and Society," *Science,* 200(4340): 388.

struction could be reduced. To explain why wind speeds increase near the storm center, we must understand the **law of conservation of angular momentum.** This law states that the product of the velocity of an object around a center of rotation (axis) and the distance squared of the object from the axis is constant. Therefore, when a parcel of air moves toward the center of a storm (the center of rotation), the product of its distance squared and velocity must remain unchanged. Consequently, if the distance decreases, the rotational velocity must increase. A common example of the conservation of angular momentum occurs when a figure skater starts whirling on the ice with both arms extended. Her arms are traveling in a circular path about an axis (her body). When the skater pulls her arms inward, she decreases the radius of the circular path of her arms. As a result, her arms go faster and the rest of her body must follow, thereby increasing her rate of spinning.

Recall that the energy for hurricanes comes from the release of latent heat in the ring of rising air, called the eye wall. The object of most hurricane modification models and experiments has been to seed portions of the storm beyond this zone of maximum wind velocity. Theoretically, at least, doing so would cause supercooled droplets to freeze, thus releasing latent heat that would then stimulate cloud growth beyond the eye wall. In this way, the inward spiraling flow of moist air would be intercepted and caused to ascend at a greater distance from the storm center. The net effect would be an outward displacement of the eye wall, that is, the eye wall would be farther from the center of the storm (Figure 16.13). Remember from the discussion of the conservation of angular momentum that if the distance between an inward moving parcel of air and the center of the storm increases, then the rotational velocity must decrease.

Thus far results are inconclusive and have not yet verified this hypothesis. It is safe to say that hurricane modification is still in its developmental stages and many questions are still unresolved.

FROST PREVENTION

Frost, the fruit-growers' plight, occurs when the air temperature falls to 0°C (32°F) or below. It may

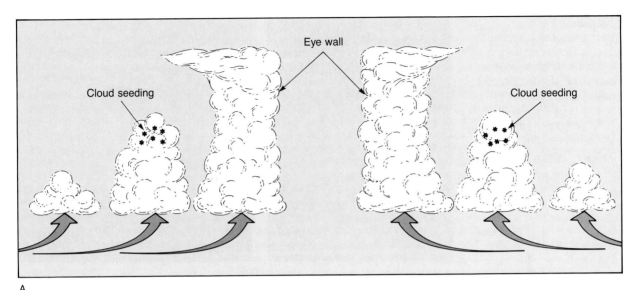

A.

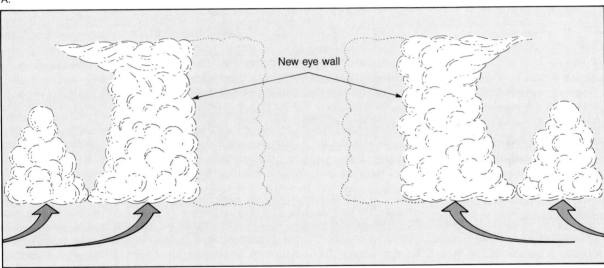

B.

FIGURE 16.13
Schematic diagram showing the proposed effect of cloud seeding in a hurricane. Seeding clouds outside of the zone of maximum winds stimulates cloud growth at a greater distance from the storm's center and leads to the decay of the old eye wall. (After Richard A. Anthes, *Tropical Cyclones, their Evolution, Structure and Effects.* Meteorological Monographs, Vol. 19. No. 41, p. 121. Boston: American Meteorological Society, 1982)

be accompanied by deposits of ice crystals commonly called *white frost*. This, however, happens only if the air becomes saturated. White frost is not a requirement for crop damage.

Frost hazards exist when a cold air mass moves into a region, or when ample radiation cooling occurs on a clear night. The conditions accompanying the invasion of cold air are characterized by

FIGURE 16.14
Two types of freeze controls used in Florida citrus groves. **A.** A wind machine mixes warmer air aloft with cooler surface air. (Courtesy of Sunkist Growers, Inc.) **B.** Orchard heaters are often effective in preventing frost damage. (Courtesy of the Florida Citrus Commission)

A.

B.

low daytime temperatures, long periods of effective frost, strong winds, and widespread damage. Frost induced by radiation loss is a nighttime phenomenon associated with a surface temperature inversion and is much easier to combat.

Several methods of frost prevention have been used with varying degrees of success. Generally these attempts are directed at reducing the amount of heat lost during the night or at adding heat to the lowermost layer of air. Heat conservation methods include covering plants with material having a low thermal conductivity, such as paper and cloth, and producing particles which, when suspended in air, reduce the radiation loss. Smudge fires have been used for particle production but have generally proven unsatisfactory. In addition to the pollution problems created by the dense clouds of black smoke, the carbon particles impede daytime warming by reducing the amount of solar radiation that can reach the surface. This

reduction in daytime warming offsets the benefits gained during the night.

Methods of warming include sprinkling, air mixing, and the use of orchard heaters (Figure 16.14). Sprinklers distribute water to the plants and add heat in two ways, first from the warmth of the water, but more importantly from latent heat of fusion, which is released when the water freezes. As long as an ice-water mixture remains on the plant, the latent heat released will keep the temperature from dropping below 0°C (32°F). Air mixing is successful when the temperature at 15 meters (50 feet) above the ground is 5°C (9°F) higher than the surface temperature. By using a wind machine, the warmer air aloft is mixed with the colder surface air. Orchard heaters are probably the most successful. As many as 30 or 40 heaters per acre are required, and the fuel cost can be significant. Nevertheless, the effectiveness of the method seems to warrant the cost.

REVIEW QUESTIONS

1 Consult Table 16.1 to answer the following.
   (a) Which source is responsible for the most pollution?
   (b) What is the single greatest air pollutant by weight?
   (c) What source produces the greatest quantity of sulfur oxides?
2 What is the difference between primary and secondary pollutants? What is a photochemical reaction?
3 What is the origin of the word "smog"? How has its meaning changed?
4 Why are air pollution episodes less likely when winds are strong?
5 How do temperature inversions form? Describe the role of inversions in air pollution episodes.
6 How much more acidic is a solution with a pH of 3 compared to a solution with a pH of 5?
7 List three effects of acid precipitation.
8 List two ways in which the difference between a "typical" rural surface and a "typical" city surface adds to the urban heat island.
9 How do each of the following factors contribute to the urban heat island: heat production, "blanket" of pollutants, three-dimensional city structure?
10 List the three factors that are likely causes of greater precipitation in and downwind of cities.
11 Describe the reasons for the following urban climatic characteristics:
   (a) Reduced solar radiation.
   (b) Reduced relative humidity.
   (c) Increased fog and cloud frequency.
   (d) Reduced wind speeds and more days when calms prevail.
12 What is a "country breeze"? Describe its cause.
13 Why has the atmosphere's carbon dioxide level been rising for more than 120 years?
14 How are temperatures in the lower atmosphere likely to change as carbon dioxide levels continue to increase? Why?
15 Is carbon dioxide the only gas contributing to a future global temperature change?
16 What are climatic feedback mechanisms? Give some examples.
17 List three possible consequences of a greenhouse warming.
18 What is the principle behind seeding supercooled clouds?
19 List three applications of cloud seeding aside from increasing precipitation.
20 How do frost and white frost differ?

KEY TERMS

primary pollutant

secondary pollutant

photochemical reaction

smog

temperature inversion

pH scale

acid precipitation

urban heat island

country breeze

climatic feedback mechanism

positive feedback mechanism

negative feedback mechanism

cloud seeding

law of conservation of angular momentum

frost

PART FOUR

ASTRONOMY

17

THE EARTH'S PLACE IN THE UNIVERSE

The earth is one of nine planets and numerous smaller bodies that orbit the sun. The sun is part of a much larger family of perhaps 100 billion stars that compose the Milky Way, which in turn is only one of billions of galaxies in an incomprehensibly large universe. This view of the earth's position in space is considerably different from that held only a few hundred years ago, when the earth was thought to occupy a privileged position as the center of the universe. This chapter examines some of the events that led to the unfolding of modern astronomy. In addition, it examines the earth's place in time and space.

Star trails around the southern celestial pole.
(Courtesy of the Anglo-Australian Observatory)

Long before recorded history, people were aware of the close relationship between events on earth and the positions of the heavenly bodies, the sun in particular. They noted that changes in the seasons and floods of great rivers like the Nile occurred when the celestial bodies reached a particular place in the heavens. Early agrarian cultures, dependent on the weather, believed that if the heavenly objects could control the seasons, they must also have a strong influence over all earthly events. This belief undoubtedly was the reason that early civilizations, particularly the Chinese, Egyptian, and Babylonian, began keeping records of the positions of celestial objects. They recorded the location of the sun, moon, and five planets visible to the unaided eye as these objects moved slowly against the background of "fixed" stars. In addition, the early Chinese kept rather accurate records of comets and "guest stars." Today, we know that a "guest star" appears when a normal star, which was too faint to be visible, increases its brightness as it explosively ejects gases from its surface, a phenomenon we call a nova. A study of Chinese archives shows that they recorded every occurrence of the famous Halley's comet for at least ten centuries. However, because this comet reappears only once every 76 years, they were unable to establish that what they saw was the same object. So they,

like most ancients, considered comets to be mystical in nature. Comets were usually believed to be bad omens and were blamed for a variety of disasters, ranging from wars to plagues (Figure 17.1).

ANCIENT ASTRONOMY

The "Golden Age" of early astronomy (600 B.C.–150 A.D.) was centered in Greece. The early Greeks have been criticized, and rightly so, for using philosophical arguments to explain natural phenomena. However, they did rely on observational data as well. The basics of geometry and trigonometry, which they had developed, were used to measure the sizes and distances of the largest-appearing bodies in the heavens—the sun and the moon.

EARLY GREEKS

Many astronomical discoveries have been credited to the Greeks. The Greeks held the **geocentric** ("earth-centered") view of the universe. The earth, they believed, was a spherical, motionless body at the center of the universe. It was, in turn, surrounded by a transparent, hollow sphere (**celestial sphere**) on which the stars were hung and carried on a daily trip around the earth (an effect that is actually caused by the earth's rotation about its axis). Some early Greeks realized that the motion of the stars could be explained as

FIGURE 17.1
Tapestry illustrating the apprehension caused by Halley's comet in 1066 A.D. (Courtesy of Wide World Photos)

452

easily by a rotating earth, but they rejected that idea, since the earth exhibited no feeling of motion and seemed too large to be moving. In fact, proof of the earth's rotation was not demonstrated until 1851.

To the Greeks, all of the stars on the celestial sphere, except seven, appeared to remain in the same relative position to one another. These seven wanderers, called planets by the Greeks, included the sun, the moon, Mercury, Venus, Mars, Jupiter, and Saturn. Each was thought to have a circular orbit around the earth. Although this system was incorrect, the Greeks were able to refine it to the point that it explained the apparent movements of the celestial bodies as seen from earth.

As early as the fifth century B.C., the Greeks understood what causes the phases of the moon. Anaxagoras reasoned that the moon shines by reflected sunlight, and because it is a sphere, only half is illuminated at one time. As the moon orbits the earth, that portion of the illuminated half that is visible from the earth is always changing. Anaxagoras also realized that an eclipse of the moon occurs when the moon moves into the shadow of the earth.

The famous Greek philosopher Aristotle (384–322 B.C.) concluded that the earth was spherical, since it always cast a curved shadow when it eclipsed the moon. Although most of the teachings of Aristotle were passed on and, in fact, considered infallible by many, his belief in a spherical earth was somehow lost.

The first Greek to profess a sun-centered, or **heliocentric,** universe was Aristarchus (312–230 B.C.). Aristarchus also used simple geometric relations to calculate the relative distances from the earth to the sun and the moon. He later used these data to calculate their relative sizes. Due to observational errors beyond his control, he came up with measurements that were much too small. However, he did learn that the sun was many times more distant than the moon and many times larger than the earth. The latter fact may have prompted him to suggest a sun-centered universe. Nevertheless, because of the strong influence of Aristotle, the earth-centered view dominated Western thought for nearly 2000 years.

The first successful attempt to establish the size of the earth is credited to Eratosthenes (276–194 B.C.). Eratosthenes observed the angles of the noonday sun at Syene (presently Aswan) and Alexandria, two cities directly north and south of each other (Figure 17.2, page 454). Having found that the angles differed by 7 degrees, or $\frac{1}{50}$ of a complete circle, he concluded that the circumference of the earth must be 50 times the distance between these two cities (5000 stadia), which gave him a figure of 250,000 stadia. Many historians believe the stadium equals 157.6 meters (517 feet), which would make Eratosthenes' calculation of the earth's circumference 39,400 kilometers (24,428 miles), a figure very close to the modern value.

Probably the greatest of the early Greek astronomers was Hipparchus (second century B.C.), who is best known for his star catalog. Hipparchus determined the location of almost 850 stars, which he divided into six groups according to their brightness. He measured the length of the year to within minutes of the modern value and developed a method for predicting the times of lunar eclipses to within a few hours.

Although many of the Greek discoveries were lost during the Middle Ages, the earth-centered view that the Greeks proposed became the established view in Europe. Presented in its finest form by Claudius Ptolemy, this geocentric outlook became known as the **Ptolemaic system.**

THE PTOLEMAIC SYSTEM

Much of our knowledge of Greek astronomy comes from a thirteen-volume treatise, *Almagest* ("the great work"), which was compiled by Ptolemy in 141 A.D. In this work, Ptolemy is credited with developing a model of the universe that accounted for the observable motions of the planets. The precision with which his model was able to predict the motions of the planets is testified to in that the model went virtually unchallenged, in principle at least, for nearly thirteen centuries.

In the Greek tradition, the Ptolemaic model had the planets moving in circular orbits around a motionless earth. (The circle was considered the pure and perfect shape by the Greeks.) Because the motion of the planets, as plotted against the background of stars, results from the combination of the motion of the earth and the planet's own motion around the sun, a rather odd apparent

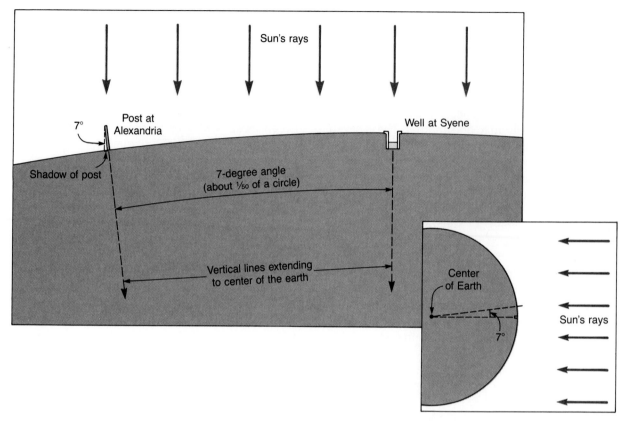

FIGURE 17.2
Orientation of the sun's rays at Syene and Alexandria on June 21 when Eratosthenes calculated the earth's circumference.

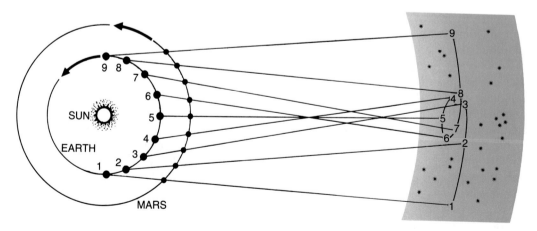

FIGURE 17.3
Retrograde (backward) motion of Mars as seen against the background of distant stars.

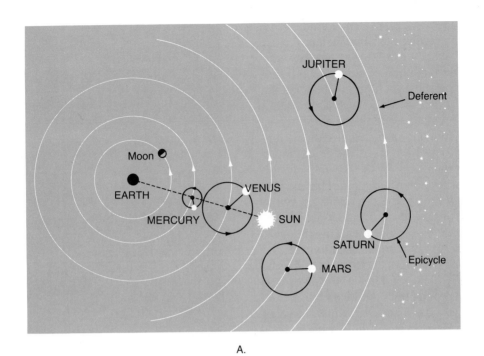

A.

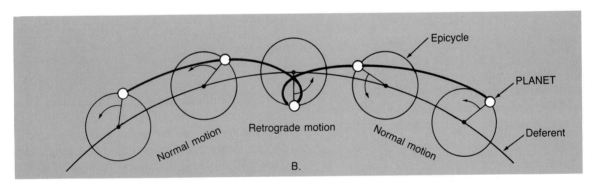

B.

FIGURE 17.4
The universe according to Ptolemy. **A.** The entire celestial sphere rotates around
a motionless earth along with the planets, which have their own orbits. **B.** Retrograde motion as explained by Ptolemy.

motion results. When viewed from the earth, the
planets move slightly eastward among the stars
each day. Periodically, each planet appears to
stop, reverse direction for a period of time, and
then resume an eastward motion. The apparent
westward drift is called **retrograde motion.** Figure
17.3 illustrates the retrograde motion of Mars. Due
to the fact that the earth has a faster orbital speed
than Mars, it overtakes Mars, and while doing so,
Mars appears to be moving backward, that is, to be

in retrograde motion. This is analogous to what a
race-car driver sees out the side window when
passing a slower car. The slower planet, like the
slower car, appears to be going backward, although its actual motion is in the same direction
as the faster-moving body.

It is much more difficult to accurately represent
retrograde motion using the incorrect earth-
centered model, but Ptolemy was able to do just
that (Figure 17.4). Rather than using one circle for

an orbit, he placed the planet on a small circle (*epicycle*), which revolved around a large circle (*deferent*). By trial and error, he was able to select just the right combinations of circles to produce the amount of retrograde motion observed for each planet. (An interesting note is that almost any closed curve can be produced by the combination of two circular motions, a fact that can be verified by persons who have used the Spirograph® design-drawing toy.)

It is a tribute to Ptolemy's genius that he was able to account for the planets' motions as well as he did, considering he used an incorrect model. Some suggest that he did not mean his model to represent reality, but only to be used to calculate the positions of the heavenly bodies. (Whether he did or not, we will never know.) However, the Church did accept Ptolemy's theory as the correct representation of the heavens, which created problems for those who found fault with it.

THE BIRTH OF MODERN ASTRONOMY

Modern astronomy was not born overnight. Its development involved a break from deeply entrenched philosophical views and the founding of a "new and greater universe" governed by discernible laws. Among the noted scientists involved in this transition were Nicolaus Copernicus, Tycho Brahe, Johannes Kepler, Galileo Galilei, and Sir Isaac Newton.

NICOLAUS COPERNICUS

For almost thirteen centuries after the time of Ptolemy, very few astronomical advances were made in Europe. The first great astronomer to emerge after the Middle Ages was the Polish astronomer Nicolaus Copernicus (1473–1543) (Figure 17.5). Copernicus became convinced that the earth is a planet, just like any of the other five then known. The daily motions of the heavens, he reasoned, could be better explained by a rotating earth. To counter the Ptolemaic objection that the earth would fly apart if it rotated, he suggested that the much larger celestial sphere, subjected to rotation, would be even more likely to fly apart.

FIGURE 17.5
Nicolaus Copernicus. (Yerkes Observatory photograph)

Having concluded that the earth is a planet, Copernicus reconstructed the solar system with the sun at the center and the planets Mercury, Venus, Earth, Mars, Jupiter, and Saturn orbiting it. This was a major break from the ancient idea that a motionless earth lies at the center of all movement. However, Copernicus retained a link to the past and used circles to represent the orbits of the planets. Unable to get satisfactory agreement between predicted locations of the planets and the observed positions, he found it necessary to add epicycles like those used by Ptolemy. (The discovery that the planets have elliptical orbits would be left to Kepler.) Also, like his predecessors, Copernicus used philosophical justifications, such as the following, to support his point of view:

> . . . in the midst of all stands the sun. For who could in this most beautiful temple place this lamp in another or better place than that from which it can at the same time illuminate the whole?

Copernicus' monumental work, *De Revolutionibus, Orbium Coelestium,* (*On the Revolution of the Heavenly Spheres*), which sets forth his controversial ideas, was published while he lay on his deathbed. Hence, he never experienced the criticisms which fell on many of his followers. The Copernican system was considered heretical, and expounding it cost at least one man his life. Giordano Bruno was seized by the Inquisition in 1600 and, refusing to denounce the Copernican theory, was burned at the stake.

TYCHO BRAHE

Tycho Brahe (1546–1601) was born of Danish nobility three years after the death of Copernicus. Reportedly Tycho became interested in astronomy while viewing a solar eclipse that had been predicted by astronomers. He persuaded King Fredrich II to establish an observatory (which he headed) on the island of Hveen, near Copenhagen. There he designed and built pointers (the telescope had not yet been invented), which he used for 20 years to systematically measure the location of the heavenly bodies. These observations, particularly those of Mars, which were far more precise than any made so far, were his legacy to astronomy.

Tycho did not believe in the Copernican (sun-centered system), because he was unable to observe an apparent shift in the position of stars caused by the earth's motion. His argument went like this: If the earth does revolve, the position of a nearby star, when observed from two points in the earth's orbit six months apart, should shift with respect to the more distant stars. This apparent shift of the stars is called *stellar parallax* (see Figure 20.1). The principle of parallax is easy to visualize: Close one eye and with your index finger in a vertical position use your eye to line your finger up with some distant object. Now, without moving your finger, view the object with your other eye and notice that its position appears to have changed. The farther away you hold your finger, the less its position seems to shift. Because the distance to even the nearest stars is so great compared with the width of the earth's orbit, the shift that occurs is too small to have been noticed using the early telescopes, let alone the unaided eye.

With the death of his patron, the King of Denmark, Tycho was forced to leave his observatory. It was probably his arrogance and extravagant nature that caused the conflict with the new ruler. Tycho moved to Prague, where in the last year of his life he acquired an able assistant by the name of Johannes Kepler. Kepler retained most of the observations made by Tycho and put them to exceptional use. Ironically, the data Tycho collected to refute the Copernican view would later be used to support it.

JOHANNES KEPLER

If Copernicus can be considered the person who ushered out the old astronomy, Johannes Kepler (1571–1630) may then be given credit for ushering in the new (Figure 17.6). Armed with Tycho's data, a good mathematical mind, and, of greater importance, a strong faith in the accuracy of Tycho's work, Kepler was able to derive three basic laws of planetary motion. The first two laws

FIGURE 17.6
Johannes Kepler. (Smithsonian Institution Photo No. 56123)

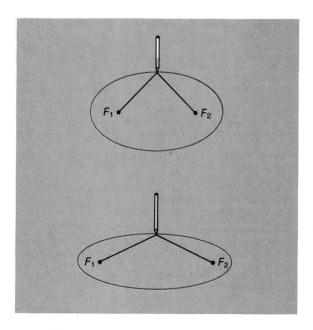

FIGURE 17.7
Drawing ellipses with various eccentricities. The farther points F_1 and F_2 (foci) are moved apart, the more flattened (greater eccentricity) is the resulting ellipse.

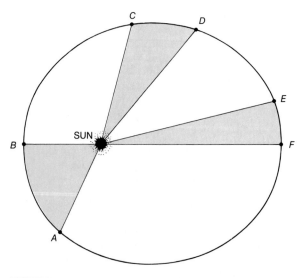

FIGURE 17.8
Law of equal areas. A line connecting a planet to the sun sweeps out an area in such a manner that equal areas (shaded) are swept out in equal times.

resulted from his inability to fit Tycho's observations of Mars to a circular orbit. Unwilling to concede that the discrepancies were due to observational error, he searched for another solution. This endeavor led him to discover that the orbit of Mars is actually elliptical (Figure 17.7). About that same time, he realized that the orbital speed of Mars varies in a predictable way. As it approaches the sun, it speeds up, and as it pulls away from the sun, it slows down. In 1609, after almost a decade of work, Kepler set forth these ideas in the form that is now known as Kepler's first two laws of planetary motion: (1) The path of each planet around the sun is an ellipse with the sun at one focus (Figure 17.7). (2) Each planet revolves so that a line connecting it to the sun sweeps over equal areas in equal intervals of time. This law of equal areas expresses the variations in orbital speeds of the planets geometrically. Figure 17.8 illustrates the second law. Note that in order for a planet to sweep equal areas in the same amount of time, it must travel more rapidly when it is nearer the sun and more slowly when it is farther from the sun.

Kepler was very religious and believed that the Creator made an orderly universe. The uniformity he tried to find eluded him for nearly a decade. Then, in 1619, he published his third law in *The Harmony of the Worlds.* Simply, this law states that the orbital periods of the planets and their distances to the sun are proportional. In its simplest form, the period of revolution is measured in years, and the planet's solar distance is expressed in terms of the earth's mean distance to the sun. The latter "yardstick" is called the **astronomical unit** (AU) and averages about 150 million kilometers (93 million miles). Using these units, Kepler's third law states that the planet's orbital period squared is equal to its mean solar distance cubed ($p^2 = d^3$). Consequently, the solar distances of the planets can be calculated when their periods of revolution are known. For example, Mars has a period of 1.88 years, which squared equals 3.54. The cube root of 3.54 is 1.52, and that is the distance to Mars in astronomical units (Table 17.1).

Kepler's laws assert that the planets revolve around the sun and therefore support the Copernican theory. Kepler, however, did fall short of deter-

TABLE 17.1
Periods and solar distances of planets.

Planet	Solar Distance (AU)	Period (years)
Mercury	0.39	0.24
Venus	0.72	0.62
Earth	1.00	1.00
Mars	1.52	1.88
Jupiter	5.20	11.86
Saturn	9.54	29.46
Uranus	19.18	84.01
Neptune	30.06	164.80
Pluto	39.44	247.70

mining the forces that act to produce the planetary motion he had so ably described. That task would remain for Galileo Galilei and Sir Isaac Newton.

GALILEO GALILEI

Galileo Galilei (1564–1642) was the greatest Italian scientist of the Renaissance (Figure 17.9). He

FIGURE 17.9
Galileo Galilei. (Yerkes Observatory photograph)

was a contemporary of Kepler and, like Kepler, strongly supported the Copernican theory of a sun-centered solar system. Galileo's greatest contributions to science were his descriptions of the behavior of moving objects. These he derived from experimentation. The method of using experiments to determine natural laws had essentially been lost since the time of the early Greeks. Legend has it that Galileo climbed to the top of the Leaning Tower of Pisa where he dropped two objects of differing masses, which to the astonishment of the observers, hit the ground at the same time.* He correctly concluded that air resistance is the cause of the variations observed for very light objects. This same experiment was performed most dramatically on the airless moon when David Scott, an *Apollo 15* astronaut, showed that a feather and a hammer fall at the same rate.

All astronomical discoveries before Galileo's time were made without the aid of a telescope. In 1609, Galileo heard that a Dutch lens maker had devised a system of lenses that magnified objects. Apparently without ever seeing one, Galileo constructed his own telescope, which had a magnification of 3. He immediately made others, his best having a magnification of about 30.

With the telescope, Galileo was able to view the universe in a completely new way. He made many important discoveries that supported the Copernican view of the universe, some of which follow:

1 He discovered four satellites or moons orbiting Jupiter and accurately determined their periods of revolution, which range from 2 to 17 days (Figure 17.10). This find dispelled the old idea that the earth is the only center of motion. It also countered the argument frequently used by those opposed to the sun-centered system that the moon would be left behind if the earth really revolves around the sun.

2 He was able to see the planets as circular disks rather than just as points of light. This indicated that they might be earthlike.

3 Venus, he learned, has phases just like the moon, proving that Venus orbits the sun. In

* Galileo probably did not actually attempt this experiment, although a colleague may have.

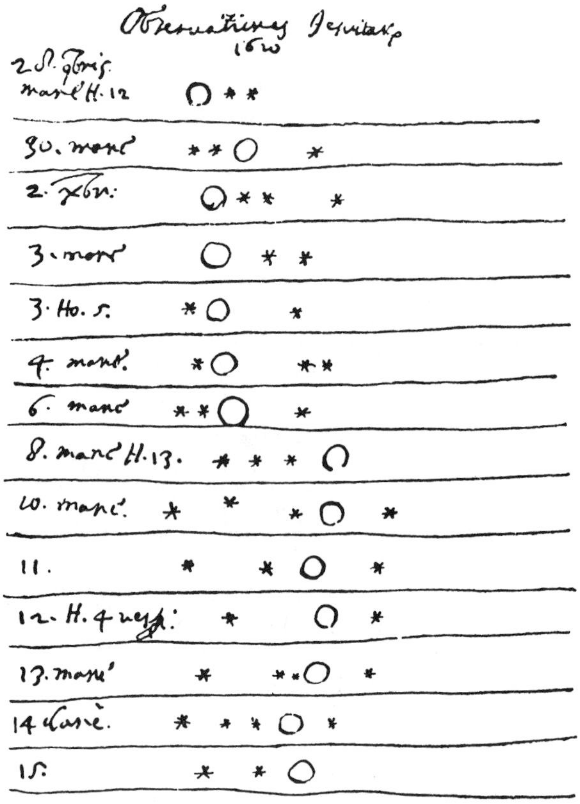

Observationes Jovialis
1610

2 d. Jovis
mane H. 12   ○ **

30. mane   ** ○   *

2. Xbr: ○ **   *

3. morr   ○ * *

3. Ho. r.   * ○   *

4. mane!   * ○   **

6. mane   *. * ○   *

8. mane H. 13.   * * * ○

10. mane!   *   *   * ○ *

11.   *   * ○ *

12. H. 4 uesp: *   ○ *

17. mane'   *   * .. ○ *

14 Lane.   * * * ○ *

15.   *   * ○

FIGURE 17.10
Drawing by Galileo of Jupiter and its four largest satellites. (Yerkes Observatory photograph)

FIGURE 17.11
Phases of Venus, illustrating that Venus appears smallest at full phase. In the Ptolemaic model, Venus is never on the opposite side of the sun and, therefore, could never appear in full phase. (Courtesy of Lowell Observatory)

the Ptolemaic system, Venus always lies between the earth and the sun, which means that only the crescent phase could be seen from earth. He also saw that Venus appears smallest when it is in full phase, and thus farthest from the earth (Figure 17.11).

4   His telescope allowed an intimate view of the moon's surface. He found that it is not a smooth glass sphere as the ancients had suspected and the Church had decreed. Rather, he saw mountains, craters, and plains. He thought the plains might be bodies of water. The idea that the moon contains seas was strongly promoted by others, as we can tell from some of the names given (Sea of Tranquility, Sea of Storms, and so forth).

5   Viewing the sun (which may have caused the eye damage that later resulted in his total blindness), Galileo saw sunspots (dark regions caused by slightly lower temperatures). He was able to track these spots and estimate the rotational period of the sun as just under a month. Hence, another heavenly body was found to have "blemishes" and a rotational motion.

In 1616 the Roman Church condemned the Copernican theory as contrary to Scripture, and Galileo was told to abandon it. Unwilling to accept this verdict, Galileo began writing his most famous work, *Dialogue of the Great World Systems*. Despite poor health, he was able to complete the project. In 1630 he went to Rome seeking permission from Pope Urban VIII to publish. Because the book was a dialogue that expounded both the Ptolemaic and Copernican systems, publication was allowed. However, Galileo's enemies were quick to realize that he was promoting the Copernican view at the expense of the Ptolemaic system. Sale of the book was quickly halted, and Galileo was called before the Inquisition. Tried and convicted of proclaiming doctrines contrary to Divine Scripture, he was sentenced to permanent house arrest, where he remained for the last ten years of his life. Despite the restrictions placed on him, and despite his age and the grief he carried after the death of his eldest daughter, Galileo continued to work. In 1637 he became totally blind, yet during the next few years he completed his finest scientific work, a book on the study of motion. When Galileo died in 1642, the Grand Duke of Tuscany wanted to erect a monument in his honor, but fear that it would offend the Holy Office still prevailed, and it was never built.

SIR ISAAC NEWTON

According to the "old" calendar, Sir Isaac Newton (1643–1727) was born in the year of Galileo's death (Figure 17.12). His many accomplishments in the fields of mathematics and physics led his successor, Lagrange, to say that, "Newton was the greatest genius that ever existed. . . . "

Although Kepler and those who followed attempted to explain the forces involved in planetary motion, they fell short of the mark. Kepler believed that some force pushes the planets along their orbits. Galileo, however, correctly reasoned that no force is required to keep an object in motion. Galileo proposed that the natural tendency for a moving object not affected by an outside force is to continue moving at a uniform speed and in a straight line. This concept was later formalized by Newton and is now known as Newton's first law of motion.

The problem, then, was not to explain the force that keeps the planets moving but rather to determine the force that keeps them from going in a straight line out into space. It was to this end that Newton conceptualized the force of gravity. At the early age of 23, he envisioned a force that extends from the earth into space and holds the moon in orbit around the earth. Although others had theorized the existence of such a force, he was the first to formulate and test the *law of universal gravitation,* which states that every body in the universe attracts every other body with a force that is proportional to their masses and inversely propor-

FIGURE 17.12
Sir Isaac Newton. (Yerkes Observatory photograph)

tional to the square of the distance between them. Thus, the gravitational force decreases with distance, so that two objects 3 kilometers apart have 3^2, or 9, times less gravitational attraction than when the same objects are 1 kilometer apart. The law of gravitation also states that the greater the mass of the object, the greater its gravitational force. The mass of an object can be considered the measurement of the total amount of matter it contains. More specifically, mass is the measurement of the resistance (or sluggishness) that an object exhibits in response to any effort made to change its state of motion. Often we confuse the concept of mass with our notion of weight. Specifically, weight is defined as the force due to gravity acting upon an object. However, unlike weight, the mass of an object does not change. For example, a person weighing 120 pounds on the earth would weigh ⅙ as much, or 20 pounds, on the moon, but the person's mass would remain unchanged.

With his laws of motion, Newton proved that the force of gravity, combined with the tendency of a planet to remain in straight-line motion, results in the elliptical orbits discovered by Kepler. The earth, for example, moves forward in its orbit about 30 kilometers (18½ miles) each second, and during the same second, the force of gravity pulls it toward the sun about ½ centimeter (⅛ inch). Therefore, as Newton concluded, it is the combination of the earth's forward motion and its "falling" motion that defines its orbit (Figure 17.13). If gravity were somehow eliminated, the earth would move in a straight line out into space. On the other hand, if the earth's forward motion were suddenly stopped, gravity would pull it toward the sun.

Up to this point, we have discussed the earth as if it were the only planet. However, all bodies in the solar system have gravitational effects on the earth. For that reason, the orbit of the earth is not the perfect ellipse determined by Kepler. Any variance in the orbit of a body from its predicted path is referred to as a **perturbation.** For example, Jupiter's gravitational pull on Saturn is enough to reduce Saturn's orbital period by nearly one week from the predicted period. As we shall see, the application of this concept led to the discovery of the planet Neptune because of Neptune's gravitational effect on the orbit of Uranus.

Newton used the law of gravitation to redefine Kepler's third law, which states the relationship between the orbital periods of the planets and their solar distances. When restated, Kepler's third law takes into account the masses of the bodies involved and thereby provides a method to determine the mass of a body when the orbit of one of its satellites is known. The mass of the sun is known from the orbit of the earth, and the mass of the earth has been determined from the orbit of the moon. In fact, the mass of any body with a satellite can be determined. The masses of bodies that do not have satellites can be determined only if the bodies noticeably affect the orbit of a neighboring body or of a nearby artificial satellite.

CONSTELLATIONS AND ASTROLOGY

At an early date, people became fascinated with the star-studded skies and began to identify the patterns they saw. These configurations, called **constellations,** were named in honor of mythological characters. It takes a good bit of imagination to make out the intended subjects, as most were probably not intended to be likenesses in the first place. Although we inherited many of the constellations from Greek mythology, it is believed that the Greeks acquired most of theirs from the Babylonians. Today, 88 constellations are recognized, and they are used to divide the sky into units, just as state boundaries divide the United States. Every star in the sky is in but is not necessarily part of one of these constellations. Constellations therefore enable astronomers to roughly identify that position of the heavens they are observing. For the student, the constellations provide a good way to become familiar with the night sky (see Appendix E).

Some of the brightest stars in the heavens were given proper names, such as Sirius, Arcturus, and Betelgeuse. In addition, the brightest stars in a constellation are generally named in order of their intensity by the letters of the Greek alphabet— alpha (α), beta (β), and so on—followed by the name of the parent constellation. For example, Sirius, the brightest star in the constellation Canis Major (Larger Dog), is called Alpha (α) Canis Majoris. Lesser stars are systematically numbered within each constellation.

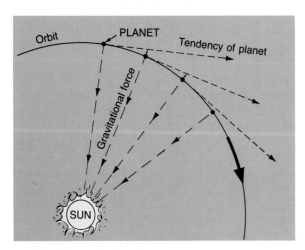

FIGURE 17.13
Orbital motion of the earth.

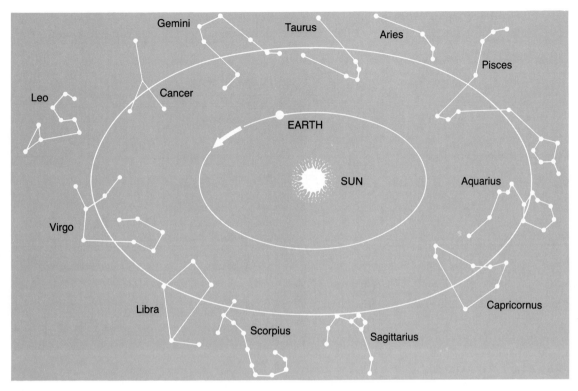

FIGURE 17.14
Constellations of the Zodiac.

The planets, moon, and sun lie along nearly the same plane. Therefore, they move along the same region of the sky, which has become known as the *Zodiac* ("zone of animals"). Because the earth's moon goes through its phases about twelve times each year, the Babylonians divided the Zodiac into twelve constellations (Figure 17.14). Thus, each successive full moon occurs in the next constellation. The twelve constellations of the Zodiac are Aries, Taurus, Gemini, Cancer, Leo, Virgo, Libra, Scorpio, Sagittarius, Capricorn, Aquarius, and Pisces. These names may be familiar to you as the astrological signs of the Zodiac. When first established, the vernal equinox (first day of spring) occurred when the sun was in the constellation Aries, and the signs of the Zodiac occurred in their respective constellations. Now, because of the earth's precession, the vernal equinox occurs when the sun is in Pisces. In several years, it will occur when the sun is in Aquarius. Hence, the "Age of Aquarius" is coming.

Although astrology is not a science and has no basis in fact, it did contribute to the science of astronomy. The positions of the moon, sun, and planets at the time of a person's birth (sign of the Zodiac) were considered to have great influence on that person's life. Even the great astronomer Kepler was required to make horoscopes as part of his duties. In order to make horoscopes for the future, astrologers attempted to predict the future positions of the celestial bodies. Consequently, some of the improvements in astronomical instruments were probably due to the desire for more accurate predictions of events such as eclipses, which were considered highly significant in a person's life.

Even prehistoric people built observatories. The structure known as Stonehenge, in England, was undoubtedly an attempt at better solar prediction (Figure 17.15). At the time of midsummer (June 21–22—the summer solstice), the rising sun emerges directly above the heel stone. Be-

FIGURE 17.15
Stonehenge, an ancient solar observatory. On June 21–22 (summer solstice), the
sun can be observed rising above the heel stone. (Photo by Robert Llewellyn)

FIGURE 17.16
Astronomical coordinate system on the celestial sphere.

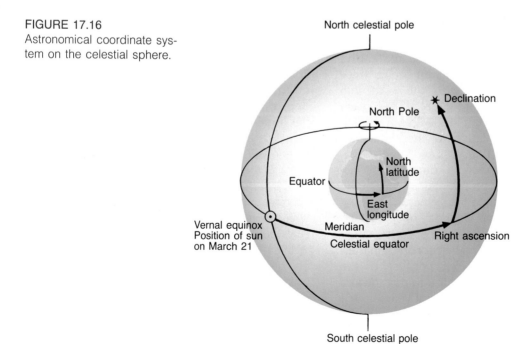

sides keeping this calendar, Stonehenge may also have provided a method of determining eclipses.

POSITIONS IN THE SKY

If you gaze away from the city lights on a clear night, you will get the distinct impression that the stars produce a spherical shell surrounding the earth. This impression seems so real that it is easy to understand why many early Greeks regarded the stars as being fixed to the clear crystalline celestial sphere. Although we realize this sphere does not exist, it is convenient to use it for locating stars.

One method for doing this, called the **equatorial system,** has the celestial sphere divided into a coordinate system very similar to that used for locations on the earth's surface (Figure 17.16). Because the celestial sphere appears to rotate around a line extending from the earth's axis, the north and south celestial poles are in line with the terrestrial North and South poles. The north celestial pole is very near the bright star Polaris ("pole star"), which is also called the North Star. To an observer in the Northern Hemisphere, the stars appear to circle Polaris, because it, like the North Pole, is in the center of motion (Figure 17.17). Figure 17.18 shows how to locate the North Star from the constellation of the Big Dipper. The intersection of a plane through the earth's equator with the celestial sphere defines the celestial equator (which is 90 degrees from the celestial poles).

The two most commonly used coordinates in the equatorial system, declination and right ascension, are analogous to latitude and longitude, respectively (Figure 17.16). Like latitude, **declination** is the angular distance north or south of the equator (celestial). **Right ascension** is the angular distance measured eastward along the celestial equator from the position of the vernal equinox. (The *vernal equinox* is at the point in the sky where the sun crosses the celestial equator, at the onset of spring.) While declination is expressed in degrees, right ascension is usually expressed in hours. This is not as complex as it first appears, since a simple relationship exists between degrees and hours. One complete rotation, or 360 degrees, takes 24 hours. Hence, 15 degrees equal 1 hour, and 1 degree equals 4 minutes. If the right

FIGURE 17.17
Star trails in the region of Polaris (north celestial pole) on a time exposure. (Courtesy of National Optical Astronomy Observatories)

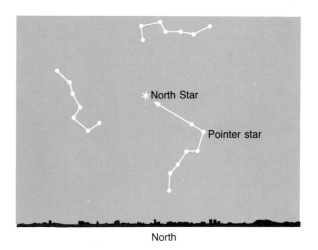

FIGURE 17.18
Locating the North Star from the Big Dipper above the northern horizon.

ascension of a star equals 2 hours, it can be expressed as 30 degrees. To visualize distances on the celestial sphere, it may be helpful to remember that the moon and sun have an apparent width of about ½ degree. The declination and right ascension of some of the brightest stars can be determined from the star charts in Appendix E.

MOTIONS OF THE EARTH

The two primary motions of the earth are rotation and revolution. **Rotation** is the turning of a body on its axis, and **revolution** is the motion of a body around some point in space. The main consequences of the earth's rotation are day and night. But, as stated earlier, day and night and the apparent motions of the stars can be accounted for equally well by a revolving sun and celestial sphere. Copernicus realized that a rotating earth greatly simplified the existing model of the universe and, therefore, strongly advocated it as the correct view. He was, however, unable to prove that the earth rotates. The first substantial proof was presented 300 years after his death, by the French physicist Jean Foucault.

In 1851 Foucault used a free-swinging pendulum to demonstrate that the earth does, in fact, turn on its axis. To envision Foucault's arguments, imagine a large pendulum swinging at the North Pole. Keep in mind that once a pendulum is put into motion, it continues swinging in the same plane unless acted upon by some outside force. A sharp stylus attached to the bottom of this pendulum marks the snow as it oscillates. When we observe the marks made by the stylus, we note that the pendulum is slowly but continually changing position. In 24 hours it has returned to the starting position (Figure 17.19). Since no outside force acted on the pendulum to change its position, what we observed must have been the earth rotating under it. Foucault conducted a similar experiment in Paris.

The earth's rotation has become a standard method of measuring time. Each rotation equals about 24 hours. We can, however, measure the earth's rotation in two ways. Consequently, there are two kinds of days. Most familiar is the **mean solar day,** the time interval from one noon to the next, which averages 24 hours. Noon is when the sun has reached its zenith (highest point). The **sidereal day,** on the other hand, is the time it takes

FIGURE 17.19
Apparent movement of a pendulum at the North Pole caused by the earth rotating beneath it.

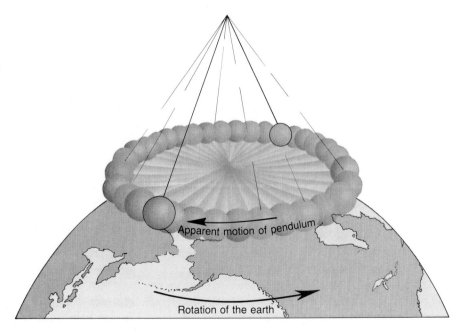

Apparent motion of pendulum

Rotation of the earth

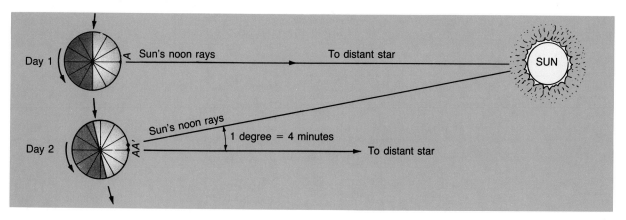

FIGURE 17.20
Illustration of a solar and sidereal day.

for the earth to make one complete rotation (360 degrees) with respect to a particular star. It is measured from the time required for a star to reappear at the identical position in the sky. The sidereal day has a period of 23 hours, 56 minutes, and 4 seconds (measured in solar time), which is almost 4 minutes shorter than the mean solar day. This difference results because the direction to the distant stars is only infinitesimally changed by the earth's motion around the sun, whereas the direction to the sun changes by almost 1 degree each day. This difference is diagrammatically shown in Figure 17.20. If it is not obvious why we use the solar day rather than the time required for one rotation as a measure of a civil day, consider the fact that in sidereal time "noon" occurs 4 minutes earlier each day. In six months it occurs at "midnight." Astronomers use sidereal time because the stars appear in the same position in the sky every 24 sidereal hours. Usually, an observatory will begin its sidereal day when the position of the vernal equinox is directly overhead, that is, over the meridian on which the observatory is located. Therefore, when the observatory's sidereal clock is the same as the star's right ascension, the star will be overhead, or at its highest point. For example, the brightest star in the heavens, Sirius, has a right ascension of 6 hours, 42 minutes, and 56 seconds and will, therefore, be overhead when the clock at the observatory indicates that time.

The earth revolves around the sun in an elliptical orbit at a speed of over 107,000 kilometers (66,000 miles) per hour and at a distance that averages 150 million kilometers (93 million miles). About January 3, the earth is closest to the sun—147 million kilometers (**perihelion** position), and around July 4, it is farthest from the sun—152 million kilometers (**aphelion** position). Due to the earth's annual motion around the sun, each day the sun appears to be displaced eastward among the constellations at a distance equal to about twice its width. The apparent annual path of the sun upon the celestial sphere is called the **ecliptic.** Most of the planets and the moon travel in nearly the same plane as the earth. Hence, their paths on the celestial sphere lie near the ecliptic. The most notable exception is Pluto, whose orbit is tilted 17 degrees to the plane of the earth's orbit.

The imaginary plane that connects the earth's orbit with the celestial sphere is called the **plane of the ecliptic.** From this reference plane, the earth's axis of rotation is tilted about 23½ degrees. The earth is not unique in this respect. Other planets have similarly tilted axes. Mars is tilted 24 degrees, and Saturn has a tilt of 27 degrees. Because of the earth's tilt, the Northern Hemisphere leans toward the sun in June and away from the sun in December (see Figure 12.6). The main consequences of this change are the seasons, which are discussed in detail in Chapter 12. Also, because of

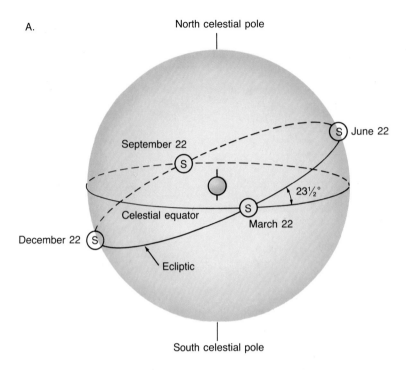

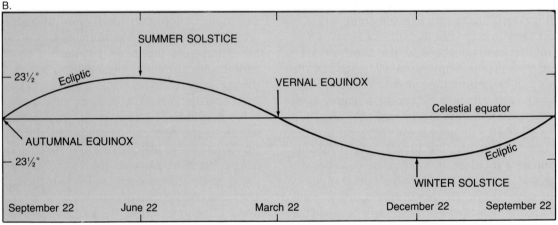

FIGURE 17.21
A. The earth's orbital motion causes the position of the sun to move about 1 degree each day on the celestial sphere. **B.** Curved path of the ecliptic as it appears plotted on the celestial sphere.

the earth's tilt, the apparent path of the sun (ecliptic) and the celestial equator intersect each other at an angle of 23½ degrees (Figure 17.21A). Therefore, when the apparent position of the sun is plotted on the celestial sphere over a period of a year's time, its path traces a curve that intersects

the celestial equator at two points (Figure 17.21B). From a Northern Hemisphere point of view, these intersections are called the vernal and autumnal equinoxes and occur about March 20–21 and September 22–23, respectively. On June 21–22, the date of the summer solstice, the sun appears 23½

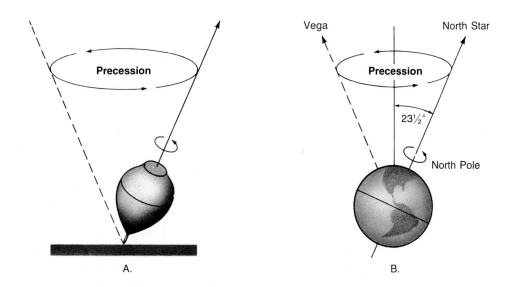

FIGURE 17.22
A. Precession illustrated by a spinning top. **B.** Precession of the earth causes the
location of the north celestial pole to change and, therefore, also the apparent
position of the North Star.

degrees north of the celestial equator, and six
months later, on December 21–22, the date of the
winter solstice, the sun appears 23½ degrees
south of the celestial equator.

A third and very slow motion of the earth is
called **precession.** Although the earth's axis maintains approximately the same angle of tilt, its direction is always changing, with the axis tracing a
circle on the sky. This movement is very similar to
the movement of a spinning top (Figure 17.22A).
At the present time, the axis points toward Polaris.
In 14,000 A.D., it will point toward the bright star
Vega, which will then be the North Star (Figure
17.22B). The period of precession is 26,000 years.
By the year 28,000, Polaris will once again be the
North Star.

Precession has only a minor effect on the seasons, because the angle of tilt changes only
slightly. It does, however, cause the positions of
the seasons (equinox and solstice) to move
slightly each year among the constellations. This
effect causes the signs of the Zodiac, measured in
30-degree segments from the position of the vernal equinox, to shift from the constellations of the
Zodiac, which are "fixed."

In addition to its own motion, the earth shares
numerous motions with the sun. It accompanies

the sun as it speeds in the direction of the bright
star Vega at 20 kilometers (12 miles) per second.
Also, the sun, like other nearby stars, revolves
around the galaxy, a trip which requires 200 million years at speeds approaching 300 kilometers
(200 miles) per second. In addition, the galaxies
themselves are in motion. We are presently approaching one of our nearest galactic neighbors,
Andromeda. In summary, the motions of the earth
are many and complex, and its speed is very great.

CALENDARS

One of the earliest responsibilities of astronomers
was the keeping of the calendar. Although the oldest known calendars date from the eighth century
B.C., others probably existed much earlier. These
calendars had as their basis the movements of the
heavenly bodies. Although the week has no astronomical significance, the seven days are clearly
named after the seven moving celestial bodies
known to the ancients: the sun, the moon, and the
five planets (Mercury, Venus, Mars, Jupiter, and
Saturn), which are easily seen with the unaided
eye. For example, Sunday is "Sun's day," Monday
is "Moon's day," and Saturday is "Saturn's day."

The remaining days of the week were named in the Romance languages after Mars, Mercury, Jupiter, and Venus, in that order. For example, in Spanish, Tuesday is *Martes.*

The 29.5-day cycle of the moon through its phases is the most noticeable heavenly phenomenon except for the day itself. It became the *moonth,* now called the month. The first Western calendars based the year on the phases of the moon. Twelve *moonths* equaled a year. Because there are actually 12.4 lunar cycles in a year, the calendar had to have a full month added every three years to keep the seasons in accord. Even with this correction, the calendar was still slowly falling behind. When Julius Caesar gained power, the calendar indicated spring, while the weather indicated the middle of winter. To correct this situation, Julius Caesar ordered that 80 days be added to the year of 46 B.C. That year, for obvious reasons, was called the "year of confusion." Also at the direction of an astronomer, Caesar ordered that the calendar be based on the tropical year* of

---

*One tropical year is the length of time required for successive passages of the sun through the vernal equinox (365.2422 days).

365.25 days. This was done by having 365 days each normal year and adding an extra day every fourth year. So began the tradition of leap year and the **Julian calendar.** However, the tropical year is slightly less than 365.25 days. The difference is only a little more than 10 minutes per year, but the Julian calendar worked just like a fast clock. Both would require adjustment if used long enough. By the sixteenth century, the Julian calendar was ahead by 10 days.

In 1582 the **Gregorian calendar,** which we presently use, was developed. The extra 10 days were eliminated by making Friday, October 15, the day after Thursday, October 4. To slow the calendar down, selected leap years were eliminated. A leap year is no longer added for centennial years except those divisible by 400. Hence, the years 1600, 2000, and 2400 are leap years, but all centennial years between them (e.g., 1900) are not. Our present calendar is accurate to within 1 day in 3000 years. Not all countries adopted the new calendar at the same time. When George Washington was born, the calendar indicated February 11, 1732, but when the colonies adopted the Gregorian calendar in 1752, his birth date became February 22, the day we now celebrate.

REVIEW QUESTIONS

1 Why did the ancients think that celestial objects had some control over their lives?

2 Describe what produces the retrograde motion of Mars. What geometric arrangement did Ptolemy use to explain this motion?

3 What major change did Copernicus make in the Ptolemaic system? Why was this change philosophically significant?

4 What was Tycho Brahe's contribution to science?

5 Does the earth move faster in its orbit near perihelion (January) or near aphelion (July)? Keeping your answer to the previous question in mind, is the solar day longest in January or in July?

6 Use Kepler's third law ($p^2 = d^3$) to determine the period of a planet whose solar distance is
  (a) 10 AU.
  (b) 1 AU.
  (c) 0.2 AU.

7 Use Kepler's third law to determine the distance from the sun of a planet whose period is
  (a) 5 years.
  (b) 10 years.
  (c) 10 days.

8   Did Galileo invent the telescope?

9   Explain how Galileo's discovery of a rotating sun supported the Copernican view of a sun-centered universe.

10 Using a diagram, explain why the fact that Venus appears full when it is smallest supports the Copernican view and is inconsistent with the Ptolemaic system.

11 Newton learned that the orbits of the planets are the result of two actions. Explain these actions.

12 Of what value are constellations to modern-day astronomers?

13 Express the declination and right ascension of Arcturus (see Appendix E).

14 Explain the difference between the solar and the sidereal day.

15 Why was the concept of a leap year introduced?

16 Why was the Julian calendar revised?

KEY TERMS

geocentric	equatorial system	aphelion
celestial sphere	declination	ecliptic
heliocentric	right ascension	plane of the ecliptic
Ptolemaic system	rotation	precession
retrograde motion	revolution	Julian calendar
astronomical unit	mean solar day	Gregorian calendar
perturbation	sidereal day	
constellation	perihelion	

18

THE MOON, CELESTIAL OBSERVATIONS, AND THE SUN

The celestial objects that have received the greatest attention are the moon and the sun. Several solar telescopes throughout the world keep the sun under constant surveillance, and the moon has inspired observers ever since Galileo first pointed his telescope skyward and revealed its beautiful crater-marked surface. Further, the lunar space probes, culminating with the *Apollo* missions, have provided first-hand information on our nearest companion in space. These data have ended years of controversy over the possible nature of the lunar surface and, as hoped, have provided clues to the early history of the solar system, for erosion has not appreciably altered the lunar surface in the last few billion years. This chapter will examine the basic features of the moon and the sun and provide a discussion of the tools used by astronomers to probe the universe.

Aerial view of the Kitt Peak Observatory, Arizona. (Association of Universities for Research in Astronomy, Inc., The Kitt Peak National Observatory)

THE MOON

Only one natural satellite, the moon, accompanies the earth on its annual journey around the sun. This planet-satellite system is unique in the solar system, because the earth's moon is unusually large compared to its parent planet. The diameter of the moon is 3475 kilometers (2150 miles), and from the calculation of its mass, its density is 3.3 times that of water. This density is comparable to that of crustal rocks on earth but a fair amount less than the earth's average density. Geologists have suggested that this difference can be accounted for if the moon's iron core is rather small. The gravitational attraction at the lunar surface is one-sixth of that experienced on the earth's surface. This difference allows an astronaut to lift a "heavy" life-support system with relative ease. When not carrying such a load, an astronaut could jump six times higher than on earth, easily clearing a one-story building.

LUNAR MOTION

Like the earth, the moon rotates on its axis but does so at a much slower rate than the earth. It also revolves around the earth with a period of about one month. The latter motion changes the relative positions of the sun, earth, and moon, producing the **phases of the moon** and eclipses of the sun and moon.

FIGURE 18.1
Phases of the moon. **A.** The outer figures show the phases as seen from the earth. **B.** Compare these photographs with the diagram. (Courtesy of Lick Observatory)

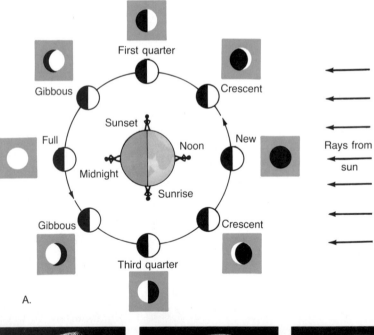

A.

B.

474

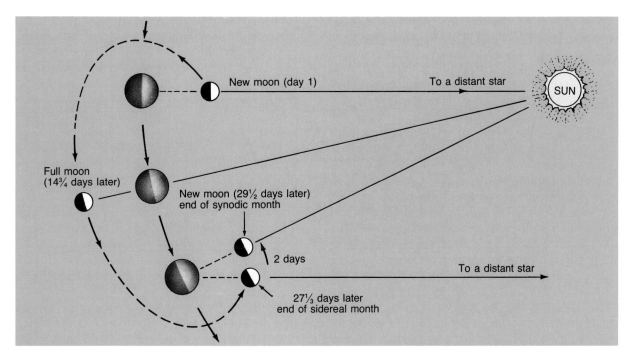

FIGURE 18.2
Sidereal and synodic months.

Phases of the Moon The first astronomical phenomenon to be understood was the cycle of the moon. On a monthly basis, we observe the phases as a systematic change in the amount of the moon that appears illuminated (Figure 18.1). About two days after a new cycle begins, a thin sliver (*crescent phase*) appears low in the western sky just after sunset. During the following week, it grows to a half circle (*first-quarter phase*), which is visible from noon to midnight. In another week, the complete disk (*full-moon phase*) can be seen rising in the east as the sun is sinking in the west. During the next two weeks, the percentage of the moon that can be seen steadily declines, until the moon disappears altogether (*new-moon phase*). The cycle soon begins anew with the reappearance of the crescent moon.

The lunar phases are a consequence of the motion of the moon and the sunlight that is reflected from its surface (Figure 18.1B). Half of the moon is illuminated at all times, but to an earthbound observer, the percentage of the bright side facing the person depends on the location of the moon with respect to the sun and the earth. When the moon lies between the sun and the earth, none of its bright side faces the earth, producing the new-moon ("no-moon") phase. Conversely, when the moon lies on the side of the earth opposite the sun, all of its lighted side faces the earth, producing the full moon. At a position between these extremes, an intermediate amount of the moon's illuminated side is visible from the earth.

Motion The cycle of the moon through its phases requires 29½ days, a time span called the **synodic month.** Recall that this cycle was the basis for the first Roman calendar. However, this is not the true period of the moon's revolution, which takes only 27⅓ days and is known as the **sidereal month.** The reason for the difference of nearly 2 days each cycle is shown diagrammatically in Figure 18.2. Note that as the moon orbits the earth, the earth-moon system also moves in an orbit around the sun. Consequently, even after the moon has made a complete revolution around the earth, it has not yet reached its starting position, which was directly between the sun and earth (new-moon phase). This additional motion takes another 2 days.

An interesting fact concerning the motions of the moon is that the periods of rotation and revolution are the same—27⅓ days. Because of this, the same lunar hemisphere always faces the earth. All of the manned *Apollo* missions have been confined to this side, although numerous photographs have been taken of the "back" side by orbiting satellites.

Since the moon rotates very slowly, any location on its surface experiences periods of daylight and darkness lasting about two weeks. This, along with the absence of an atmosphere, accounts for the high surface temperature of 127°C (261°F) on the day side of the moon and the low surface temperature of −173°C (−280°F) on its night side.

Eclipses Along with understanding the moon's phases, the early Greeks also realized that eclipses are simply shadow effects (Figure 18.3). When the moon moves directly between the earth and the sun (new-moon phase), it casts a dark shadow on the earth, producing a **solar eclipse.** On the other hand, the moon is eclipsed **(lunar eclipse)** when it moves within the shadow of the earth, a situation that is only possible during the full-moon phase. However, if this is the case, why does a solar eclipse not occur with each new-moon phase and a lunar eclipse with each full-moon phase? They would if the orbit of the moon lay along the plane of the earth's orbit. However, the moon's orbit is inclined about 5 degrees to the plane of the

ecliptic. Thus, the shadow of the moon (or earth) can pass above or below the earth (or moon). Only if a new- or full-moon phase occurs when the moon lies in the plane of the ecliptic can an eclipse take place. The maximum number of eclipses in one year is seven (more of which are solar), and the minimum number is two (both solar).

During a total lunar eclipse, the circular shadow of the earth can be seen moving across the disk of the full moon. When totally eclipsed, the moon will be completely within the earth's shadow but will still be visible as a coppery disk, because the earth's atmosphere bends and transmits some long-wavelength light (red) into its shadow. A total eclipse of the moon can last up to four hours and is visible to anyone on the side of the earth facing the moon.

During a total solar eclipse, the moon casts a shadow that is never wider than 275 kilometers (170 miles). Anyone observing in this region will see the moon slowly block the sun from view and the sky darken. Near totality, a sharp drop in temperature is experienced. The solar disk is blocked for at most seven minutes, and then one edge reappears. At totality, the dark moon is seen covering the complete solar disk, and only the sun's brilliant white outer atmosphere is visible (see Figure 18.26). Total solar eclipses are only visible to people in the dark part of the moon's shadow (*umbra*), while a partial eclipse can be seen by those in the light portion (*penumbra*) (Figure 18.3). Partial solar eclipses are relatively common in the polar regions, for it is there that the penum-

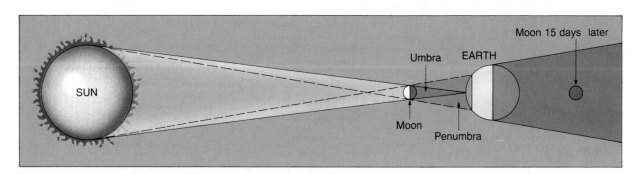

FIGURE 18.3
Positions of the sun, moon, and earth during a solar eclipse and a lunar eclipse (not to scale).

bra hits when the dark shadow of the moon just misses the earth. A total solar eclipse is a rare event at any given location. The next one that will be visible from the contiguous United States will take place on August 21, 2017.

THE LUNAR SURFACE

When Galileo first pointed his telescope toward the moon, he saw two different types of terrain (Figure 18.4). The dark areas he observed are now known to be fairly smooth lowlands, while the bright regions are densely cratered highlands. Because the dark regions resembled seas on earth, they were later named **maria** (singular, *mare:* Latin for "sea"). This name is unfortunate because the moon's surface is totally void of water.

Today we know that the moon has no atmosphere and lacks water as well. Therefore, the processes of weathering and erosion which continually modify the earth are virtually lacking. In addition, tectonic events such as earthquakes and volcanic eruptions no longer occur on the moon. However, since the moon is unprotected by an atmosphere, tiny particles (micrometeorites) continually bombard its surface and ever so gradually smooth the landscape. Rocks, for example, can become

FIGURE 18.4
Telescopic view of the lunar surface. (Courtesy of Lick Observatory)

slightly rounded on top if exposed at the lunar surface for a long enough period. Nevertheless, it is unlikely, except for the addition of a few large craters, that the moon has changed appreciably in the last 3 billion years.

The most obvious features of the lunar surface are craters. They are so profuse that craters within craters within craters are the rule. The larger ones seen in the lower portion of Figure 18.4 are about 250 kilometers (155 miles) in diameter, and these often overlap. Most craters were produced by the impact of rapidly moving debris (meteoroids), a situation that was considerably more common in the early history of the solar system than it is today. By contrast, the earth has only about a dozen recognized impact craters. This difference can be attributed to the earth's atmosphere, which burns up small debris before it reaches the ground. In addition, evidence for most of the craters which did form early in the earth's history has since been destroyed by erosion or tectonic processes.

The formation of an impact crater is illustrated in Figure 18.5. Upon impact, the high-speed meteoroid compresses the material it strikes, then almost instantaneously the compressed rock rebounds, ejecting material from the crater. This process is analogous to the splash that occurs when a rock is dropped into water, and it often results in the formation of a central peak as seen in the large crater in Figure 18.6. Most of the ejected material (*ejecta*) lands near the crater, building a rim around it. The heat generated by the impact is sufficient to melt some of the impacted rock. Astronauts have brought back samples of glass beads produced in this manner, as well as rock formed when angular fragments and

FIGURE 18.5
Formation of an impact crater. The energy of the rapidly moving meteoroid is transferred into heat energy and compressional waves. The rebound of the compressed rock causes debris to be ejected from the crater, and the heat melts some material, producing glass beads. Small secondary craters are formed by the material "splashed" from the impact crater. (After E. M. Shoemaker)

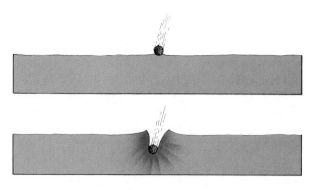
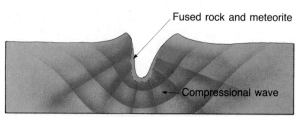
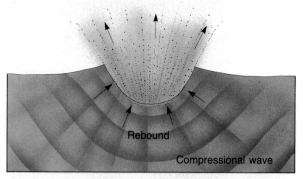

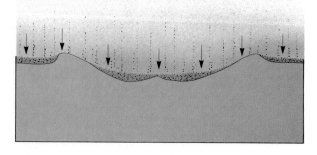

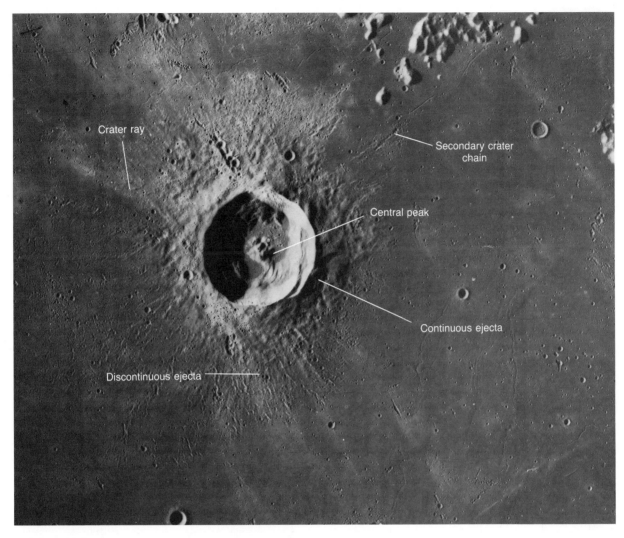

FIGURE 18.6
The 20-kilometer-wide lunar crater Euler located in the southwestern Mare Imbrium. Clearly visible are the bright rays, central peak, secondary craters, and the large accumulation of ejecta near the crater rim. (Courtesy of NASA)

dust were welded together by the impact. The latter material is called **lunar breccia.**

A meteoroid only 3 meters (10 feet) wide can blast out a 150-meter (500-foot) crater. A few of the large craters such as Kepler and Copernicus, shown in Figure 18.4, formed from the impact of bodies one kilometer, or more, in diameter. These two large craters are thought to be relatively young because of the bright **rays** ("splash" marks) that radiate outward for hundreds of kilometers.

These bright rays consist of fine debris ejected from the primary crater, including impact-generated glass beads, as well as material displaced during the formation of smaller, secondary craters.

Densely pock-marked highland areas make up most of the lunar surface. In fact, all of the "back" side of the moon is characterized by such topography. Within the highland regions are mountain ranges that have been named for mountainous

FIGURE 18.7
Formation of lunar maria. **A.** Impact of an asteroid-sized mass produced a huge crater hundreds of kilometers in diameter and disturbed the lunar crust beyond the crater. **B.** Filling of the impact area with fluid basalts, perhaps derived from partial melting deep within the lunar mantle.

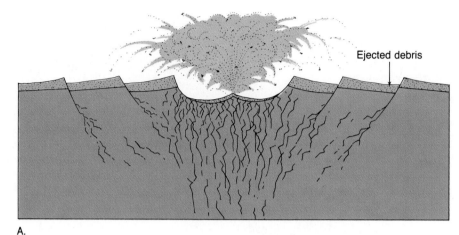

Ejected debris

A.

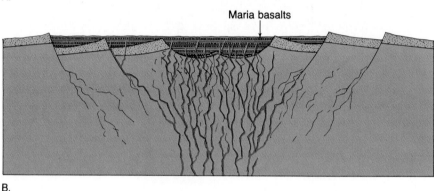

Maria basalts

B.

terrains on earth. The highest lunar peaks reach elevations approaching 8 kilometers, only 1 kilometer lower than Mount Everest.

Although highlands predominate, the less rugged maria have attracted most of the interest. The origin of maria basins as enormous impact craters produced by the violent impact of at least a dozen asteroid-sized bodies was hypothesized before the turn of the century by the noted American geologist G. K. Gilbert (Figure 18.7A). However, it remained for the *Apollo* missions to determine what filled these depressions to produce the relatively flat topography. Apparently the craters were flooded, layer upon layer, by very fluid basaltic lava, in which case they somewhat resemble the Columbia Plateau in the northwestern United States (Figure 18.7B). The photograph in Figure 18.8 reveals the forward edge of a lava flow that is "frozen" in place. Astronauts have also viewed and photographed the layered nature of maria. The

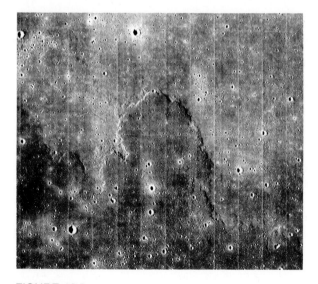

FIGURE 18.8
Margin of a lava flow on the surface of Mare Imbrium. (Courtesy of National Space Data Center)

FIGURE 18.9
Astronaut Harrison Schmitt sampling the lunar surface. Notice the footprints left in the lunar "soil." (Courtesy of NASA)

layers are often over 30 meters (100 feet) thick, and the total thickness of the material that fills the maria must approach thousands of meters.

On several occasions the lava flowed beyond the impact crater, engulfing the surrounding lowlands. If the rim of a remnant crater can be seen above the lava, an estimate of the flow's thickness can be made. Many geologists believe that the impacts that produced maria basins were great enough to fracture the lunar crust some distance away. Examples of basins in which all of the disturbed regions were filled to overflowing include Mare Tranquillitatis (Sea of Tranquility), the site where astronaut Neil Armstrong was the first person to step onto the lunar surface, and Mare Imbrium (Sea of Rains). Some maria basalts fill only the central crater; these appear as dark, smooth crater floors in Figure 18.4.

All lunar terrains are mantled with a layer of gray, unconsolidated debris derived from a few billion years of meteoric bombardment (Figure 18.9). This soil-like layer, properly called **lunar regolith,** is composed of igneous rocks, breccia, glass beads, and fine particles commonly called *lunar dust.* As meteoroid after meteoroid collided with the lunar surface, the thickness of the lunar regolith increased, while the size of the bombarded debris diminished. In the maria that have been explored by *Apollo* astronauts, the lunar regolith is apparently just over 3 meters (10 feet) thick, but it is believed to form a thicker mantle upon the older highlands.

LUNAR HISTORY

Although the moon is our nearest planetary neighbor and astronauts have sampled its surface, much is still unknown about is origin. Until recently, the most widely held hypothesis argued that the formation of the moon paralleled that of the earth and the other planets. That is, the moon formed from minute rock fragments and gases which composed a disk-shaped structure that orbited the protosun. Debris from this disk collided and accumulated into larger masses which, in turn, accreted into planetary-sized bodies.

A new hypothesis, which has recently gained support from many scientists, suggests that a giant body collided with the earth to produce the moon. The explosion caused by the impact of a Mars-sized body with a semimolten Earth is thought to have ejected huge quantities of mantle rock from the primordial Earth. A portion of this ejecta remained in orbit around the earth, while the remainder either escaped or impacted upon the earth's surface. In a manner similar to that proposed in the earlier hypothesis, the material orbiting the earth then began to accumulate, eventually producing the moon. Though the giant-impact hypothesis provides a plausible mechanism for the moon's formation, many questions must be answered before this proposal can be considered viable.

Despite the fact that the origin of the moon is still debated, planetary geologists have been able to work out some of the basic details of the moon's history, using among other things variations in crater density (quantity per unit area). Simply stated, the higher the crater density, the longer the topographic feature has existed. During its early history the moon was continually impacted as it swept up debris from the solar nebula. This continuous bombardment and perhaps radioactive decay generated enough heat to melt the moon's outer shell, and quite possibly the rest of the moon as well.

When a large percentage of the debris had been gathered, the outer layer of the moon began to cool and form a crystalline crust. From samples obtained by *Apollo* astronauts, the rocks of the primitive lunar crust are thought to be composed of a high percentage of a calcium-rich feldspar (anorthosite). This feldspar mineral crystallized early and, because it was less dense than the remaining melt, floated to the top and formed a surface scum. While this process was taking place, iron and other heavy metals probably sank to form a small central core. Even after the crust had solidified, its surface was continually bombarded. Remnants of the original crust occupy the densely cratered highlands, which have been estimated to

be as much as 4.5 billion years old.

The last period of heavy bombardment recorded in the lunar highlands occurred almost 500 million years after the crust had formed. It is not known with certainty whether this final episode of bombardment was simply a clean-up phase where the remaining large particles in the earth-moon orbit were swept up or whether it was an influx of bodies from farther out in the solar system.

The next major event in the moon's evolution was the formation of maria basins (see Figure 18.7). The meteoroids that produced these huge pits ejected mountainous quantities of lunar rock into piles rising 5 kilometers or more. The Apennine mountain range, which typifies such an accumulation, was produced in conjunction with the formation of the Imbrium basin, the site explored by the *Apollo 15* astronauts. The crater density of the ejected material is greater than that of the surface of the associated mare, confirming that an appreciable time elapsed between the formation and filling of these basins. Radiometric dating of the maria basalts puts their age between 3.2 and 3.8 billion years, somewhat younger than the initial crust. In places, the lava flows overlap the highlands, another testimonial to the lesser age of the maria deposits.

The last prominent features to form on the lunar surface were the rayed craters as exemplified by the crater Copernicus (see Figure 18.4). Material ejected from these "young" depressions is clearly seen blanketing the surface of the maria and many older rayless craters. By contrast, the older craters have rounded rims, and their rays have been erased by the impact of small debris. However, even a relatively young crater like Copernicus must be millions of years old. Had it formed on the earth, erosional forces would have long since obliterated it.

Evidence carried back from the lunar landings indicates that most, if not all, of the moon's tectonic activity ceased about 3 billion years ago. The youngest maria lava flows are about equivalent in age to the oldest rocks so far discovered on the earth. If photos of the moon taken several hundreds of millions of years ago were available, they would reveal that the moon has changed little in

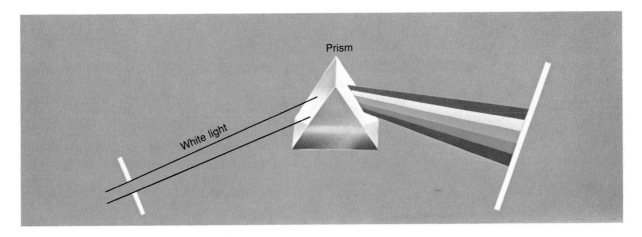

FIGURE 18.10
A spectrum is produced when sunlight is passed through a prism, which bends
each wavelength differently.

the intervening years. By all standards of measure the moon is a dead body wandering through space and time.

THE STUDY OF LIGHT

The vast majority of our information about the universe is obtained from the study of the light emitted from celestial bodies. Although visible light is most familiar to us, it constitutes only a small part of an array of energy generally referred to as **electromagnetic radiation.** Included in this array are gamma rays, x rays, ultraviolet light, visible light, infrared light, and radio waves (see Figure 12.8). All radiant energy travels through the vacuum of space in a straight line at the rate of 300,000 kilometers (186,000 miles) per second, which equals a staggering 26 billion kilometers per day.*

NATURE OF LIGHT

The properties of light do not conform to our common knowledge. In some instances, light behaves like waves, and in others, like discrete particles. In the wave sense, light is analogous to swells in the

ocean. This motion is characterized by the wavelength, which is the distance from one crest to the next. Wavelengths vary, from several kilometers in length for radio waves to less than a billionth of a centimeter for gamma rays. Most of these waves are either too long or too short for our eyes to detect. The small band of electromagnetic radiation we can see is often called **white light.** However, even white light consists of an array of waves having various wavelengths, a fact that can easily be demonstrated with a prism (Figure 18.10). As white light is passed through a prism, violet, the color with the shortest wavelength, is bent more than blue, which is bent more than green, and so forth (Table 18.1). Thus, white light can be

TABLE 18.1
Colors and corresponding wavelengths.

Color	Wavelength (angstroms*)
Violet	3800–4400
Blue	4400–5000
Green	5000–5600
Yellow	5600–5900
Orange	5900–6400
Red	6400–7500

*One angstrom equals 10^{-8} centimeter.

*Light rays are "bent" slightly when they pass nearby a very massive object such as the sun.

separated into its component colors in the order of their wavelengths, producing the familiar rainbow of colors.

Nevertheless, some effects of light cannot be explained using the wave theory. In these cases, light acts like a stream of particles, analogous to infinitesimally small bullets fired from a machine gun. These particles, called **photons,** can exert a pressure ("push") on matter, which is called **radiation pressure.** The photons from the sun are re-sponsible for "pushing" material away from a comet to produce its tail. Each photon has a spe-cific amount of energy, which is related to its wavelength in a simple way: Shorter wavelengths correspond to more energetic photons. Thus, blue light has more energetic photons than red light.

Which theory of light—the wave theory or the particle theory—is correct? Both, since each will predict the behavior of light for certain phenom-ena. As George Abell, a prominent astronomer,

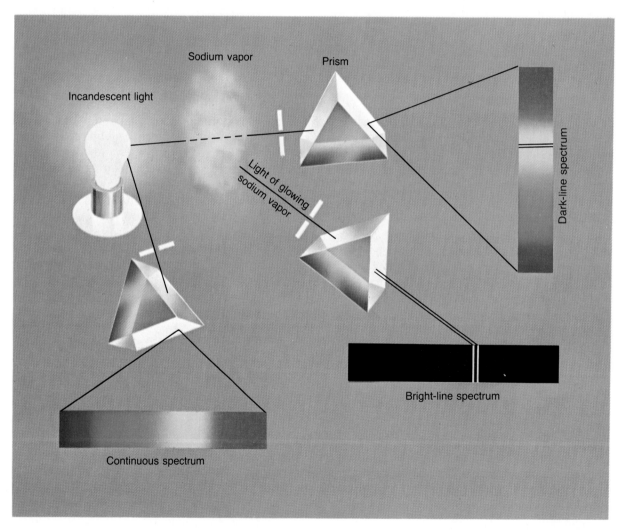

FIGURE 18.11
Formation of the three types of spectra.

stated about all scientific laws, "The mistake is only to apply them to situations that are outside their range of validity."

SPECTROSCOPY

When Sir Isaac Newton used a prism to disperse white light into its component colors, he unknowingly initiated the field of **spectroscopy,** which is the study of wavelength-dependent properties of light. The rainbow of colors Newton produced is called a continuous spectrum, because all wavelengths of light are included. It was later learned that two other types of spectra exist and that all three are generated under somewhat different conditions (Figure 18.11).

1. A **continuous spectrum** is produced by an incandescent solid, liquid, or gas under high pressure. It consists of all visible wavelengths. One example would be light generated by a common light bulb.
2. A **bright-line,** or **emission, spectrum** is produced by an incandescent gas under low pressure. It consists of a series of bright lines of particular wavelengths, which are dependent on the gas that produces them.
3. A **dark-line,** or **absorption, spectrum** is produced when white light is passed through a gas under low pressure. Since the gas will absorb selected wavelengths of light, the spectrum that is produced appears as a continuous spectrum with a series of dark lines running through it. These dark lines appear in the exact location as the bright lines that are produced by this gas in an emission spectrum.

The spectra of most stars are of the dark-line type. The importance of these spectra is that each element or compound in the gaseous form (in a star, material is usually in the gaseous form) produces a unique set of spectral lines. When the spectrum of a star is studied, the spectral lines act as "fingerprints," which identify the elements present. In an admittedly oversimplified manner, we can imagine how the sun and other stars create their dark-line spectrum. A continuous spectrum is produced in the interior of the sun, where the gases are under very high pressure. When this light passes through the less dense gases of the solar atmosphere, they extract selected wavelengths, producing the dark lines.

When Newton studied solar light, he obtained a continuous spectrum. However, when a prism is used in conjunction with lenses, the solar spectrum can be dispersed even further. An instrument that does this is called a **spectroscope.** The spectrum of the sun contains thousands of dark lines. Over 60 elements have been identified by matching these lines with those of elements known on earth.

Two other facts concerning a radiating body are important. First, if the temperature of a radiating surface is increased, the total amount of energy emitted increases at a rate known as the Stefan-Boltzmann law. Simply stated, the total amount of energy radiated by a body is directly proportional to the fourth power of its absolute temperature. For example, if the temperature of a star is doubled, the total radiation emitted increases by 2^4 ($2 \times 2 \times 2 \times 2$), or 16 times. Second, as the temperature of an object increases, a larger proportion of its energy is radiated at shorter wavelengths. To illustrate this, imagine a metal that is heated slowly. The rod first appears dull red (long wavelengths), and later bluish white (short wavelengths). From this, it follows that blue stars are hotter than yellow stars, which are hotter than red stars.

THE DOPPLER EFFECT

You may have heard the change in pitch of a train whistle as a train passes by. When the train is approaching, the sound seems to have a higher than normal pitch, and when it is moving away, the pitch appears lower than normal. This effect, which occurs for both sound and light waves, was first explained by Christian Doppler in 1842 and is called the **Doppler effect.** The reason for the difference is that it takes some time for the wave to

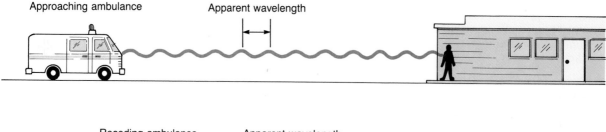

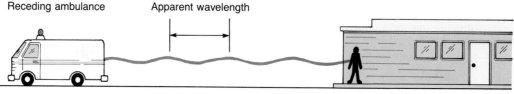

FIGURE 18.12
The Doppler effect, illustrating the apparent lengthening and shortening of wavelengths caused by the relative motion of a source and an observer.

be emitted. If the source is moving away, the beginning of the wave is emitted nearer to you than the end, which tends to "stretch" the wave, that is,

give it a longer wavelength (Figure 18.12). The opposite would be true for an approaching source. In the case of light, when a source is moving away, its light will appear redder than it actually is, since its waves appear lengthened, while objects approaching will have their light waves shifted toward the blue (shorter wavelength). Thus, if a source of red light approached you at a very high speed (near the speed of light), it would actually appear blue. The same effect would be produced if you moved and the light was stationary.

Therefore, the Doppler effect reveals whether the earth is approaching or receding from a star or another celestial body. In addition, the amount of shift allows us to calculate the rate at which the relative movement is occurring. Larger Doppler shifts indicate higher velocities and vice versa. Doppler shifts are generally measured from the dark lines in the spectra of stars by making a comparison with a standard spectrum produced in the laboratory (Figure 18.13).

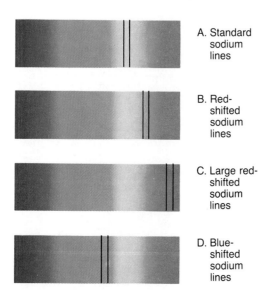

A. Standard sodium lines

B. Red-shifted sodium lines

C. Large red-shifted sodium lines

D. Blue-shifted sodium lines

FIGURE 18.13
Illustration of standard sodium lines as produced in a laboratory compared to sodium lines as they would appear when the source is receding (red shift) or approaching (blue shift).

ASTRONOMICAL TOOLS

Having examined the nature of light, we will now turn our attention to the tools astronomers use to intercept and study the energy emitted by distant objects in the universe. Since the basic principles of detecting radiation were originally developed through visual observations, we will consider optical telescopes first.

REFRACTING TELESCOPES

Galileo is considered the first person to use telescopes for astronomical observations. Having learned about the newly invented instrument, Galileo built one of his own design that was capable of magnifying objects 30 times. Because this early instrument, as well as its modern counterparts, used a lens to bend or refract light, it is know as a **refracting telescope.** The most important lens in a refracting telescope, the **objective lens,** produces the image by bending light from a distant object in such a way that the light converges at an area called the **focus** (Figure 18.14). For an object such as a star, the image appears as a point of light, but for a nearby object, it appears as an inverted replica of the original. You can easily demonstrate the latter case by holding a lens in one hand and with the other hand placing a white card behind the lens. Now, vary the distance between them until an image appears on the card. The distance between the focus (place where the image appears) and the lens is called the **focal length** of the lens.

Astronomers usually study an image from a telescope by first photographing the image. If a telescope is used to examine an image directly, a second lens, called an **eyepiece,** is required (Figure 18.14). The eyepiece magnifies the image produced by the objective lens. In this respect, it is similar to a magnifying glass. Thus, the objective lens produces a very small, bright image of an object and the eyepiece enlarges the image so that details can be seen.

Although used extensively in the nineteenth century, refracting telescopes suffer a major optical defect. As light passes through any lens, the shorter wavelengths of light are bent more than the longer wavelengths. Recall the effect of a prism in separating the colors of the spectrum. Consequently, when a refracting telescope is in focus for red light, blue and violet light are out of focus. This troublesome effect, known as **chromatic** ("color") **aberration,** weakens the image and produces a halo of color around it. When blue light is in focus, a reddish halo appears, and vice versa. Although this effect cannot be eliminated completely, it is reduced by using a second lens made of a different type of glass.

REFLECTING TELESCOPES

Newton was so bothered by chromatic aberration that he built and used telescopes that reflected light from a shiny surface (mirror). Because reflected light is not dispersed into its component colors, the problem is avoided. **Reflecting telescopes** use a concave mirror that focuses the light

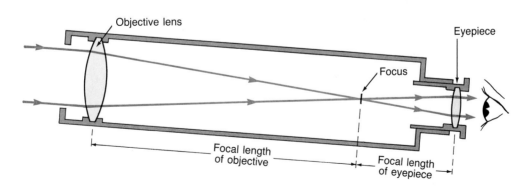

FIGURE 18.14
Simple refracting telescope.

FIGURE 18.15
A. Sketch of a mirror used to gather light. **B.** Preparation of the 5-meter mirror for the Hale Telescope. (Hale Observatories photo. Copyright by the California Institute of Technology and the Carnegie Institution of Washington.)

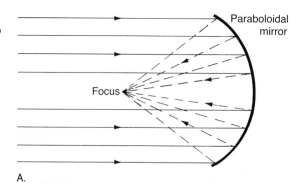

A.

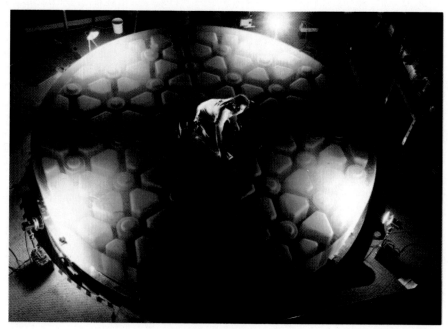

B.

in front of the objective rather than behind it like a lens (Figure 18.15). The mirror is generally made of glass that is finely ground. In the case of the 5-meter (200-inch) Hale Telescope, the grinding is accurate to about 1 millionth of a centimeter. The surface is then coated with a highly reflective material, usually an aluminum-based compound. In order to focus parallel incoming light into one spot, a special curved surface called a *paraboloid* is used. This is the same shape as that used for the reflector in the headlights of automobiles. In the case of the auto bulb, however, the light source is at the focus, and the light goes out in parallel rays rather than coming in.

Due to the fact that the focus of a reflecting telescope is in front of the mirror, provisions have to be made to view the image without blocking too much of the incoming light. Figure 18.16 illustrates the most commonly used arrangements. Most large telescopes employ more than one type. When using a very large reflecting telescope, the observer can actually get inside a viewing cage positioned at the focus to make the observations. The viewing cage blocks only about 10 percent of

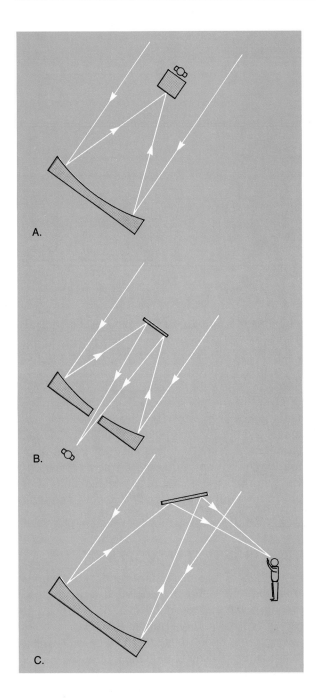

FIGURE 18.16
Viewing methods used with reflecting telescopes.
A. Prime focus method (only used in very large telescopes). **B.** Cassegrain method (most common).
C. Newtonian method.

the total incoming light, and this is more than compensated for by the large objectives that are used.

Most modern telescopes have supplemental devices that enhance the image. A simple, but important, example is a photographic plate that can be exposed for long periods of time, thereby collecting enough light from a star to make an image that otherwise would be undetectable.

PROPERTIES OF OPTICAL TELESCOPES

Telescopes have three properties that aid astronomers in their work. They provide the observer with light-gathering power, resolving power, and magnifying power.

Since most celestial objects are very faint sources of light, astronomers are most interested in improving the *light-gathering* power of their instruments. As seen in Figure 18.17, a telescope with a large lens (or mirror) intercepts more light from distant objects and thereby produces brighter images. Since very distant stars appear very dim as well, a great deal of light must be collected before the image will be bright enough to be seen. Consequently, telescopes with large objectives "see" farther into space than those with small objectives.

Another advantage of telescopes with large-diameter objectives is their high *resolving power* (Figure 18.18), which allows for sharper images and finer detail. For example, with the unaided eye, the Milky Way appears as a band of light, but even a small telescope is capable of resolving (separating it into) individual stars. Even so, the condition of the earth's atmosphere (which is known as *seeing*) greatly limits the resolving power of earthbound telescopes. On a night when the stars twinkle, the seeing is poor, because the air is moving rapidly. This causes the image to move about, blurring the photograph. Conversely, when the stars shine steadily, the seeing is described as good. Even under ideal conditions, however, some blurring occurs, eliminating the fine details. Thus, even the largest telescopes cannot obtain photographs of lunar features less than 0.5 kilometer (0.3 mile) in size.

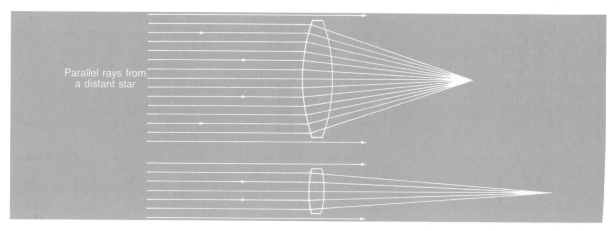

FIGURE 18.17
Comparison of the light-gathering ability of two lenses.

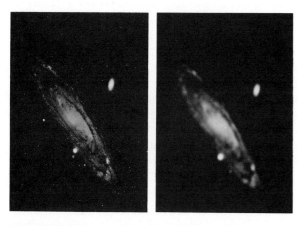

FIGURE 18.18
How the galaxy in Andromeda would appear using telescopes with decreasing resolution. (Courtesy of Leiden Observatory)

To eliminate the problems of earthbound viewing, the United States is currently building the Hubble Space Telescope, which it hopes to put into earth orbit by late 1988. When operational, this 2.4-meter (94-inch) space telescope will have 10 billion times more light-gathering power than the human eye.

When you think of the power of a telescope, you probably think of its *magnifying power,* the ability to make an object larger. Although the magnification of a telescope can be changed by simply changing the eyepiece, increased magnification does not necessarily improve the clarity of the image. What can be viewed telescopically is limited by atmospheric conditions and the resolving power of the telescope. Any part of an image that is not clear at low magnification will only appear as a larger blur at high magnification. Thus, astronomers describe telescopes not in terms of their magnification but by the diameter of the objective mirror or lens, because it is this factor that determines both the light-gathering power and the resolving power of a telescope.

THE LARGEST TELESCOPES

For a variety of reasons, all large optical telescopes built today are reflectors. Included among these reasons is the monumental task of producing a large piece of quality, bubble-free glass for refracting telescopes. In addition, both sides of a lens must be polished, and when a double lens is employed to reduce the chromatic aberration, a total of four sides must be polished to the highest optical standards. However, only one side of a reflector must be polished. In addition, a lens can be supported only around the edge and suffers the effects of sagging. Mirrors, on the other hand, can be fully supported from behind.

The largest telescope in the world today is the 6-meter (236-inch) reflector located in the Caucasus Mountains of the Soviet Union. However, since this telescope is not in a particularly good location, it has not been as useful as many smaller telescopes. Just behind it in size is the 5-meter (200-inch) Hale Telescope on Mount Palomar, in California. The Hale Telescope is a 500-ton steel and glass marvel which floats into position on oil bearings. To ready the instrument for the night's vigil, the astronomer uses an elevator to reach the 2-meter-wide viewing cage, located inside the telescope almost 24 meters above the floor of the dome. From this perch, the astronomer can see the 5-meter Pyrex mirror shimmering below. In the "good old days" an astronomer would spend the night outdoors in the cold mountain air, but advances in photographic techniques and computer-enhanced pictures allow indoor work in a heated room below and reduce the time required to expose the photographic film. Other large reflecting telescopes of about 4-meters (150-inches) are located at Kitt Peak, Arizona; Mauna Kea, Hawaii; Cerro Tololo, Chile; and Siding Spring, Australia (Figure 18.19). By comparison, the largest refractor in the world is the 1-meter (40-inch) telescope at Yerkes Observatory in Williams Bay, Wisconsin. This instrument was built before the turn of the century.

Despite the expense and difficulty of building ever larger optical telescopes, their greater light-gathering and resolving power has led to the development of new technologies in telescope building. A recently completed telescope called the Multiple Mirror Telescope uses six 1.8-meter mirrors that are linked together to focus at a single point. The result is a telescope with the light-gathering capacity of a mirror 4.5 meters in diameter. An even more ambitious endeavor is the New Technology Telescope, which is being designed for use in the United States. When completed, this instrument will use four 8-meter mirrors in tandem to form one truly large telescope. The technology of casting an 8-meter mirror is currently being developed under the football stadium at the University of Arizona. If successful, several 8-meter telescopes may eventually be produced.

Another project in the developmental stage is the 10-meter Keck Telescope. Rather than casting a single large mirror, this instrument will consist

FIGURE 18.19
The 4-meter Mayall Telescope at Kitt Peak, Arizona, is one of the world's largest. (Courtesy of National Optical Astronomy Observatories)

of a mosaic of 36 small segments that will be carefully positioned by computer to give the optical effect of a 10-meter mirror.

DETECTING INVISIBLE RADIATION

Sunlight is made up of more than just the radiation that is visible to our eyes. As stated earlier, gamma rays, x rays, ultraviolet radiation, infrared radiation, and radio waves are also produced by celestial bodies. Photographic film that is sensitive to ultraviolet and infrared radiation has been developed, thereby extending the limits of our vision. However, most of this radiation cannot penetrate our atmosphere, so balloons, rockets, and satellites must be used to record it. Of great importance is a narrow band of radio radiation that does penetrate the atmosphere. One particular

A.

FIGURE 18.20
A. The 43-meter (140-foot) steerable radio telescope at Green Bank, West Virginia. **B.** Twenty-seven identical antennas operate together to form the Very Large Array located near Socorro, New Mexico. (Courtesy of National Radio Astronomy Observatory)

B.

wavelength is the 21-centimeter (8-inch) line produced by neutral hydrogen. Measurement of this radiation has permitted us to map the galactic distribution of hydrogen—the material from which stars are made.

The detection of radio waves is accomplished by "big dishes" called **radio telescopes** (Figure 18.20A). In principle, the dish of one of these telescopes operates in the same manner as the mirror of an optical telescope. It is parabolic in shape and focuses the radio waves on an antenna, which "hears" the source. Because radio waves are on the order of 100,000 times longer than visible radiation, the accuracy to which the surface of the dish has to be built is greatly reduced. However, this is more than offset by the fact that radio energy from celestial sources is weak, making large dishes a requirement. The largest radio telescope is a bowl-shaped antenna hung in a natural depression in Puerto Rico (Figure 18.21). It is 300 meters (1000 feet) in diameter and has some directional flexibility in its movable antenna. The largest steerable types have about 100-meter (330-foot) dishes like that at the National Radio Astronomy Observatory, in Green Bank, West Virginia. Radio telescopes also have rather poor resolution, making it difficult to pinpoint the radio source. Pairs or groups of them are used to reduce this problem. When several radio telescopes are wired together, the resulting network is called a **radio interferometer** (Figure 18.20B).

Radio telescopes do have a few advantages over optical telescopes. They are much less affected by turbulence in the atmosphere, clouds, and the weather in general. No protective dome is required, which reduces the cost of construction, and viewing is possible 24 hours a day. They are, however, hindered by human-made radio interference. Thus, while optical telescopes are placed on a mountaintop, radio telescopes are often hidden in a valley, or at least behind some obstruction.

Radio telescopes have revealed such spectacular events as the collision of two galaxies, but of even greater interest was the discovery of *quasars* (quasi-stellar radio sources). These perplexing objects may be the most distant things in the universe.

FIGURE 18.21
The 300-meter (1000-foot) telescope at Arecibo, Puerto Rico. (Courtesy of National Astronomy and Ionosphere Center's Arecibo Observatory, operated by Cornell University under contract with the National Science Foundation)

THE SUN

The sun is one of billions of stars that make up the Milky Way galaxy, and although it is of no significance to the universe as a whole, to us on earth, it is the primary source of energy. Everything from the fossil fuels we burn in our automobiles and power plants to the food we eat is ultimately derived from solar energy. The sun is also important to astronomers, since it is the only star whose surface can be observed. Even with the largest telescopes, other stars can only be resolved as points of light.

Due to the sun's brightness and its damaging radiation, it is not safe to observe it directly. However, a small telescope will project an image on a piece of cardboard held behind its eyepiece, and the sun may be studied from that. Several specially built telescopes around the world keep a constant vigil of the sun. One of the finest is at the Kitt Peak National Observatory in southern

FIGURE 18.22
Robert J. McMath solar telescope at Kitt Peak, Arizona. (Courtesy of National Optical Astronomy Observatories)

Arizona (Figure 18.22). It consists of a 150-meter sloped enclosure that directs sunlight to a 150-centimeter-long focal length mirror situated below ground. From the mirror, an 85-centimeter image of the sun is projected to an observing room, where it can be studied.

Compared to the other stars of the universe, the sun can be considered an "average star." However, on the scale of our solar system, it is truly gigantic, having a diameter equal to 109 earth diameters (1.35 million kilometers) and a volume 1¼ million times as great as that of the earth. Yet, because of its gaseous nature, its density is only ¼ that of the solid earth, very closely approximating the density of water.

STRUCTURE OF THE SUN

For convenience of discussion, we divide the sun into four parts: the solar interior; the visible surface, or photosphere; and the two layers of its atmosphere, the chromosphere and the corona (Figure 18.23). Since the sun is gaseous throughout, no sharp boundaries exist between these layers. The sun's interior makes up all but a tiny fraction of the solar mass, and unlike the outer three layers, it is not accessible to direct observation. We shall discuss the visible layers first.

The **photosphere** ("sphere of light") is aptly named, since it is the layer that radiates most of

the sunlight we see and therefore appears as the bright disk of the sun. Although it is considered to be the sun's surface, it is unlike most surfaces we are accustomed to. The photosphere consists of a layer of incandescent gas 300 kilometers (200 miles) thick having a pressure less than 1/100 of our atmosphere. Furthermore, it is not smooth or uniformly bright as the ancients had imagined. It has numerous blemishes. When viewed through a telescope under ideal conditions, a grainy texture is apparent. This is the result of numerous relatively small, bright markings called **granules,** which are surrounded by narrow, dark regions (Figure 18.24). Granules are typically 1000 kilometers (620 miles) in diameter and owe their brightness to hotter gases that are rising from below. As this gas spreads laterally, cooling causes it to darken and sink back into the interior. Although each granule lasts only a few minutes, the combined motion of all granules gives the photosphere the appearance of boiling. This up-and-down movement of gas is called convection and, besides producing the grainy appearance of the photosphere, is believed to be responsible for the transfer of energy in the uppermost part of the sun's interior (Figure 18.23).

The composition of the photosphere is revealed by the dark lines of its absorption spectrum (see Figure 18.11). When these "fingerprints" are compared to the spectrum of known elements, they

FIGURE 18.23
Diagram of solar structure in
cross section.

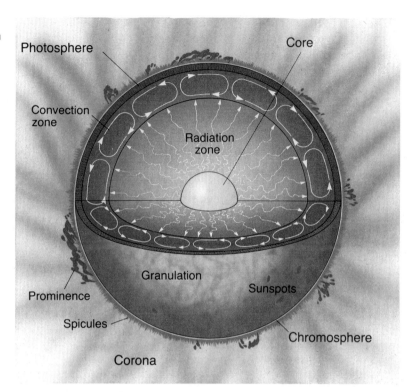

FIGURE 18.24
Granules of the solar photosphere as photographed
from a balloon at 24,000 meters (80,000 feet). (Courtesy of Project Stratoscope of Princeton University,
sponsored by NASA, NSF, and ONR)

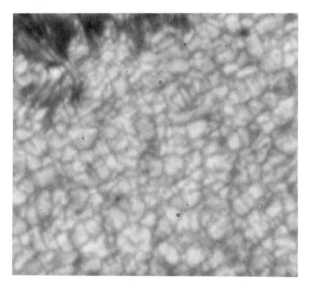

indicate that most of the elements found on earth
also occur on the sun. When the strengths of the
absorption lines are analyzed, the relative abundance of the elements can be determined. These
studies reveal that 90 percent of the sun's surface
atoms are hydrogen, almost 10 percent are helium, and only minor amounts of the other detectable elements are present. Other stars also indicate similar disproportionate percentages of these
two lightest elements, a fact we shall consider
later.

Just above the photosphere lies the **chromosphere** ("color sphere"), a relatively thin layer of
hot, incandescent gases a few thousand kilometers thick. The chromosphere is observable for a
few moments during a total solar eclipse or by
using a special instrument which blocks out the
light from the photosphere. On such occasions, it
appears as a thin red rim around the sun. Because
the chromosphere consists of hot, incandescent
gases under low pressure, it produces a bright-
line spectrum which is nearly the reverse of the

dark-line spectrum of the photosphere. One of the bright lines of hydrogen contributes a good portion of its total light and accounts for this sphere's red color. A study of the chromospheric spectrum conducted in 1868 revealed the existence of the heretofore unknown element helium (from *helios,* the Greek word for "sun"). Originally, helium was thought to be an element unique to the stars, but 27 years later, it was discovered in natural-gas wells on earth. The top of the chromosphere contains numerous **spicules,** which extend upward into the lower corona, almost like the trees which reach into our atmosphere (Figure 18.25). The spicules may be continuations of the turbulent motion of the granules below.

FIGURE 18.26
Solar corona photographed during a total eclipse. (Courtesy of Hale Observatories)

FIGURE 18.25
Spicules of the chromosphere as seen on the edge of the solar disk. (Courtesy of Sacramento Peak Observatory, Air Force Cambridge Research Laboratories)

The outermost portion of the solar atmosphere, the **corona,** is very tenuous and, as with the chromosphere, is visible only when the brilliant photosphere is covered (Figure 18.26). This envelope of ionized gases normally extends a million kilometers from the sun and produces a glow about half as bright as the full moon. At the outer fringe of the corona, the ionized gases have acquired speeds great enough to escape the gravitational pull of the sun. The streams of protons and electrons that "boil" from the corona constitute the **solar wind.** They travel outward through the solar system at very high speeds and eventually are lost to interstellar space. During their journey, the solar winds interact with the bodies of the solar system, continually bombarding lunar rocks and altering their appearance. Although the earth's magnetic field prevents the solar winds from reaching our surface, these winds do affect our atmosphere, as we shall discuss later.

Studies of the energy emitted from the photosphere indicate that its temperature averages about 6000 K (10,000°F). Upward from the photosphere, the temperature unexpectedly increases, exceeding 1 million K at the top of the corona. It should be noted that although the coronal temperature exceeds that of the photosphere many times, it radiates much less energy because of its very low density. The high temperature of the corona is probably caused by sound waves generated by the convective motion of the photosphere. Just as boiling water makes noise, the supersonic booms similarly generated in the photosphere are believed to be absorbed by the gases of the corona and thereby raise their temperatures.

THE ACTIVE SUN

The most conspicuous features on the surface of the sun are the dark blemishes called **sunspots** (Figure 18.27A). Although sunspots were occasionally observed before the advent of the telescope, they were generally regarded as opaque objects located somewhere between the sun and the earth. In 1610 Galileo concluded that they were definitely residents of the solar surface,

and from their motion, he deduced that the sun rotates on its axis about once a month. Later observations indicated that not all parts of the sun rotate at the same speed. The sun's equator rotates once in 25 days, while a place located 70 degrees from the solar equator, either north or south, requires 33 days for one rotation. If the earth rotated in a similar manner, imagine the consequences! The sun's nonuniform rotation is a testimonial to its gaseous nature.

Sunspots begin as small dark pores about 1600 kilometers (1000 miles) in diameter. While most pores last for only a few hours, some grow into blemishes many times larger than the earth and last for a month or more. The largest spots often occur in pairs surrounded by several smaller spots. An individual spot contains a black center, the **umbra,** which is rimmed by a lighter region, the **penumbra** (Figure 18.27B). Sunspots appear dark only by contrast with the brilliant photosphere, a fact accounted for by their temperature,

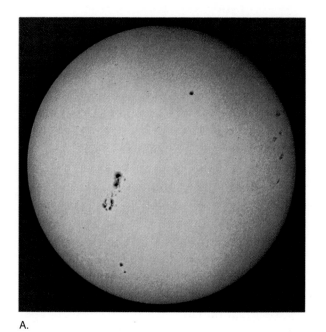

A.

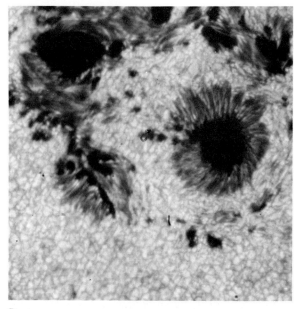

B.

FIGURE 18.27
A. Large sunspot group on the solar disk. (Celestron 8 photo courtesy of Celestron International) B. View of sunspots whose umbra and penumbra are visible. (Courtesy of Project Stratoscope of Princeton University, sponsored by NASA, NSF, and ONR)

FIGURE 18.28
Solar disk photographed in hydrogen alpha light, showing the manifestations of the active sun. (This composite courtesy of Hale Observatories and Sacramento Peak Observatory, Air Force Cambridge Research Laboratories)

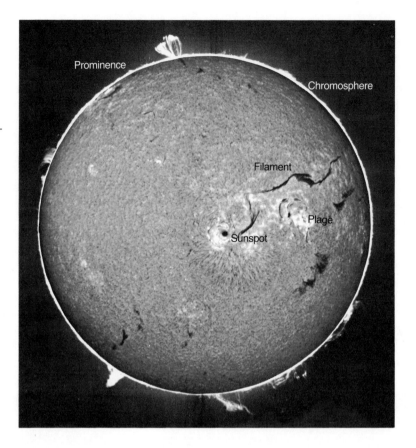

which is about 1500 K less than that of the solar surface. If these dark spots could be observed away from the sun, they would appear many times brighter than the full moon.

During the search conducted in the early 1800s for the planet Vulcan, believed to orbit between Mercury and the sun, an accurate record of sunspot occurrences was kept. Although the planet was never found, the sunspot data collected did reveal that the number of sunspots observable on the solar disk varies in an 11-year cycle. First, the number of sunspots increases to a maximum, when perhaps a hundred or more are visible at a given time, and then over a period of 5–7 years, their numbers decline to a minimum, when only a few, or possibly none, are visible. At the beginning of each cycle, the first sunspots form about 30 degrees from the solar equator, but as the cycle progresses and their numbers increase, they form nearer the equator. During the period when sun-

spots are most abundant, the majority form about 15 degrees from the equator. They rarely occur more than 40 degrees away from the sun's equator or within 5 degrees of it.

Another interesting characteristic of sunspots was discovered by George Hale. Hale deduced that the large spots are strongly magnetized, and when they occur in pairs, they have opposite magnetic poles. For instance, if one member of the pair is a north pole, then the other member will be a south pole, as with the north and south poles of the earth's magnetic field. Also, every pair located in the same hemisphere will be magnetized in the same manner. However, all pairs in the other hemisphere will be magnetized in the opposite manner. At the beginning of each sunspot cycle, the situation reverses, and the polarity of these sunspot pairs is opposite those of the previous cycle. The cause of this change in polarity, in fact the cause of sunspots themselves, is not fully ex-

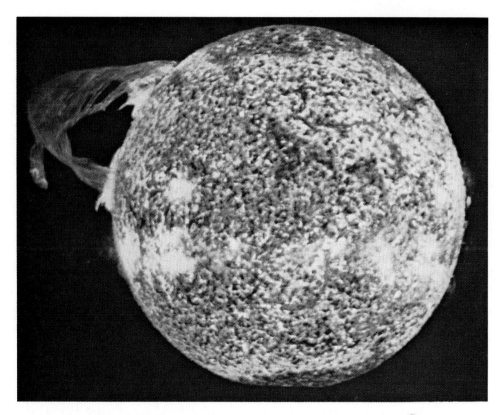

FIGURE 18.29
A huge solar prominence. (Courtesy of NASA)

plained. However, other solar activity varies in the same cyclic manner as sunspots do, indicating a common origin.

The brilliance of the photosphere makes viewing the activity occurring above the solar surface very difficult. To overcome this problem, numerous photographic methods are employed to filter solar radiation so that light of a particular spectral region (color) can be viewed. Using such techniques, large "clouds" can be seen in the chromosphere directly above sunspot clusters (Figure 18.28). These bright centers of solar activity are called **plages** and occasionally can even be viewed before or after the sunspot occurrences. Among the more spectacular features of the active sun are the **prominences.** These huge cloudlike structures are best observed when they are on the edge, or limb, of the sun, where they often appear as great arches that extend well into the corona

(Figure 18.29). Many prominences have the appearance of a fine tapestry and seem to hang motionless for days at a time, but motion pictures reveal that the material within them is continually falling like luminescent rain. Apparently these quiet prominences are condensations of coronal material which are gracefully "sliding down" the lines of magnetic force back into the chromosphere. More rarely, the material within a prominence rises almost explosively away from the sun. These active prominences reach velocities up to 1000 kilometers (620 miles) per second and may leave the sun entirely. Prominences can also be seen against the bright disk of the sun, in which case they appear as dark, thin streaks called **filaments** (Figure 18.28).

Solar flares are the most explosive events associated with sunspots. These brief outbursts normally last an hour or so and appear as a sudden

FIGURE 18.30
Aurora borealis (northern lights) as seen from Alaska. (Courtesy of Gustav Lamprecht)

brightening of the region above a sunspot cluster. During the period of their existence, enormous quantities of energy are released, much of it in the form of ultraviolet, radio, and x-ray radiation. Simultaneously, fast-moving atomic particles are ejected, causing the solar winds to intensify noticeably. Although a major flare could conceivably endanger a manned space flight, they are relatively rare. About a day after a large outburst, the ejected particles reach the earth and disturb the ionosphere,* affecting long-distance radio communications. Their most spectacular effects, however, are the **auroras,** also called the northern and southern lights (Figure 18.30). Following a strong solar flare the earth's upper atmosphere near its magnetic poles is set aglow for several nights. The auroras appear in a wide variety of forms. Sometimes the displays consist of vertical streamers in which there can be considerable movement. At other times, the auroras appear as a series of luminous expanding arcs or as a quiet glow that has an almost foglike character. Auroral displays, like other solar activities, vary in intensity with the 11-year sunspot cycle.

---

*The ionosphere is a complex atmospheric zone of ionized gases that extends between about 80 and 400 kilometers (50 and 250 miles) above the earth's surface.

THE SOLAR INTERIOR

The interior of the sun cannot be observed directly. For that reason, all we know about it is based on information acquired from the energy it radiates and from theoretical studies. The source of the sun's energy, **nuclear fusion,** was not discovered until the late 1930s. Deep in its interior, a nuclear reaction called the **proton-proton chain** converts four hydrogen nuclei (protons) into a nucleus of helium. The energy from the proton-proton reaction results because some of the matter involved is actually converted to energy. This can be illustrated by noting that four hydrogen atoms have a combined atomic mass of 4.032 (4×1.008), whereas the atomic mass of helium is 4.003, or 0.029 less than the combined mass of the hydrogen. The missing mass is emitted as energy according to Einstein's formula $E = mc^2$, where m equals mass and c equals the speed of light. Because the speed of light is very great, the amount of energy released from even a small amount of mass is enormous. The conversion of just one pinhead's worth of hydrogen to helium can generate more energy than burning thousands of tons of coal. Most of this energy is in the form of light photons which work their way toward the solar surface, being absorbed and re-emitted many times until they reach a rather opaque layer just below the photosphere. Here, convection currents transport this energy to the solar surface, where it can radiate through the transparent chromosphere and corona (see Figure 18.23).

Only a small percentage (0.7 percent) of the hydrogen involved in the proton-proton reaction is actually converted to energy. Nevertheless, the sun is consuming an estimated 600 million tons of hydrogen each second, with about 4 million tons of that amount being converted to energy. As hydrogen is consumed, the product of this reaction, helium, forms the solar core, which continually grows in size. Just how long can the sun produce energy at its present rate before all of its fuel (hydrogen) is consumed? Even at the enormous rate at which the sun engulfs its fuel, it has enough to easily last another 100 billion years. However, evidence from other stars indicates that the sun will change dramatically long before all of its hydrogen is gone. It is thought that a star the size of the sun can exist in its present form for 10 billion years. since the sun is already 5 billion years old, it may be considered "middle-aged."

To initiate the proton-proton reaction, the sun's internal temperature must have reached several million degrees. What was the source of this heat? As previously noted, the solar system is believed to have formed from an enormous cloud of dust and gases (mostly hydrogen) which condensed gravitationally. The consequence of squeezing (compressing) a gas is to increase its temperature.

Although all of the bodies in the solar system were compressed, the sun was the only one, because of its size, that became hot enough to trigger the proton-proton reaction. Astronomers currently estimate its internal temperature at 15 million K. If Jupiter were about ten times more massive, it too might have become a star.

The idea of one star orbiting another may seem unusual, but recent evidence indicates that about 50 percent of the stars in the universe probably occur in pairs or multiples.

REVIEW QUESTIONS

1 How much would you weigh on the moon?

2 Draw the earth-moon system to scale (approximate the diameters of the moon and the earth—3500 and 13,000 kilometers, respectively).

3 What is different about the crescent phase that precedes the new-moon phase and that which follows the new-moon phase?

4 What phase of the moon occurs approximately two weeks after the new moon?

5 On your drawing from Question 2, show how the 5-degree inclination of the moon's orbit prevents an eclipse from occurring each month.

6 Solar eclipses are more common than lunar eclipses. Why, then, is it more likely that your region of the country will experience a lunar eclipse?

7 Why are large-rayed craters considered relatively young features on the lunar surface?

8 Briefly outline the history of the moon.

9 How are the maria of the moon thought to be similar to the Columbia Plateau?

10 Which color has the longest wavelength? The shortest?

11 Explain how each of the three types of spectra is produced.

12 How can astronomers determine whether a star is moving toward or away from the earth?

13 What three properties (powers) do telescopes have that aid astronomers?

14 What is the advantage of a space telescope over a similar earthbound instrument?

15 Why do astronomers seek to design telescopes with larger and larger objectives?

16 Why do all large optical telescopes use mirrors to collect light rather than lenses?

17 Why are special viewing systems needed with reflecting telescopes?

18 Explain the statement "Photography has extended the limits of our vision."

19 Why are radio telescopes built much larger than optical telescopes?

20 What are some of the advantages of radio telescopes over optical telescopes?

21 Describe the photosphere, chromosphere, and corona.

22 List the features associated with the active sun and describe each.

23 Explain how a sunspot can be very hot and yet appear dark.

24 What is the sun's source of energy?

KEY TERMS

phases of the moon

synodic month

sidereal month

solar eclipse

lunar eclipse

maria

lunar breccia

rays

lunar regolith

electromagnetic
 radiation

white light

photon

radiation pressure

spectroscopy

continuous spectrum

bright-line (emission)
 spectrum

dark-line (absorption)
 spectrum

spectroscope

Doppler effect

refracting telescope

objective lens

focus

focal length

eyepiece

chromatic aberration

reflecting telescope

radio telescope

radio interferometer

photosphere

granules

chromosphere

spicule

corona

solar wind

sunspot

umbra

penumbra

plage

prominence

filament

solar flare

aurora

nuclear fusion

proton-proton chain

19

THE SOLAR SYSTEM

When people first recognized that the other planets of our solar system are "worlds" much like the earth, a great deal of interest was generated. The primary concern has always been the possibility of intelligent life existing elsewhere in the universe. Recent unmanned space explorations have renewed this interest. Since all the planets probably have formed from the same primeval cloud of dust and gases, they should provide valuable information concerning the earth's formation and early history.

Saturn photographed from 18 million kilometers
(11 million miles) by *Voyager 1.* (Courtesy of NASA)

The sun is the hub of a huge rotating system consisting of nine planets, their satellites, and numerous small, but nonetheless interesting bodies, including asteroids, comets, and meteoroids. An estimated 99.85 percent of the mass of our solar system is contained within the sun, while the planets collectively make up most of the remaining 0.15 percent. The planets, in order from the sun, are Mercury, Venus, Earth, Mars, Jupiter, Saturn, Uranus, Neptune, and Pluto (Figure 19.1). Under the control of the sun's gravitational force, each planet maintains an elliptical orbit and travels in a counterclockwise direction. The nearest planet to the sun, Mercury, has the fastest orbital motion, 48 kilometers per second, and the shortest period of revolution, 88 days; whereas the most distant planet, Pluto, has an orbital speed of 5 kilometers per second and requires 248 years to complete one revolution. Further, the orbits of all the planets lie within 3 degrees of the plane of the sun's equator, except for those of Mercury and Pluto, which are inclined 7 and 17 degrees, respectively.

THE PLANETS: AN OVERVIEW

Careful examination of Table 19.1 shows that the planets fall quite nicely into two groups: the **terrestrial** (earthlike) **planets** (Mercury, Venus, Earth, and Mars); and the **Jovian** (Jupiterlike) **planets** (Jupiter, Saturn, Uranus, and Neptune). Pluto is not included in either category, since its position at the far edge of the solar system and its small size make this planet's true nature a mystery. The most obvious difference between these groups is the size of their members (Figure 19.2, page 508). The largest terrestrial planet (Earth) has a diameter only one-quarter as great as the diameter of the smallest Jovian planet (Neptune), and its mass is only one-seventeenth as great. Hence, the Jovian planets are often called *giants*. Also, because of their relative locations, the four Jovian planets are referred to as the *outer planets*, while the terrestrial planets are called the *inner planets*. As we shall see, there appears to be a correlation between the locations of these planets and their sizes.

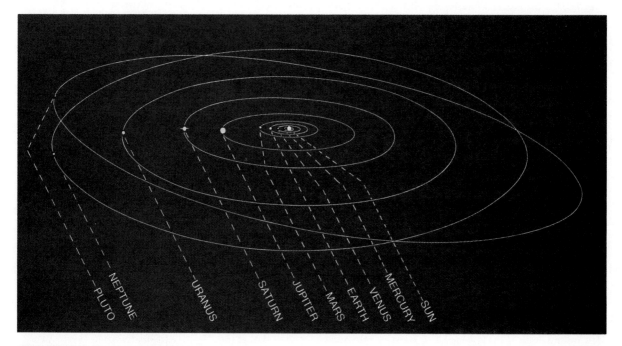

FIGURE 19.1
Orbits of the planets to scale.

TABLE 19.1
Planetary data.

Planet	Symbol	Mean Distance from Sun			Period of Revolution	Inclination to Ecliptic	Orbital Velocity	
		AU	Millions of Miles	Millions of Kilometers			mi/s	km/s
Mercury	☿	0.387	36	58	88^d	7°00′	29.5	47.5
Venus	♀	0.723	67	108	225^d	3°24′	21.8	35.0
Earth	⊕	1.000	93	150	365.25^d	0°00′	18.5	29.8
Mars	♂	1.524	142	228	687^d	1°51′	14.9	24.1
Jupiter	♃	5.203	483	778	12^{yr}	1°19′	8.1	13.1
Saturn	♄	9.539	886	1427	29.5^{yr}	2°30′	6.0	9.6
Uranus	♅	19.180	1780	2866	84^{yr}	0°46′	4.2	6.8
Neptune	♆	30.060	2790	4492	165^{yr}	1°46′	3.3	5.3
Pluto	♇	39.440	3670	5909	248^{yr}	17°12′	2.9	4.7

Planet	Period of Rotation	Diameter		Relative Mass (Earth = 1)	Average Density (g/cm^3)	Polar Flattening (%)	Eccentricity	Number of Known Satellites
		Miles	Kilometers					
Mercury	59^d	3015	4854	0.056	5.1	0.0	0.206	0
Venus	243^d	7526	12,112	0.82	5.3	0.0	0.007	0
Earth	$23^h56^m04^s$	7920	12,751	1.00	5.52	0.3	0.017	1
Mars	$24^h37^m23^s$	4216	6788	0.108	3.94	0.5	0.093	2
Jupiter	$\sim 9^h50^m$	88,700	143,000	318.000	1.34	6.5	0.048	16
Saturn	$\sim 10^h25^m$	75,000	121,000	95.200	0.70	10.5	0.056	17
Uranus	10^h45^m	29,000	47,000	14.600	1.55	7.0	0.047	15
Neptune	18^h(?)	28,900	46,529	17.300	2.27	2.5	0.008	2
Pluto	6.4^d	~1500	~2400	~0.01(?)	~1.5(?)	?	0.250	1

Other dimensions along which the two groups markedly differ include density, composition, and rate of rotation. The densities of the terrestrial planets average about 5 times the density of water, whereas the Jovian planets have densities that average only 1.5 times that of water. One of the outer planets, Saturn, has a density only 0.7 that of water. Variations in the compositions of the planets is largely responsible for these differences.

The substances of which both groups of planets are composed can be divided into three groups based upon their melting points. They are called *gases*, *rocks*, and *ices*. The gases are those materials with melting points near absolute zero,* −273°C, and consist of hydrogen and helium. The rocky materials are made principally of silicate minerals and metallic iron, which have melting points exceeding 700°C. The ices have intermediate melting points and include ammonia (NH_3), methane (CH_4), carbon dioxide (CO_2), and water (H_20).

The terrestrial planets are composed mostly of dense rocky and metallic material with minor amounts of gases. The Jovian planets, on the other hand, contain a large percentage of hydrogen and helium, with varying amounts of ices (mostly water, ammonia, and methane), which accounts for their low densities. The outer planets are also thought to contain as much rocky and metallic material as the terrestrial planets, and this

*Absolute zero (zero kelvin) is the lowest possible (theoretical) temperature. All molecular motion ceases at that point.

FIGURE 19.2 The planets drawn to scale.

material may be concentrated in a small central core.

The Jovian planets have very thick atmospheres consisting of varying amounts of hydrogen, helium, methane, and ammonia. By comparison, the terrestrial planets have meager atmospheres at best. A planet's ability to retain an atmosphere depends on its temperature and mass. Simply stated, a gas molecule can "evaporate" from a planet if it reaches a speed known as the **escape velocity.** For the earth, this velocity is 11 kilometers (7 miles) per second. Any material, including a rocket, must reach this speed before it can leave the earth and go into space. The Jovian planets, because of their greater mass, have higher escape velocities than the terrestrial planets. Consequently, it is more difficult for gases to "evaporate" from them. Also, because the molecular motion of a gas is temperature dependent, at the low temperatures of the Jovian planets even the lightest gases are unlikely to acquire the speed needed to escape. On the other hand, a comparatively warm body with a small mass, like our moon, is unable to hold even the heaviest gas and thus lacks an atmosphere. The slightly larger terrestrial planets of Earth, Venus, and Mars retain some heavy gases (as compared to hydrogen), but even their atmospheres make up only an infinitesimally small portion of their total mass.

It is hypothesized that the primordial cloud of dust and gas from which all the planets are thought to have condensed had a composition somewhat similar to that of Jupiter. However, unlike Jupiter, the terrestrial planets are nearly void of light gases and ices. Were the terrestrial planets once much larger? Did they contain these materials but lose them because of their close proximity to the sun? In the following section we will consider the evolutionary histories of these two diverse groups of planets in an attempt to answer these questions.

ORIGIN OF THE SOLAR SYSTEM

The orderly nature of our solar system leads most astronomers to conclude that the planets formed at essentially the same time and from the same primordial material as the sun. This **nebular hypothesis** suggests that all bodies of the solar sys-

FIGURE 19.3

Nebular hypothesis. **A.** A huge rotating cloud of dust and gases (nebula) begins to contract. **B.** Most of the material is gravitationally swept toward the center, producing the sun. However, due to rotational motion some dust and gases remain orbiting the central body as a flattened disk. **C.** The planets begin to acrete from the material that is orbiting within the flattened disk. **D.** In time most of the remaining debris was either collected into the nine planets and their moons or swept out into space by the solar wind.

tem formed from an enormous nebular cloud consisting of approximately 80 percent hydrogen, 15 percent helium, and a few percent of all the other heavier elements known to exist. The heavier substances in this frigid cloud of dust and gases consisted mostly of elements such as silicon, aluminum, iron, and calcium—the substances of the common rocky materials. Also prevalent were the other familiar elements, including oxygen, carbon, and nitrogen.

About 5 billion years ago, and for reasons not yet fully understood, this huge cloud of minute rocky fragments and gases began to contract under its own gravitational influence (Figure 19.3). The contracting clump of material is assumed to have had some component of rotational motion, which rotated faster and faster as it gravitationally contracted. This rotation caused the nebular cloud to assume a disk-like shape. Within this rotating disk, relatively small eddy-like contractions formed the nuclei from which the planets would eventually develop. However, the greatest concentration of material was gravitationally pulled toward the center, forming the *protosun.*

As more and more of this gas plunged inward, the temperature of the central mass continued to increase. The nebular material located near the protosun reached temperatures of several thousand degrees and was completely vaporized. However, at distances beyond the orbit of Mars, the temperatures probably always remained very low. Here, at $-200°C$, the dust fragments were most likely covered with a thick layer of water ice, and ices of carbon dioxide, ammonia, and methane. The disk-shaped cloud also contained appreciable amounts of the lighter gases, namely hydrogen and helium, which had not been consumed by the protosun.

In a relatively short time after the protosun formed, the temperature in the inner portion of the nebula dropped significantly. This temperature decrease caused those substances with high melting points to condense into perhaps sand-sized particles. Materials such as iron and nickel solidified first. Next to condense were the elements of which the rock-forming minerals are composed. As these fragments collided they joined into larger asteroid-sized objects which in a few tens of millions of years accreted into the four inner planets we call Mercury, Venus, Earth, and Mars. As more and more of the nebular debris was swept up by these *protoplanets*, the inner solar system began to clear, allowing sunlight to heat the planets' surfaces. Due to their relatively high temperatures and weak gravitational fields, the inner planets were unable to accumulate an appreciable amount of the lighter components of this nebular cloud. These materials, namely hydrogen, ammonia, methane, and water, were eventually wisked from the inner solar system by the solar winds.

Shortly after the four terrestrial planets formed, the decay of radioactive isotopes within them plus the heat from the colliding particles produced at least some melting of the planets' interiors. Melting, in turn, caused the heavier elements, principally iron and nickel, to sink, while the lighter silicate minerals floated upward. During this period of chemical differentiation, gaseous materials were allowed to escape from the planets' interiors, much like what happens during a volcanic event on Earth. The hottest and second smallest planet, Mercury, was unable to retain even the heaviest of these gases. Mars, on the other hand, being slightly larger and cooler than Mercury, retained a

thin layer of carbon dioxide and some water in the form of ice. The largest of the terrestrial planets, Venus and Earth, have surface gravitations strong enough to retain a substantial amount of the heavier gases. However, when compared to the four Jovian planets, even the atmospheres of these planets must be looked upon as meager at best.

At the same time that the terrestrial planets were forming, the larger Jovian planets, along with their extensive satellite systems, were also developing. However, because of the frigid temperatures existing far from the sun, the fragments out of which these planets formed contained a high percentage of ices—water, carbon dioxide, ammonia, and methane. Perhaps by random chance, two of the outer planets, Jupiter and Saturn, grew many times larger (by mass) than Uranus and Neptune. For comparison, Jupiter is 318 and Saturn 95 times more massive than the earth. However, Uranus and Neptune have masses about 14 and 17 times greater, respectively, than the earth. When Jupiter and Saturn reached a certain size, estimated to be about 10 earth masses, their surface gravitation was sufficient to attract and hold even the lightest materials—hydrogen and helium. It is thought that these gases gravitationally collapsed onto these large protoplanets as they swept through their region of the solar system. Thus, much of their size is attributable to the large envelope of light elements, which exists as a dense liquid below a thick hydrogen-rich atmosphere. Jupiter and Saturn therefore consist of a central core of ices and rock, and a much larger outer envelope containing mostly hydrogen and helium.

By contrast, the smaller Jovian planets, Uranus and Neptune, grew more slowly and contain proportionately much smaller amounts of hydrogen and helium. Nevertheless, hydrogen, methane, and ammonia are still the major constituents of their dense atmospheres and perhaps a thin outer ocean of hydrogen exists on these planets as well. Thus, Uranus and Neptune are proposed to have a small rocky-iron core and a large mantle of water, ammonia, and methane surrounded by a thin ocean of liquid hydrogen. Consequently, these planets structurally resemble Jupiter and Saturn without their large hydrogen-helium envelopes.

In many respects the development of the outer

planets with their large satellite systems roughly parallels the events which formed the solar system as a whole. Like their parents, the satellites of the outer planets are composed primarily of icy materials with lesser amounts of rocky substances. However, because of their small size, they could not contain appreciable amounts of hydrogen and helium.

In the remainder of this chapter, we will consider each planet in more detail. However, we will first consider some small but nevertheless important members of the solar system.

THE MINOR MEMBERS OF THE SOLAR SYSTEM

ASTEROIDS

When Uranus was discovered in 1781, the distance from the sun to all the known planets fit a simple arithmetic progression known as Bode's law, which is determined by adding 4 to the series of numbers 0, 3, 6, 12, 24, . . . (found by doubling the previous number) and dividing the sum by 10. The sequence of numbers so generated is given in Table 19.2, where it is compared to the actual distances from the sun to the planets in astronomical units (AU). In 1781, the only major discrepancy in Bode's law was a "gap" at 2.8 AU. For that reason, a search was begun for a planet which occupied that orbit. In 1801 the elusive "missing planet" was discovered by an Italian monk, Giuseppe Piazzi, who named it Ceres after the protective god of Sicily. But a year later, another moving body, Pallas, was discovered at the same distance from the sun. Then, a search began for other similar small bodies, called **asteroids.** By the year 1900, over 300 asteroids had been sighted, and today the number exceeds 50,000. The orbits of 1800 of the larger asteroids have been calculated. After each orbit is established, the asteroid is given a number and the discoverer is given the opportunity to name it. At first, mythological names were used, but they were soon exhausted, and names such as Hungaria, Hilda, and Chicago became common.

The orbits of most asteroids lie between Mars and Jupiter and have periods ranging from three to six years. Some asteroids with eccentric orbits

TABLE 19.2
Bode's law compared with actual distances.

Planet	Actual Distance from Sun (AU)	Sequence
Mercury	0.39	$(0 + 4) \div 10 = 0.4$
Venus	0.72	$(3 + 4) \div 10 = 0.7$
Earth	1.00	$(6 + 4) \div 10 = 1.0$
Mars	1.52	$(12 + 4) \div 10 = 1.6$
		$(24 + 4) \div 10 = 2.8$
Jupiter	5.20	$(48 + 4) \div 10 = 5.2$
Saturn	9.54	$(96 + 4) \div 10 = 10.0$
Uranus	19.18	$(192 + 4) \div 10 = 19.6$
Neptune*	30.06	$(384 + 4) \div 10 = 38.8$
Pluto*	39.44	$(768 + 4) \div 10 = 77.2$

*Unknown at the time Bode's law was formulated.

travel very near the sun, and a few large ones regularly pass close to the earth and moon. Some of the larger impact craters on the moon were probably caused by collisions with asteroids.

Asteroids are rather small, rocky bodies that have been likened to "flying mountains." The largest, Ceres, is only about 1000 kilometers (620) miles in diameter, but most of those that have been observed are about one kilometer or so across. The smallest asteroids are no larger than grains of dust. Due to the irregular shapes of asteroids, astronomers first speculated that they may have formed from a catastrophic breakup of a planet that once occupied an orbit at 2.8 AU. However, the total mass of the asteroids is estimated to be only 1/1000 that of the earth, itself not a large planet. What, then, happened to the remainder of the original planet? Other astronomers have hypothesized that several larger bodies once existed and that collisions produced numerous smaller bodies. The existence of several "families" of asteroids has been used to support this latter explanation. Nevertheless, no conclusive evidence has been found for either hypothesis.

COMETS

Comets are among the most spectacular and unpredictable bodies in the solar system (Figure 19.4). They have been compared to large, dirty

FIGURE 19.4
Comet West as photographed by a Celestron 20-centimeter (8-inch) Schmidt telescope. (Courtesy of Celestron International)

snowballs, since they are made of frozen gases (water, ammonia, methane, and carbon dioxide) which hold together small pieces of rocky and metallic materials. Many comets travel along very elongated orbits that carry them beyond Pluto (Figure 19.4). On their return, the comets are visible only after they have moved within the orbit of Saturn.

When first observed, comets appear very small, but as they approach the sun, solar energy begins to vaporize the frozen gases, producing a glowing head called the **coma.** The size of the coma varies greatly from one comet to another. Some exceed the size of the sun, but most approximate the size of Jupiter. Within the coma, a small glowing nucleus with a diameter of only a few kilometers can sometimes be detected. As they approach the sun, some, but not all, comets develop a tail that extends for millions of kilometers. Despite the enormous size of their tail and coma, comets are

thought to have an insignificant mass. They are generally less than one billionth as massive as the earth.

The tail of a comet points away from the sun in a slightly curved manner (Figure 19.5). This fact led early astronomers to propose that the sun had a repulsive force that pushed the particles of the coma, thus forming the tail. Today, two solar forces are known to contribute to this formation. One, called radiation pressure, pushes dust particles away from the coma, and the second, known as solar wind, is responsible for moving the ionized gases, particularly carbon monoxide. Usually a single tail composed of both types of materials is produced, but two somewhat separate tails like those shown in Figure 19.5 can sometimes form.

As a comet moves away from the sun, the gases begin to condense, the tail disappears, and the comet once again returns to "cold storage." The material that was blown from the coma to form the tail is lost from the comet forever. Consequently, it is believed that most comets cannot survive more than 100 close orbits of the sun. Once all the gases are expended, the remaining material—a swarm of unconnected metallic and stony particles—continues the orbit without a coma or a tail.

Little is known about the origin of comets. The most widely accepted proposal considers them to be members of the solar system that formed at great distances from the sun. Accordingly, millions of comets are believed to orbit beyond Pluto with periods measured in hundreds of years. It is proposed that the gravitational effect of stars passing nearby sends some of them into highly eccentric orbits, which carry them toward the center of our solar system. Here, the gravitation of the larger planets, particularly Jupiter, alters a comet's orbit and reduces its period of revolution. Many short-period comets of this type have been discovered. However, since they have a short life expect-

ancy, we can be fairly certain that they are always being replaced by other long-period comets which are deflected toward the sun.

The most famous short-period comet is Halley's comet. Its orbital period averages 76 years, and every one of its appearances since 240 B.C. has been recorded by the Chinese. This record is a testimonial to their dedication as astronomical observers. When seen in 1910, Halley's comet had developed a tail nearly 1.6 million kilometers (1 million miles) long and was visible during the daylight hours. Although the 1986 visit of Halley's comet was a disappointment to those of us in the Northern Hemisphere, its next visit should be truly spectacular.

One of the most spectacular events of modern times occurred in Siberia in 1908. A brilliant fireball exploded with a force strong enough to create a shock wave that rattled windows and was heard hundreds of kilometers away. Trees were scorched

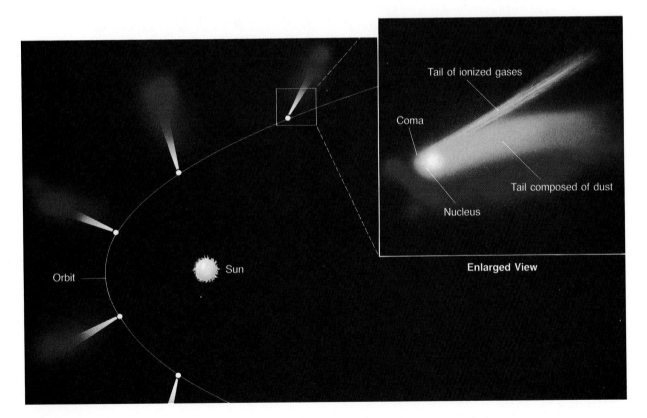

FIGURE 19.5
Orientation of a comet's tail as it orbits the sun.

and flattened for a distance of 30 kilometers from the impact zone. Although numerous craters up to 50 meters wide resulted, no large metallic fragments were found. The notable lack of impact fragments points strongly to the explosion of a comet nucleus which was rapidly heated as it penetrated our atmosphere.

METEOROIDS

Nearly everyone has seen a **meteor,** popularly called a "shooting star." This streak of light lasts for, at most, a few seconds and occurs when a small solid particle, a **meteoroid,** enters the earth's atmosphere from interplanetary space. The friction between the meteoroid and the air heats both and produces the light we see. Most meteors are about the size of sand grains and weigh less than $1/100$ gram. Consequently, they vaporize before reaching the earth's surface. Some, called *micrometeorites,* are so tiny that their rate of fall becomes too slow to cause them to burn up, so they drift down like dust. Each day, the total number of meteoroids that enter the earth's atmosphere must number into the millions. From a single spot on earth, a half dozen or more are bright enough to be seen with the unaided eye each hour.

Occasionally, the number of meteor sightings increases dramatically to 60 or more per hour. These spectacular displays, called **meteor showers,** result when the earth encounters a swarm of

FIGURE 19.6
Meteor Crater, northern Arizona, about 32 kilometers (20 miles) west of Winslow.
(Courtesy of Meteor Crater Enterprises, Inc.)

meteoroids traveling in the same direction and at nearly the same speed as the earth. The close association of these swarms to the orbits of some short-term comets strongly suggests that they represent material lost by these comets. Some swarms not associated with orbits of known comets are probably the remains of the nucleus of a defunct comet. The meteor showers that occur regularly each year around August 12 are believed to be the remains of the tail of Comet 1862 III, which has a period of 110 years. Meteoroids associated with comets are small and not known to reach the ground. Most meteoroids large enough to survive the fall are thought to originate in the belt of asteroids, where a chance collision sends them toward the earth. The earth's gravitational force does the rest.

The remains of meteoroids, when found on the earth, are referred to as **meteorites.** Several very large meteorites have blasted out craters on the earth's surface that are not much different in appearance from those found on the lunar surface. The most famous is Meteor Crater in Arizona (Figure 19.6). This crater is about 1.2 kilometers (0.75 mile) across, 170 meters (560 feet) deep, and has an upturned rim that rises 50 meters (165 feet) above the surrounding countryside. Over 30 tons of iron fragments have been found in the immediate area, but attempts to locate the main body have been unsuccessful. As evidenced from the amount of erosion, it appears that the impact occurred within the last 20,000 years.

Prior to lunar exploration, meteorites were the only samples of extraterrestrial material that could be directly examined (Figure 19.7). Depending upon their composition, they can generally be put into one of three categories: (1) **Irons**—mostly iron with 5–20 percent nickel; (2) **Stony**—silicate minerals with inclusions of other minerals; and (3) **Stony-irons**—mixtures. Although stony meteorites are probably more common, most meteorite finds are irons. This is understandable, since irons tend to withstand the impact, weather more slowly, and are much easier for a lay person to distinguish from terrestrial rocks than are stony meteorites. One rare kind of meteorite, called a *carbonaceous chondrite,* was found to contain some simple amino acids, which are the basic building blocks to life. This discovery confirms

FIGURE 19.7
Meteorite found near Meteor Crater, Arizona.

similar findings in observational astronomy which indicate that numerous organic compounds exist in the frigid realm of outer space.

If the composition of meteorites is representative of the material that makes up the Earthlike planets, as some planetary geologists believe, then the earth must contain a much larger percentage of iron than is indicated by the surface rock. This is one of the reasons why geologists suggest that the core of the earth may be mostly iron and nickel. In addition, the dating of meteorites has indicated that our solar system has an age which certainly exceeds 4 billion years. This "old age" has also been confirmed by data obtained from lunar samples.

THE PLANETS: AN INVENTORY

MERCURY: THE INNERMOST PLANET

Mercury, the innermost and swiftest planet, has a diameter of 4878 kilometers, not much larger than Earth's moon. Also, like the moon, it absorbs most of the sunlight that strikes it, reflecting only 6 percent into space. This is characteristic of terrestrial bodies that have no atmosphere. By way of comparison, the earth reflects more than 30 percent of the light which strikes it, most of it from clouds.

Mercury's close proximity to the sun makes viewing from earthbound telescopes difficult at best. The first good glimpse of this planet came in the spring of 1974, when *Mariner 10* passed within

800 kilometers (500 miles) of its surface (Table 19.3). Its striking resemblance to the moon was immediately evident from the high-resolution images that were radioed back. In fact, the similarity was so great that a project scientist remarked that these photos could be substituted for

TABLE 19.3
Summary of significant space probes and planned space probes.

Mariner 2	1962	Fly-by of Venus (first to any planet)
Mariner 4, 6, 7	1965 1969 1969	Fly-by missions to Mars
Mariner 9	1971	Orbiter of Mars
Apollo 8	1968	Astronauts circled the moon and returned to the earth
Apollo 11	1969	First astronaut landed on the moon
Apollo 17	1972	Last of six manned *Apollo* missions to the moon
Mariner 10	1974	Orbited the sun, allowing it to pass Mercury several times; fly-by mission to Venus
Pioneer 10, 11	1973 1974	First close-up views of Jupiter
Venera 8, 9, 10	1972 1975	Soviet landers on Venus (operative about one hour each)
Venera 13, 14	1982	First color images of Venus
Viking 1, 2	1976	Orbiters and landers of Mars
Voyager 1	1979 1980	Fly-by of Jupiter Fly-by of Saturn
Voyager 2	1979 1981 1986 1989	Fly-by of Jupiter Fly-by of Saturn Fly-by of Uranus Fly-by of Neptune
Ulysses	1987*	Fly-by of Jupiter
Galileo	1988*	Orbiter of Jupiter
Magellan	1988*	Venus Radar Mapper (VRM)
Mars Observer	1990*	Designed to determine the role of water in the Martian climate

*Proposed launch date.

ones of the back side of the moon and most non-scientists would not know the difference.

Mercury has densely cratered highlands much like the moon and vast smooth terrains which resemble maria. However, unlike the moon, Mercury is a very dense planet, which implies that it contains an iron core perhaps larger than the earth's. Also, Mercury has very long scarps that cut across the plains and craters alike. One proposal is that these scarps resulted from crustal shortening as the planet cooled and shrank.

A typical period of daylight and a typical period of darkness on Mercury both last 88 days. The night temperatures drop as low as −173°C (−280°F) and the noontime temperatures exceed 427°C (800°F), hot enough to melt tin and lead. Mercury has the greatest temperature extremes of any planet. The odds of life as we know it existing on Mercury are nil.

VENUS: THE VEILED PLANET

Venus, second only to the moon in brilliance in the night sky, is named for the goddess of love and beauty. It orbits the sun in a nearly perfect circle once every 225 days. Venus is similar to the earth in size, density, and mass, and is also located in the inner portion of the solar system. Thus, it has been referred to as "Earth's twin." Due to these similarities, it is hoped a detailed study of Venus will provide geologists with a better understanding of the earth's evolutionary history.

Although Venus is shrouded in a thick cloud cover, radar mapping by the *Pioneer/Venus* orbiter, as well as by earthbound instruments, has revealed a rather varied topography with features somewhat between those of Earth and Mars (Figure 19.8). Simply stated, radar impulses are sent toward the Venusian surface and the heights of features such as plateaus and mountains are measured by timing the return of the radar echo. A high percentage of the Venusian surface consists of a rather subdued rolling topography. Within this area of rolling plains are several large valleys believed to have a tectonic origin. This region also has the highest density of circular features, which most probably are impact craters. Only 8 percent of the surface consists of highlands that may be likened to continental areas on Earth. The high-

FIGURE 19.8
Venus as photographed from *Mariner 10*. (Courtesy of NASA)

lands of Venus are dominated by large plateau-like structures; however, discrete mountain ranges, including Maxwell Montes, which rises 11 kilometers above the lowlands, also exist. In addition, two topographic features rising nearly 5 kilometers above the adjacent terrain resemble large shield volcanoes like those found on Earth and Mars.

Before the advent of space vehicles, Venus and Mars were considered the most hospitable sites in the solar system (Earth excluded) and the logical places to expect living organisms. Unfortunately, evidence from *Mariner* fly-by space probes and Soviet unmanned landings on Venus indicates differently. The surface of Venus reaches temperatures of 480°C (900°F), and the Venusian atmosphere is composed mostly of carbon dioxide (97 percent). Only minor amounts of water vapor and nitrogen have been detected. Its atmosphere contains an opaque cloud deck about 25 kilometers thick, which begins approximately 70 kilometers from the surface. Although the unmanned Soviet *Venera 8* survived less than one hour on the Venusian surface, it determined that the atmospheric

pressure on that planet is 90 times greater than that found on the earth's surface. This hostile environment makes it unlikely that life as we know it exists on Venus and makes manned space flights to Venus improbable in the future.

Later *Venera* missions transmitted a few images of Venus before they too succumbed to the tremendous heat and pressure. The landing site of *Venera 9* showed numerous angular rocks, a setting similar to what the *Viking* lander found on Mars. *Venera 10*, on the other hand, photographed rather flat, rounded rocks, possibly the result of longer exposure to erosion than those found by *Venera 9*. Soil measurements taken by the *Venera* landers indicated a composition similar to the basalts found on the earth, the moon, and Mars.

Carl Sagan, one of the foremost experts on extraterrestrial life, in a description of planetary environments, called Earth "the Heaven of the solar system" and Venus "the Hell." Why should the environments of two planets that are nearly the same size, which evolved in a similar manner, and are located in close proximity to one another, be so dramatically different? The primary reason is the blistering temperature of the Venusian atmosphere, which is caused by a runaway greenhouse effect. The **greenhouse effect** is also experienced on Earth on a sunny day by anyone entering a greenhouse or a car with its windows rolled up. The dense Venusian atmosphere acts like the glass in a greenhouse, allowing visible solar energy to penetrate to the surface. Planets typically lose energy at nearly the same rate as it is received by reradiating it back to space. However, the dense Venusian atmosphere, like the glass in the greenhouse, is quite opaque to the outgoing radiation. Thus, the surface of Venus is very hot because the heat is temporarily trapped.

Carbon dioxide, which makes up 97 percent of the Venusian atmosphere, is the main contributor to the greenhouse effect. The primary source of carbon dioxide is outgassing during volcanic eruptions. Why does the earth, which has numerous active volcanoes, have an atmosphere which consists of only a small fraction of one percent carbon dioxide? The answer can be found in the earth's abundant plant life. Plants use carbon dioxide and water in photosynthesis to generate organic matter, while oxygen is released as a by-product. Thus, over millions of years plant life has altered our atmosphere, making it carbon dioxide-poor and oxygen-rich. In addition, carbon dioxide dissolved in seawater is used by a vast number of organisms for the production of carbonate shells. These shells are eventually deposited as sediment on the ocean floor. Consequently, huge quantities of carbon dioxide from the earth's atmosphere are continually being converted into organic matter and carbonate sediments. Apparently, Venus does not possess living organisms or a chemical mechanism capable of removing atmospheric carbon dioxide. Hence, our neighboring planet has a carbon dioxide concentration that has reached extreme proportions.

The discovery of a hostile Venusian atmosphere provides a good example of the practical benefits that humans might derive from space exploration. These data clearly reveal what can happen to a planet's environment when the greenhouse effect runs rampant. Some climatologists believe a similar situation may be in the making on earth. Through the burning of fossil fuels, we are con-

verting increasingly larger amounts of oxygen and organic materials into carbon dioxide and water. For that reason, we are altering the earth's atmosphere by rapidly undoing what has taken plant life millions of years to accomplish.*

We should expect that studies of the Venusian atmosphere will provide insight to how the earth's atmosphere evolved and how it will respond to changes induced by human activity. This insight could keep our planet from becoming, as some have suggested, "hothouse Earth."

MARS: THE RED PLANET

Mars has evoked greater interest than any other planet, for astronomers as well as for nonscientists. When we think of intelligent life on other worlds, "little green Martians" may come to mind. Interest in Mars stems mainly from this planet's accessibility to observation. All other planets within telescopic range have their surfaces hidden

*An in-depth discussion of the possible impact of additional carbon dioxide on global climate is found in Chapter 16.

FIGURE 19.9
The boulder-strewn Martian landscape as viewed from *Viking 2*. The red color of the Martian surface and the salmon color of the sky are believed to be accurately reproduced. The color of the sky results from dust particles suspended in the atmosphere. (Courtesy of NASA)

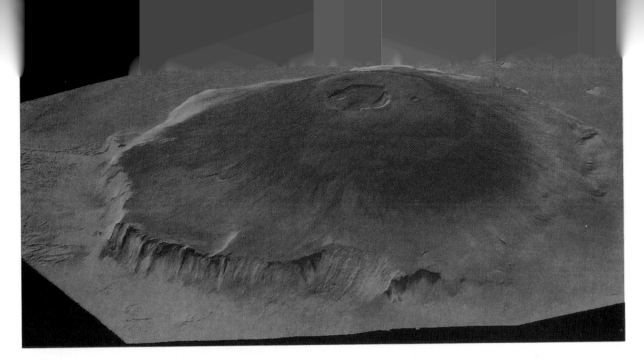

FIGURE 19.10
Image of Mons Olympus, an inactive shield volcano on Mars that covers an area about the size of the state of Ohio. (Courtesy of the U.S. Geological Survey)

by clouds, except for Mercury, whose nearness to the sun makes viewing difficult. Through the telescope Mars appears as a reddish ball interrupted by some permanent dark regions that change intensity during the Martian year. The most prominent telescopic features of Mars are its brilliant white polar caps.

The Martian atmosphere is only 1 percent as dense as that of the earth and is composed primarily of carbon dioxide with only very small amounts of water vapor. Data radioed back to Earth confirmed speculations that the polar caps of Mars are made of water ice covered by a relatively thin layer of frozen carbon dioxide. As the winter nears in a given hemisphere, we see the equatorward growth of that hemisphere's ice cap as additional carbon dioxide is deposited. This finding is compatible with the temperatures observed at the polar caps. Here temperatures reach a frigid $-125°C$ $(-193°F)$, cold enough to solidify carbon dioxide.

Although the atmosphere of Mars is very thin, extensive dust storms do occur and may be responsible for the color changes observed from earth-based telescopes. Winds up to 270 kilometers (170 miles) per hour can persist for weeks. Images radioed back by *Viking 1* revealed a Martian landscape remarkably similar to a rocky desert on Earth (Figure 19.9). Sand dunes are abundant, and many Martian impact craters have

flat bottoms because they are partially filled with dust. Thus, unlike the craters of the moon, the oldest craters on Mars are completely obscured by deposits of windblown material.

When *Mariner 9*, the first artificial satellite to orbit another planet, reached Mars in 1971, a large dust storm was raging, and only the ice caps were initially observable. It was spring in the southern hemisphere, and that polar cap was receding rapidly. About mid-summer, when the rate of evaporation should have been greatest, the size of the polar cap showed little change. A residual cap remained through the summer, providing strong evidence that the residual polar cap is made of water ice. (Water ice has a much lower rate of evaporation than frozen carbon dioxide.)

When the dust cleared, images of the northern hemisphere revealed numerous large volcanoes. The largest, called Mons Olympus, covers an area the size of the state of Ohio and is no less than 23 kilometers (75,000 feet) in elevation. This gigantic volcano and others that were observed closely, resemble the shield volcanoes found on Earth, being similar to those of Hawaii (Figure 19.10). Their extreme size is thought to result from the absence of plate movements on Mars. Therefore, rather than a chain of volcanoes forming as we find in Hawaii, single, larger cones developed.

Impact craters are notably less abundant in the

region where the volcanic activity appears most prevalent. This indicates that at least some of the volcanic topography formed more recently in that planet's history, somewhat after the early period of heavy bombardment. Nevertheless, age determinations based on crater densities obtained from

Viking photographs indicate that most Martian surface features are old by Earth standards. The highly cratered Martian southern hemisphere is probably similar in age to comparable lunar highlands (3.5–4.5 billion years old). The discovery of several highly cratered and weathered volcanoes

FIGURE 19.11
Vast chasm with branching tributaries located 500 kilometers south of the Martian equator. (Courtesy of NASA)

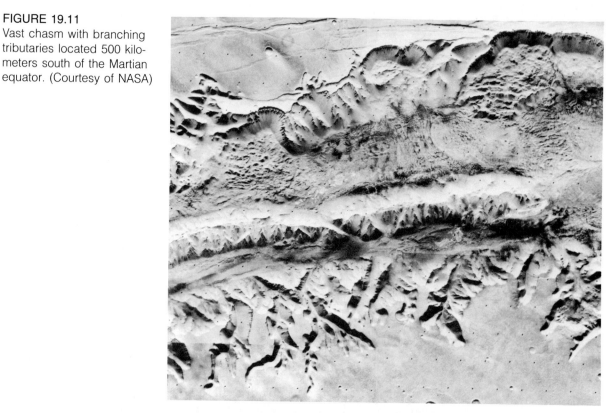

FIGURE 19.12
Extending across the middle of this photograph is a "stream" channel that most probably formed as a result of the melting of subsurface ice. As the ice melted, the unsupported land above collapsed to produce the hummocky topography to the right. The flood of meltwater carved the channel to the left. (Courtesy of NASA)

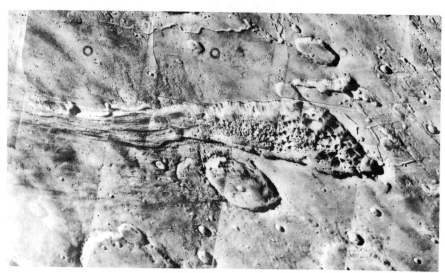

on Mars further indicates that volcanic activity began early and had a long history. However, even the relatively fresh-appearing volcanic features of the northern hemisphere may be older than 1 billion years. This fact, coupled with the absence of "marsquake" recordings by *Viking* seismographs, points to a tectonically dead planet.

Another surprising find made by *Mariner 9* was the existence of several canyons that dwarf even the Grand Canyon of the Colorado River. One of the largest, Valles Marineris, is roughly 6 kilometers deep, up to 160 kilometers wide, and extends for almost 5000 kilometers along the Martian equator (Figure 19.11). This vast chasm is thought to have formed by slippage of crustal material along huge faults in the crustal layer. In this respect, it would be comparable to the rift valleys of Africa.

Not all valleys on Mars have a tectonic origin. Many Martian valleys have tributaries exhibiting drainage patterns similar to those of stream valleys found on Earth. In addition, *Viking* orbiter photographs have revealed streamlined features that are unmistakably ancient islands located in what is now a dry stream bed. When these streamlike channels were first discovered, some observers speculated that a thick water-laden atmosphere capable of generating torrential downpours once existed on Mars. What happened to this water? The present Martian atmosphere contains only trace amounts of water. Moreover, the environment of Mars is far too harsh to allow water to exist as a liquid. Despite these difficulties, the work of flowing water still remains the most acceptable explanation for many of the Martian channels.

Many planetary geologists do not accept the premise that Mars once had an active water cycle similar to that on Earth. Rather they believe that many of the large streamlike valleys were created by the collapse of surface material caused by the slow melting of subsurface ice (Figure 19.12). If this is the case, these large valleys would be more akin to features formed by mass wasting processes on the earth. Some of the smaller channels, they contend, have been cut by the gradual release of subsurface water, which flowed out of the ground like springs. Further, the heat from volcanic activity or meteoroid impact may have rap-idly melted subsurface ice which, in turn, caused local flooding. These floods may have carved some of the channels. An occasional flood would also account for the streamlined features found in what are presently dry river beds. The question of whether rainfall, gradual seepage of groundwater, or the collapse of surface material created the branching valleys of Mars will be debated for some time. The answer might not come until we obtain images of Mars which reveal greater detail than those presently available.

Due to their small size, Phobos and Deimos, the two satellites of Mars, were not discovered until 1877. Phobos is nearer to its parent than any other natural satellite (Figure 19.13). Only 5500 kilometers from the Martian surface, Phobos requires just 7 hours and 39 minutes for one revolution.

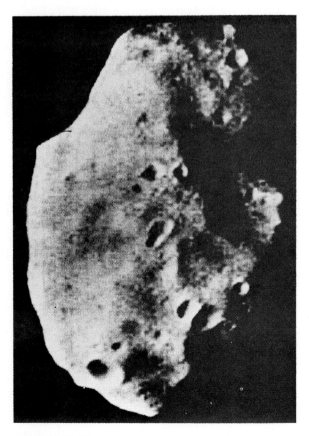

FIGURE 19.13
Phobos, the larger of the two Martian satellites, photographed by *Mariner 9* from 5500 kilometers. (Courtesy of NASA)

Deimos, which is smaller and 20,000 kilometers away, revolves in 30 hours and 18 minutes. *Mariner 9* revealed that both satellites are irregularly shaped and have numerous impact craters, much like their parent. The maximum diameter of Phobos is 24 kilometers and the maximum diameter of Deimos is only about 15 kilometers. Undoubtedly, these moons are asteroids which were captured by Mars. One of the most interesting coincidences in astronomy is the close resemblance between Phobos and Deimos and the two fictional satellites of Mars described by Jonathan Swift in

Gulliver's Travels, written about 150 years before they were actually discovered.

JUPITER: THE LORD OF THE HEAVENS

Jupiter, the largest planet in our solar system, has a mass 2½ times greater than the combined mass of all the remaining planets, satellites, and asteroids. It is truly a giant among planets. In fact, had Jupiter been about ten times larger, it would have evolved into a small star. Despite its great size, however, it is only 1/1000 as massive as the sun.

FIGURE 19.14
A mosaic of Jupiter and its four Galilean satellites as seen from *Voyager 1.*
Ganymede is located at lower left, Callisto at lower right. Europa is adjacent to
Jupiter, and Io appears in the distance. (Courtesy of NASA)

Jupiter also rotates more rapidly than any other planet, completing one rotation in slightly less than ten hours. The effect of Jupiter's rapid rotation causes the equatorial region to be slightly bulged and the polar dimension to be flattened.

When viewed through a telescope or binoculars, Jupiter appears to be covered with alternating bands of multicolored clouds aligned parallel to its equator. The most striking feature on the disk of Jupiter is the great red spot (Figure 19.14). Although its color varies greatly in intensity, it has been an ever-present feature since it was first discovered more than three centuries ago. When *Voyager 2* swept by Jupiter in July, 1979, the size of the red spot was 11,000 kilometers by 22,000 kilometers, which equals two earth-sized circles placed side by side. On occasion it has grown even larger. Although the great red spot varies in size, it does remain the same distance from the Jovian equator. The cause of the spot has been attributed to everything from volcanic activity to a large cyclonic storm. Images obtained by *Pioneer 11* as it moved

to within 42,000 kilometers of Jupiter's cloud tops in December, 1974, support the latter view. The great red spot apparently is a counterclockwise-rotating storm caught between two jetstream-like bands of atmosphere flowing in opposite directions. This huge hurricane-like storm rotates once every 12 Earth days. Although several smaller storms have been observed in other regions of Jupiter's atmosphere, none has survived for more than a few days.

Structure of Jupiter Jupiter's atmosphere is composed primarily of hydrogen and helium, with methane, ammonia, water, and sulfur compounds as minor constituents. The wind systems generate the light- and dark-colored bands that encircle this giant. Unlike the winds on Earth, which are driven by solar energy, Jupiter gives off nearly twice as much heat as it receives from the sun. Thus, it is the heat emanating from Jupiter's interior that produces huge convection currents in the atmosphere (Figure 19.15). The light-colored zones are

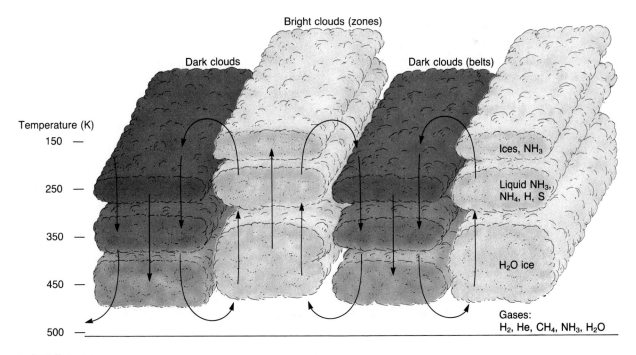

FIGURE 19.15
The structure of Jupiter's atmosphere. The bright zones are clouds composed of ice particles located higher in the atmosphere than the relatively warm clouds of the dark bands. The dark color of the belts may be caused by sulfur compounds or perhaps organic molecules.

regions where gases are ascending and cooling. The light color is thought to be produced by ammonia crystallizing into "snowflakes" near the cloud tops where temperatures of $-120°C$ have been measured. The cold material from the light zones spills over onto the lower dark belts, where air is descending and heating. The reason for the rusty-brown color of the dark belts is not known for certain; however, they may be colored by sulfur compounds.

Atmospheric pressure at the top of the clouds is equal to sea-level pressure on Earth, but because of Jupiter's immense gravity, the pressure increases rapidly toward its surface. At 1000 kilometers below the clouds, the pressure is great enough to liquify hydrogen. Consequently, the surface of Jupiter is thought to be a gigantic ocean of liquid hydrogen. Less than halfway into Jupiter's interior, pressures of unimaginable magnitude cause liquid hydrogen to turn into liquid metallic hydrogen. Although scientists have never seen this unusual material, its properties are thought to be predictable. Jupiter is also believed to contain as much rocky and metallic material as is found in the terrestrial planets, such as Earth. Whether this "earthly" material is scattered as small bits or makes up a central core is not known with certainty, but the latter idea is generally accepted.

Jupiter's Moons Jupiter's satellite system, consisting of at least sixteen moons, resembles a miniature solar system. The four largest satellites were discovered by Galileo and travel in nearly circular orbits around the parent with periods of from two to seventeen days (Figure 19.14). The two largest Galilean satellites, Callisto and Ganymede, surpass Mercury in size, while the two smaller ones, Europa and Io, are about the size of Earth's moon. These Galilean moons can be observed with a small telescope and are interesting in their own right. Because their orbits are along Jupiter's equatorial plane and since they all have the same orbital direction, these moons most probably formed from "leftover" debris in much the same way as the planets did. By contrast, the four outermost satellites are very small (20 kilometers in diameter), revolve in a direction that is

opposite the other moons, and have orbits that are steeply inclined to the Jovian equator. These satellites appear to be asteroids that passed near enough to be captured gravitationally by Jupiter.

The images obtained by *Voyagers 1* and *2* in 1979, revealed to the surprise of almost everyone that each of the four Galilean satellites has a character all its own. The entire surface of Callisto, the outermost of the Galilean satellites is densely cratered, much like the surfaces of Mercury and the moon. However, in this instance the impacts occurred in a crust which appears to be a dirty, frozen ocean of water ice. The largest features discovered were sets of bright concentric rings surrounding impact craters. These bright rings are thought to resemble the ripples produced when a pebble is dropped into calm water. They are probably composed of blocks of ice, which were deformed and uplifted by the impact that generated the central crater.

Ganymede, the largest Jovian satellite, contains the most diverse terrain. Like our moon, it has densely cratered regions, and other very smooth areas, where a younger icy layer covers the older cratered surface. In addition, Ganymede has numerous parallel grooves. This suggests some type of tectonic activity has occurred in the distant past. In some places structural features show evidence of lateral displacement along strike-slip faults, a phenomenon previously found only on the earth.

Europa, smallest of the Galilean satellites, has an icy surface that is crisscrossed by many linear features. Although some of these linear markings are thousands of kilometers in length, they have very little surface expression. Further, this satellite is notably void of large impact craters. Europa is as smooth as a billiard ball of the same scale would be. Therefore, the present surface of Europa must have formed sometime after the early period of bombardment, when rocky chunks were far more numerous in the solar system. The crust of Europa may be a thick, frozen ice layer which caps a slushy ocean.

The innermost of the Galilean moons, Io, is the only volcanically active body other than Earth so far discovered in our solar system (Figure 19.16). To date, eight active sulfurous volcanic centers

FIGURE 19.16
Io, a satellite of Jupiter, is the only body in the solar system other than Earth known to have active volcanism. Image obtained in 1979 by *Voyager 1*. (Courtesy of NASA)

have been discovered. Umbrella-shaped plumes have been seen rising from the surface of Io to heights approaching 200 kilometers (125 miles). The surface of Io is very colorful, a result of its sulfurous "rocks" which change colors from red to yellow, and from black to white depending upon temperature. The heat source for this volcanic activity is thought to be tidal energy generated by a "tug of war" between Jupiter and the Galilean satellites. Since Io is gravitationally locked to Jupiter, the same side always faces the giant planet. The gravitational influence of Jupiter and the other nearby satellites pulls and pushes on Io's tidal bulge as its slightly eccentric orbit takes it alternately closer and farther from Jupiter. This gravitational flexing of Io is transformed into heat energy.

One of the most interesting discoveries made by *Voyager 1* is the ring system of Jupiter. Thought to be less than 30 kilometers thick, the apparently continuous ring may extend outward from the surface of Jupiter to a distance equal to twice the diameter of the planet. This ring system is not believed to be like that encircling Saturn. Rather than particles held in planetary-type orbits, the particles composing Jupiter's ring appear to be temporarily entrapped by the planet's intense magnetic field. The source of the ring material may be sulfur from the volcanoes of Io, which is ionized and pulled toward Jupiter by the planet's magnetic field.

SATURN: THE ELEGANT PLANET

Requiring 29½ years to make one revolution, Saturn is almost twice as far from the sun as Jupiter, yet its atmosphere, composition, and internal structure are thought to be remarkably similar to Jupiter's. The most prominent feature of Saturn is its system of rings (Figure 19.17). These rings were discovered by Galileo and appeared to him as two smaller bodies adjacent to the planet because he could not resolve them with his primitive telescope. Their ring nature was revealed 50 years later by the Dutch astronomer Christian Huygens. Until the recent discovery that Jupiter and Uranus have very faint ring systems, this spectacular phenomenon was thought to be unique to Saturn.

When viewed from Earth, Saturn's rings appear to consist of three rather distinct, concentric bands, which have classically been called the *A*, *B*, and *C* rings (Figure 19.18). The *A* ring is the outermost of the bright rings and is separated from the brightest ring (*B* ring) by a large gap. This space, called the **Cassini gap,** is easily seen in a photo of Saturn (see chapter-opening photo). Having a width of 5000 kilometers, the Cassini gap would be large enough to accommodate our moon. During this century evidence has been gathered for the existence of four other very faint bands; the *D* ring is located inside and the *F*, *G*, and *E* rings are located outside the classically known rings (Figure 19.18). Even the Cassini gap is not empty, but rather contains a number of very faint bands.

FIGURE 19.17
A view of Saturn's rings taken from a distance of 8 million kilometers by *Voyager 1*. At this distance almost 100 individual rings can be seen, including several narrow bands in the dark Cassini division. (Courtesy of NASA)

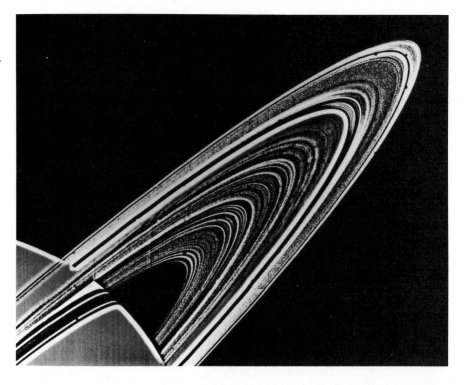

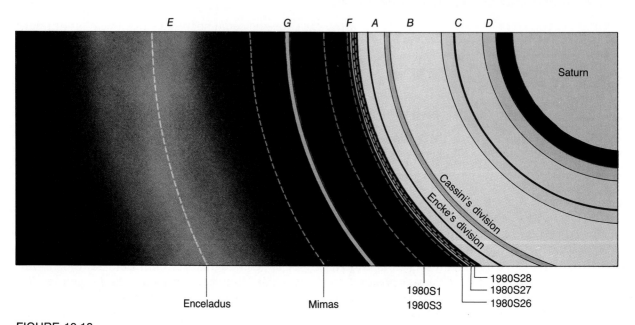

FIGURE 19.18
A polar view of the ring system of Saturn is shown diagrammatically to scale. (From "Rings in the Solar System," by James B. Pollack and Jeffrey N. Cuzzi. Copyright © 1981 by Scientific American, Inc. All rights reserved.)

From the earth, Saturn's rings can be viewed on edge once every 15 years and appear as an extremely fine line.

In 1980 and 1981, fly-by missions of the nuclear-powered *Voyagers 1* and *2* unmanned space vehicles came within one hundred thousand kilometers of the surface of Saturn. More information on Saturn and its satellite system was gained in a few days than had been acquired since Galileo first viewed this elegant planet telescopically in 1610. Some of the information acquired by these space probes is as follows:

1 Saturn's atmosphere is very dynamic with winds roaring up to 1500 kilometers per hour.
2 Large cyclonic "storms" similar although much smaller than Jupiter's great red spot occur in Saturn's atmosphere.
3 Three additional moons were discovered, bringing the total so far detected to seventeen.
4 Verification was made of the existence of the sixth and innermost ring system, and evidence was gathered for a seventh ring system.
5 The icy rings of Saturn were discovered to be more complex than expected. Each of the seven rings is made of numerous ringlets resembling the grooves on a phonograph record, while the rings in the faint *F* ring are intertwined in stable kinked and braidlike configurations. Further, the *B* ring develops perplexing outwardly radiating spokes which survive for hours at a time.
6 Satellite images reveal the thickness of the ring system to be no more than a few hundred meters while its lateral extent exceeds 200,000 kilometers.

Although none of the images obtained so far has the resolution needed to "see" the fine structures of the rings, the rings are undoubtedly composed of relatively small particles (moonlets) which orbit the planet much like any other satellite. Radar observations indicate that most of the rings' particles are no larger than 10 meters and the more abundant particles are perhaps as small as 10 centimeters. Further, it has been determined that these moonlets are good reflectors of visible light and poor reflectors of near-infrared wavelengths, both characteristic of water ice. Hence, it is thought that at least the outer surfaces of these particles are coated with a highly reflective layer of water ice. Saturn's rings therefore are actually swarms of small ice-covered debris that orbit the planet in nearly circular paths that are aligned with the equatorial plane of the planet. Most probably they are composed of material which failed to accrete into moons at the time the Saturnian System was forming.

The origin of the rings is believed to be related to their distance from the surface of Saturn. Since they are very close, the disruptive gravitational force of the parent prevented the individual particles from accreting into a larger satellite. Stated another way, objects cannot be held together by self-gravitation when they are within the influence of a stronger gravitational field. In fact, should one of Saturn's large icy satellites approach closer than the outer edge of the bright rings, it would be destroyed and its remains distributed among the rings. The gravitational (tidal) force of Saturn pulling on the near side of such a satellite would be enough greater than the force pulling on its far side that it would tear the satellite apart. Although the rings could be the remains of a satellite destroyed in this fashion, these moonlets are more likely part of the primordial material from which Saturn formed.

Beyond the outermost bright ring (*A* ring), some moonlets have accreted to form very small satellites having diameters on the order of 100 kilometers. Five of these asteroid-sized moons have been discovered orbiting within the faint outer rings, and others probably exist. The gravitational influence of these so-called shepherding satellites is believed to be responsible for keeping the moonlets confined within the rings, thereby producing the sharp edges which are observed. Planetary geologists are very interested in the gravitational interaction of the objects that comprise Saturn's ring system. It is hoped that this information will reveal how material from the primordial cloud of dust and gases condensed to produce the planets. This information would be valuable in reconstructing the earth's early history.

FIGURE 19.19
Montage of the Saturnian System. Dione foreground; Tethys and Mimas distant
right; Enceladus and Rhea off ring's left; Titan, upper right. (Courtesy of NASA)

The Saturnian satellite system consists of at
least seventeen known bodies, all but two of which
have nearly circular, counterclockwise orbits
along Saturn's equatorial plane (Figure 19.19).
The largest Saturnian moon is the only satellite in
the solar system known to have a substantial atmosphere. Due to its dense gaseous cover, the
atmospheric pressure at the surface of Titan is
about one and one-half times that experienced at
the earth's surface. Although Titan's atmosphere
was predicted to be composed largely of methane,
data from *Voyager 1* revealed that this was not the
case. Rather, scientists discovered that as much as
80 percent of Titan's atmosphere is nitrogen, with
methane probably accounting for less than 6 percent. The orange color of Titan's atmosphere may
result from photochemical "smog" composed of
hydrocarbon molecules, including ethylene and
hydrogen cyanide. Further, this planet-sized moon

appears to have polar ice caps, which show seasonal variations in size. Its surface, if unfrozen,
would be an ocean of liquid nitrogen.

URANUS: THE GREEN PLANET

The honor of discovering the first planet that was
unknown to the ancients belongs to William
Herschel, the great German-born English astronomer. Herschel, a musician by training, originally
made telescopes and studied astronomy as hob-
bies. On March 13, 1781, using a 7-inch reflecting
telescope of his own making, he came upon a dim
"star," which appeared as a small disk. Believing it
was a comet, he recorded its position. After several
months, a calculation of its orbit revealed that it
was definitely a planet located beyond Saturn. For
this discovery, King George III awarded Herschel a
stipend, which allowed him to pursue his interest

in astronomy full-time. In response, Herschel named the planet "Georgium Sidus" (George's Star). In keeping with tradition, however, the name Uranus, a mythological Greek god and the grandfather of Jupiter, was adopted. Later, a review made of earlier records indicated that Uranus had been observed and recorded as a star on many occasions prior to Herschel's discovery.

The unique feature of Uranus is that its axis of rotation lies only 8 degrees from the plane of its orbit. Its rotational motion, therefore, has the appearance of rolling, rather than spinning like a top as do the other planets. Because Uranus is inclined almost 90 degrees, the sun is nearly overhead at one of the poles once each revolution, and then half a revolution later, it is nearly overhead at the other.

A surprise discovery in 1977 revealed that Uranus is surrounded by rings, much like those found encircling Jupiter. This find occurred as Uranus passed in front of a distant star and blocked its view, a process called **occultation.** Observers saw the star "wink" briefly five times before the primary occultation and again five times afterward. Later studies have indicated that Uranus has at least nine distinct belts of debris orbiting its equatorial region.

The first close-up views of Uranus and its satellites were transmitted to Earth by *Voyager 2* in January, 1986 (Figure 19.20). Preliminary studies of

FIGURE 19.20
Painting of *Voyager 2* as it might have appeared when it encountered Uranus on January 24, 1986. (By Don Davis, courtesy of NASA)

FIGURE 19.21
This computer-assembled mosaic of the Uranian satellite Miranda was obtained by *Voyager 2* in 1986. In addition to its crater marked areas, Miranda contains regions where folded ridges produce curvilinear patterns unique to this satellite. (Courtesy of NASA)

these images revealed ten previously unknown satellites that are much smaller than the five moons known from Earth-based observations. Two additional rings were also discovered, along with evidence that others may encircle the planet.

Spectacular views of the five largest moons of Uranus showed quite varied terrains. Some contain long, deep canyons and linear scars, whereas others possess large, smooth areas on otherwise crater-ridden surfaces. A spokesperson for the Jet Propulsion Laboratory described Miranda, the innermost of the five largest moons, as having a greater variety of landforms than any other body yet examined in the solar system. In particular, Miranda contains three vast areas that exhibit families of grooves that encircle one another much like the lanes of a gigantic racetrack (Figure 19.21). Careful study of the data obtained by *Voyager 2* should give planetary geologists greater insight into the processes that have shaped this planet and its satellites. Further, *Voyager 2* has also been programmed for a fly-by probe of Neptune in 1989 before it follows its predecessor, *Voyager 1,* out of the solar system.

NEPTUNE: A TWIN

If any two planets in the solar system can be considered "twins," Neptune and Uranus can. Besides being similar in size, they appear a pale green color, attributable to the methane in their atmospheres. Their structure and composition are believed to be similar, too, but because of its greater distance, Neptune experiences somewhat colder temperatures.

Whereas the discovery of Uranus was an accidental sighting by a very alert observer, the discovery of Neptune was truly a triumph for modern astronomy. A short time after an orbit had been calculated for Uranus, its motion began to show irregularities. Prior to 1822, it moved slightly ahead of its predicted position, and after 1822, it began lagging behind. Could it be that Newton's law of universal gravitation was not applicable for such great distances? Some people thought so.

A young Englishman, John Adams, approached the problem by considering the gravitational effect of a yet unknown planet. If this planet lay beyond Uranus, it would, in accordance with Kepler's laws, travel at a slower velocity. Consequently, Uranus would approach and pass this unknown planet during a revolution and be influenced by its gravity. When Uranus approached the unknown planet, its orbital speed would increase slightly, and as it receded from the unknown planet, its orbital speed would be reduced slightly. The amount of influence by the unknown planet would be dependent on its mass and its distance to Uranus. Adams made tedious calculations using differently-sized planets at various distances until he found a combination that worked. In October of 1845 he sent the location he had predicted for the unknown planet to the observatory at Greenwich, England. Undoubtedly, the Astronomer Royal had little faith in Adam's prediction, since a thorough search was never conducted. At this same time, a Frenchman, Urbain Leverrier, unaware of Adam's work, was tackling the same problem in a similar manner. Leverrier sent his data to the Berlin Observatory, and during the first night of observation, a likely candidate for the unknown planet was located. Since Adams and Leverrier independently predicted the position of Neptune, they are both credited with its discovery.

PLUTO: PLANET X

After the discovery of Neptune, there still remained discrepancies in the orbit of Uranus. This led some astronomers to suggest that another planet might lie beyond the orbit of Neptune. A vigorous search for this planet was begun in 1906 by Percival Lowell at the observatory in Flagstaff, Arizona, that now bears his name. He unsuccessfully searched for the object he called "Planet X" until his death ten years later. Some years afterward, a new photographic telescope was installed at the Lowell Observatory and the search was resumed full scale. Two photographs of the same portion of the sky were taken at intervals of a week or so and then compared. In such short time intervals, stars appear stationary, but a planet shifts its position relative to its background. Twenty-five years had lapsed since Lowell's death, and 2 million stars were examined before Clyde Tombaugh discovered a body which had a shift that was about right for an object 1 billion miles beyond the orbit of Neptune (Figure 19.22). This newly discovered planet was named Pluto, for the Greek god of the underworld, because the first two letters were Percival Lowell's initials. Although the planet was located in the constellation Gemini as Lowell had calculated, its mass was much less than he had estimated. It seems unlikely that a planet of Pluto's size could have altered the orbit of Uranus measurably. Most probably, the discovery of Pluto can be credited to an extensive search of the correct portion of the sky, although for the wrong reasons.

Pluto lies on the fringe of the solar system almost 40 times farther from the sun than the earth. Due to this great distance and Pluto's slow orbital speed, Pluto requires 248 years to orbit the sun. Since its discovery in 1930, it has completed slightly more than 1/5 of a revolution. Because of Pluto's elongated (highly eccentric) orbit, it is presently inside the orbit of Neptune, where it will remain until 1999. There is little likelihood that Pluto and Neptune will ever collide, because their orbits are inclined to each other and do not actually cross. Pluto's eccentric orbit and small size suggest to some astronomers that it might originally have been a satellite of Neptune.

In June of 1978 a moon was discovered orbiting Pluto. Although this satellite is too small to be observed visually, it appears as an elongated bulge on photographic plates taken of Pluto. The satellite is about 20,000 kilometers (12,400 miles) from the parent planet, which is 20 times closer than our moon is to the earth. This discovery greatly altered early estimations of the size of Pluto. Current data suggest that Pluto has a diameter of less

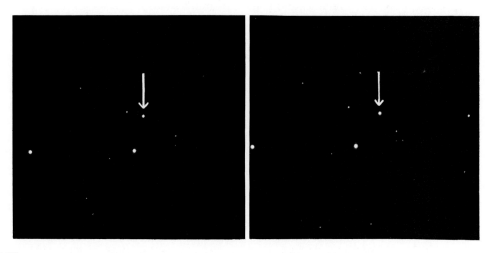

FIGURE 19.22
Pluto appears starlike on a photograph but is revealed by its motion to be a planet. (Courtesy of Hale Observatories)

than 3000 kilometers, less than that of our moon. Evidence from other studies indicates that its density is slightly greater than that of water. Consequently, Pluto might best be described as a large, dirty iceball made up of a mixture of frozen gases and rock.

Other than its orbital and rotational motions, little else is known about Pluto. It is 10,000 times too dim to be visible with the unaided eye. We do know that Pluto fluctuates slightly in brightness with a period of 6.4 days. This variation is believed to result from its rotation, which causes regions of its surface that have different reflectivities to periodically face the earth. The average temperature of Pluto is estimated, considering its solar distance, at −210°C (−350°F), which is cold enough to liquify or solidify any gas that may be present. Thus, Pluto could not have an atmosphere.

After the discovery of Pluto, Tombaugh started a systematic search of the heavens for a tenth planet. His inability to find one makes it unlikely that any undiscovered large planets orbit the sun near the plane of the ecliptic. However, a planet with a very inclined orbit may exist.

REVIEW QUESTIONS

1 Compare the rotational periods of the terrestrial and Jovian planets.

2 How fast does a location on the equator of Jupiter rotate? Note that Jupiter's circumference equals 416,000 kilometers (260,000 miles).

3 By what criteria are the planets placed into either the Jovian or terrestrial group?

4 What are the three types of materials thought to make up the planets? How are they different? How does their distribution account for the density differences between the terrestrial and Jovian planetary groups?

5 What fact led to the search for asteroids? Why were they difficult to find?

6 What do you think would happen if the earth passed through the tail of a comet?

7 Describe the origin of comets according to the most widely accepted hypothesis.

8 Compare a meteoroid, meteor, and meteorite.

9 Why are meteorite craters more common on the moon than on the earth, even though the moon is much smaller?

10 Why has Mars been the planet most studied telescopically?

11 What surface features does Mars have that are also common on Earth?

12 Although Mars has valleys that appear to be products of stream erosion, what fact makes it unlikely that Mars has had a water cycle like that on Earth?

13 The two "moons" of Mars were once suggested to be artificial. What characteristics do they have that would cause such speculation?

14 What is thought to be the nature of Jupiter's great red spot?

15 Why are the Galilean satellites of Jupiter so named?

16 What is unique about Jupiter's satellite Io?

17 Describe the nature of the surface of Callisto.

18 Why are the four outer satellites of Jupiter thought to have been captured?

19 What evidence indicates that Saturn's rings are composed of individual moonlets rather than consisting of solid disks?

20 Explain why a large satellite cannot exist closer to Saturn than the outer edge of its ring system.

21 What is unique about Titan?

22 Which of the five largest moons of Uranus has the most varied terrain?

23 Why was the name Pluto appropriate for "Planet X"?

KEY TERMS

terrestrial planet	coma	stony meteorite
Jovian planet	meteor	stony-iron meteorite
escape velocity	meteoroid	greenhouse effect
nebular hypothesis	meteor shower	Cassini gap
asteroid	meteorite	occultation
comet	iron meteorite	

20

BEYOND OUR SOLAR SYSTEM

The star Proxima Centauri is about 4.3 light-years away, roughly 100 million times farther than the moon, yet other than our own sun it is the closest star. The universe this fact suggests is incomprehensibly large. What is the nature of this vast cosmos beyond our solar system? Are the stars distributed randomly, or are they organized into distinct clusters? Do stars move, or are they permanently fixed features, like lights strung out against the black cloak of outer space? Does the universe extend infinitely in all directions, or is it bounded? To consider these questions, this chapter will examine the universe by taking a census of the stars—the most numerous objects in the night sky.

Veil Nebula in Cygnus is the remnant of an ancient supernova explosion. (Hale Observatories photo. Copyright by the California Institute of Technology and Carnegie Institution of Washington)

lthough the sun is the only star whose surface can be observed, a great deal is known about the universe beyond our solar system. In fact, more is known about the stars than about our outermost planet, Pluto. This knowledge hinges on the fact that stars, and even interstellar gases, radiate energy in all directions into space. The trick is to collect this radiation and unravel the secrets it holds. Astronomers have devised many ingenious methods to do just that. We will begin our discussion of the universe by examining some intrinsic properties of individual stars, including luminosity, color, mass, temperature, and size.

PROPERTIES OF STARS

DISTANCES TO THE STARS

As stated in Chapter 17, the annual back-and-forth shift of nearby stars with respect to very distant stars was considered proof of the earth's orbital motion. The measurement of this shift, called **stellar parallax,** is the only direct method of determining distances to stars. Figure 20.1 illustrates these shifts and the parallax angles determined from them. The nearest stars have the greatest parallax, while that of distant stars is imperceptible. Because the distance to even the nearest stars is thousands of times greater than the earth-sun distance, the triangles that astronomers work with are very "long and narrow," making the angles that are measured very small. The parallax angle of the nearest star, Proxima Centauri, is less than 1 second of arc, which equals ⅟₃₆₀₀ of a degree. It should be apparent why Tycho Brahe was unable to observe stellar parallax without a telescope and why he therefore concluded that the earth does not revolve.

Astronomers generally express stellar distances in units called **parsecs.** One parsec is the distance at which a star would have a parallax of 1 second of arc. Simply stated, the distance in parsecs is equal to 1 divided by the parallax angle. For example, a star with a parallax of 0.5 second of arc is 1/0.5, or 2, parsecs away, and one with a parallax of 0.1 is 1/0.1, or 10, parsecs away. Another unit popularly used to express stellar distance is the **light-year,** which is the distance light travels in a year. For purposes of conversion, 1 parsec equals 3.26 light-years. The distances to some nearby stars are given in all three units in Table 20.1.

In principle, the method used to measure stellar distances is rather elementary and was known to the ancient Greeks. In practice, the small angles involved and the fact that the sun, as well as the designated star, have actual motion greatly complicate the measurements. The first accurate stellar parallax was not determined until 1838, and today, parallaxes for only a few thousand of the nearest stars are known with certainty.

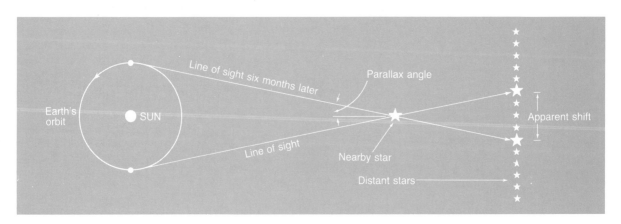

FIGURE 20.1
Geometry of stellar parallax.

TABLE 20.1
Distance to some of the nearest stars.

Name	Par-secs	Parallax Angle	Light-years
Proxima Centauri	1.31	0.763	4.27
Alpha Centauri	1.33	0.752	4.34
Barnard's Star	1.83	0.545	5.97
Sirius	2.67	0.377	8.70
61 Cygni	3.42	0.292	11.15
Kapteyn's Star	3.98	0.251	12.97

Stars more distant than 50 parsecs have such small shifts that accurate measurement is not possible. Fortunately, a few indirect methods exist so that distances to these stars can be estimated; however, it is important to check the accuracy of these methods on stars whose parallax is known.

STELLAR MAGNITUDES

Stars have been classified according to their brightness since the second century B.C., when Hipparchus placed about 1000 of them into six categories. The measure of a star's brightness is called its **magnitude.** Some stars may appear dimmer than others only because they are farther away. Therefore, a star's brightness, as it appears when viewed from the earth, has been termed its **apparent magnitude.** When numbers are employed to designate relative brightness, the larger the magnitude number, the dimmer the star. Stars that appear the brightest are of the first magnitude, while the faintest stars visible to the unaided eye are of the sixth magnitude. With the advent of the telescope, many stars fainter than the sixth magnitude were discovered.

In the middle of the nineteenth century, a method was developed whereby magnitudes could be used to quantitatively compare the brilliance of stars. So, just as we can compare the brightness of a 50-watt light bulb to that of a 100-watt bulb, we can compare the brightness of stars having different magnitudes. It was determined that a first-magnitude star was about 100 times brighter than a sixth-magnitude star, so on the scale adopted, two stars which differ by 5 magni-

tudes have a ratio in brightness of 100 to 1. Hence, a seventh-magnitude star is 100 times brighter than a twelfth-magnitude star. It follows, then, that the brightness ratio of two stars differing by one magnitude is the fifth root of 100, or 2.512. Thus, a star of the first magnitude is 2.512 times brighter than a star of the second magnitude and $(2.512)^4$, or 40, times brighter than a star of the fifth magnitude (Table 20.2). Due to the fact that some stars are brighter than the so-called first-magnitude stars, negative magnitudes were introduced. The brightest star in the night sky, Sirius, has an apparent magnitude of −1.4, about 10 times brighter than a first-magnitude star. On this scale, the sun has an apparent magnitude of −26.7, and when brightest, Venus has a magnitude of −4.3. At the other end of the scale, the 5-meter (200-inch) Hale telescope can view stars with an apparent magnitude of +23, which are approximately 100 million times dimmer than stars that are visible with the unaided eye.

Astronomers are also interested in the "true" brightness of stars, called their **absolute magnitudes.** Stars of the same luminosity, or brightness, usually do not have the same apparent magnitude, because their distances from us are not equal. In order to compare their true, or intrinsic, brightness, astronomers determine what magnitude the star would have if it were at a standard distance of 10 parsecs, or about 32.6 light-years. For example, the sun, which has an apparent magnitude of −26.7, would, if located at a distance of 32.6 light-years, have an absolute magnitude of

TABLE 20.2
Ratios of stellar luminosity.

Difference in Magnitude	Brightness Ratio
0.5	1.6:1
1	2.5:1
2	6.3:1
3	16:1
4	40:1
5	100:1
10	10,000:1
20	100,000,000:1

TABLE 20.3
Distance, absolute magnitude, and apparent magnitude of some stars.

Name	Distance (Parsecs)	Apparent Magnitude	Absolute Magnitude
Sun		−26.7	5.0
Alpha Centauri A	1.3	0.0	4.4
Sirius A	2.7	−1.4	1.5
Epsilon Indi	3.4	4.7	7.0
Kruger 60B	3.9	11.2	13.2
Arcturus	11.0	−0.1	−0.3
Antares	120.0	1.0	−4.5
Betelgeuse	150.0	0.8	−5.5
Deneb	430.0	1.3	−6.9

about 5. Thus, stars with absolute magnitudes greater than 5 (smaller numerical value) are intrinsically brighter than the sun, but because of their distance, they appear much dimmer. Table 20.3 lists the absolute and apparent magnitudes of some stars as well as their distances from the earth. Most stars have an absolute magnitude between −5 and +15, which puts the sun near the middle of this range.

For a star too distant for parallax measurements, knowing its absolute and apparent brightness provides astronomers with a tool for determining its distance. The apparent magnitude can be determined relatively easily with a *photometer* (light meter) attached to a telescope. If we also know a star's true brightness, we can determine just how far away that star would have to be in order to have the apparent brightness we observe. You use this same principle when you drive at night and estimate the distance to an oncoming car simply from the brightness of its headlights. You know the true brightness of an automobile headlight, but how do astronomers determine the intrinsic brightness of a star? Fortunately, some stars have characteristics that provide the necessary data. One important group of these is called **cepheid variables.** These are pulsating stars that get brighter and fainter in a rhythmic fashion. The interval between two successive occurrences of maximum brightness of a pulsating variable is called its **light period.** Most cepheid variables pulsate with periods of between 2 and 50 days. For example, the North Pole Star (Polaris) varies about 10 percent in brightness over a period of 4 days. In general, the longer the light period of a cepheid, the greater is its absolute magnitude (Figure 20.2). So, by determining the light period of a cepheid, its absolute magnitude can be calculated. When this absolute magnitude is compared to the apparent magnitude, a good approximation of its distance can be made.

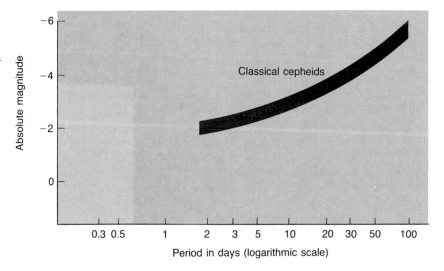

FIGURE 20.2
Relationship between the light period and absolute magnitude of pulsating variable stars (cepheids).

COLOR

The next time you are outdoors on a clear night, take a good look at the stars and note their color. Some that are quite colorful can be found in the constellation Orion (see Appendix E). One of the two brightest stars in Orion, Betelgeuse (α Orionis), is definitely red, whereas the other, Rigel (β Orionis), appears blue.

Very hot stars with surface temperatures above 30,000 K emit most of their energy in the form of short-wavelength light and therefore appear blue. Red stars, on the other hand, are much cooler, generally less than 3000 K, and most of their energy is emitted as longer-wavelength red light. Stars with temperatures between 5000 and 6000 K appear yellow like the sun. Some stars are so cool that they radiate mostly in the infrared and are not visible. However, these can sometimes be detected by using infrared-sensitive photography. Since a star's color is primarily a manifestation of its temperature, it does more for the astronomer than just decorate the sky.

BINARY STARS

Persons with good eyesight can resolve the second star in the handle of the Big Dipper (Mizar) as two stars. Numerous star pairs were discovered by astronomers during the eighteenth century with their new tool, the telescope. One of the stars in the pair is usually fainter than the other, and for this reason, was considered to be farther away. In other words, the stars were not considered pairs, but were only thought to lie in the same line of sight. In the early nineteenth century, careful examination of numerous pairs by William Herschel revealed that these stars actually orbit one another. The two stars are in fact united by their mutual gravitation. Newton's law of gravitation had been extended to the stars. These pairs of stars, in which the members are far enough apart to be resolved telescopically, are called **visual binaries.**

The idea of one star orbiting another may seem unusual, but evidence indicates that more than 50 percent of the stars in the universe occur in pairs or multiples.

One of the more interesting groups of nonvisual binaries is detectable by changes in brightness. This change occurs when the plane of the orbits of two stars lies in our view such that the stars eclipse each other (Figure 20.3). When the fainter of the two stars eclipses the brighter, a major drop in intensity is noted, and when the brighter star eclipses the fainter one, a lesser drop is observed.

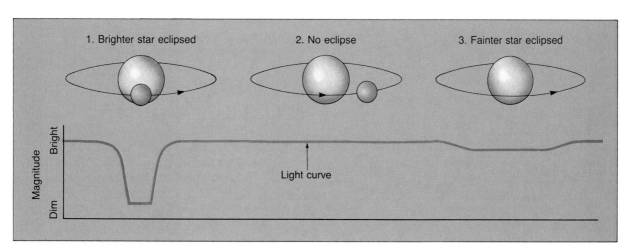

FIGURE 20.3
Idealized light curve and geometry of an eclipsing binary star.

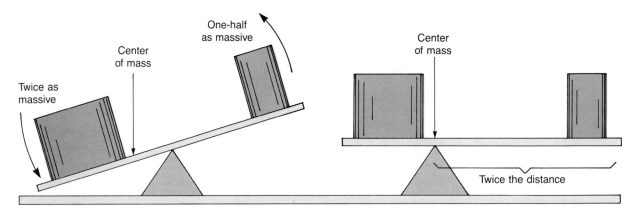

FIGURE 20.4
Illustration of the center of mass using a seesaw.

Binary stars are used to determine the property of a star most difficult to calculate—its mass. Recall that the mass of a body can be established if it is gravitationally attached to a partner, which is the case for any binary star system. Using Kepler's third law as redefined by Newton, the sum of the masses of two stars equals the cube of their mean distance (measured in AU) divided by their period squared. However, this gives information only about the combined masses of the two stars, not about the individual masses. Fortunately, binary stars orbit each other around a common point called the *center of mass* in a manner dependent on their individual masses. For stars of equal mass, the center of mass lies exactly halfway between them. If one star is more massive than its partner, their common center will be located closer to the more massive one. Thus, if the sizes of their orbits can be observed, a determination of their individual masses can be made. You can experience this relationship on a seesaw by trying to balance a person having a much greater mass (Figure 20.4). For illustration, when one star has an orbit half the size of its companion, it is twice as massive. If their combined masses as determined by Kepler's law are equal to 3 times the mass of the sun, then the larger will be twice as massive as the sun, and the smaller will have a mass equal to that of the sun. Most stars have masses that range between $1/10$ and 50 times the mass of the sun.

STELLAR CLASSIFICATION

The vast majority of stars have continuous spectra on which a number of dark absorption lines are superimposed. Recall from our discussion of light that a given set of dark lines can be attributed to the presence of a particular element. Early studies of stellar spectra revealed that an element such as calcium or helium produces strong absorption lines in the spectrum of some stars, but only weak lines in others, and no lines in still others. After the spectra of numerous stars were obtained, it was inevitable that someone would try to group stars according to their spectra. The first classification scheme was based on the relative strengths of selected absorption lines, and thus stars with similarly-appearing spectra were placed in the same **spectral class.** This scheme assumed that stars with an abundance of a particular element, for instance calcium, would produce strong calcium lines. Elements for which no absorption lines appeared were considered absent or rare.

Later, it was learned that the spectral variations observed were primarily the result of temperature, not compositional, differences. For example, very cool red stars (class M) have strong absorption lines of molecular titanium oxide (TiO). However, in the spectrum of stars whose temperatures are above 5000 K, these lines do not appear. At these temperatures, the titanium oxide molecules are unstable and are split into atomic titanium and atomic oxygen. Hence, hotter stars probably contain as much titanium as cooler stars, but the absorption lines of the titanium oxide molecule cannot be produced. Similarly, the presence of most absorption lines was found to be dependent on the star's surface temperature, not its composition.

The original spectral classes which had been labeled with letters of the alphabet were rearranged in order of decreasing temperatures (Table 20.4). This order is O, B, A, F, G, K, M. The O-type stars are hot, blue stars with temperatures up to 50,000 K, and the coolest red stars (temperatures less than 3500 K) belong to the M spectrum class. The sun is a yellow G-type star, and the somewhat cooler K-type stars appear orange. By now it should be obvious that once a star's spectrum has been classified, its surface temperature has also been established. As we shall see, a star's temperature is very useful in determining its size.

HERTZSPRUNG-RUSSELL DIAGRAM

Early this century, two men, Einar Hertzsprung and Henry Russell, independently studied the relationship between the brightness (absolute magnitude) and temperature (spectral class) of stars. From this research each developed a graph, now called a **Hertzsprung-Russell diagram,** or simply, **H-R diagram,** which exhibits these intrinsic stellar properties. By studying H-R diagrams, we learn a great deal about stars.

TABLE 20.4
Characteristics of spectral classes.

Spectral Class	Color	Approximate Temperature (K)	Principal Spectral Characteristics	Approximate Mass of Main-Sequence Star (Sun = 1)	Stellar Example
O	Blue	>30,000	Strong lines of ionized helium; weak hydrogen lines.	40.0	10 Lacertae
B	Blue-White	10,500–30,000	Stronger hydrogen lines than 0 type; lines of neutral helium dominate.	10.0	Rigel
A	White	7500–10,500	Hydrogen lines reach maximum strength; lines of ionized metals strong.	3.0	Vega
F	Yellow-White	6000–7500	Hydrogen lines weaken; lines of neutral metals strengthen; lines of ionized metals weaken.	1.5	Canopus
G	Yellow	5000–6000	Strong calcium lines; strong lines of neutral and ionized metals; weak hydrogen lines.	1.0	Sun
K	Orange	3500–5000	Lines of neutral metals dominate. Hydrogen lines weak but detectable.	0.8	Arcturus
M	Red	< 3500	Numerous lines of neutral metals; band of TiO molecules	0.2	Antares

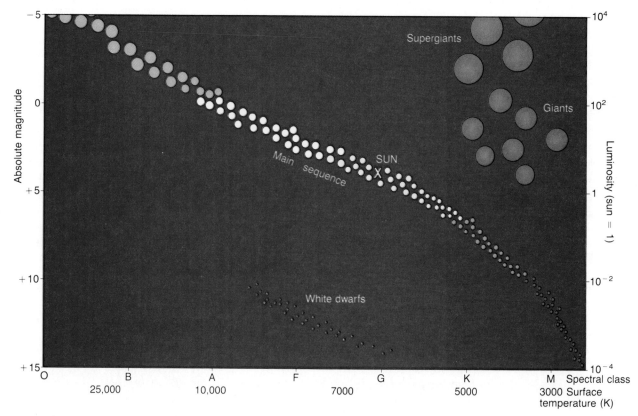

FIGURE 20.5

Idealized Hertzsprung-Russell diagram, on which stars are plotted according to temperature and absolute magnitude.

To produce an H-R diagram, astronomers survey a portion of the sky and plot each star according to its true brightness and temperature (Figure 20.5). Notice that the plots in Figure 20.5 are not uniformly distributed, rather, most fall along a band that runs from the upper-left corner to the lower-right corner. Stars whose plots lie in this portion of an H-R diagram are called **main-sequence stars.** As shown in Figure 20.5, the hottest main-sequence stars are intrinsically the brightest.

Although discovered independently, the luminosity of the main-sequence stars is also related to their mass. The hottest (blue) stars are about 50 times more massive than the sun, while the coolest (red) stars are only 1/10 as massive. Therefore, on the H-R diagram, the main-sequence stars appear in a decreasing order, from hotter, more

massive blue stars to cooler, less massive red stars. The position of the sun is shown by the X. The sun is a G-type, main-sequence star with an absolute magnitude of about +5. Because the magnitudes of a vast majority of main-sequence stars lie between −5 and +15, and since the sun falls midway in this range, the sun is often considered an average star. However, more main-sequence stars are cooler and less massive than our sun.

Above and to the right of the main sequence in the H-R diagram lies a group of very luminous stars called **giants** or, on the basis of their color, **red giants.** The size of these giants can be approximated by comparing them with stars of known size that belong to the same spectral class and, therefore, have the same surface temperature. According to the Stefan-Boltzmann law (Chapter

18), stars having equal surface temperatures radiate the same amount of energy per unit area. Therefore, any difference in the brightness (absolute magnitude) of stars belonging to the same spectral class is attributable to their relative sizes. As an example, let us compare the sun, which has an absolute magnitude of about +5, with another, more luminous G-type star that has an absolute magnitude of 0. Since their magnitudes differ by 5, their brightness ratio is 100:1. Because both stars radiate the same amount of energy per unit area, it follows that in order for the more luminous star to be 100 times brighter than the sun, it must have about 100 times more surface area. Note the position of a G-type star with a magnitude of 0 on the H-R diagram in Figure 20.5. It should be clear why stars whose plots fall in this portion of the graph are called giants.

Some stars are so large that they are called **super-giants.** Betelgeuse, a bright-red supergiant in the constellation Orion, has a radius about 800 times that of the sun. If this voluminous star were at the center of the solar system, it would extend beyond the orbit of Mars, and the earth would find itself inside the star. Other red giants which are easy to locate in our sky are Arcturus in the constellation Boötes and Antares in Scorpius (see the listing of Star Charts in Appendix E).

In the lower-left portion of the H-R diagram, the opposite situation occurs. These stars are much fainter than main-sequence stars of the same temperature, and by using the same reasoning as before, they must be much smaller. Some probably approximate the earth in size. This group has come to be called **white dwarfs,** although not all are white in color.

Soon after the first H-R diagrams were developed, astronomers realized their importance in interpreting stellar evolution. Just like living things, a star is born, ages, and dies. Due to the fact that almost 90 percent of the stars lie on the main sequence, we can be relatively certain that stars spend most of their active years as main-sequence stars. Only a few percent are giants, and perhaps 10 percent are white dwarfs. After considering some variable stars and some "unusual" stars, we will return to the topic of stellar evolution.

VARIABLE STARS

Stars that fluctuate in brightness are known as variables. Some, called **pulsating variables,** fluctuate regularly in brightness by expanding and contracting in size. The importance of one member of this group (cepheid variables) in determining stellar distances was discussed earlier. Other pulsating variables have irregular periods and are of no value in this respect.

The most spectacular variables belong to a group known as **eruptive variables.** When one of these explosive events occurs, it appears as a sudden brightening of a star, called a **nova** (Figure 20.6). (The term *nova*, meaning "new," was used by the ancients since these stars were unknown to them before increasing in luminosity.) During the outburst, the outer layer of the star is ejected at high speeds. The "cloud" of ejected material is occasionally revealed photographically (Figure 20.7). A nova generally reaches maximum brightness in one day, remains bright for only a few weeks, then slowly returns in a year or so to its original brightness. Since the star returns to its prenova brightness, we can assume that only a small amount of its mass is lost during the flare-up. Some stars have experienced more than one such event. In fact, the process probably occurs over and over again.

The modern explanation of novae proposes that they occur in binary systems consisting of an expanding red giant and a hot white dwarf which are in close proximity. Hydrogen-rich gas from the oversized giant encroaches near enough to the white dwarf to be gravitationally transferred. Eventually, enough of the hydrogen-rich gas is added to the dwarf to cause it to ignite explosively. Such a thermonuclear reaction rapidly heats and expands the outer gaseous envelope of the hot dwarf to produce a nova event. In a relatively short time, the white dwarf returns to its prenova state where it remains inactive until the next buildup occurs.

Possibly the most cataclysmic act of nature is the **supernova.** While a nova may increase its brightness several thousand times, a supernova becomes millions of times brighter than its prenova stage. If one of the nearer stars to the earth produced such an outburst, its brilliance

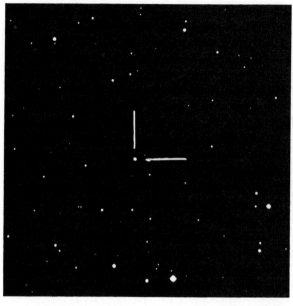

FIGURE 20.6
Photographs of Nova Herculis taken about two months apart showing the decrease in brightness. (Courtesy of Lick Observatory)

FIGURE 20.7
Expanding "cloud" around Nova Persei. (Courtesy of Hale Observatories)

would surpass that of the sun. Supernovae are rare; none have been observed in our galaxy since the advent of the telescope, but Tycho Brahe and Galileo each recorded one about 30 years apart. Probably an even larger supernova was recorded in 1054 A.D. by the Chinese. The remnant of this great outburst is the Crab nebula, shown in Figure 20.8. The amount of material ejected surely must represent a considerable portion of the original star.

The current consensus of opinion is that supernovae occur when stars more massive than the sun exhaust their nuclear fuel. Then, because they are no longer able to oppose the grip of their own gravity, they collapse. Because of their great mass, the resulting implosion is of cataclysmic proportions, releasing enormous amounts of gravitational energy, which eject the outer shell of the star. Theoretical work predicts that the star's interior condenses into a very hot object possibly no larger than 20 kilometers in diameter. These incomprehensibly dense bodies have been named

FIGURE 20.8
Crab nebula in Taurus; the remains of the supernova of 1054 A.D. (Courtesy of Lick Observatory)

neutron stars. Some supernovae events are thought to produce even smaller and most intriguing objects called black holes. In the following sections, we will consider the nature of neutron stars and black holes.

UNUSUAL STARS

DWARFS

White dwarfs are extremely small stars with densities greater than that of any known terrestrial material. It is believed that white dwarfs were once normal-sized stars whose internal heat energy was able to keep these gaseous masses from shrinking

under their own gravitational force. Eventually, however, these stars depleted their nuclear fuel and collapsed to planetary size. Although some white dwarfs are no larger than the earth, their mass is equal to that of the sun. Thus, their densities may be a million times greater than water. A spoonful of such matter would weigh several tons. Densities this great are possible only when electrons are displaced from their regular orbits and pushed closer to the nucleus, allowing the atoms to take up less space. Material in this state is called **electron degenerate matter.**

In order to better visualize electron degenerate matter, consider that in ordinary atoms the electrons circle the nucleus at distances proportional to the orbits of the planets around the sun. Thus, like our solar system, atoms are made up mostly of empty space. You might envision an atom as having a very small nucleus surrounded by speedy electrons which produce a comparatively large electrically-charged cloud. In an ordinary gas, such as that found in an automobile tire, it is the motion of these atoms which causes them to collide with each other and with the walls of the tire to provide the air pressure to keep the tire inflated. In a similar manner, gas pressure supports the outer envelope of a star and keeps it from gravitational collapse. When a gas cools, the atoms slow, thus reducing the pressure they exert. In the envelope of a star, cooling results in a reduced volume until finally, at absolute zero, the electron clouds of the atom are touching. However, if the gravitational force of the star is great enough the atoms themselves will collapse, producing electron degenerate matter. Recall that in electron degenerate matter the atoms have been squeezed together so tightly that the electrons are displaced much nearer to the nucleus. Degenerate matter uses electrical repulsion rather than molecular motion to support itself from total collapse. Consequently, pressure exists no matter what the temperature becomes. Although atomic particles in degenerate matter are much closer together than in normal earth matter, they are not yet packed as tightly as possible. Stars made of matter that has even greater densities are thought to exist.

As a star contracts into a white dwarf, its surface becomes very hot, often exceeding 50,000 K. Even so, without a source of energy, it can only become cooler and dimmer. Although none has been observed, the terminal stage of a white dwarf must be a small, cold, nonluminous body called a **black dwarf.**

NEUTRON STARS

A study of white dwarfs produced what might at first appear to be a surprising conclusion. The smallest white dwarfs are the most massive, and the largest are the least massive. The reason for this is that a more massive star, because of its greater gravitational force, is able to squeeze itself into a smaller, more densely packed object than can a less massive star. Thus, the smallest white dwarfs were produced from the collapse of larger, more massive stars than were the larger white dwarfs.

Theoretical calculations indicate that stars of greater than three solar masses would not become white dwarfs. Rather, they would form into the even smaller **neutron stars.** In a white dwarf, the electrons are pushed close to the nucleus, whereas in a neutron star, the electrons, under much greater pressure, are forced to combine with protons to produce neutrons (hence, the name). If the earth were to collapse to the density of a neutron star, it would have a diameter equivalent to the length of a football field. A pea-sized sample of this matter would weigh more than 130,000,000 tons.

Recall that a neutron star is thought to be a remnant of a supernova implosion. During such an event, the envelope of the star is ejected, while the core collapses into a very hot star about 20 kilometers (12.4 miles) in diameter. Although neutron stars have high surface temperatures, their small size would greatly limit their luminosity. Consequently, locating one visually would be extremely difficult. However, theory predicts that a neutron star would have a very strong magnetic field. Further, because of the law of conservation of angular momentum, we would expect neutron stars to develop a high rotational velocity during their collapse. Radio waves generated by these rotating stars would be concentrated into two narrow zones which would align with the magnetic poles. Consequently, these stars would resemble a rapidly rotating search beacon omitting strong radio waves. If the earth happened to be in the path of these beacons, the star would appear to blink on and off as the waves swept past. In the early 1970s, a source that radiates short pulses of radio energy called a **pulsar** (pulsating radio source) was discovered in the Crab nebula. Visual inspection of this radio source revealed it to be a small star centered in the nebula. The pulsar found in the Crab nebula is undoubtedly the remains of the supernova of 1054 A.D. Thus, the first neutron star had been discovered.

BLACK HOLES

Are neutron stars made of the most dense materials possible? No. During a supernova event, stars greater than five solar masses apparently collapse into objects even smaller and denser than neutron stars. Even though these objects would be very hot, their immense gravitation would prevent light from leaving their surface, and consequently, they would literally disappear from sight. These interesting bodies have appropriately been named **black holes.** Anything which moved too near a black hole would be devoured and lost forever.

How can astronomers locate an object whose gravitational field prevents the escape of all matter and energy? It so happens that astronomers have already found several likely candidates, one of which is named Cygnus X-1. These apparent black holes are located in binary systems in conjunction with large red giants (Figure 20.9). Gases are pulled from their giant companion and spiral into the black hole at great speeds. Theory predicts that these whirling gases would be heated to very high temperatures and thus emit a flood of x rays before being engulfed. Although we think of x rays as unusual phenomena, they are simply electromagnetic waves quite similar to visible light, but much shorter. Because x rays do not penetrate our atmosphere efficiently, the existence of black holes was not confirmed until recently. The first x ray sources were discovered in 1970 by detectors placed aboard satellites.

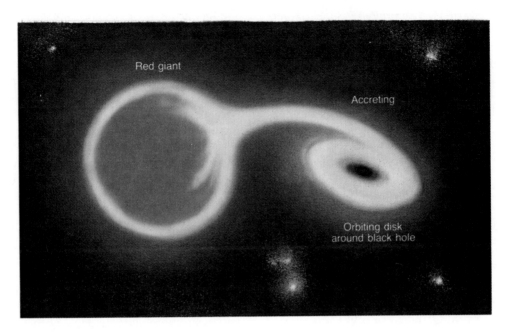

FIGURE 20.9
Red giant with rotating disk around a black hole.

A recent proposal suggests that a large number of very small black holes formed early in the history of the universe. The hypothesis further states that these black holes may be able to slowly "leak" radiation and gradually shrink in size. A black hole which emits matter and energy might be called a "white hole." However, whether "white holes" really exist remains to be documented.

STELLAR EVOLUTION

The idea of describing how a star is born, ages, and then dies may seem a bit presumptuous, since these objects have life spans that surely exceed tens of billions of years. However, by studying stars of different ages, astronomers have been able to piece together an acceptable model for stellar evolution. The method that was used to create this model is analogous to what an alien being might do upon reaching the earth, in order to determine the development of human life. By examining a large number of humans, this distant traveler would be able to observe the birth of human life, the activities of children and adults,

and the death of other humans. From this information, the alien would attempt to put the stages of human development into their proper sequence. Based on the relative abundance of humans in each stage of development, it would even be possible to conclude that humans spend more of their lives as adults than as toddlers.

The birthplaces of stars are dark, cool, interstellar clouds which are comparatively rich in dust and gases. In the neighborhood of the Milky Way, these large **globules**, as they are called, consist of 92 percent hydrogen, 7 percent helium, and less than 1 percent of the remaining heavier elements. For reasons not yet fully understood, these gaseous clouds become concentrated enough to begin contracting by self-gravitation. One event proposed to explain the triggering of stellar formation is a shock wave traveling from a catastrophic explosion (supernova) of a nearby star. Regardless of whatever force initially concentrates interstellar matter, once it is accomplished, mutual gravitational attraction begins to cause material from this large cloud to accrete (collect together). As the matter contracts, gravitational energy (potential energy) is converted into energy of motion, or heat energy.

The initial contraction spans a period of a million years or so. With the passage of time, the temperature of this gaseous body will slowly rise, eventually reaching a temperature that is sufficiently high to cause it to radiate energy from its surface in the form of long-wavelength visible light (red). Because this large red object is not yet a star, the name **protostar** is generally applied to it.

During the protostar phase, contraction continues, slowly at first, and then much more rapidly. This contraction causes the core of the developing star to heat much more intensely than the outer envelope. When the core has reached 15 million K, the pressure within is so great that four hydrogen nuclei are fused together into a single helium nucleus. Astronomers refer to this nuclear reaction as **hydrogen burning*** because of the energy released. However, keep in mind that thermonuclear burning is not burning in the usual chemical sense.

Heat from hydrogen burning, as well as heat released by further gravitational contraction, cause the stellar gases to increase their motion. This in turn results in an increase in the outward gas pressure. At some point this outward pressure will exactly balance the inward force of gravity. When this balance is reached, the star becomes a stable main-sequence star. From this point in the star's evolution until its death, its internal gas pressure struggles to offset the unrelenting force of gravity. Typically, hydrogen burning continues for a few billion years, providing the outward pressure required to support the star. Most stars spend as much as 90 percent of their active lives as stable main-sequence stars.

Different stars do not age at the same rate. Hot, massive, blue stars radiate energy at such an enormously high rate that they substantially deplete their hydrogen fuel in only a few million years. By contrast, the very smallest main-sequence stars may remain stable for hundreds of

billions of years. A yellow star, such as the sun, remains a main-sequence star for about 10 billion years. Once the hydrogen in the sun's core has been converted to helium, it will become a red giant and grow large enough to engulf the earth. Since the solar system is 5 billion years old, it is comforting to know that the sun will remain stable for another 5 billion years.

The evolution to the red giant stage results because the zone of hydrogen burning continually migrates outward, leaving behind an ever larger helium core. Eventually, much of the hydrogen which exists at proper temperatures and pressures is consumed. Although the star's outer envelope can be supported by the reduced level of hydrogen burning, the core, which is much more dense, begins to contract. The gravitational energy released by this contraction is transmitted to the layer surrounding the core. This additional heat initiates a more vigorous level of hydrogen burning in this shell. This energy in turn heats and enormously expands the star's outer envelope. As the star expands, its surface cools, and becomes a bloated, cool, red body hundreds to thousands of times its main-sequence size. Eventually the star's gravitational force will stop this outward flight. Once again these two opposing forces, gravity and gas pressure, will be in balance, and this gaseous mass will be a stable, but much larger, star. Some red giants overshoot the equilibrium point and then rebound like an overcompressed spring. These stars continue to oscillate in size, becoming variable stars. The cepheid variables discussed earlier would be examples of variable giant stars.

Although the envelope of a red giant expands, the core continues to collapse and heat until it reaches 100 million K. At this temperature, it is hot enough to initiate a nuclear reaction in which helium is converted to carbon. Thus, a red giant consumes both hydrogen and helium for a source of energy. In stars more massive than the sun, still other thermonuclear reactions occur which generate all the elements on the periodic table up to iron. Nuclear burning of elements heavier than iron requires an additional source of energy to keep the reaction progressing. Hence, these elements are not formed by "normal" nuclear burning.

---

*This reaction is also called the proton-proton reaction and is discussed in more detail in Chapter 18 in the subsection "The Solar Interior." See page 500.

Eventually, all the star's nuclear fuel will be consumed. As is the case with main-sequence stars, red giants are only temporary. The force of gravity again controls the star's destiny as it attempts to squeeze it into the smallest, most dense piece of matter possible.

Most of the events of stellar evolution discussed thus far are well documented. What happens to a star after the red-giant phase is much more speculative. However, astronomers do have a widely accepted model of stellar evolution which predicts still further changes. Data supporting this evolutionary model are being discovered almost daily; however, some changes in the model are inevitable. We know that a star, regardless of its size, must eventually exhaust all its nuclear fuel and collapse under the force of its immense gravitational field. The type of death each star experiences is determined primarily by its initial mass. A star about as massive as the sun will slowly shrink to an earth-sized star of great density called a white dwarf. The density of a white dwarf is as great as nature will allow without totally destroying the existence of protons and electrons. The gravitational energy supplied to these collapsing stars is reflected in their very high surface temperatures. However, without a source of nuclear energy, a white dwarf continually radiates its remaining thermal energy, becoming cooler and dimmer until it dies as a black dwarf. These non-luminous earth-sized ash piles are undoubtedly scattered through space.

A star having a mass greater than 1.5 times that of the sun probably goes through additional phases before reaching its terminal state. As previously mentioned, such a star is too massive to become a white dwarf directly. It may lose mass during its red-giant phase, much as the sun does, in the form of solar wind, but at a much higher rate. Several white dwarfs have been observed with slowly expanding shells of gas surrounding a hot core. These objects, called **planetary nebulae,** suggest that some stars do, in fact, eject large quantities of material as they move toward the white dwarf stage. A good example of a planetary nebula is the Ring nebula in the constellation Lyra (Figure 20.10). This nebula appears as a ring because our line of sight through the center tra-

verses less gaseous material than at its edge. It is nevertheless spherical in shape.

Theory predicts that stars of over three solar masses are not able to lose enough mass by "normal" ejections to allow them to evolve into white dwarfs. Rather, supernovae events are thought to mark the end of these massive stars. During a supernova, the star is torn apart. The burned-out core is believed to implode, forming a very dense neutron star or possibly a black hole. During the supernova implosion, the star's internal temperatures may reach 1 billion K, at which time very heavy elements such as gold and uranium are thought to be produced. These heavy elements,

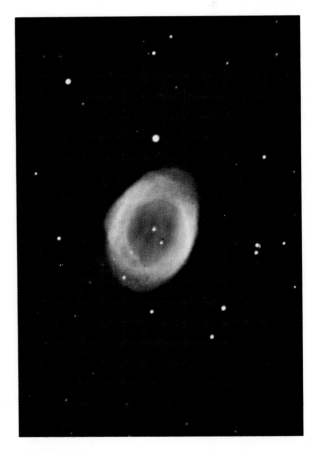

FIGURE 20.10
Ring nebula, a planetary nebula in Lyra. (Hale Observatories photo. Copyright by the California Institute of Technology and Carnegie Institution of Washington)

plus the debris of novae and the planetary nebulae are continually returning to interstellar space as dust and gas, which would then be available for the formation of other stars.

Astronomers believe that the earliest stars were made of nearly pure hydrogen. These in turn produced the heavy elements, some of which were returned to space. Because the sun contains some heavy elements but has not yet reached the stage in its evolution where it could have produced them, it must be a second-generation star. Thus, our sun as well as the rest of the solar system is believed to have formed, at least in part, from debris scattered from pre-existing stars. If this is the case, the atoms in your body were produced billions of years ago inside a star, and the gold in your ring perhaps formed during a supernova event that occurred trillions of kilometers away.

Without these catastrophic events, life on earth would not be possible.

The Hertzsprung-Russell diagrams have been very helpful in developing and testing models of stellar evolution. They are also useful for illustrating the changes that take place in an individual star during its life span. Figure 20.11 shows the evolution of a star about the size of the sun on an H-R diagram. Keep in mind that the star does not physically move along this path, but rather that its position on the H-R diagram represents the color (temperature) and absolute magnitude (brightness) of the star at various stages in its evolution. For example, on the H-R diagram, a protostar would be located to the right and above the main sequence (Figure 20.11). It is found to the right because of its relatively cool surface temperature (red color) and above because it would be more

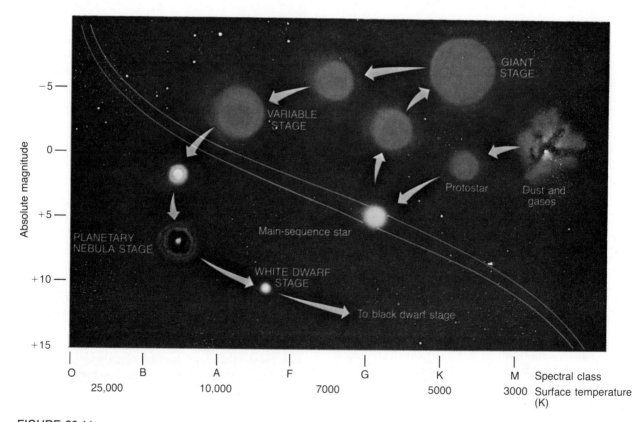

FIGURE 20.11
Diagram of stellar evolution on H-R diagram for a star about as massive as the sun.

TABLE 20.5
Summary of stellar evolution.

Initial Mass of Interstellar Cloud (Sun = 1)	Main-Sequence Stage	Giant Phase	Evolution After Giant Phase	Terminal State (Final mass*)
0.001	None (Planet)	No	None	Planet (0.001)
0.1	Red (Type M)	No	None	Black dwarf (0.09)
1	Yellow (Type G)	Yes	Planetary nebula	White dwarf (0.7)
6	White (Type A)	Yes	Supernova	Neutron star (3)
20	Blue (Type O)	Yes Supergiant	Supernova	Black hole (10)

*These figures are estimates.

luminous than a main-sequence star of the same color, a fact attributable to its large size. Careful examination of Figure 20.11 should help you visualize the evolutionary changes experienced by a star the size of the sun. In addition, Table 20.5 provides a summary of the evolutionary history of stars having various masses.

STAR CLUSTERS

Many stars, like the members of ancient nomadic tribes, travel in groups called **clusters.** Two common types are **open clusters,** like Pleiades in the constellation Taurus (Figure 20.12) and **globular clusters** such as the cluster in the constellation Hercules (Figure 20.13). As their names suggest, the members of a globular cluster are arranged in a spherical shape, while the members of an open cluster are less well organized. Other differences between these groups include size, star density, and location within the galaxy.

A large globular cluster may contain 100,000 member stars, which are packed near enough together to give the appearance of a solid when photographed. However, there is plenty of space, so

the chance of stellar collision is almost nonexistent. Globular clusters are located in all directions from the center of our galaxy but appear in greater numbers nearest the center. The typical open clusters, on the other hand, contain tens to hundreds of members and are almost exclusively confined to the plane of our galaxy.

It is assumed that the members of each cluster have similar origins and compositions, and nearly the same "birthday." Consequently, they are very beneficial in testing concepts of stellar evolution. At birth, each star cluster presumably contained members of every spectral type (color). Today, however, some clusters, such as Pleiades, are notably missing the hot 0- and B-type main-sequence stars, and others are nearly depleted of all members hotter than the G type. These differences can be explained if the clusters are of differing ages and the 0-type stars do, in fact, use their fuel most rapidly, followed by B-type stars, and so on. Thus, studies of clusters support the proposal that the more massive (brighter) stars evolve to another stage, leaving the main sequences before less massive (dimmer) stars. In a similar manner, other aspects of stellar evolution have been tested using star clusters.

FIGURE 20.12
The Pleiades, an open star cluster in Taurus shown embedded in the interstellar clouds, which reflect the light of the brighter cluster members. (Hale Observatories photo. Copyright by the California Institute of Technology and Carnegie Institution of Washington)

FIGURE 20.13
Globular star cluster in Hercules. (Courtesy of U.S. Naval Observatory)

INTERSTELLAR MATTER

The structure of the Milky Way galaxy was rather difficult to determine, since large amounts of interstellar dust and gas are found in its arms, obstructing the view. The name applied to concentrations of this intervening material is **nebula** (Latin for "cloud"), a term originally used for any fuzzy region in the heavens, including what we now know to be other galaxies. If this interstellar matter is in close proximity to a hot star, usually an 0- or B-type star, it will glow and is then called a **bright nebula**. The two main types of bright nebulae are known as **emission nebulae** and **reflection nebulae**. Emission nebulae are gaseous in nature, containing a large percentage of hydrogen, which absorbs the *ultraviolet radiation* emitted by an

FIGURE 20.14
Orion nebula. (Courtesy of Lick Observatory)

embedded or nearby hot star. Because these gases are under very low pressure, they reradiate this energy in the form of *visible light*. The process of converting ultraviolet light to visible light is known as **fluorescence,** an effect you have surely observed. A well-known emission nebula that can easily be observed with binoculars is located in the sword of the hunter in the constellation Orion (Figure 20.14). Reflection nebulae, as the name implies, merely reflect the light of nearby stars. This fact was discovered when the spectra of these nebulae were found to closely match those of the benevolent stars. Reflection nebulae are thought to be composed of rather dense clouds of large particles called **interstellar dust.** This belief is based on the fact that atomic gases with low densities could not reflect light sufficiently to produce the glow observed.

When a dense cloud of interstellar material is not close enough to a bright star to be illuminated, it

FIGURE 20.15
Horsehead nebula. (Hale Observatories photo. Copyright by the California Institute of Technology and Carnegie Institution of Washington)

is referred to as a **dark nebula.** Exemplified by the Horsehead nebula in Orion, dark nebulae appear as opaque objects silhouetted against a bright background (Figure 20.15, page 553). Dark nebulae can also easily be seen as starless regions—"holes in the heavens"—when viewing the Milky Way.

Although nebulae appear very dense, they actually approximate a good artificially-made vacuum. Even so, because of their enormous size, their total mass may be many times that of the sun. Interstellar matter is of great interest to astronomers. It is from this material that the stars and the planets were formed. Although the original composition of this material is not known, it was probably mostly hydrogen. Heavier elements (dust) were added by supernovae and, less dramatically, by all stars.

THE MILKY WAY

On a clear, moonless night away from city lights, you can see a truly marvelous sight—the Milky Way (Figure 20.16). With his telescope, Galileo discovered that this band of light was produced by countless individual stars which the unaided eye is unable to resolve. Today, we realize that the sun is actually part of this vast system of stars, which number about one hundred billion. The "milky" appearance of our galaxy results because the solar system is located within the flat *galactic disk.* Thus, when it is viewed from the "inside," a higher concentration of stars appears in the direction of the galactic plane than in any other direction.

When astronomers began to telescopically survey the stars located along the plane of the Milky Way, it appeared that equal numbers lay in every direction. Could the earth, which had recently been dethroned from its position at the center of the solar system, actually be at the center of the galaxy? A better explanation was put forth. Imagine that the trees in an enormous forest represent the stars in the galaxy. After hiking into this forest a short distance, you look around. What you see is an equal number of trees in every direction. Are you really in the center of the forest? Not necessarily; anywhere in the forest, except at the very edge, you will seem to be in the middle.

Attempts to visually inspect the Milky Way are hindered by the large quantities of interstellar

FIGURE 20.16
Panorama of the Milky Way. Notice the dark bands caused by the presence of interstellar nebula. (Lund Observatory photograph)

FIGURE 20.17
Structure of the visible portion of the Milky Way Galaxy.

matter which lie in our line of sight. Nevertheless, with the aid of radio telescopes, the gross structure of our galaxy has been determined. The Milky Way is a rather large spiral galaxy whose disk spans some 100,000 light-years and is about 10,000 light-years thick at the nucleus (Figure 20.17). As viewed from the earth, the center of the galaxy lies beyond the constellation Sagittarius. Radio telescopes reveal the existence of four, or possibly five, distinct *spiral arms,* with some showing splintering. The sun is believed positioned in one of these arms about two-thirds of the way from the center, at a distance of 30,000 light-years. The stars in the arms of the Milky Way rotate around the *galactic nucleus,* with the most outward ones moving the slowest such that the ends of the arms appear to trail. The sun requires

about 200 million years for each orbit. Surrounding the galactic disk is a nearly spheroidal *halo* made of very tenuous gas and numerous globular clusters. These star clusters do not participate in the rotating motion of the arms but rather have their own orbits that carry them through the disk. Although some clusters are very dense, they may pass among the stars of the arms with plenty of room to spare.

It was recently demonstrated that in addition to its nucleus, disk, and halo, our galaxy contains a large outer shell called the *galactic corona.* This feature is believed to contain much more material than is contained in the visible part of the galaxy. However, because the corona does not emit any form of radiation, it cannot be detected by direct observation. Astronomers know of its existence only by its gravitational properties. Additional research should provide more information about the nature of this mysterious material.

The Milky Way was probably formed from an enormous hydrogen cloud, which began to condense some 10 billion years ago. As it contracted, its density increased, eventually becoming sufficient for stellar formation to begin. Initially, stars developed in scattered globular-type clusters. The globular clusters of our galaxy all lack hot 0- or B-type stars, which testifies to their age. Astronomers call old stars like these, which formed early from nearly pure hydrogen, **population II stars.** At the time of the formation of population II stars, the galaxy had the spheroidal shape and size which is presently outlined by the farthest of the globular clusters (Figure 20.17). As this gaseous mass continued to condense, its rotational speed increased, causing it to flatten. While this was taking place, the original stars were generating heavy elements, and more and more of this matter was being added to the condensing gases. Therefore, the stars that formed later in the disk of the galaxy are richer in heavy elements. These younger stars are said to be **population I stars.** The sun is a population I star about 5 billion years old found in the galactic disk. The arms of the galaxy still contain large quantities of dust and gases, and stars are still forming there. These new stars will contain an even higher percentage of heavy elements.

GALAXIES

In the mid-1700s Immanuel Kant proposed that the telescopically visible fuzzy patches of light that appeared scattered among the stars were actually distant galaxies like the Milky Way. Kant described them as "island universes." Each, he felt, contained billions of stars and, as such, was a universe in itself. The weight of opinion, however, favored the hypothesis that they were dust and gas clouds (nebulae) within our galaxy. This matter was not resolved until the 1920s when Edwin Hubble was able to identify some cepheid variables in the Great Galaxy in Andromeda (Figure 20.18). Hubble realized that since these intrinsically very bright stars had apparent magnitudes of +18 (very

dim), they must lie outside the Milky Way. Recall that cepheid variables can be used to determine stellar distances. When they were used to find the distance to Andromeda, astronomers determined it to be over a million light-years away. Hubble had extended the universe far beyond the limits of our imagination to consist of hundreds of billions of galaxies, each of which contains hundreds of billions of stars. It has been said that a million galaxies are found in that portion of the sky bounded by the cup of the Big Dipper. There are more stars in the heavens than grains of sand on all the beaches on earth.

From the hundreds of billions of galaxies, four basic types have been identified: elliptical, spiral, barred spiral, and irregular. Only 10 percent of the known galaxies lack symmetry and are classified as **irregular.** The best known are the Large and Small Magellanic Clouds, which are seen as the two bright areas in the lower half of Figure 20.16. They are easily visible with the unaided eye in the Southern Hemisphere and were named after the explorer Ferdinand Magellan, who observed them when he circumnavigated the earth. They are also our nearest galactic neighbors, only 150,000 light-years away.

The Milky Way and the famous Great Galaxy in Andromeda are examples of fairly large **spiral gal-**

FIGURE 20.18
Great Galaxy in the constellation Andromeda. (Hale Observatories photo. Copyright by the California Institute of Technology and Carnegie Institution of Washington)

FIGURE 20.19
Two views illustrating the idealized structure of spiral galaxies. (Courtesy of U.S. Naval Observatory)

axies (Figure 20.19). Andromeda can be seen with the unaided eye as a fuzzy fifth-magnitude star. Typically, spiral galaxies are disk-shaped with a somewhat greater concentration of stars near their center, but there are numerous variations. Viewed broadside, arms are often seen extending from the central nucleus and sweeping gracefully away. The outermost stars of these arms rotate most slowly, giving the galaxy the appearance of a fireworks pinwheel. However, one type of spiral galaxy has the stars arranged in the shape of a bar, which rotates as a rigid system. This requires that the outer stars move faster than the inner ones, a fact not easy for astronomers to reconcile with the laws of motion. Attached to each end of these bars are curved spiral arms. These have become known as **barred spiral galaxies** (Figure 20.20). Spiral galaxies are generally quite large, ranging from 20,000 to about 125,000 light-years in diameter. About 10 percent of all galaxies are thought to be barred spirals and another 20 percent are regular spiral galaxies like the Milky Way.

The most abundant group, making up 60 percent of the total, is the **elliptical galaxies.** These are generally smaller than spiral galaxies. Some are so much smaller, in fact, that the term *dwarf* has been applied. Since these dwarf galaxies are not visible at great distances, a survey of the sky reveals more of the conspicuous large spiral gal-

FIGURE 20.20
Barred spiral galaxy. (Courtesy of Hale Observatories)

axies. Although most elliptical galaxies are small, the very largest known galaxies (200,000 light-years in diameter) are also elliptical. As their name implies, elliptical galaxies have an ellipsoidal shape that ranges to nearly spherical, and they lack spiral arms. The two dwarf companions of Andromeda shown in Figure 20.18 are elliptical galaxies. Some resemble the dense nucleus of a spiral galaxy, a fact which generated speculation that they may be the remains of a spiral galaxy whose arms wound around themselves. However, this has not been shown to be the case.

One of the major differences among the galactic types is the age of the stars which make them up. The irregular galaxies are composed mostly of young population I stars, whereas the elliptical galaxies contain old population II stars. The Milky Way and other spiral galaxies consist of both populations, with the youngest stars located in the arms.

Once astronomers discovered that stars were associated in groups, they set out to determine whether galaxies were also grouped or just randomly distributed throughout the universe. They found that, like stars, galaxies are grouped in **galactic clusters** (Figure 20.21). Some abundant clusters contain thousands of galaxies, while our own, called the **Local Group,** contains at least 28

galaxies. Of these, three are spirals, eleven are irregulars, and fourteen are ellipticals. Also, galactic clusters reside in huge swarms called superclusters. From visual observations, it appears that superclusters are spread randomly throughout the universe.

RED SHIFTS

One of the most important discoveries of modern astronomy was made in 1929 by Edwin Hubble. Observations completed several years earlier revealed that most galaxies have Doppler shifts in their spectral lines toward the red end of the spectrum. Recall that red shifts occur because the waves are "stretched," which indicates the source is moving away (see Chapter 18). Hubble set out to explain the predominance of red shift which had been observed in the spectra of galaxies.

Recall that Hubble established a method of estimating the distance to galaxies by using stars called cepheid variables. He also realized that the dimmer galaxies were probably farther away. Thus, he tried to determine if there is a relationship between the distances to galaxies and their red shifts. Using these estimated distances and the observed Doppler red shifts, Hubble discovered that galaxies which exhibit the greatest red shifts are the most distant. A very important benefit of this discovery was that it gave astronomers another method of determining distances in the universe. Since many galaxies are too far away to observe individual stars, such as cepheid variables, the amount of red shift in their spectra can be used to estimate their distance.

Another consequence of the universal red shift is that it predicts that most galaxies (except for a few nearby) are receding from us. Recall that the amount of Doppler red shift is dependent on the velocity at which the object is moving away. Greater red shifts indicate faster recessional velocities. Because more distant galaxies have greater red shifts, Hubble concluded that they must be retreating from us at greater velocities. This idea is currently called **Hubble's law,** and states that galaxies are receding from us at a speed that is proportional to their distance.

FIGURE 20.21
Numerous galaxies grouped in Hercules. (Courtesy of Hale Observatories)

Hubble was surprised at this discovery because it implied that the most distant galaxies are moving away from us many times faster than those nearby. What type of cosmological theory can explain this fact? It was soon realized that an expanding universe can adequately account for the observed red shifts.*

To help visualize the nature of this expanding universe, we will employ a popularly used analogy. Imagine a loaf of raisin bread dough which has been set out to rise for a few hours. As the dough doubles in size, so does the distance between all of the raisins. However, the raisins which were originally farther apart traveled a greater distance in the same time span than those located closer together. We therefore conclude that in an expanding universe, as in our analogy, those objects located farther apart recede at greater velocities.

Another feature of the expanding universe can be demonstrated using the raisin bread analogy. Let the raisins represent clusters of galaxies. During expansion, the size of the raisins, like the space occupied by the clusters, does not get larger, only the space separating them increases. It is believed that the galaxies within each cluster are held together by their mutual gravitational attraction. Within each cluster, galaxies have random motion in addition to the universal expansion. This explains why some of the galaxies in the Local Group exhibit a blue shift, indicating they are moving toward us. In addition, no matter which raisin you select, it will move away from all the other raisins. Likewise, no matter where one is located in the universe, every other galaxy (except those in the same cluster) will be receding. Hubble had indeed advanced our understanding of the universe.

THE BIG BANG

The universe—did it have a beginning? Will it have an end? Cosmologists are trying to answer

---

*One other reason for a red shift is known. An immense gravitational field can stretch light waves. However, a gravitational red shift would produce other visible effects that are not observed.

this question, and that makes them a rare breed.

First and foremost, any viable theory regarding the origin of the universe must account for the fact that all galaxies (except for the very nearest) are moving away from us. Since all galaxies appear to be moving away from the earth, is the earth in the center of the universe? If we are not in the center of the solar system and the solar system is not in the center of the galaxy, it seems unlikely that we should be in the center of the universe. A more probable explanation exists for this fact. Imagine a rubber balloon with paper punch "dots" glued to its surface. When the balloon is inflated, each dot spreads apart from every other dot. Similarly, if the universe is expanding, every galaxy would be moving away from every other galaxy.

This belief in the expanding universe led to the widely accepted **Big Bang** theory. According to this theory, the entire universe was at one time confined to a dense, hot, supermassive ball. Then about 20 billion years ago, a cataclysmic explosion occurred, hurling this material in all directions. The big bang marks the inception of the universe; all matter and space was created at this instant. The ejected masses of gas cooled and condensed, forming the stellar systems we now observe fleeing from their birthplace.

Will the stars eventually dim from view and invisible galaxies travel on forever? It has been suggested that after a certain point, perhaps 20 billion years in the future, the galaxies will slow and eventually stop their outward flight. Gravitational contraction would follow. The galaxies would then collide and coalesce, and a new fireball would be born. For this event to occur, the universe is required to have an average density of one atom for every cubic meter of space. But present estimates indicate that the concentration of matter in the universe is far less than this amount.

However, this subject does not end here. It has been proposed that heretofore undetected matter exists in large quantities in the universe. For example, numerous black holes may occupy many of the voids in the universe. If this is true, the galaxies could, in fact, collapse upon themselves.

"Absence of evidence is not evidence of absence." (Anonymous)

REVIEW QUESTIONS

1 What is the distance in parsecs to a star that has a parallax angle of 0.25 second of arc? What is its distance in light-years?

2 Explain the difference between a star's apparent and absolute magnitudes. Which one is an intrinsic property of a star?

3 What is the ratio of brightness between a twelfth-magnitude and fifteenth-magnitude star?

4 What information about a star can be determined from its color?

5 Why were the original spectral classes rearranged?

6 Make a generalization relating the mass and luminosity of main-sequence stars.

7 The disk of a star cannot be resolved telescopically. Explain the method that astronomers have used to estimate the diameter of stars.

8 Compare a black dwarf to a black hole.

9 Where on an H-R diagram does a star spend most of its lifetime?

10 What causes a star to become a giant?

11 Why are less massive stars thought to age more slowly than more massive stars, even though they have much less "fuel"?

12 Enumerate the steps thought to be involved in the evolution of stars more massive than the sun.

13 How has the word *nebula* taken on different meanings?

14 Describe the different types of nebulae.

15 Describe the general structure of the Milky Way.

16 Why is the sun considered a second-generation star?

17 Compare the three general types of galaxies.

18 Explain why astronomers consider elliptical galaxies more abundant than spiral galaxies, even though more spiral galaxies have been sighted.

19 How did Hubble determine that Andromeda is located beyond our galaxy?

20 What evidence supports the Big Bang theory?

KEY TERMS

stellar parallax

parsec

light-year

magnitude

apparent magnitude

absolute magnitude

cepheid variable

light period

visual binaries

spectral class

Hertzsprung-Russell
  (H-R) diagram

main sequence

giant

red giant

supergiant

white dwarf

pulsating variable

eruptive variable

nova

supernova

electron degenerate
  matter

black dwarf

neutron star

pulsar

black hole

globule

protostar

hydrogen burning

planetary nebula

cluster

open cluster

globular cluster

nebula

bright nebula

emission nebula

reflection nebula

fluorescence

interstellar dust

dark nebula

population II stars

population I stars

irregular galaxy

spiral galaxy

barred spiral galaxy

elliptical galaxy

galactic cluster

Local Group

Hubble's law

Big Bang

APPENDIX A
Metric and English Units Compared

UNITS

1 kilometer (km)	= 1000 meters (m)
1 meter (m)	= 100 centimeters (cm)
1 centimeter (cm)	= 0.39 inches (in.)
1 mile (mi)	= 5280 feet (ft)
1 foot (ft)	= 12 inches (in.)
1 inch (in.)	= 2.54 centimeters (cm)
1 square mile (mi^2)	= 640 acres (a)
1 kilogram (kg)	= 1000 grams (g)
1 pound (lb)	= 16 ounces (oz)
1 fathom	= 6 feet (ft)

CONVERSIONS

When you want to convert:	Multiply by:	To find:
LENGTH		
inches	2.54	centimeters
centimeters	0.39	inches
feet	0.30	meters
meters	3.28	feet
yards	0.91	meters
meters	1.09	yards
miles	1.61	kilometers
kilometers	0.62	miles

When you want to convert:	Multiply by:	To find:
AREA		
square inches	6.45	square centimeters
square centimeters	0.15	square inches
square feet	0.09	square meters
square meters	10.76	square feet
square miles	2.59	square kilometers
square kilometers	0.39	square miles
VOLUME		
cubic inches	16.38	cubic centimeters
cubic centimeters	0.06	cubic inches
cubic feet	0.028	cubic meters
cubic meters	35.3	cubic feet
cubic miles	4.17	cubic kilometers
cubic kilometers	0.24	cubic miles
liters	1.06	quarts
liters	0.26	gallons
gallons	3.78	liters
MASSES AND WEIGHTS		
ounces	28.35	grams
grams	0.035	ounces
pounds	0.45	kilograms
kilograms	2.205	pounds

TEMPERATURE

When you want to convert degrees Fahrenheit (°F) to degrees Celsius (°C), subtract 32 degrees and divide by 1.8.

When you want to convert degrees Celsius (°C) to degrees Fahrenheit (°F), multiply by 1.8 and add 32 degrees.

When you want to convert degrees Celsius (°C) to kelvins (K), delete the degree symbol and add 273.

When you want to convert kelvins (K) to degrees Celsius (°C), add the degree symbol and subtract 273.

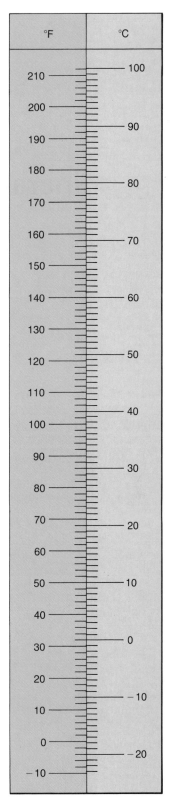

FIGURE A.1

APPENDIX B
Mineral Identification Key

GROUP I METALLIC LUSTER

Hardness	Streak	Other Diagnostic Properties	Chemical Composition
Harder than glass	Black streak	Black; magnetic; hardness = 6; specific gravity = 5.2; often granular	Magnetite (Fe_3O_4)
	Greenish-black streak	Brass yellow; hardness = 6; specific gravity = 5.2; generally an aggregate of cubic crystals	Pyrite (FeS_2)— fool's gold
	Reddish streak	Gray or reddish brown; hardness = 5–6; specific gravity = 5; platy appearance	Hematite (Fe_2O_3)
Softer than glass	Greenish-black streak	Brass yellow; hardness = 4; specific gravity = 4.2; massive	Chalcopyrite ($CuFeS_2$)
	Gray-black streak	Silvery gray; hardness = 2.5; specific gravity = 7.6 (very heavy); good cubic cleavage	Galena (PbS)
	Yellow-brown streak	Yellow brown to dark brown; hardness variable (1–6); specific gravity = 3.5–4; often found in rounded masses; earthy appearance	Limonite ($Fe_2O_3H_2O$)
	Gray-black streak	Black to bronze; tarnishes to purples and greens; hardness = 3; specific gravity = 5; massive	Bornite (Cu_5FeS_4)
Softer than your fingernail	Dark gray streak	Silvery gray; hardness = 1 (very soft); specific gravity = 2.2; massive to platy; writes on paper (pencil lead); feels greasy	Graphite (C)

GROUP II   NONMETALLIC LUSTER (DARK COLORED)

Hardness	Cleavage	Other Diagnostic Properties	Chemical Composition
Harder than glass	Cleavage present	Black to greenish black; hardness = 5–6; specific gravity = 3.4; fair cleavage, two planes at nearly 90 degrees	Augite (Ca, Mg, Fe, Al silicate)
		Black to greenish black; hardness = 5–6; specific gravity = 3.2; fair cleavage, two planes at nearly 60 degrees and 120 degrees	Hornblende (Ca, Na, Mg, Fe, Al silicate)
	Cleavage not prominent	Red to reddish brown; hardness = 6.5–7.5; conchoidal fracture; glassy luster	Garnet (Fe, Mg, Ca, Al silicate)
		Gray to brown; harness = 9; specific gravity = 4; hexagonal crystals common	Corundum (Al_2O_3)
		Dark brown to black; hardness = 7; conchoidal fracture; glassy luster	Smoky quartz (SiO_2)
		Olive green; hardness = 6.5–7; small grains	Olivine ($(Mg, Fe)_2SiO_4$)
Softer than glass	Cleavage present	Yellow brown to black; hardness = 4; good cleavage in six directions, light yellow streak that has the smell of sulfur	Sphalerite (ZnS)
	Cleavage absent	Generally tarnished to brown or green; hardness = 2.5; specific gravity = 9; massive	Native copper (Cu)
Softer than your fingernail	Cleavage present	Dark brown to black; hardness = 1; excellent cleavage in one direction; elastic in thin sheets; black mica	Biotite (K, Mg, Fe, Al silicate)
	Cleavage not prominent	Reddish brown; hardness = 1–5; specific gravity = 4–5; red streak; earthy appearance	Hematite (Fe_2O_3)
		Yellow brown; hardness = 1–3; specific gravity = 3.5; earthy appearance; powders easily	Limonite ($Fe_2O_3 \cdot H_2O$)

GROUP III   NONMETALLIC LUSTER (LIGHT COLORED)

Hardness	Cleavage	Other Diagnostic Properties	Chemical Composition
Harder than glass	Cleavage present	Flesh colored or white to gray; hardness = 6; specific gravity = 2.6; two planes of cleavage at nearly right angles	Potassium feldspar $(KAlSi_3O_8)$ Plagioclase feldspar $(NaAlSi_3O_8$ to $CaAl_2Si_2O_8)$
	Cleavage absent	Any color; hardness = 7; specific gravity = 2.65; conchoidal fracture; glassy appearance; varieties; milky, rose, smoky, amethyst (violet)	Quartz (SiO_8)
Softer than glass	Cleavage present	White, yellowish to colorless; hardness = 3; three planes of cleavage at 75 degrees (rhombohedral); effervesces in HCl; often transparent	Calcite $(CaCO_3)$
		White to colorless; hardness = 2.5; three planes of cleavage at 90 degrees (cubic); salty taste	Halite $(NaCl)$
		Yellow, purple, white; hardness = 4; white streak; translucent to transparent; four planes of cleavage	Fluorite (CaF_2)
Softer than your fingernail	Cleavage present	Colorless; hardness = 2; transparent and elastic in thin sheets; excellent cleavage in one direction; light mica	Muscovite $(K, Al$ silicate)
		White to transparent; hardness = 2; when in sheets, is flexible but not elastic; varieties: selenite (transparent, three planes of cleavage); satin spar (fibrous, silky luster); alabaster (aggregate of small crystals)	Gypsum $(CaSO_4 \cdot 2H_2O)$
		White, pink, green; hardness = 1–2; forms in thin plates; soapy feel; pearly luster	Talc $(Mg$ silicate)
		Yellow; hardness = 1–2.5	Sulfur (S)
	Cleavage not prominent	White; hardness = 1 (very soft); smooth feel; earthy odor; when moistened, has typical clay texture	Kaolinite $(Al$ silicate)
		Green; hardness = 2.5; fibrous; variety of serpentine	Asbestos $(Mg, Al$ silicate)
		Pale to dark reddish brown; hardness = 1–3; dull luster; earthy; often contains spheroidal-shaped particles; not a true mineral	Bauxite (Hydrous Al oxide)

APPENDIX C
The Earth's Grid System

A glance at any globe reveals a series of north-south and east-west lines that together make up the earth's grid system, a universally used scheme for locating points on the earth's surface. The north-south lines of the grid are called **meridians** and extend from pole to pole (Figure C.1). All are halves of great circles. A **great circle** is the largest possible circle that may be drawn on a globe; if a globe were sliced along one of these circles, it would be divided into two equal parts called **hemispheres.** By viewing a globe or Figure C.1, it can be seen that meridians are spaced farthest apart at the equator and converge toward the poles. The east-west lines (circles) of the grid are known as **parallels.** As their name implies, these circles are parallel to one another (Figure C.1). While all meridians are parts of great circles, all parallels are not. In fact, only one parallel, the equator, is a great circle.

LATITUDE AND LONGITUDE

Latitude may be defined as distance, measured in degrees, *north* and *south* of the equator. Parallels are used to show latitude. Since all points that lie along the same parallel are an identical distance from the equator, they all have the same latitude designation. The latitude of the equator is 0 degrees, while the north and south poles lie 90 degrees N and 90 degrees S, respectively.

Longitude is defined as distance, measured in degrees, *east* and *west* of the zero or prime meridian. Since all meridians are identical, the choice of a zero line is obviously arbitrary. However, the meridian that passes through the Royal Observatory at Greenwich, England,

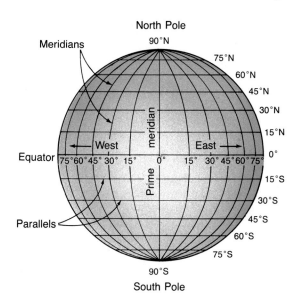

FIGURE C.1

The earth's grid system. (After J. B. Hoyt, *Man and the Earth* 3rd ed., © 1966. Adapted by permission of Prentice-Hall, Inc., Englewood Cliffs, N.J.)

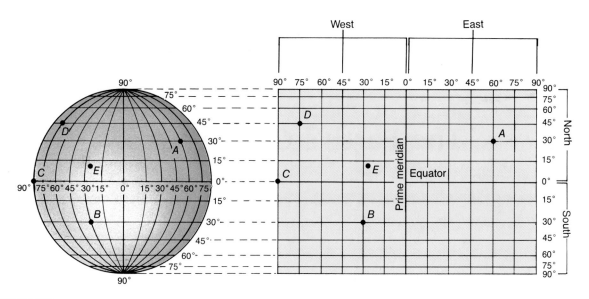

FIGURE C.2

Locating places using the grid system. For both diagrams: Point *A* is latitude 30 degrees N, longitude 60 degrees E; Point *B* is latitude 30 degrees S, longitude 30 degrees W; Point *C* is latitude 0 degrees, longitude 90 degrees W; Point *D* is latitude 45 degrees N, longitude 75 degrees W; Point *E* is approximately latitude 10 degrees N, longitude 25 degrees W. (After J. B. Hoyt, *Man and the Earth* 3rd ed., © 1966. Adapted by permission of Prentice-Hall, Inc., Englewood Cliffs, N.J.)

is universally accepted as the reference meridian. Thus, the longitude for any place on the globe is measured east or west from this line. Longitude can vary from 0 degrees along the prime meridian to 180 degrees, halfway around the globe.

It is important to remember that when a location is specified, directions must be given, that is, north or south latitude and east or west longitude (Figure C.2). If this is not done, more than one point on the globe is being designated. The only exceptions, of course, are places that lie along the equator, the prime meridian, or the 180-degree meridian. It should also be noted that while it is not incorrect to use fractions, a degree of latitude or longitude is usually divided into minutes and seconds. A minute (′) is ⅟₆₀ of a degree, and a second (″) is ⅟₆₀ of a minute. When locating a place on a map, the degree of exactness will depend upon the scale of the map. When using a small-scale world map or globe, it may be difficult to estimate latitude and longitude to the nearest whole degree or two. On the other hand, when a large-scale map of an area is used, it is often possible to estimate latitude and longitude to the nearest minute or second.

DISTANCE MEASUREMENT

The length of a degree of longitude depends upon where the measurement is taken. At the equator, which is a great circle, a degree of east-west distance is equal to approximately 111 kilometers (69 miles). This figure is found by dividing the earth's circumference—40,075 kilometers (24,900 miles)—by 360. However, with an increase in latitude, the parallels become smaller, and the length of a degree of longitude diminishes (see Table C.1). Thus, at about latitude 60 degrees N and S, a degree of longitude has a value equal to about half of what it was at the equator.

Since all meridians are halves of great circles, a degree of latitude is equal to about 111 kilometers (69 miles), just as a degree of longitude along the equator is. However, the earth is not a perfect sphere but is slightly flattened at the poles and bulges slightly at the equator. Because of this, there are small differences in the length of a degree of latitude.

Determining the shortest distance between two points on a globe can be done easily and fairly accurately using the "globe and string" method. It should be noted here that the arc of a great circle is the shortest distance between two points on a sphere. In order to determine the great circle distance (as well as observe the great circle route) between two places, stretch the string between the locations in question. Then, measure the length of the string along the equator (since it is a great circle with degrees marked on it) to determine the number of degrees between the two points. To calculate the distance in kilometers or miles, simply multiply the number of degrees by 111 or 69, respectively.

TABLE C.1

Longitude as distance.

°Lat.	Length of 1° Long.		°Lat.	Length of 1° Long.		°Lat.	Length of 1° Long.	
	km	miles		km	miles		km	miles
0	111.367	69.172	30	96.528	59.955	60	55.825	34.674
1	111.349	69.161	31	95.545	59.345	61	54.131	33.622
2	111.298	69.129	32	94.533	58.716	62	52.422	32.560
3	111.214	69.077	33	93.493	58.070	63	50.696	31.488
4	111.096	69.004	34	92.425	57.407	64	48.954	30.406
5	110.945	68.910	35	91.327	56.725	65	47.196	29.314
6	110.760	68.795	36	90.203	56.027	66	45.426	28.215
7	110.543	68.660	37	89.051	55.311	67	43.639	27.105
8	110.290	68.503	38	87.871	54.578	68	41.841	25.988
9	110.003	68.325	39	86.665	53.829	69	40.028	24.862
10	109.686	68.128	40	85.431	53.063	70	38.204	23.729
11	109.333	67.909	41	84.171	52.280	71	36.368	22.589
12	108.949	67.670	42	82.886	51.482	72	34.520	21.441
13	108.530	67.410	43	81.575	50.668	73	32.662	20.287
14	108.079	67.130	44	80.241	49.839	74	30.793	19.126
15	107.596	66.830	45	78.880	48.994	75	28.914	17.959
16	107.079	66.509	46	77.497	48.135	76	27.029	16.788
17	106.530	66.168	47	76.089	47.260	77	25.134	15.611
18	105.949	65.807	48	74.659	46.372	78	23.229	14.428
19	105.337	65.427	49	73.203	45.468	79	21.320	13.242
20	104.692	65.026	50	71.727	44.551	80	19.402	12.051
21	104.014	64.605	51	70.228	43.620	81	17.480	10.857
22	103.306	64.165	52	68.708	42.676	82	15.551	9.659
23	102.565	63.705	53	67.168	41.719	83	13.617	8.458
24	101.795	63.227	54	65.604	40.748	84	11.681	7.255
25	100.994	62.729	55	64.022	39.765	85	9.739	6.049
26	100.160	62.211	56	62.420	38.770	86	7.796	4.842
27	99.297	61.675	57	60.798	37.763	87	5.849	3.633
28	98.405	61.121	58	59.159	36.745	88	3.899	2.422
29	97.481	60.547	59	57.501	35.715	89	1.950	1.211
30	96.528	59.955	60	55.825	34.674	90	0.000	0.000

APPENDIX D
Topographic Maps

A map is a representation on a flat surface of all or a part of the earth's surface drawn to a specific scale. Maps are often the most effective means for showing the locations of both natural and cultural features, their sizes, and their relationships to one another. Like photographs, maps readily display information that would be impractical to express in words.

While most maps show only the two horizontal dimensions, geologists, as well as other map users, often require that the third dimension, elevation, be shown on maps. Maps that show the shape of the land are called **topographic maps.** Although various techniques may be used to depict elevations, the most accurate method involves the use of contour lines.

CONTOUR LINES

A **contour line** is a line on a map representing a corresponding imaginary line on the ground that has the same elevation above sea level along its entire length. While many map symbols are pictographs resembling the objects they represent, a contour line is an abstraction that has no counterpart in nature. It is, however, an accurate and effective device for representing the third dimension on paper.

Some useful facts and rules concerning contour lines are listed as follows. This information should be studied in conjunction with Figure D.1.

1 Contour lines bend upstream or upvalley. The contours form Vs that point upstream, and in the upstream direction the successive contours represent higher elevations. For example, if you were standing on a stream bank and wished to get to the point at the same elevation directly opposite you on the other bank, without stepping up or down, you would need to walk upstream along the contour at that elevation to where it crosses the stream bed, cross the stream, and then walk back downstream along the same contour.

2 Contours near the upper parts of hills form closures. The top of a hill is higher than the highest closed contour.

3 Hollows (depressions) without outlets are shown by closed, hatched contours. Hatched contours are contours with short lines on the inside pointing downslope.

4 Contours are widely spaced on gentle slopes.

5 Contours are closely spaced on steep slopes.

6 Evenly spaced contours indicate a uniform slope.

7 Contours usually do not cross or intersect each other, except in the rare case of an overhanging cliff.

8 All contours eventually close, either on a map or beyond its margins.

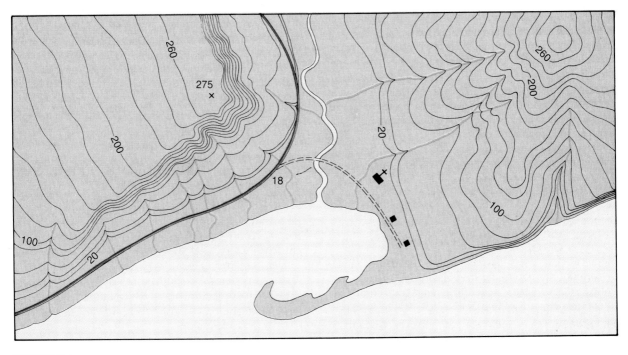

FIGURE D.1

Perspective view of an area and a contour map of the same area. These illustrations show how features are depicted on a topographic map. The upper illustration is a perspective view of a river valley and the adjoining hills. The river flows into a bay, which is partly enclosed by a hooked sandbar. On either side of the valley are terraces through which streams have cut gullies. The hill on the right has a smoothly eroded form and gradual slopes, whereas the one on the left rises abruptly in a sharp precipice, from which it slopes gently, and forms an inclined plateau traversed by a few shallow gullies. A road provides access to a church and two houses situated across the river from a highway that follows the seacoast and curves up the river valley. The lower illustration shows the same features represented by symbols on a topographic map. The contour interval (vertical distance between adjacent contours) is 20 feet. (After U.S. Geological Survey)

9 A single high contour never occurs between two lower ones, and vice versa. In other words, a change in slope direction is always determined by the repetition of the same elevation either as two different contours of the same value or as the same contour crossed twice.

10 Spot elevations between contours are given at many places, such as road intersections, hill summits, and lake surfaces. Spot elevations differ from control elevation stations, such as bench marks, in not being permanently established by permanent markers.

RELIEF

Relief refers to the difference in elevation between any two points. Maximum relief refers to the difference in elevation between the highest and lowest points in the area being considered. Relief determines the **contour interval,** which is the difference in elevation between succeeding contour lines that is used on topographic maps. Where relief is low, a small contour interval, such as 10 or 20 feet, may be used. In flat areas, such as wide river valleys or broad, flat uplands, a contour interval of 5 feet is often used. In rugged mountainous terrain, where relief is many hundreds of feet, contour intervals as large as 50 or 100 feet are used.

SCALE

Map **scale** expresses the relationship between distance or area on the map to the true distance or area on the earth's surface. This is generally expressed as a ratio or fraction, such as 1:24,000 or 1/24,000. The numerator, usually 1, represents map distance, and the denominator, a large number, represents ground distance. Thus, 1:24,000 means that a distance of 1 unit on the map represents a distance of 24,000 such units on the surface of the earth. It does not matter what the units are.

Often, the graphic, or bar, scale is more useful than the fractional scale, because it is easier to use for measuring distances between points. The graphic scale (Figure D.2) consists of a bar that is divided into equal segments, which represent equal distances on the map. One segment on the left side of the bar is usually divided into smaller units to permit more accurate estimates of fractional units.

Topographic maps, which are also referred to as quadrangles, are generally classified according to publication scale. Each series is intended to fulfill a specific type of map need. To select a map with the proper scale for a particular use, remember that large-scale-maps show more detail and small-scale maps show less detail. The sizes and scales of topographic maps published by the U.S. Geological Survey are shown in Table D.1.

COLOR AND SYMBOL

Each color and symbol used on a U.S. Geological Survey topographic map has significance. Common topographic map symbols are shown in Figure D.3 on page 574. The meaning of each color is as follows:

Blue—water features
Black—construction works, such as homes, schools, churches, roads, and so forth
Brown—contour lines
Green—woodlands, orchards, and so forth
Red—Urban areas, important roads, public land subdivision lines

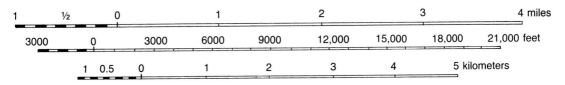

FIGURE D.2
Graphic scale.

TABLE D.1
National topographic maps.

Series	Scale	1 inch Represents	Standard Quadrangle Size (latitude-longitude)	Quadrangle Area (square miles)	Paper Size E-W N-S Width Length (inches)
7½-minute	1:24,000	2000 feet	7½′ × 7½′	49–70	22 × 27*
Puerto Rico 7½-minute	1:20,000	about 1667 feet	7½ × 7½′	71	29½ × 32½
15-minute	1:62,500	nearly 1 mile	15′ × 15′	197–282	17 × 21*
Alaska 1:63,360	1:63,360	1 mile	15′ × 20′–36′	207–281	18 × 21**
U.S. 1:250,000	1:250,000	nearly 4 miles	1° × 2°†	4580–8669	34 × 22††
U.S. 1:1,000,000	1:1,000,000	nearly 16 miles	4° × 6°†	73.734–102,759	27 × 27

SOURCE: U.S. Geological Survey.
*South of latitude 31 degrees, 7½-minute sheets are 23 × 27 inches; 15-minute sheets are 18 × 21 inches.
**South of latitude 62 degrees, sheets are 17 × 21 inches.
†Maps of Alaska and Hawaii vary from these standards.
††North of latitude 42 degrees, sheets are 29 × 22 inches; Alaska sheets are 30 × 23 inches.

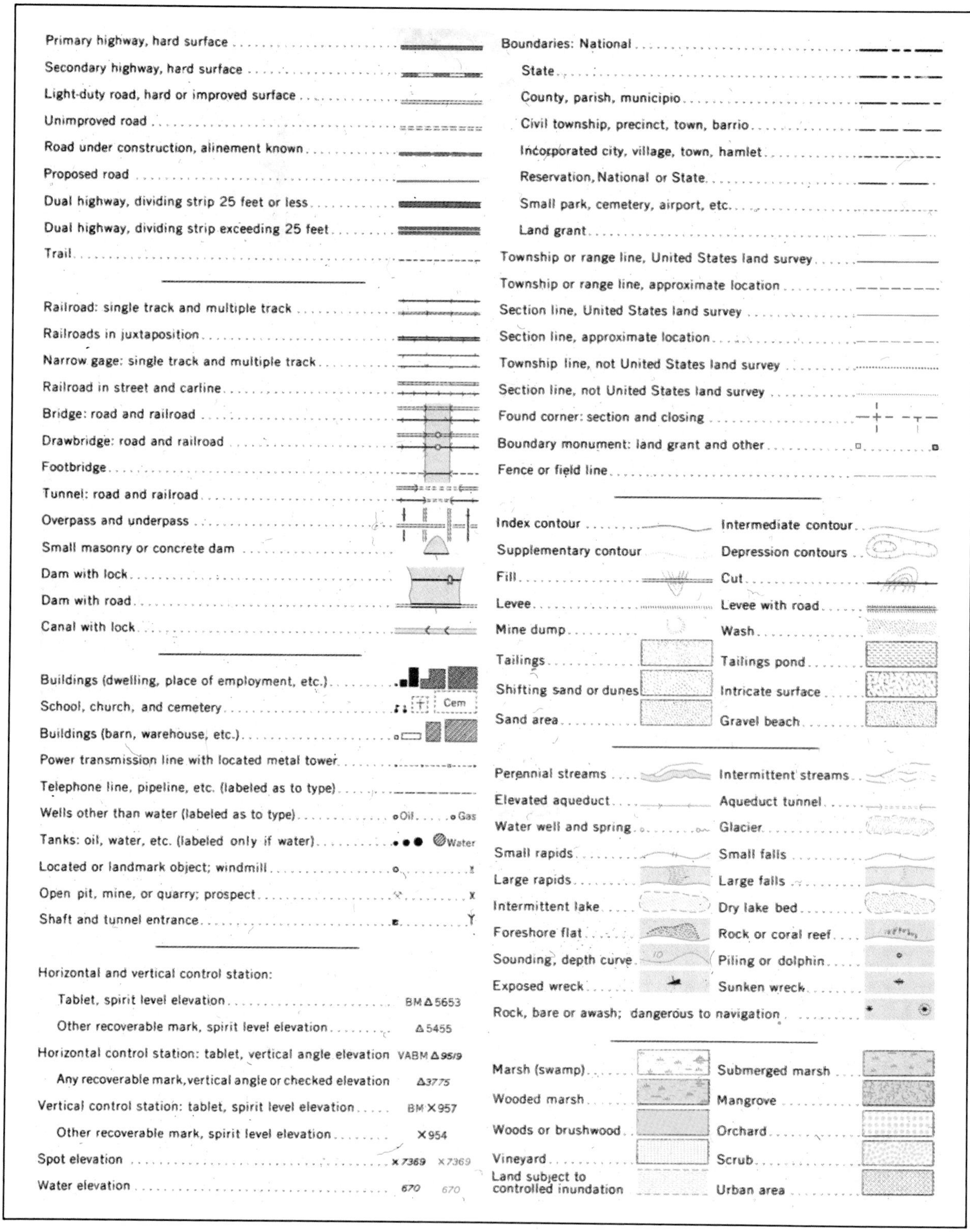

Primary highway, hard surface . . . .

Secondary highway, hard surface . . . .

Light-duty road, hard or improved surface . . . .

Unimproved road . . . .

Road under construction, alinement known . . . .

Proposed road . . . .

Dual highway, dividing strip 25 feet or less . . . .

Dual highway, dividing strip exceeding 25 feet . . . .

Trail . . . .

Railroad: single track and multiple track . . . .

Railroads in juxtaposition . . . .

Narrow gage: single track and multiple track . . . .

Railroad in street and carline . . . .

Bridge: road and railroad . . . .

Drawbridge: road and railroad . . . .

Footbridge . . . .

Tunnel: road and railroad . . . .

Overpass and underpass . . . .

Small masonry or concrete dam . . . .

Dam with lock . . . .

Dam with road . . . .

Canal with lock . . . .

Buildings (dwelling, place of employment, etc.) . . . .

School, church, and cemetery . . . . Cem

Buildings (barn, warehouse, etc.) . . . .

Power transmission line with located metal tower . . . .

Telephone line, pipeline, etc. (labeled as to type) . . . .

Wells other than water (labeled as to type) . . . . oOil oGas

Tanks: oil, water, etc. (labeled only if water) . . . . ● ● Water

Located or landmark object; windmill . . . . o

Open pit, mine, or quarry; prospect x

Shaft and tunnel entrance . . . . Y

Horizontal and vertical control station:

    Tablet, spirit level elevation . . . . BM △ 5653

    Other recoverable mark, spirit level elevation . . . . △ 5455

Horizontal control station: tablet, vertical angle elevation VABM △ 95/9

    Any recoverable mark, vertical angle or checked elevation △3775

Vertical control station: tablet, spirit level elevation . . . . . BM × 957

    Other recoverable mark, spirit level elevation . . . . × 954

Spot elevation . . . . × 7369 × 7369

Water elevation 670 670

Boundaries: National . . . .

    State . . . .

    County, parish, municipio . . . .

    Civil township, precinct, town, barrio . . . .

    Incorporated city, village, town, hamlet . . . .

    Reservation, National or State . . . .

    Small park, cemetery, airport, etc. . . . .

    Land grant . . . .

Township or range line, United States land survey . . . . . .

Township or range line, approximate location . . . .

Section line, United States land survey . . . .

Section line, approximate location . . . .

Township line, not United States land survey . . . .

Section line, not United States land survey . . . .

Found corner: section and closing . . . .

Boundary monument: land grant and other . . . .

Fence or field line . . . .

Index contour	Intermediate contour . . .
Supplementary contour	Depression contours . . .
Fill	Cut . . . .
Levee	Levee with road . . . .
Mine dump	Wash . . . .
Tailings	Tailings pond . . . .
Shifting sand or dunes	Intricate surface . . . .
Sand area	Gravel beach . . . .

Perennial streams	Intermittent streams . . .
Elevated aqueduct	Aqueduct tunnel . . . .
Water well and spring	Glacier . . . .
Small rapids	Small falls . . . .
Large rapids	Large falls . . . .
Intermittent lake	Dry lake bed . . . .
Foreshore flat	Rock or coral reef . . . . .
Sounding, depth curve 10	Piling or dolphin . . . . . .
Exposed wreck	Sunken wreck . . . .
Rock, bare or awash; dangerous to navigation . . . .

Marsh (swamp)	Submerged marsh . . . .
Wooded marsh	Mangrove . . . .
Woods or brushwood	Orchard . . . .
Vineyard	Scrub . . . .
Land subject to controlled inundation	Urban area . . . .

FIGURE D.3

U.S. Geological Survey topographic map symbols. (Variations will be found on older maps.)

APPENDIX E
Star Charts

The star charts on the next five pages can be used to locate and identify some of the very bright stars in the sky.* The months along the bottom of each map indicate the stars which will be over the observer's meridian in the early evening during that month. The stars with a declination equal to the observer's latitude will be directly overhead (zenith).

---

*Reproduced from J. A. Ripley, Jr. and R. C. Whitten, *The Elements and Structure of the Physical Sciences,* 2nd ed. Copyright © 1969 by John Wiley & Sons, Inc. Reprinted by permission of John Wiley & Sons, Inc.

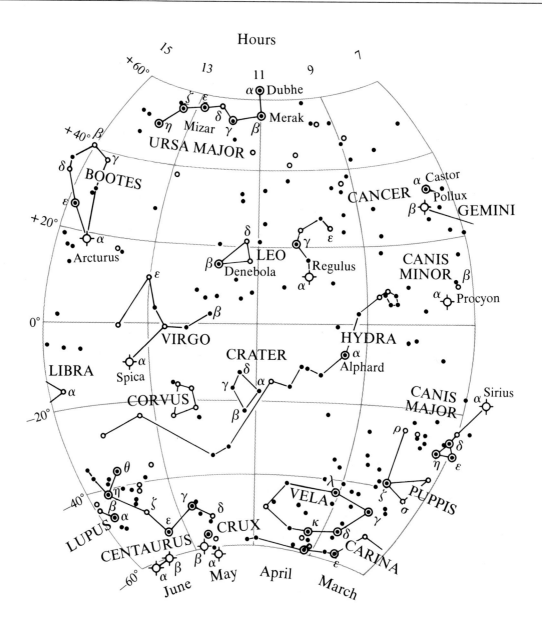

Hours

⟡ Brighter than magnitude 1.5

◉ Between magnitudes 1.5 and 2.5

○ Between magnitudes 2.5 and 3.5

• Fainter than magnitude 3.5

Hours

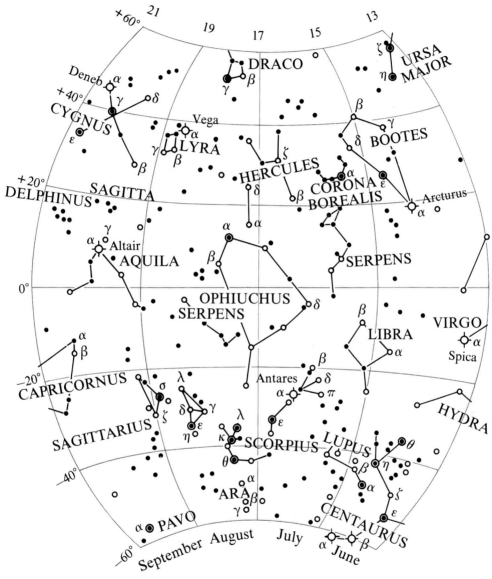

◇   Brighter than magnitude 1.5

◉   Between magnitudes 1.5 and 2.5

○   Between magnitudes 2.5 and 3.5

•   Fainter than magnitude 3.5

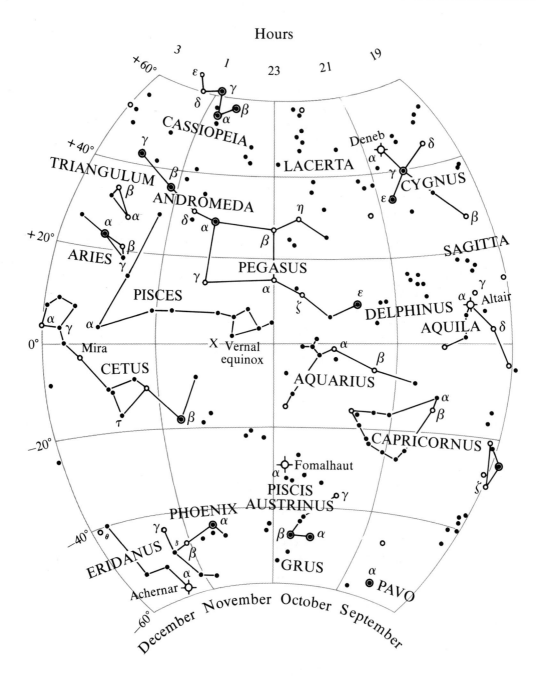

Hours

Brighter than magnitude 1.5

Between magnitudes 1.5 and 2.5

Between magnitudes 2.5 and 3.5

Fainter than magnitude 3.5

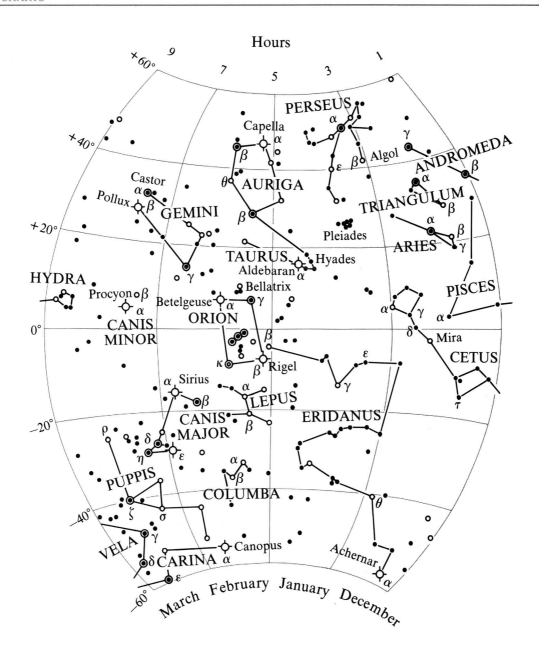

Hours

+60°   9   7   5   3   1

+40°

+20°

0°

−20°

−40°

−60°   March February January December

PERSEUS
Capella
α
β
θ   AURIGA
β

Castor
α
Pollux β   GEMINI
γ

HYDRA
Procyon β
α
CANIS
MINOR

α
ANDROMEDA
γ
Algol β
ε   β
α
TRIANGULUM
β
α   β
ARIES   γ

Pleiades
TAURUS Hyades
Aldebaran
α

PISCES
α   γ
α
δ   Mira
CETUS

Bellatrix
Betelgeuse α   γ
ORION
β
κ   β   Rigel
ε
γ

α Sirius
β
CANIS
MAJOR
α LEPUS
β

ERIDANUS

τ

ρ
δ
η ε
PUPPIS
ζ   σ

α
β
COLUMBA

θ

VELA   γ
δ CARINA α
ε
Canopus

Achernar
α

⬦   Brighter than magnitude 1.5

●   Between magnitudes 1.5 and 2.5

○   Between magnitudes 2.5 and 3.5

·   Fainter than magnitude 3.5

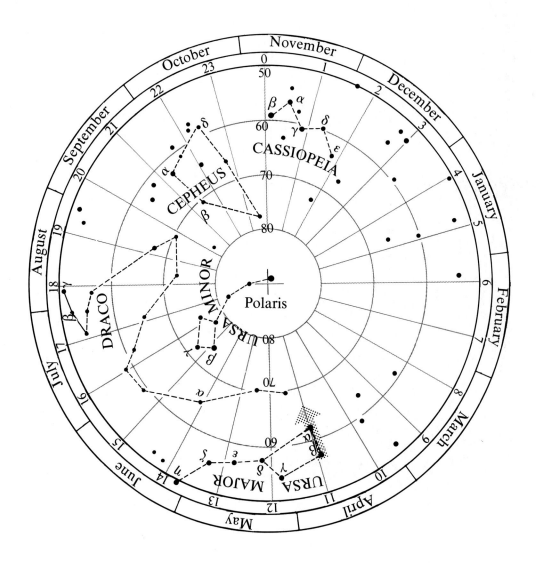

APPENDIX F
World Climates

The distribution of major atmospheric elements is very complex. Due to the many variations from place to place and from time to time at the same place, it is unlikely that any two sites on the earth's surface experience exactly the same weather. Since the number of places on earth is almost infinite, the number of different climates must also be extremely large. In order to cope with such variety, it is essential to devise a means of classifying this vast array of data. By establishing groups consisting of items that have certain important characteristics in common, order and simplicity are introduced. Organizing large amounts of information not only aids comprehension and understanding but also facilitates analysis and explanation.

We should remember that the classification of climates (or of anything else) is not a natural phenomenon, but the product of human ingenuity. The value of any particular classification is determined largely by its intended use. Here we shall use a system devised by the Russian-born German climatologist Wladimir Koeppen (1846–1940). As a tool for presenting the general world pattern of climates, it has been the best-known and most used system for more than fifty years. Koeppen believed that the distribution of natural vegetation was the best expression of the totality of climate. Therefore, the boundaries he chose were based largely on the limits of certain plant associations.

Five principal groups are recognized, with each group designated by a capital letter as follows:

A Humid tropical: winterless climates; all months have a mean temperature above 18°C.

B Dry: climates where evaporation exceeds precipitation; there is a constant water deficiency.

C Humid middle-latitude: mild winters; the average temperature of the coldest month is below 18°C but above −3°C.

D Humid middle-latitude: severe winters; the average temperature of the coldest month is below −3°C and the warmest monthly mean exceeds 10°C.

E Polar: summerless climates; the average temperature of the warmest month is below 10°C.

Notice that four of the major groups (A, C, D, E) are defined on the basis of temperature characteristics, and the fifth, the B group, has precipitation as its primary criterion. Each of the five groups is further subdivided by using the criteria and letter symbols presented in Table F.1* on page 582.

What follows is a brief summary of the major climatic types. While reading this summary, you should refer to Figure F.1 on pages 584–85 and Table F.2 on page 583, which shows data for representative places of each climatic type.

---

*When classifying climatic data by using Table F.1, you should first determine whether or not the data meet the criteria for the E climates. If the station is not a polar climate, proceed to the criteria for B climates. If your data do not fit into either E or B groups, check the data against the criteria for A, C, and D climates, in that order.

TABLE F.1
Koeppen system of climatic classification.

	Letter Symbol		
1st	2nd	3rd	
A			Average temperature of the coldest month is 18°C or higher.
	f		Every month has 6 centimeters of precipitation or more.
	m		Short dry season; precipitation in driest month less than 6 centimeters but equal to or greater than 10 − (R/25) (R is annual rainfall in centimeters).
	w		Well-defined winter dry season; precipitation in driest month less than 10 − (R/25).
	s		Well-defined summer dry season (rare).
B			Potential evaporation exceeds precipitation. The dry/humid boundary is defined by the following formulas: (NOTE: R is average annual precipitation in centimeters and T is average annual temperature in °C) $R < 2T + 28$ when 70% or more of rain falls in warmer 6 months. $R < 2T$ when 70% or more of rain falls in cooler 6 months. $R < 2T + 14$ when neither half year has 70% or more of total annual rain.
	S		Steppe—The BS/BW boundary is 1/2 the dry/humid boundary.
	W		Desert
		h	Average annual temperature is 18°C or greater.
		k	Average annual temperature is less than 18°C.
C			Average temperature of the coldest month is below 18°C and above −3°C.
	w		At least ten times as much precipitation in a summer month as in the driest winter month.
	s		At least three times as much precipitation in a winter month as in the driest summer month; precipitation in driest summer month less than 4 centimeters.
	f		Criteria for w and s cannot be met.
		a	Warmest month is over 22°C; at least 4 months over 10°C.
		b	No month above 22°C; at least 4 months over 10°C.
		c	One to 3 months above 10°C.
D			Average temperature of coldest month is −3°C or below; average temperature of warmest month is greater than 10°C.
	s		Same as under C.
	w		Same as under C.
	f		Same as under C.
		a	Same as under C.
		b	Same as under C.
		c	Same as under C.
		d	Average temperature of the coldest month is −38°C or below.
E			Average temperature of the warmest month is below 10°C.
	T		Average temperature of the warmest month is greater than 0°C and less than 10°C.
	F		Average temperature of the warmest month is 0°C or below.

TABLE F.2
Climatic data for representative stations.

	J	F	M	A	M	J	J	A	S	O	N	D	Year
Iquitos, Peru (Af); lat. 3°39′S; 115 m													
Temp. (°C)	25.6	25.6	24.4	25.0	24.4	23.3	23.3	24.4	24.4	25.0	25.6	25.6	24.7
Precip. (mm)	259	249	310	165	254	188	168	117	221	183	213	292	2619
Rio de Janeiro, Brazil (Aw); lat. 22°50′S; 26 m													
Temp. (°C)	25.9	26.1	25.2	23.9	22.3	21.3	20.8	21.1	21.5	22.3	23.1	24.4	23.2
Precip. (mm)	137	137	143	116	73	43	43	43	53	74	97	127	1086
Faya, Chad (BWh); lat. 18°00′N; 251 m													
Temp. (°C)	20.4	22.7	27.0	30.6	33.8	34.2	33.6	32.7	32.6	30.5	25.5	21.3	28.7
Precip. (mm)	0	0	0	0	0	2	1	11	2	0	0	0	16
Salt Lake City, Utah (BSk); lat. 40°46′N; 1288 m													
Temp. (°C)	−2.1	0.9	4.7	9.9	14.7	19.4	24.7	23.6	18.3	11.5	3.4	−0.2	10.7
Precip. (mm)	34	30	40	45	36	25	15	22	13	29	33	31	353
Washington, D.C. (Cfa); lat. 38°50′N; 20 m													
Temp. (°C)	2.7	3.2	7.1	13.2	18.8	23.4	25.7	24.7	20.9	15.0	8.7	3.4	13.9
Precip. (mm)	77	63	82	80	105	82	105	124	97	78	72	71	1036
Brest, France (Cfb); lat. 48°24′N; 103 m													
Temp. (°C)	6.1	5.8	7.8	9.2	11.6	14.4	15.6	16.0	14.7	12.0	9.0	7.0	10.8
Precip. (mm)	133	96	83	69	68	56	62	80	87	104	138	150	1126
Rome, Italy (Csa); lat. 41°52′N; 3 m													
Temp. (°C)	8.0	9.0	10.9	13.7	17.5	21.6	24.4	24.2	21.5	17.2	12.7	9.5	15.9
Precip. (mm)	83	73	52	50	48	18	9	18	70	110	113	105	749
Peoria, Illinois (Dfa); lat. 40°45′N; 180 m													
Temp. (°C)	−4.4	−2.2	4.4	10.6	16.7	21.7	23.9	22.7	18.3	11.7	3.8	−2.2	10.4
Precip. (mm)	46	51	69	84	99	97	97	81	97	61	61	51	894
Verkhoyansk, U.S.S.R. (Dfd); lat. 67°33′N; 137 m													
Temp. (°C)	−46.8	−43.1	−30.2	−13.5	2.7	12.9	15.7	11.4	2.7	−14.3	−35.7	−44.5	−15.2
Precip. (mm)	7	5	5	4	5	25	33	30	13	11	10	7	155
Ivigtut, Greenland (ET); lat. 61°12′N; 129 m													
Temp. (°C)	−7.2	−7.2	−4.4	−0.6	4.4	8.3	10.0	8.3	5.0	1.1	−3.3	−6.1	0.7
Precip. (mm)	84	66	86	62	89	81	79	94	150	145	117	79	1132
McMurdo Station, Antarctica (EF); lat. 77°53′S; 2 m													
Temp. (°C)	−4.4	−8.9	−15.5	−22.8	−23.9	−24.4	−26.1	−26.1	−24.4	−18.8	−10.0	−3.9	−17.4
Precip. (mm)	13	18	10	10	10	8	5	8	10	5	5	8	110

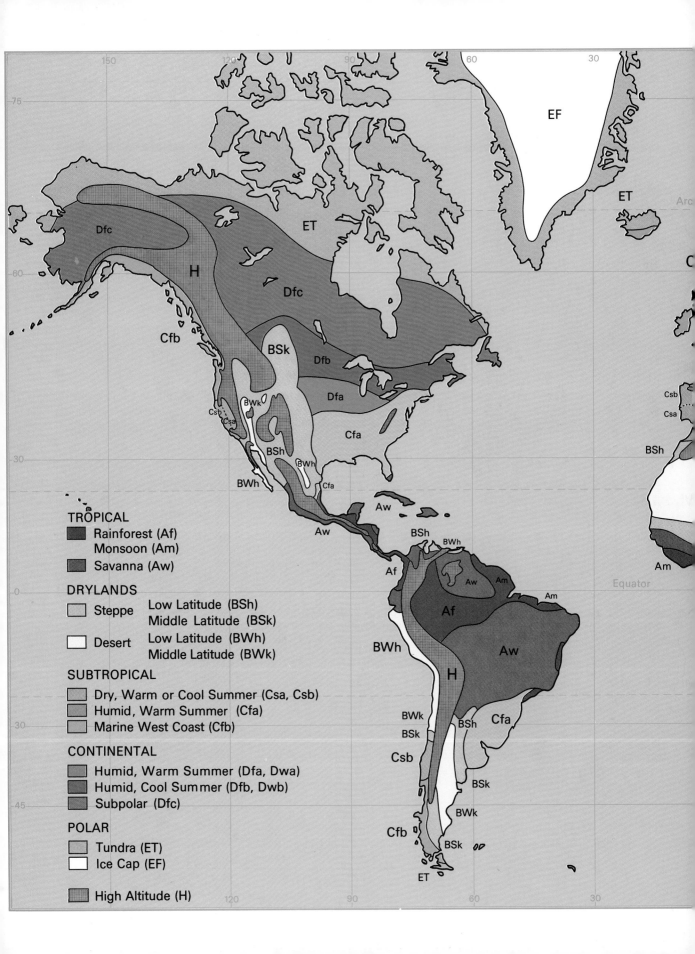

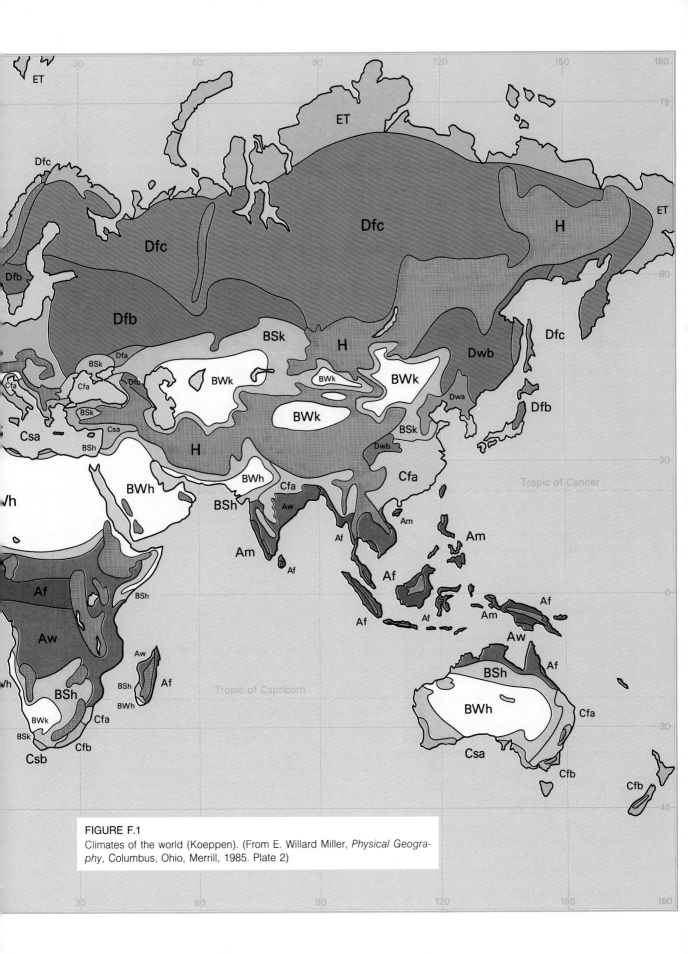

FIGURE F.1
Climates of the world (Koeppen). (From E. Willard Miller, *Physical Geography*, Columbus, Ohio, Merrill, 1985. Plate 2)

HUMID TROPICS (A CLIMATES)

Within the A group of climates, two main types are recognized—wet tropical climates (Af and Am) and tropical wet and dry (AW).

WET TROPICAL CLIMATES (Af, Am)

Since places with an Af or Am designation are lowland areas that lie near the equator, temperatures are consistently high. Consequently, not only is the annual mean high, but the annual range is very small. Total annual precipitation is high, often exceeding 200 centimeters; and although precipitation is not evenly distributed throughout the year, places with this climate are generally wet in all months. If a dry season exists, it is very short. In response to the constantly high temperatures and year-round rainfall, the wet tropics are dominated by the most luxuriant vegetation found in any climatic realm—the tropical rain forest.

TROPICAL WET AND DRY (Aw)

In the latitude zone poleward of the wet tropics and equator-ward of the tropical deserts lies a transitional climatic region called tropical wet and dry. Here the rain forest gives way to the savanna, a tropical grassland with scattered drought-tolerant trees. Since temperature characteristics among all A climates are quite similar, the primary factor which distinguishes the Aw climate from Af and Am is precipitation. Although the overall amount of precipitation in the tropical wet and dry realm is often considerably less than in the wet tropics, the most distinctive feature of this climate is not the annual rainfall total but the markedly seasonal character of the rainfall. As the equatorial low advances poleward in summer, the rainy season commences and features weather patterns typical of the wet tropics. Later, with the retreat of the equatorial low, the subtropical high advances into the region and brings with it intense drought conditions. In some Aw regions such as India, Southeast Asia, and portions of Australia, the alternating periods of rainfall and drought are associated with a pronounced monsoon circulation (see Chapter 14).

DRY (B) CLIMATES

It is important to realize that the concept of dryness is a relative one and refers to any situation in which a water deficiency exists. Climatologists define a dry climate as one in which the yearly precipitation is not as great as

the potential loss of water by evaporation. Thus, dryness is not only related to annual rainfall totals, but it is also a function of evaporation, which in turn is closely dependent upon temperature. To establish the boundary between dry and humid climates, the Koeppen classification uses formulas that involve three variables: average annual precipitation, average annual temperature, and seasonal distribution of precipitation. The use of average annual temperature reflects its importance as an index of evaporation. The amount of rainfall defining the humid-dry boundary increases as the annual mean temperature increases. The use of seasonal precipitation as a variable is also related to this idea. If rain is concentrated to the warmest months, loss to evaporation is greater than if the precipitation were concentrated in the cooler months.

Within the regions defined by a general water deficiency there are two climatic types: arid or desert (BW) and semiarid or steppe (BS). These two groups have many features in common; their differences are primarily a matter of degree. The semiarid is a marginal and more humid variant of the arid and represents a transition zone that surrounds the desert and separates it from the bordering humid climates.

The heart of the low-latitude dry climates (BWh and BSh) lies in the vicinity of the Tropic of Cancer and the Tropic of Capricorn. The existence and distribution of this rather extensive dry tropical realm is primarily a consequence of the subsidence and marked stability of the subtropical highs. Unlike their low-latitude counterparts, middle-latitude deserts and steppes are not controlled by the subsiding air masses of the subtropical anticyclones. Instead, these dry lands exist principally because of their position in the deep interiors of large landmasses far removed from the oceans. In addition, the presence of high mountains across the paths of the prevailing winds further acts to separate these areas from water-bearing maritime air masses.

HUMID MIDDLE-LATITUDE CLIMATES WITH MILD WINTERS (C CLIMATES)

Although the term *subtropical* is often used for the C climates, it can be misleading. While many areas with C climates do indeed possess some near-tropical characteristics, other regions do not. For example, we would be stretching the use of the term *subtropical* to describe the climates of coastal Alaska and Norway, which belong to the C group. Within the C group of climates, several subgroups are recognized.

HUMID SUBTROPICS (Cfa)

Located on the eastern sides of the continents, in the 25- to 40-degree latitude range, this climatic type dominates the southeastern United States, as well as other similarly situated areas around the world. In the summer the humid subtropics experience hot, sultry weather of the type one expects to find in the rainy tropics. Daytime temperatures are generally high, and since both specific and relative humidities are high, the night brings little relief. An afternoon or evening thunderstorm is also possible, for these areas experience such storms on an average of 40 to 100 days each year, the majority during the summer months. As summer turns to autumn, the humid subtropics lose their similarity to the rainy tropics. Although winters are mild, frosts are common in the higher-latitude Cfa areas and occasionally plague the tropical margins as well. The winter precipitation is also different in character from the summer. Some is in the form of snow, and most is generated along fronts of the frequent middle-latitude cyclones that sweep over these regions.

MARINE WEST COAST CLIMATE (Cfb, Cfc)

Situated on the western (windward) side of continents, from about 40 to 60 degrees north and south latitude, is a climatic region dominated by the onshore flow of oceanic air. The prevalence of maritime air masses means that mild winters and cool summers are the rule, as is an ample amount of rainfall throughout the year. Although there is no pronounced dry period, there is a drop in monthly precipitation totals during the summer. The reason for the reduced summer rainfall is the poleward migration of the oceanic subtropical highs. Although the areas of marine west coast climate are situated too far poleward to be dominated by these dry anticyclones, their influence is sufficient to cause a decrease in warm season rainfall.

DRY-SUMMER SUBTROPICS (Csa, Csb)

This climate is typically located along the west sides of continents, between latitudes of 30 and 45 degrees. It is unique because it is the only humid climate that has a strong winter rainfall maximum, a feature that reflects its intermediate position between the marine west coast on the poleward side and the tropical steppes on the equatorward side. In summer the region is dominated by the stable eastern side of the oceanic subtropical highs. In winter, as the wind and pressure systems follow the sun equatorward, it is within range of the cy-

clonic storms of the polar front. Thus, during the course of a year these areas alternate between being part of the dry tropics and being an extension of the humid middle latitudes. While middle-latitude changeability characterizes the winter, tropical constancy describes the summer.

HUMID MIDDLE-LATITUDE CLIMATES WITH SEVERE WINTERS (D CLIMATES)

The D climates are land-controlled climates, the result of broad continents in the middle latitudes. Because continentality is a basic feature, D climates are absent in the Southern Hemisphere where the middle-latitude zone is dominated by the oceans.

HUMID CONTINENTAL (Dfa, Dfb, Dwa, Dwb)

This climate is located in the central and eastern portions of North America and Eurasia in the latitude range between approximately 40 and 50 degrees north latitude. Both winter and summer temperatures may be characterized as relatively severe. Consequently, annual temperature ranges are high throughout the climate. Precipitation is characteristically greatest in summer. A portion of the winter precipitation is in the form of snow, the proportion increasing with latitude. Although the precipitation is generally considerably less during the cold season, it is usually much more conspicuous than the greater amounts which characterize the summer. An obvious reason is that snow remains on the ground, often for extended periods, and rain, of course, does not. Furthermore, while summer rains are often in the form of relatively short showers, winter snows generally occur over a prolonged period.

SUBARCTIC CLIMATE (Dfc, Dfd, Dwc, Dwd)

Situated north of the humid continental climate and south of the polar tundra is an extensive subarctic region. It is often referred to as the taiga climate because its extent closely corresponds to the northern coniferous forest of the same name. The outstanding feature in this realm is the dominance of winter. Not only is it long, but temperatures are also bitterly cold. By contrast, summers in the subarctic are remarkably warm, despite their short duration. However, when compared with regions farther south, this short season must be characterized as cool. The extremely cold winters and relatively warm summers combine to produce the highest annual temperature ranges on earth. Since these far northerly continental interiors are the source regions for cP air masses, there is very limited moisture available throughout the year. Precipitation totals are therefore small, with a maximum occurring during the warmer summer months.

POLAR (E) CLIMATES

Polar climates are those in which the mean temperature of the warmest month is below 10°C. Thus, just as the tropics are defined by their year-round warmth, the polar realm is known for its enduring cold. Since winters are periods of perpetual night, or nearly so, temperatures at most polar locations are understandably bitter. During the summer months temperatures remain cool despite the long days, because the sun is so low in the sky that its oblique rays are not effective in bringing about a genuine warming. Although polar climates are classified as humid, precipitation is generally meager. Evaporation, of course, is also limited. The scanty pre-

cipitation totals are easily understood in view of the temperature characteristics of the region. The amount of water vapor in the air is always small because low specific humidities must accompany low temperatures. Usually precipitation is most abundant during the warmer summer months when the moisture content of the air is highest.

Two types of polar climates are recognized. The tundra climate (ET) is a treeless climate found almost exclusively in the Northern Hemisphere. Due to the combination of high latitude and continentality, winters are severe, summers are cool, and annual temperature ranges are high. Further, yearly precipitation is small, with a modest summer maximum. The ice cap climate (EF) does not have a single monthly mean above 0°C. Consequently, since the average temperature for all months is below freezing, the growth of vegetation is prohibited and the landscape is one of permanent ice and snow. This climate of perpetual frost covers a surprisingly large area—more than 15.5 million kilometers, or about 9 percent of the earth's land area. Aside from scattered occurrences in high mountain areas, it is confined to the ice caps of Greenland and Antarctica.

APPENDIX G
World Soils

Figure G.1, on pages 590–591, shows the generalized pattern of global soil orders according to the *Comprehensive Soil Classification System* (CSCS). It should be examined in conjunction with Table G.1 which briefly describes each of the soil orders depicted on the map. To avoid subjective decisions as to classification (a problem that plagued earlier systems), the CSCS defined its classes strictly in terms of soil characteristics. That is, it is based upon features that can be observed or inferred.

The CSCS uses a hierarchy of six categories, or levels. The system recognizes 10 major global *orders* which can be further subdivided into *suborders, great groups, subgroups, families,* and *series*. Note, however, that on the scale of a world map such as Figure G.1, only the largest units (soil orders) can be shown and then only in an extremely generalized way. Although the distribution pattern of major soil orders is more complex than can be shown in Figure G.1, the major distinguishing regional properties of world soils is depicted.

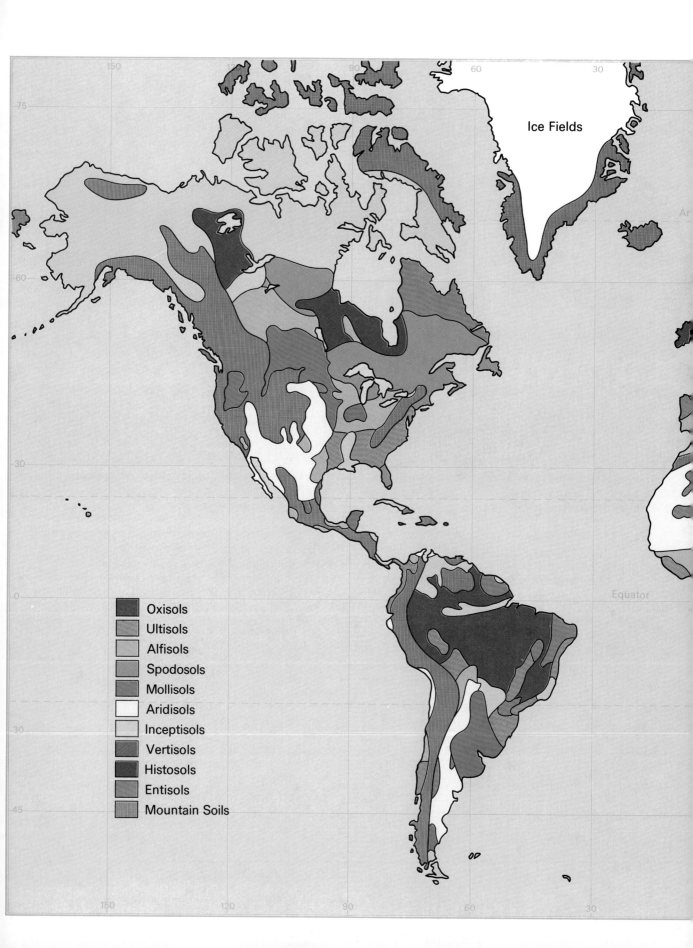

Ice Fields

Oxisols
Ultisols
Alfisols
Spodosols
Mollisols
Aridisols
Inceptisols
Vertisols
Histosols
Entisols
Mountain Soils

Equator

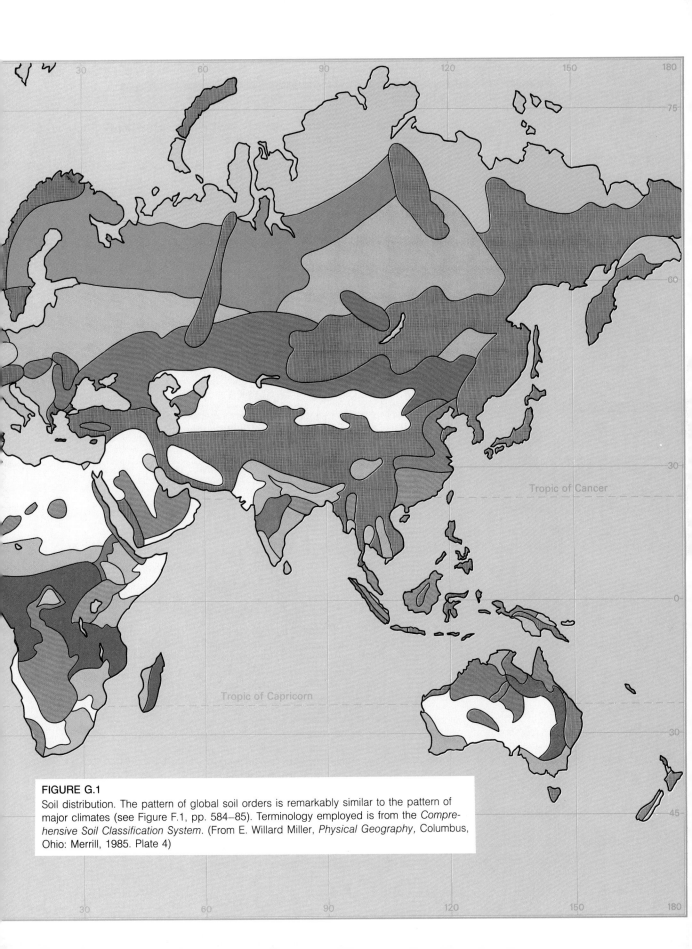

FIGURE G.1
Soil distribution. The pattern of global soil orders is remarkably similar to the pattern of major climates (see Figure F.1, pp. 584–85). Terminology employed is from the *Comprehensive Soil Classification System*. (From E. Willard Miller, *Physical Geography,* Columbus, Ohio: Merrill, 1985. Plate 4)

TABLE G.1
World soil orders.

Entisols	Youngest soils on the earth. Just beginning to develop in response to the weathering phenomena in the environment. Do not display natural horizons. Found in all climates. They weather slowly over thousands of years; consequently, volcanic ash deposits or sand deposits form the basis for entisols.
Vertisols	Soils containing large amounts of clay, which shrink upon drying and swell with the addition of water. Found in subhumid to arid climates, provided that adequate supplies of water are available to saturate the soil after periods of drought. Soil expansion and contraction exert stresses on human structures.
Inceptisols	Young soils that reveal developmental characteristics (horizons) in response to climate and vegetation. Exist from the Arctic to the tropics on young land surfaces. Common in alpine areas, on river floodplains, in stables and dune areas, and in areas once glaciated.
Aridisols	Soils that develop in dry places, such as the desert, where water—precipitation and groundwater—is insufficient to remove soluble minerals. Frequently irrigated for intensive agricultural production, although salt accumulation poses a problem.
Mollisols	Dark, soft soils that have developed under grass vegetation, generally found in prairie areas. Soil fertility is excellent because potential evaporation generally exceeds precipitation. Also found in hard-wood forests with significant earthworm activity. Climatic range is boreal or alpine to tropical. Dry seasons are normal.
Spodosols	Soils found only in humid regions on sandy material. Range from the boreal coniferous forests into tropical forests. Beneath the dark upper horizon of weathered organic material lies a light-colored horizon of leached material, the distinctive property of this soil.
Alfisols	Mineral soils that form under boreal forests or broadleaf deciduous forests, rich in iron and aluminum. Clay particles accumulate in a subsurface layer in response to leaching in moist environments. Fertile, productive soils, because they are neither too wet nor too dry.
Ultisols	Soils that represent the products of long periods of weathering. Water percolating through the soil concentrates clay particles in the lower horizons (argillic horizons). Restricted to humid climates in the temperate regions and the tropics where the growing season is long. Abundant water and a long frost-free period contribute to extensive leaching, hence poorer soil quality.
Oxisols	Soils that occur on old land surfaces unless parent materials were strongly weathered before they were deposited. Generally found in the tropics and subtropical regions. Rich in iron and aluminum oxides, oxisols are heavily leached; hence are poor soils for agricultural activity. Few, if any, exist in the United States.
Histosols	Organic soils with little or no climatic implications. Can be found in any climate where organic debris can accumulate to form a "bog soil." Dark, partially decomposed organic material commonly referred to as *peat*.

SOURCE: Robert E. Norris, et al. *Geography: An Introductory Perspective*, Columbus, Ohio: Merrill, 1982.

GLOSSARY

Aa A type of lava flow that has a jagged blocky surface.

Ablation A general term for the loss of ice and snow from a glacier.

Absolute dating Determination of the number of years since the occurrence of a given geologic event.

Absolute humidity The weight of water vapor in a given volume of air (usually expressed in grams/m^3).

Absolute instability Air that has a lapse rate greater than the dry adiabatic rate.

Absolute magnitude The apparent brightness of a star if it were viewed from a distance of 10 parsecs (32.6 light-years). Used to compare the true brightness of stars.

Absolute stability Air with a lapse rate less than the wet adiabatic rate.

Absorption spectrum A continuous spectrum with dark lines superimposed.

Abyssal plain Very level area of the deep ocean floor, usually lying at the foot of the continental rise.

Accretionary prism A large wedge-shaped mass of sediment that accumulates in subduction zones. Here sediment is scraped from the subducting oceanic plate and accreted to the overriding crustal block.

Acid precipitation Rain or snow with pH values of less than 5.6, the pH of unpolluted precipitation.

Adiabatic temperature change Cooling or warming of air caused when air is allowed to expand or is compressed, not because heat is added or subtracted.

Advection Horizontal convective motion, such as wind.

Advection fog A fog formed when warm, moist air is blown over a cool surface.

Aftershocks Smaller earthquakes that follow the main earthquake.

Air mass A large body of air that is characterized by a sameness of temperature and humidity.

Albedo The reflectivity of a substance, usually expressed as a percentage of the incident radiation reflected.

Alluvial fan A fan-shaped deposit of sediment formed when a stream's slope is abruptly reduced.

Alluvium Unconsolidated sediment deposited by a stream.

Alpine glacier A glacier confined to a mountain valley, which in most instances had previously been a stream valley.

Altitude (of the sun) The angle of the sun above the horizon.

Anemometer An instrument used to determine wind speed.

Angular unconformity An unconformity in which the strata below dip at an angle different from that of the beds above.

Annual mean An average of the twelve monthly temperature means.

Annual temperature range The difference between the highest and lowest monthly means.

Anthracite A hard, metamorphic form of coal that burns clean and hot.

Anticline A fold in sedimentary strata that resembles an arch.

Anticyclone A high-pressure center characterized by a clockwise flow of air in the Northern Hemisphere.

Aphelion The place in the orbit of a planet where the planet is farthest from the sun.

Apparent magnitude The brightness of a star when viewed from the earth.

Aquicludes Impermeable beds that hinder or prevent groundwater movement.

Aquifer Rock or soil through which groundwater moves easily.

Arête A narrow knifelike ridge separating two adjacent glaciated valleys.

Arkose A feldspar-rich sandstone.

Artesian well A well in which the water rises above the level where it was initially encountered.

Asteroids Thousands of small planetlike bodies, ranging in size from a few hundred kilometers to less than a kilometer, whose orbits lie mainly between those of Mars and Jupiter.

Asthenosphere Layer of the earth found below the lithosphere that is thought to be partially molten.

Astronomical theory A theory of climatic change first developed by the Yugoslavian astronomer Milankovitch. It is based upon changes in the shape of the earth's orbit, variations in the obliquity of the earth's axis, and the wobbling of the earth's axis.

Atoll A continuous or broken ring of coral reef surrounding a central lagoon.

Atom The smallest particle that exists as an element.

Atomic number The number of protons in the nucleus of an atom.

Atomic weight The average of the atomic masses of isotopes for a given element.

Aurora A bright display of ever-changing light caused by solar radiation interacting with the upper atmosphere in the region of the poles.

Azoic zone A well-known but incorrect theory formulated about 1850 by Edward Forbes stating that no life existed in the ocean below a depth of about 550 meters.

Back swamp A poorly drained area on a floodplain that results when natural levees are present.

Barchan dune Solitary sand dune shaped like a crescent with its tips pointing downwind.

Barometer An instrument that measures atmospheric pressure.

Barred spiral A galaxy having straight arms extending from its nucleus.

Barrier island A low, elongate ridge of sand that parallels the coast.

Basalt A fine-grained igneous rock of mafic composition.

Base level The level below which a stream cannot erode.

Basin A circular downfolded structure.

Batholith A large mass of igneous rock that formed when magma was emplaced at depth, crystallized, and was subsequently exposed by erosion.

Baymouth bar A sandbar that completely crosses a bay, sealing it off from the open ocean.

Beach drift The transport of sediment in a zig-zag pattern along a beach caused by the uprush of water from obliquely breaking waves.

Beach nourishment Large quantities of sand are added to the beach system to offset losses caused by wave erosion.

Bed load Sediment that is carried by a stream along the bottom of its channel.

Benioff zone Zone of inclined seismic activity that extends from a trench downward into the asthenosphere.

Big Bang theory The theory that proposes that the universe originated as a single mass, which subsequently exploded.

Binary stars Two stars revolving around a common center of mass under their mutual gravitational attraction.

Biogenous sediment Sea-floor sediments consisting of material of marine-organic origin.

Bituminous The most common form of coal, often called soft, black coal.

Blowout (deflation hollow) A depression excavated by the wind in easily eroded deposits.

Bode's law A sequence of numbers that approximates the mean distances of the planets from the sun.

Body waves Seismic waves that travel through the earth's interior.

Braided stream A stream consisting of numerous intertwining channels.

Breakwater A structure protecting a near-shore area from breaking waves.

Breccia A sedimentary rock composed of angular fragments that were lithified.

Bright-line spectrum The bright lines produced by an incandescent gas under low pressure.

Cactolith A quasi-horizontal chonolith composed of anastomosing ductoliths, whose distal ends curl like a harpolith, thin like a sphenolith, or bulge discordantly like an akmolith or ethmolith.

Caldera A large depression typically caused by collapse or ejection of the summit area of a volcano.

Calorie The amount of heat required to raise the temperature of one gram of water 1°C.

Calving Wastage of a glacier that occurs when large pieces of ice break off into water.

Capacity The total amount of sediment a stream is able to transport.

Catastrophism The concept that the earth was shaped by catastrophic events of a short-term nature.

Celestial sphere An imaginary hollow sphere upon which the ancients believed the stars were hung and carried around the earth.

Cenozoic era A time span on the geologic calendar beginning about 65 million years ago following the Mesozoic era.

Cepheid variable A star whose brightness varies periodically because it expands and contracts. A type of pulsating star.

Chemical weathering The processes by which the internal structure of a mineral is altered by the removal and/or addition of elements.

Chinook A wind blowing down the leeward side of a mountain and warming by compression.

Chromatic aberration The property of a lens whereby light of different colors is focused at different places.

Chromosphere The first layer of the solar atmosphere found directly above the photosphere.

Cinder cone A rather small volcano built primarily of pyroclastics ejected from a single vent.

Circle of illumination The great circle that separates daylight from darkness.

Cirque An amphitheater-shaped basin at the head of a glaciated valley produced by frost wedging and plucking.

Clastic rock A sedimentary rock made of broken fragments of pre-existing rock.

Cleavage The tendency of a mineral to break along planes of weak bonding.

Climate A description of aggregate weather conditions; the sum of all statistical weather information that helps describe a place or region.

Climatic feedback mechanism One of the several different outcomes that may result if one of the many elements in the atmosphere's extremely complex interactive system is altered.

Col A pass between mountain valleys where the headwalls of two cirques intersect.

Cold front A front along which a cold air mass thrusts beneath a warmer air mass.

Column A feature found in caves that is formed when a stalactite and stalagmite join.

Columnar joints A pattern of cracks that forms during cooling of molten rock generating columns.

Comet A small body which generally revolves about the sun in an elongated orbit.

Competence A measure of the largest particle a stream can transport; a factor dependent on velocity.

Composite cone A volcano composed of both lava flows and pyroclastic material.

Compound A substance formed by the chemical combination of two or more elements in definite proportions and usually having properties different than those of its constituent elements.

Condensation The change of state from a gas to a liquid.

Condensation nuclei Tiny bits of particulate matter that serve as surfaces on which water vapor condenses.

Conditional instability Moist air with a lapse rate between the dry and wet adiabatic rates.

Conduction The transfer of heat through matter by molecular activity. Energy is transferred through collisions from one molecule to another.

Cone of depression A cone-shaped depression in the water table immediately surrounding a well.

Conformable Layers of rock that were deposited without interruption.

Conglomerate A sedimentary rock composed of rounded gravel-sized particles.

Constellation An apparent group of stars originally named for mythical characters. The sky is presently divided into 88 constellations.

Contact metamorphism Changes in rock caused by the heat from a nearby magma body.

Continental drift theory A theory that originally proposed that the continents are rafted about. It has essentially been replaced by the plate tectonics theory.

Continental glacier A massive accumulation of ice that covers extensive land areas and whose flow is not usually controlled by the underlying typography.

Continental rise The gently sloping surface at the base of the continental slope.

Continental shelf The gently sloping submerged portion of the continental margin extending from the shoreline to the continental slope.

Continental slope The steep gradient that leads to the deep ocean floor and marks the seaward edge of the continental shelf.

Continuous spectrum An uninterrupted band of light emitted by an incandescent solid, liquid, or gas under pressure.

Convection The transfer of heat by the movement of a mass or substance. It can only take place in fluids.

Convergence zone An area where plates move together and either collide or one is subducted.

Coriolis force (effect) The deflective force of the earth's rotation on all free-moving objects, including the atmosphere and oceans. Deflection is to the right in the Northern Hemisphere and to the left in the Southern Hemisphere.

Corona The outer, tenuous layer of the solar atmosphere.

Correlation Establishing the equivalence of rocks of similar age in different areas.

Country breeze A circulation pattern characterized by a light wind blowing into a city from the surrounding countryside. It is best developed on clear and otherwise calm nights when the urban heat island is most pronounced.

Crater The depression at the summit of a volcano, or that which is produced by a meteorite impact.

Creep The slow downhill movement of soil and regolith.

Crevasse A deep crack in the brittle surface of a glacier.

Cross-cutting A principle of relative dating. A rock or fault is younger than any rock (or fault) through which it cuts.

Crust The very thin outermost layer of the earth.

Crystal An orderly arrangement of atoms.

Crystal form The external appearance of a mineral as determined by its internal arrangement of atoms.

Curie point The temperature above which a material loses its magnetization.

Cyclone A low-pressure center characterized by a counter-clockwise flow of air in the Northern Hemisphere.

Daily mean The mean temperature for a day that is determined by averaging the 24 hourly readings or, more commonly, by averaging the maximum and minimum temperatures for a day.

Daughter product An isotope resulting from radioactive decay.

Declination (stellar) The angular distance north or south of the celestial equator denoting the position of a celestial body.

Deep-sea fan A cone-shaped deposit at the base of the continental slope. The sediment is transported to the fan by turbidity currents that follow submarine canyons.

Deflation The lifting and removal of loose material by wind.

Delta An accumulation of sediment formed where a stream enters a lake or ocean.

Dendritic pattern A stream system that resembles the pattern of a branching tree.

Density The weight per unit volume of a particular material.

Deposition The process by which water vapor is changed directly to a solid without passing through the liquid state.

Desalination The removal of salts and other chemicals from seawater.

Desert pavement A layer of coarse pebbles and gravel created when wind removed the finer material.

Detrital sedimentary rock Rock formed from the accumulation of material that originated and was transported in the form of solid particles derived from both mechanical and chemical weathering.

Dew point The temperature to which air has to be cooled in order to reach saturation.

Diffused light Solar energy scattered and reflected in the atmosphere that reaches the earth's surface in the form of diffuse blue light from the sky.

Dike A tabular-shaped intrusive igneous feature that cuts through the surrounding rock.

Discharge The quantity of water in a stream that passes a given point in a period of time.

Disconformity A type of unconformity in which the beds above and below are parallel.

Dissolved load That portion of a stream's load carried in solution.

Distributary A section of a stream that leaves the main flow.

Diurnal tide Tides characterized by a single high and low water height each tidal day.

Divergent zone A region where the rigid plates are moving apart, typified by the mid-oceanic ridges.

Divide An imaginary line that separates the drainage of two streams; often found along a ridge.

Dome A roughly circular upfolded structure similar to an anticline.

Doppler effect The apparent change in wavelength of radiation caused by the relative motions of the source and the observer.

Doppler radar In addition to the tasks performed by conventional radar, this new generation of weather radar can detect motion directly and hence greatly improve tornado and severe storm warnings.

Drift The general term for any glacial deposit.

Drumlin A streamlined asymmetrical hill composed of glacial till. The steep side of the hill faces the direction from which the ice advanced.

Dune A hill or ridge of wind-deposited sand.

Earthflow The downslope movement of water-saturated, clay-rich sediment. Most characteristic of humid regions.

Earthquake The vibration of the earth produced by the rapid release of energy.

Ebb tide The movement of a tidal current away from the shore.

Eccentricity The variation of an ellipse from a circle.

Echo sounder An instrument used to determine the depth of water by measuring the time interval between emission of a sound signal and the return of its echo from the bottom.

Eclipse The cutting off of the light of one celestial body by another passing in front of it.

Ecliptic The yearly path of the sun plotted against the background of stars.

Elastic rebound The sudden release of stored strain in rocks that results in movement along a fault.

Electromagnetic spectrum The distribution of electromagnetic radiation by wavelength.

Element A substance that cannot be decomposed into simpler substances by ordinary chemical or physical means.

Emergent coast A coast where land that was formerly below sea level has been exposed either because of crustal uplift or a drop in sea level or both.

End moraine A ridge of till marking a former position of the front of a glacier.

Entrenched meander A meander cut into bedrock when uplifting rejuvenated a meandering stream.

Epicenter The location on the earth's surface that lies directly above the focus of an earthquake.

Epoch A unit of the geologic calendar that is a subdivision of a period.

Equinox The time when the vertical rays of the sun are striking the equator. The length of daylight and darkness is equal at all latitudes at equinox.

Era A major division on the geologic calendar; eras are divided into shorter units called periods.

Erosion The incorporation and transportation of material by a mobile agent, such as water, wind, or ice.

Esker Sinuous ridge composed largely of sand and gravel deposited by a stream flowing in a tunnel beneath a glacier near its terminus.

Estuary A funnel-shaped inlet of the sea that formed when a rise in sea level or subsidence of land caused the mouth of a river to be flooded.

Evaporation The process of converting a liquid to a gas.

Evaporite A sedimentary rock formed of material deposited from solution by evaporation of the water.

Exfoliation A mechanical weathering process characterized by the splitting off of slablike sheets of rock.

Exotic stream A permanent stream that traverses a desert and has its source in well-watered areas outside the desert.

Extrusive Igneous activity that occurs outside the crust.

Eye A zone of scattered clouds and calm averaging about 20 kilometers in diameter at the center of a hurricane.

Fault A break in a rock mass along which movement has occurred.

Fault-block mountain A mountain formed by the displacement of rock along a fault.

Felsic The group of igneous rocks composed primarily of feldspar and quartz.

Fetch The distance that the wind has traveled across the open water.

Fiord A steep-sided inlet of the sea formed when a glacial trough was partially submerged.

Flare A sudden brightening of an area on the sun.

Floodplain The flat, low-lying portion of a stream valley subject to periodic inundation.

Flood tide The tidal current associated with the increase in the height of the tide.

Flood basalts Flows of basaltic lava that issue from numerous cracks or fissures and commonly cover extensive areas to thicknesses of hundreds of meters.

Fluorescence The absorption of ultraviolet light, which is re-emitted as visible light.

Focus (earthquake) The zone within the earth where rock displacement produces an earthquake.

Focus (light) The point where a lens or mirror causes light rays to converge.

Foliated A texture of metamorphic rocks that gives the rock a layered appearance.

Foreshocks Small earthquakes that often precede a major earthquake.

Fossils The remains or traces of organisms preserved from the geologic past.

Fractional crystallization The process that separates magma into components having varied compositions and melting points.

Front The boundary between two adjoining air masses having contrasting characteristics.

Frontal wedging Lifting of air resulting when cool air acts as a barrier over which warmer, lighter air will rise.

Frost wedging The mechanical breakup of rock caused by the expansion of freezing water in cracks and crevices.

Geocentric The concept of an earth-centered universe.

Geostrophic wind A wind, usually above a height of 600 meters (2000 feet), that blows parallel to the isobars.

Geothermal energy Natural steam used for power generation.

Geyser A fountain of hot water ejected periodically.

Glacial erratic An ice-transported boulder that was not derived from bedrock near its present site.

Glacial striations Scratches and grooves on bedrock caused by glacial abrasion.

Glacial trough A mountain valley that has been widened, deepened, and straightened by a glacier.

Glacier A thick mass of ice originating on land from the compaction and recrystallization of snow that shows evidence of past or present flow.

Glassy A term used to describe the texture of certain igneous rocks, such as obsidian, that contain no crystals.

Glaze A coating of ice on objects formed when supercooled rain freezes on contact.

Globular cluster A nearly spherically shaped group of densely packed stars.

Gondwanaland The southern portion of Pangaea consisting of South America, Africa, Australia, India, and Antarctica.

Graben A valley formed by the downward displacement of a fault-bounded block.

Graded bed A sediment layer that is characterized by a decrease in sediment size from bottom to top.

Graded stream A stream that has the correct channel characteristics to maintain exactly the velocity required to transport the material supplied to it.

Gradient The slope of a stream; generally measured in feet per mile.

Groin A short wall built at a right angle to the shore to trap moving sand.

Ground moraine An undulating layer of till deposited as the ice front retreats.

Groundwater Water in the zone of saturation.

Guyot A submerged flat-topped seamount.

Gyre The large circular surface current pattern found in each ocean.

Hail Nearly spherical ice pellets having concentric layers and formed by the successive freezing of layers of water.

Half-life The time required for one-half of the atoms of a radioactive substance to decay.

Halocline A layer of water in which there is a high rate of change in salinity in the vertical dimension.

Hanging valley A tributary valley that enters a glacial trough at a considerable height above its floor.

Hardness The resistance a mineral offers to scratching.

Heliocentric The view that the sun is at the center of the solar system.

Horn A pyramidlike peak formed by glacial action in three or more cirques surrounding a mountain summit.

Horst An elongate, uplifted block of crust bounded by faults.

H-R diagram A plot of stars according to their absolute magnitudes and spectral types.

Humidity A general term referring to water vapor in the air but not to liquid droplets of fog, cloud, or rain.

Humus Organic matter in soil produced by the decomposition of plants and animals.

Hurricane A tropical cyclonic storm having winds in excess of 119 kilometers (74 miles) per hour.

Hydrogenous sediment Sea-floor sediments consisting of minerals that crystallize from seawater. An important example is manganese nodules.

Hygroscopic nuclei Condensation nuclei having a high affinity for water, such as salt particles.

Igneous rock A rock formed by the crystallization of molten magma.

Immature soil A soil lacking horizons.

Inclination of the axis The tilt of the earth's axis from the perpendicular to the plane of the earth's orbit.

Inclusion A piece of one rock unit contained within another. Inclusions are used in relative dating. The rock mass adjacent to the one containing the inclusion must have been there first in order to provide the fragment.

Index fossil A fossil that is associated with a particular span of geologic time.

Inertia A property of matter that resists a change in its motion.

Infiltration The movement of surface water into rock or soil through cracks and pore spaces.

Inner core The solid innermost layer of the earth, about 1300 kilometers (800 miles) in radius.

Inselberg An isolated mountain remnant characteristic of the late stage of erosion in an arid region.

Intermediate composition The composition of igneous rocks lying between felsic and mafic.

Interstellar matter Dust and gases found between stars.

Intrusive rock Igneous rock that formed below the earth's surface.

Ion An atom or molecule that possesses an electrical charge.

Ionosphere A complex zone of ionized gases that coincides with the lower portion of the thermosphere.

Island arc A group of volcanic islands formed by the subduction and partial melting of oceanic lithosphere.

Isobar A line drawn on a map connecting points of equal atmospheric pressure, usually corrected to sea level.

Isostacy The concept that the earth's crust is "floating" in gravitational balance upon the material of the mantle.

Isotherms Lines connecting points of equal temperature.

Isotopes Varieties of the same element that have different mass numbers; their nuclei contain the same number of protons but different numbers of neutrons.

Jet stream Swift (120–240-kilometer per hour), high-altitude winds.

Jetties A pair of structures extending into the ocean at the entrance to a harbor or river that are built for the purpose of protecting against storm waves and sediment deposition.

Joint A fracture in rock along which there has been no movement.

Kame A steep-sided hill composed of sand and gravel originating when sediment collected in openings in stagnant glacial ice.

Karst A topography consisting of numerous depressions called sinkholes.

Kettle holes Depressions created when blocks of ice became lodged in glacial deposits and subsequently melted.

Laccolith A massive igneous body intruded between pre-existing strata.

Lahar Mudflows on the slopes of volcanoes that result when unstable layers of ash and debris become saturated and flow downslope, usually following stream channels.

Land breeze A local wind blowing from land toward the water during the night in coastal areas.

Lapse rate (normal) The average drop in temperature (6.5°C per kilometer; 3.5°F per 1000 feet) with increased altitude in the troposphere.

Latent heat The energy absorbed or released during a change in state.

Lateral moraine A ridge of till along the sides of an alpine glacier composed primarily of debris that fell to the glacier from the valley walls.

Laterite A red, highly leached soil type found in the tropics that is rich in oxides of iron and aluminum.

Laurasia The northern portion of Pangaea consisting of North America and Eurasia.

Lava Magma that reaches the earth's surface.

Law of conservation of angular momentum The product of the velocity of an object around a center of rotation (axis) and the distance squared of the object from the axis is constant.

Law of superposition In any undeformed sequence of sedimentary rocks, each bed is older than the one above it and younger than the one below.

Light-year The distance light travels in a year; about 6 trillion miles.

Lithification The process, generally cementation and/or compaction, of converting sediments to solid rock.

Lithogenous sediment Sea-floor sediments having their source as products of weathering on the continents.

Lithosphere The rigid outer layer of the earth, including the crust and upper mantle.

Loess Deposits of wind-blown silt, lacking visible layers, generally buff colored, and capable of maintaining a nearly vertical cliff.

Longitudinal (seif) dunes Long ridges of sand oriented parallel to the prevailing wind; these dunes form where sand supplies are limited.

Longshore current A near-shore current that flows parallel to the shore.

Low velocity zone See Asthenosphere.

Luminosity The brightness of a star. The amount of energy radiated by a star.

Luster The appearance or quality of light reflected from the surface of a mineral.

Mafic Igneous rocks with a low silica content and a high iron-magnesium content.

Magma A body of molten rock found at depth, including any dissolved gases and crystals.

Magnitude (earthquake) The total amount of energy released during an earthquake.

Magnitude (stellar) A number given to a celestial object to express its relative brightness.

Manganese nodules Rounded lumps of hydrogenous sediment scattered on the ocean floor consisting mainly of manganese and iron, and usually containing small amounts of copper, nickel, and cobalt.

Mantle The 2900-kilometer-(1800-mile-) thick layer of the earth located below the crust.

Maria The Latin name for the smooth areas of the moon formerly thought to be seas.

Mass number The number of neutrons and protons in the nucleus of an atom.

Mass wasting   The downslope movement of rock, regolith, and soil under the direct influence of gravity.

Meander   A looplike bend in the course of a stream.

Mechanical weathering   The physical disintegration of rock, resulting in smaller fragments.

Medial moraine   A ridge of till formed when lateral moraines from two coalescing alpine glaciers join.

Mélange   A highly deformed mixture of rock material formed in areas of plate convergence.

Melt   The liquid portion of magma excluding the solid crystals.

Mesocyclone   An intense, rotating wind system in the lower part of a thunderstorm that precedes tornado development.

Mesopause   The boundary between the mesosphere and the thermosphere.

Mesosphere   The layer of the atmosphere immediately above the stratosphere and characterized by decreasing temperatures with height.

Mesozoic era   A time span on the geologic calendar between the Paleozoic and Cenozoic eras—from about 225 to 65 million years ago.

Metamorphism   The changes in mineral composition and texture of a rock subjected to high temperature and pressure within the earth.

Meteoroid   Small solid particles that have orbits in the solar system.

Mid-ocean ridge   A continuous mountainous ridge on the floor of all the major ocean basins and varying in width from 500–5000 kilometers (300–3000 miles). The rifts at the crests of these ridges represent divergent plate boundaries.

Mineral   A naturally occurring, inorganic crystalline material with a unique chemical composition.

Mixed tide   A tide characterized by a large inequality in successive high water heights, low water heights, or both.

Moho   The boundary separating the crust from the mantle, discernible by an increase in seismic velocity.

Mohs scale   A series of ten minerals used as a standard in determining hardness.

Monsoon   Seasonal reversal of wind direction associated with large continents, especially Asia. In winter, the wind blows from land to sea; in summer, from sea to land.

Mudflow   The flowage of debris containing a large amount of water; most characteristic of canyons and gullies in dry, mountainous regions.

Nansen bottle   A device used by oceanographers to obtain subsurface seawater samples.

Natural levees   The elevated landforms that parallel some streams and act to confine their waters, except during floodstage.

Nebula   A cloud of interstellar gas and/or dust.

Neap tide   Lowest tidal range, occurring near the times of the first- and third-quarter phases of the moon.

Neutron   A subatomic particle found in the nucleus of an atom. The neutron is electrically neutral and has a mass approximately that of a proton.

Nonconformity   An unconformity in which older metamorphic or intrusive igneous rocks are overlain by younger sedimentary strata.

Normal fault   A fault in which the rock above the fault plane has moved down relative to the rock below.

Nova   A star that explosively increases in brightness.

Nucleus   The small heavy core of an atom that contains all of its positive charge and most of its mass.

Nuée ardente   Incandescent volcanic debris buoyed up by hot gases that moves downslope in an avalanche fashion.

Oblique fault   A fault having both vertical and horizontal movement.

Obliquity   The angle between the planes of the earth's equator and orbit.

Obsidian   A volcanic glass of felsic composition.

Occluded front   A front formed when a cold front overtakes a warm front. It marks the beginning of the end of a middle-latitude cyclone.

Open cluster   A loosely formed group of stars of similar origin.

Orbit   The path of a body in revolution around a center of mass.

Orogenesis   The processes that collectively result in the formation of mountains.

Orographic lifting   Mountains acting as barriers to the flow of air force the air to ascend. The air cools adiabatically, and clouds and precipitation may result.

Outer core   A layer beneath the mantle about 2200 kilometers (1364 miles) thick that has the properties of a liquid.

Outwash   Stratified deposits dropped by glacial meltwater.

Oxbow lake   A curved lake produced when a stream cuts off a meander.

Ozone   A molecule of oxygen containing three oxygen atoms.

Pahoehoe   A lava flow with a smooth-to-ropy surface.

Paleontology   The systematic study of fossils and the history of life on earth.

Paleozoic era A time span on the geologic calendar between the Precambrian and Mesozoic eras—from about 600 million to 225 million years ago.

Pangaea The proposed supercontinent which 200 million years ago began to break apart and form the present land masses.

Parabolic dunes The shape of these dunes resembles barchans except their tips point into the wind; they often form along coasts that have strong onshore winds, abundant sand, and vegetation that partly covers the sand.

Parallax The apparent shift of an object when viewed from two different locations.

Parent material The material upon which a soil develops.

Parsec The distance at which an object would have a parallax angle of 1 second of arc (3.26 light-years).

Pedalfer Soil of humid regions characterized by the accumulation of iron oxides and aluminum-rich clays in the B horizon.

Pediment A sloping bedrock surface fringing a mountain base in an arid region, formed when erosion causes the mountain front to retreat.

Pedocal Soil associated with drier regions and characterized by an accumulation of calcium carbonate in the upper horizons.

Peneplain In the idealized cycle of landscape evolution, an undulating plain near base level associated with old age.

Perched water table A localized zone of saturation above the main water table created by an impermeable layer (aquiclude).

Peridotite An igneous rock of ultramafic composition thought to be abundant in the upper mantle.

Perihelion The point in the orbit of a planet where it is closest to the sun.

Period A basic unit of the geologic calendar that is a subdivision of an era. Periods may be divided into smaller units called epochs.

Permeability A measure of a material's ability to transmit water.

Perturbation The gravitational disturbance of the orbit of one celestial body by another.

pH scale A common measure of the degree of acidity or alkalinity of a solution, it is a logarithmic scale ranging from 0 to 14. A value of 7 denotes a neutral solution, values below 7 indicate greater acidity and numbers above 7 indicate greater alkalinity.

Phenocryst Conspicuously large crystals imbedded in a matrix of finer-grained crystals.

Photochemical reaction A chemical reaction in the atmosphere that is triggered by sunlight, often yielding a secondary pollutant.

Photon A discrete amount (quantum) of electromagnetic energy.

Photosphere The region of the sun that radiates energy to space. The visible surface of the sun.

Plage A bright region in the solar atmosphere located above a sunspot.

Planetary nebula A shell of incandescent gas expanding from a star.

Plate One of numerous rigid sections of the lithosphere that moves as a unit over the material of the asthenosphere.

Plate tectonics The theory which proposes that the earth's outer shell consists of individual plates which interact in various ways and thereby produce earthquakes, volcanoes, mountains, and the crust itself.

Playa A flat area on the floor of an undrained desert basin. Following heavy rain, the playa becomes a lake.

Plucking (quarrying) The process by which pieces of bedrock are lifted out of place by a glacier.

Pluvial lake A lake formed during a period of increased rainfall. During the Pleistocene epoch this occurred in some nonglaciated regions during periods of ice advance elsewhere.

Porphyry An igneous texture consisting of large crystals embedded in a matrix of much smaller crystals.

Porosity The volume of open spaces in rock or soil.

Pothole A depression formed in a stream channel by the abrasive action of the water's sediment load.

Precambrian All geologic time prior to the Paleozoic era.

Precession A slow motion of the earth's axis which traces out a cone over a period of 26,000 years.

Precipitation fog Fog formed when rain evaporates as it falls through a layer of cool air.

Pressure gradient The amount of pressure change occurring over a given distance.

Primary pollutants Those pollutants emitted directly from identifiable sources.

Principle of faunal succession Fossil organisms succeed one another in a definite and determinable order, and any time period can be recognized by its fossil content.

Principle of original horizontality Layers of sediment are generally deposited in a horizontal or nearly horizontal position.

Prominence A concentration of material above the solar surface that appears as a bright archlike structure.

Psychrometer A device consisting of two thermometers (wet bulb and dry bulb) that is rapidly whirled and, with the use of tables, yields the relative humidity and dew point.

Pulsating star A star that fluxuates in brightness by periodically expanding and contracting.

P wave The fastest earthquake wave, which travels by compression and expansion of the medium.

Pyroclastic material The volcanic rock ejected during an eruption, including ash, bombs, and blocks.

Radial drainage A system of streams running in all directions away from a central elevated structure, such as a volcano.

Radiation The transfer of energy (heat) through space by electromagnetic waves.

Radiation fog Fog resulting from radiation heat loss by the earth.

Radioactivity The spontaneous decay of certain unstable atomic nuclei.

Radiocarbon (carbon-14) The radioactive isotope of carbon, which is produced continuously in the atmosphere and is used in dating events as far back as 40,000 years.

Radiometric dating The procedure of calculating the absolute ages of rocks and minerals that contain radioactive isotopes.

Rainshadow A dry area on the lee side of a mountain range.

Recessional moraine An end moraine formed as the ice front stagnated during glacial retreat.

Reflecting telescope A telescope that concentrates light from distant objects by using a concave mirror.

Refracting telescope A telescope that employs a lens to bend and concentrate the light from distant objects.

Refraction The process by which the portion of a wave in shallow water slows, causing the wave to bend and tend to align itself with the underwater contours.

Regional metamorphism Metamorphism associated with the large-scale mountain-building processes.

Regolith The layer of rock and mineral fragments that nearly everywhere covers the earth's land surface.

Rejuvenation A change, often caused by regional uplift, that causes the forces of erosion to intensify.

Relative dating Rocks are placed in their proper sequence or order. Only the chronologic order of events is determined.

Relative humidity The ratio of the air's water vapor content to its water vapor capacity.

Residual soil Soil developed directly from the weathering of the bedrock below.

Resolving power The ability of a telescope to separate objects that would otherwise appear as one.

Retrograde motion The apparent westward motion of the planets with respect to the stars.

Reverse fault A fault in which the material above the fault plane moves up in relation to the material below.

Revolution The motion of one body about another, as the earth about the sun.

Richter scale A scale of earthquake magnitude based on the motion of a seismograph.

Rift A region of the earth's crust along which divergence is taking place.

Right ascension An angular distance measured eastward along the celestial equator from the vernal equinox. Used with declination in a coordinate system to describe the position of celestial bodies.

Rime A thin coating of ice on objects produced when supercooled fog droplets freeze on contact.

Rock A consolidated mixture of minerals.

Rock flour Ground-up rock produced by the grinding effect of a glacier.

Rockslide The rapid slide of a mass of rock downslope along planes of weakness.

Rotation The spinning of a body, such as the earth, about its axis.

Runoff Water that flows over the land rather than infiltrating into the ground.

Salinity The proportion of dissolved salts to pure water, usually expressed in parts per thousand (‰).

Saltation Transportation of sediment through a series of leaps or bounces.

Scoria Hardened lava which has retained the vesicles produced by escaping gases.

Sea arch An arch formed by wave erosion when caves on opposite sides of a headland unite.

Sea breeze A local wind blowing from the sea during the afternoon in coastal areas.

Sea-floor spreading The process of producing new sea floor between two diverging plates.

Seamount An isolated volcanic peak that rises at least 1000 meters (3000 feet) above the deep ocean floor.

Sea stack An isolated mass of rock standing just offshore, produced by wave erosion of a headland.

Seawall A barrier constructed to prevent waves from reaching the area behind the wall. Their purpose is to defend property from the force of breaking waves.

Secondary pollutants Pollutants that are produced in the atmosphere by chemical reactions that occur among primary pollutants.

Sedimentary rock Rock formed from the weathered

products of pre-existing rocks that have been transported, deposited, and lithified.

Seismic sea wave A rapidly moving ocean wave generated by earthquake activity which is capable of inflicting heavy damage in coastal regions.

Seismograph An instrument that records earthquake waves.

Semidiurnal tide The predominant type of tide throughout the world, with two high waters and two low waters each tidal day.

Shadow zone The zone between 104 and 143 degrees distance from an earthquake epicenter in which direct waves do not arrive because of refraction by the earth's core.

Shelf break The point where a rapid steepening of the gradient occurs marking the outer edge of the continental shelf and the beginning of the continental slope.

Shield volcano A broad, gently sloping volcano built from fluid basaltic lavas.

Silicate Any one of numerous minerals that have the oxygen and silicon tetrahedron as their basic structure.

Silicon-oxygen tetrahedron A structure composed of four oxygen atoms surrounding a silicon atom that constitutes the basic building block of silicate minerals.

Sill A tabular igneous body that was intruded parallel to the layering of pre-existing rock.

Sink hole A depression produced in a region where soluble rock has been removed by groundwater.

Sleet Frozen or semifrozen rain formed when raindrops freeze as they pass through a layer of cold air.

Slip face The steep, leeward slope of a sand dune; it maintains an angle of about 34 degrees.

Slump The downward slipping of a mass of rock or unconsolidated material moving as a unit along a curved surface.

Smog Originally used to describe a combination of smoke and fog, today the term is used as a synonym for general air pollution.

Snow A solid form of precipitation produced by sublimation of water vapor.

Snowfield An area where snow persists year-round.

Snowline Lower limit of perennial snow.

Soil A combination of mineral and organic matter, water, and air; that portion of the regolith that supports plant growth.

Soil horizon A layer of soil that has identifiable characteristics produced by chemical weathering and other soil-forming processes.

Soil profile A vertical section through a soil showing its succession of horizons and the underlying parent material.

Solar constant The rate at which solar radiation is received outside the earth's atmosphere on a surface perpendicular to the sun's rays when the earth is at an average distance from the sun.

Solar winds Subatomic particles ejected at high speed from the solar corona.

Solifluction Slow, downslope flow of water-saturated materials common to permafrost areas.

Solstice The time when the vertical rays of the sun are striking either the Tropic of Cancer or the Tropic of Capricorn. Solstice represents the longest or shortest day (length of daylight) of the year.

Source region The area where an air mass acquires its characteristic properties of temperature and moisture.

Specific gravity The ratio of a substance's weight to the weight of an equal volume of water.

Specific humidity The weight of water vapor compared with the total weight of the air, including the water vapor.

Spheroidal weathering Any weathering process that tends to produce a spherical shape from an initially blocky shape.

Spit An elongate ridge of sand that projects from the land into the mouth of an adjacent bay.

Spring A flow of groundwater that emerges naturally at the ground surface.

Spring tide Highest tidal range that occurs near the times of the new and full moons.

Stalactite The iciclelike structure that hangs from the ceiling of a cavern.

Stalagmite The columnlike form that grows upward from the floor of a cavern.

Steam fog Fog having the appearance of steam; produced by evaporation from a warm water surface into the cool air above.

Strata Parallel layers of sedimentary rock.

Stratopause The boundary between the stratosphere and the mesosphere.

Stratosphere The layer of the atmosphere immediately above the troposphere, characterized by increasing temperatures with height due to the concentration of ozone.

Stratovolcano See Composite cone.

Streak The color of a mineral in powdered form.

Striations (glacial) Scratches or grooves in a bedrock surface caused by the grinding action of a glacier and its load of sediment.

Strike-slip fault A fault along which the movement is horizontal.

Subduction   The process of thrusting oceanic lithosphere into the mantle along a convergent zone.

Sublimation   The conversion of a solid directly to a gas without passing through the liquid state.

Submarine canyon   A seaward extension of a valley that was cut on the continental shelf during a time when sea level was lower, or a canyon carved into the outer continental shelf, slope, and rise by turbidity currents.

Submergent coast   A coast with a form that is largely the result of the partial drowning of a former land surface either because of a rise of sea level or subsidence of the crust or both.

Subsoil   A term applied to the *B* horizon of a soil profile.

Sunspot   A dark spot on the sun, which is cool by contrast to the surrounding photosphere.

Supernova   An exploding star that increases in brightness many thousands of times.

Surf   A collective term for breakers; also the wave activity in the area between the shoreline and the outer limit of breakers.

Surface soil   The uppermost layer in a soil profile: the *A* horizon.

Surface waves   Seismic waves that travel along the outer layer of the earth.

Suspended load   The fine sediment carried within the body of flowing water.

S wave   An earthquake wave, slower than a P wave, that travels only in solids.

Swells   Wind-generated waves that have moved into an area of weaker winds or calm.

Syncline   A linear downfold in sedimentary strata; the opposite of anticline.

Talus   An accumulation of rock debris at the base of a cliff.

Tarn   A small lake in a cirque.

Tectonics   The study of the large-scale processes that collectively deform the earth's crust.

Temperature inversion   A layer in the atmosphere of limited depth where the temperature increases rather than decreases with height.

Temporary (local) base level   The level of a lake, resistant rock layer, or any other base level that stands above sea level.

Terminal moraine   The end moraine marking the farthest advance of a glacier.

Terrace   A flat, benchlike structure produced by a stream, which was left elevated as the stream cut downward.

Terrane   A crustal block bounded by faults, whose geologic history is distinct from the histories of adjoining crustal blocks.

Texture   The size, shape, and distribution of the particles that collectively constitute a rock.

Thermal gradient   The increase in temperature with depth. It averages 1°C per 30 meters (1–2°F per 100 feet) in the crust.

Thermocline   A layer of water in which there is a rapid change in temperature in the vertical dimension.

Thermohaline circulation   Movements of ocean water caused by density differences brought about by variations in temperature and salinity.

Thermosphere   The region of the atmosphere immediately above the mesosphere and characterized by increasing temperatures due to absorption of very short-wave solar energy by oxygen.

Thrust fault   A low-angle reverse fault.

Tidal current   The alternating horizontal movement of water associated with the rise and fall of the tide.

Tidal flat   A marshy or muddy area that is covered and uncovered by the rise and fall of the tide.

Tide   Periodic change in the elevation of the ocean surface.

Till   Unsorted sediment deposited directly by a glacier.

Tombolo   A ridge of sand that connects an island to the mainland or to another island.

Tornado   A small, very intense cyclonic storm with exceedingly high winds, most often produced along cold fronts in conjunction with severe thunderstorms.

Tornado warning   A warning issued when a tornado has actually been sighted in an area or is indicated by radar.

Tornado watch   A forecast issued for areas of about 65,000 square kilometers (25,000 square miles) indicating that conditions are such that tornadoes may develop; they are intended to alert people to the possibility of tornadoes.

Transform fault   A boundary where plates slide past each other, such as the San Andreas fault.

Transpiration   The release of water vapor to the atmosphere by plants.

Transported soil   Soils that form on unconsolidated deposits.

Transverse dunes   A series of long ridges oriented at right angles to the prevailing wind; these dunes form where vegetation is sparse and sand is very plentiful.

Travertine   A form of limestone ($CaCO_3$) that is deposited by hot springs or as a cave deposit.

Trellis drainage   A system of streams in which nearly parallel tributaries occupy valleys cut in folded strata.

Trench An elongate depression in the sea floor produced by bending of oceanic crust during subduction.

Tropic of Cancer The parallel of latitude, 23½ degrees north latitude, marking the northern limit of the sun's vertical rays.

Tropic of Capricorn The parallel of latitude, 23½ degrees south latitude, marking the southern limit of the sun's vertical rays.

Tropical depression By international agreement, a tropical cyclone with maximum winds that do not exceed 61 kilometers (38 miles) per hour.

Tropical storm By international agreement, a tropical cyclone with maximum winds between 61 and 119 kilometers (38 and 74 miles) per hour.

Tropopause The boundary between the troposphere and the stratosphere.

Troposphere The lowermost layer of the atmosphere. It is generally characterized by a decrease in temperature with height.

Tsunami The Japanese word for a seismic sea wave.

Turbidite Turbidity current deposit characterized by graded bedding.

Turbidity current A downslope movement of dense, sediment-laden water created when sand and mud on the continental shelf and slope are dislodged and thrown into suspension.

Ultimate base level Sea level; the lowest level to which stream erosion could lower the land.

Ultramafic Igneous rocks composed mainly of iron and magnesium-rich minerals.

Unconformity A surface that represents a break in the rock record, caused by erosion or nondeposition.

Uniformitarianism The concept that the processes that have shaped the earth in the geologic past are essentially the same as those operating today.

Upslope fog Fog created when air moves up a slope and cools adiabatically.

Upwelling The rising of cold water from deeper layers to replace warmer surface water that has been moved away.

Urban heat island Refers to the fact that temperatures within a city are generally higher than in surrounding rural areas.

Vapor pressure That part of the total atmospheric pressure attributable to water vapor content.

Ventifact A cobble or pebble polished and shaped by the sandblasting effect of wind.

Vesicular A term applied to igneous rocks that contain small cavities called vesicles, which are formed when gases escape from lava.

Viscosity A measure of a fluid's resistance to flow.

Volcanic bomb A streamlined pyroclastic fragment ejected from a volcano while molten.

Volcano A mountain formed of lava and/or pyroclastics.

Warm front A front along which a warm air mass overrides a retreating mass of cooler air.

Wash A common term for a desert stream course which is typically dry except for brief periods immediately following a rain.

Water table The upper level of the saturated zone of groundwater.

Wave-cut cliff A seaward-facing cliff along a steep shoreline formed by wave erosion at its base and mass wasting.

Wave-cut platform A bench or shelf in the bedrock at sea level, cut by wave erosion.

Wave height The vertical distance between the trough and crest of a wave.

Wave length The horizontal distance separating successive crests or troughs.

Wave period The time interval between the passage of successive crests at a stationary point.

Weather The state of the atmosphere at any given time.

Weathering The disintegration and decomposition of rock at or near the surface of the earth.

Welded tuff A pyroclastic rock composed of particles that have been fused together by the combination of heat still contained in the deposit after it has come to rest and the weight of overlying material.

Well An opening bored into the zone of saturation.

Wind vane An instrument used to determine wind direction.

Yazoo tributary A tributary that flows parallel to the main stream because a natural levee is present.

Zodiac A band along the ecliptic containing the twelve constellations of the zodiac.

Zone of aeration Area above the water table where openings in soil, sediment, and rock are not saturated, but filled mainly with air.

Zone of fracture The upper portion of a glacier consisting of brittle ice.

Zone of saturation Zone where all open spaces in sediment and rock are completely filled with water.

INDEX